Drought and Water Crises

Integrating Science, Management, and Policy

Drought and Water Crises

Series Editor: Donald A. Wilhite

Published Titles:

Drought and Water Crises

Integrating Science, Management, and Policy

Second Edition

Donald A. Wilhite
Roger S. Pulwarty

CRC Press
Taylor & Francis Group
Boca Raton London New York

CRC Press is an imprint of the
Taylor & Francis Group, an **informa** business

CRC Press
Taylor & Francis Group
6000 Broken Sound Parkway NW, Suite 300
Boca Raton, FL 33487-2742

First issued in paperback 2020

© 2018 by Taylor & Francis Group, LLC
CRC Press is an imprint of Taylor & Francis Group, an Informa business

No claim to original U.S. Government works

ISBN 13: 978-0-367-57285-3 (pbk)
ISBN 13: 978-1-138-03564-5 (hbk)

Library of Congress Cataloging-in-Publication Data

Names: Wilhite, Donald A. | Pulwarty, Roger S., 1960-
Title: Drought and water crises : integrating science, management, and policy / [edited by] Donald A. Wilhite and Roger S. Pulwarty.
Description: Second edition. | Boca Raton : CRC Press, 2018. | 1st edition published in 2005. | Includes bibliographical references.
Identifiers: LCCN 2017017957 | ISBN 9781138035645 (hardback : alk. paper)
Subjects: LCSH: Droughts. | Water-supply--Risk assessment.
Classification: LCC QC929.24 .D75 2018 | DDC 363.34/9297--dc23
LC record available at https://lccn.loc.gov/2017017957

Visit the Taylor & Francis Web site at
http://www.taylorandfrancis.com

and the CRC Press Web site at
http://www.crcpress.com

To our wives, Myra and Susan, for their ongoing support and patience

Contents

Section I   Overview

Section II   Drought Risk Reduction: Shifting the Paradigm from Managing Disasters to Managing Risk

Section III   Advances in Tools for Drought Prediction, Early Warning, Decision Support, and Management

Section IV   Case Studies in Integrated Drought and Water Management: The Role of Science, Technology, Management, and Policy

Section V   Integration and Conclusions

Preface

Drought and Water Crises: Integrating Science, Management, and Policy, 2nd edition, documents the remarkable progress in drought management since the publication of the first edition in 2005. Significant advances in the past decade have enhanced our ability to monitor and detect drought and its severity, and to communicate this information to decision makers at all levels. Many decision makers are now using this information for risk mitigation. Progress has been made on improving the reliability of seasonal drought forecasts for some regions in order to better serve decision makers in the management of water and other natural resources. Efforts are being directed at expressing these forecasts in ways that better meet the needs of end users. Researchers and practitioners have developed improved impact assessment tools that can help to determine the true economic, social, and environmental costs associated with drought. Planning and mitigation tools have been developed that can assist government and others in the development of drought mitigation plans. New tools for vulnerability assessment have been developed and tested in various settings, although much work remains on this complex process. New water-conserving technologies have been developed and are being applied in both the agricultural and urban sectors, which provide opportunities for continued improvement in the efficiency of water use.

Although progress on operationalizing these improved drought management techniques has been significant, much of this progress was sporadic prior to 2013 and restricted to specific geographic settings. Especially deficient was progress on the development of national drought policies that emphasize risk reduction as a strategy to shift the paradigm for drought management from the traditional crisis management approach, that is, the "hydro-illogical" cycle. The High-level Meeting on National Drought Policy (HMNDP), held in March 2013, served as a major stimulus to promote the development of national drought policies and many of the concepts associated with integrated drought management. This meeting brought together three major organizations of the United Nations, all with a strong interest in improving drought management and impact reduction. This collaboration facilitated a global dialogue on the importance of formulating national drought policies with a goal of reducing the risks and, therefore, the vulnerability of all nations to this insidious natural hazard. The World Meteorological Organization (WMO) served as the catalyst for the HMNDP with substantial input and support from the Food and Agricultural Organization (FAO) and the UN Convention to Combat Desertification (UNCCD). The goals of HMNDP and its outcomes are described in detail in Chapter 2. The declaration that was an outcome of the conference and endorsed unanimously by the 87 nations that participated in this conference represents a dramatic shift away from crisis management. That declaration called for all nations to develop a national drought policy that focuses on risk reduction. Since 2013, the attention that has been drawn to this topic has been significant and the progress, described in detail in this book, outstanding.

The chapters in this book are organized into several parts in order to highlight both the progress that has been made in drought risk assessment and management in recent years on the development of tools and methodologies, and how these tools and methodologies have been applied in various settings around the world. Part I discusses drought as a hazard and the complexities associated with its management. It lays the foundation for subsequent chapters. Part II details the advances made in moving the conversation forward on drought risk management to shift the paradigm from managing disasters to managing risk, and the difficulties and opportunities in making that transition. Part III focuses on the advances in drought monitoring, early warning, and prediction and management. Part IV has been expanded considerably from the first edition of this book to provide relevant case studies and advances in drought and water management that demonstrate the progress made on the process of integrating science, technology, and management to improve the policies associated with drought management. Part V is a capstone chapter that succinctly integrates and summarizes the progress made to date and, hopefully, the way forward.

As coeditors of this volume, it is our intent that the chapters contained in the second edition will better equip readers with both the knowledge and appropriate examples that illustrate how these tools can be applied in actions to reduce societal vulnerability to drought through proactive drought risk management.

Donald A. Wilhite
University of Nebraska-Lincoln

Roger S. Pulwarty
NOAA

Acknowledgments

Drought and Water Crises: Integrating Science, Management, and Policy, Second edition, is the result of the efforts of many persons who have been working diligently over the past year to bring this volume to fruition. This edition of the book provides an update on the progress made in drought management over the past decade. That progress has emphasized a proactive, risk-reduction approach rather than the traditional, crisis management approach. The decision to pursue the publication of a second edition of this book in 2016 was the result of discussions with Irma Shagla Britton, Senior Editor, Environmental Sciences, GIS and Remote Sensing, CRC Press/Taylor & Francis Group. It has been a pleasure to work with Irma and the entire CRC team in the production of this book.

We would especially like to thank the contributors to this volume. These colleagues were chosen for their expertise, the quality of their research throughout their professional careers, and the contribution of their research efforts and experiences to the theme of this book. We appreciate their responsiveness to the short deadlines given to prepare their chapter submissions and their receptivity to suggested edits and modifications.

We also thank Deborah Wood of the National Drought Mitigation Center (NDMC) for her many contributions to the preparation of the final manuscript for the book. Deb's exceptional editing skills and attention to detail are highly valued. Without her diligence in finalizing the book chapters, we would not have been able to complete this project on schedule. This book is just one of many manuscripts that Deb has contributed her many talents and skills to through the years.

Finally, we would like to thank the staff and partners of the NDMC at the University of Nebraska, Lincoln, and the National Oceanic and Atmospheric Administration's National Integrated Drought Information System for inspiring much of the material in this volume through their work in proactive drought risk management.

Editors

Donald A. Wilhite is a professor and director emeritus of applied climate science in the School of Natural Resources at the University of Nebraska–Lincoln, USA. Prior to August 2012, he served as director of the School of Natural Resources, a position he held from 2007 to 2012. Donald was founding director of the NDMC in 1995 and the International Drought Information Center in 1989 at the University of Nebraska–Lincoln. He was elected fellow of the American Meteorological Society in 2013. Donald's research and outreach activities have focused on issues of drought monitoring, planning, mitigation, and policy; the use of climate information in decision-making; and climate change. In 2013, he chaired the International Organizing Committee for the HMNDP, sponsored by WMO, FAO, and UNCCD. Donald chairs the management and advisory committees of the Integrated Drought Management Program (IDMP), launched in 2013 by WMO and the Global Water Partnership. In 2014, he authored *National Drought Management Policy Guidelines: A Template for Action* for the IDMP. Donald has authored or coauthored more than 150 journal articles, monographs, book chapters, and technical reports. He is editor or coeditor of numerous books on drought and drought management, including *Drought and Water Crises* (CRC Press, 2005); *From Disaster Response to Risk Management: Australia's National Drought Policy* (Springer, 2005); *Drought: A Global Assessment* (Routledge, 2000); *Drought Assessment, Management, and Planning: Theory and Case Studies* (Kluwer Publishers, 1993); and *Planning for Drought: Toward a Reduction of Societal Vulnerability* (Westview Press, 1987). Donald is also coauthor of two reports on the implications of climate change for Nebraska, published in 2014 and 2016, and coeditor of the *Atlas of Nebraska* (University of Nebraska Press, 2017).

Roger S. Pulwarty is the senior science advisor for climate research at the National Oceanic and Atmospheric Administration (NOAA) Climate Program and the Earth System Research Laboratory in Boulder, Colorado. Roger's publications focus on climate and risk management in the United States, Latin America, and the Caribbean. Throughout his career, he has helped develop and lead widely recognized programs on climate science, adaptation, and services, including the Regional Integrated Sciences and Assessments, the US National Integrated Drought Information System, and the Mainstreaming Adaptation to Climate Change project in the Caribbean. Roger is a convening lead author on the Intergovernmental Panel on Climate Change, UN International Strategy for Disaster Reduction, and US National Climate Assessment reports. He serves on advisory committees of the National Academy of Sciences, provides testimonies before the US Congress, and acts as an advisor on climate risk management to the Western Governors Association, Organization of American States, UNDP, the InterAmerican Bank, and World Bank, among others. He chairs the WMO Climate Services Information System implementation team. Roger's work has received US government, international, and professional society awards for integrating scientific research into decision-making. Roger is professor-adjunct at the University of Colorado and the University of the West Indies.

Contributors

Amir AghaKouchak is an associate professor of civil and environmental engineering at the University of California, Irvine. His research focuses on climate extremes and crosses the boundaries between hydrology, climatology, and remote sensing. Amir's group has developed models for monitoring and assessing climatic extremes, including the global integrated drought monitoring and prediction system (GIDMaPS). The long-term goal of his research group is to utilize the continuously growing satellite data along with ground-based observations to develop/improve integrated drought, flood and landslide modeling, drought prediction, and decision support systems. He has led several research grants funded by the National Science Foundation, National Oceanic and Atmospheric Administration, United States Bureau of Reclamation, and the National Aeronautics and Space Administration.

Martha C. Anderson received a BA in physics from Carleton College, Northfield, Minnesota, and a PhD in astrophysics from the University of Minnesota, Minneapolis. She is a research physical scientist for the USDA Agricultural Research Service in the Hydrology and Remote Sensing Laboratory in Beltsville, Maryland. Martha's research interests focus on mapping water, energy, and carbon land–surface fluxes at field to continental scales using thermal remote sensing, with applications in drought monitoring and yield estimation. She is currently a member of the Landsat and ECOSTRESS Science Teams, and the HysPIRI Science Working Group.

Felipe I. Arreguín Cortés holds master's and doctor's degrees in hydraulics from the National Autonomous University of Mexico (UNAM). Since 1989, he has served as a professor of the Postgraduate Studies Coordination, UNAM. Felipe was deputy technical director at the National Water Commission from 2002 to 2015 and president of the 30th Board of Directors of the Mexican Hydraulics Association (AMH). He has authored more than 240 publications and delivered more than 200 lectures in several countries, including Mexico. Honors and awards include the AMH's Enzo Levi Prize for Research and Hydraulics Pedagogy, the Javier Barrios Prize for the Best Book on Civil Engineering in 2002 (Spillways) and the 2002 Best Theoretical-Oriented Paper Award by the ASCE *Journal of Water Resources Planning and Management*, among others.

Paulo Barbosa is a senior scientist at the European Commission's Joint Research Centre in Ispra, Italy. He has a PhD in forest engineering from the Technical University of Lisbon (Portugal) and an MS in rural management in relation to the environment from the University of Lerida and the Agronomic Mediterranean Institute of Zaragoza (Spain). Paulo has worked for more than 25 years in the area of environmental sustainability and natural hazards, developing remote sensing applications in the areas of agriculture, forestry, and land cover classification, and developing early warning systems for forest fires and droughts. In the last few years, he has focused on the topic of climate change adaptation and its link to disaster risk reduction.

Dan Barrie is a program manager in the NOAA's Climate Program Office (CPO). Within CPO, he manages the modeling, analysis, predictions, and projections (MAPP) program, which focuses on model development, improvements to predictions and projections of climate conditions, and analysis of the climate system toward improved modeling and predictions. The MAPP program works extensively on supporting the transition of research to operations, and Dan leads the drought task force, which focuses on research and transition to advance capabilities. His background is in climate and energy modeling, with a degree in physics from Colgate University and a PhD in atmospheric and oceanic science from the University of Maryland. In addition to working at NOAA, Dan teaches climate and energy classes at Johns Hopkins University.

Mohamed Bazza, currently working in the FAO of the United Nations, obtained a BS in general agriculture from Hassan II Agronomic and Veterinary Medicine Institute in Rabat, Morocco, and an MS and a PhD in land and water sciences and an MS in statistics from the University of California-Davis, USA. Mohamed has over 35 years of hands-on experience in water resources planning and management, with a focus on agriculture water governance and management under water scarcity conditions, irrigation modernization, and drought policy. He led FAO's work at the global level in the areas of drought policy and water and groundwater governance, and supported field projects in several countries from 2011 to 2017. Mohamed led FAO's program on water resources and irrigation in the Near East and North Africa region from 2000 to 2011 and a World Bank financed project in Yemen in 1998–1999. He was a professor and researcher in agricultural engineering and provider of consulting services to UN organizations, financing institutions, and the public sector. Mohamed has work experience in many countries worldwide and is the author of numerous publications.

Courtney Black, PE, has been a consultant primarily in municipal and watershed water resources planning, stakeholder outreach, water rights litigation, environmental impact statements, and wetland restoration projects in Colorado and other areas of the United States. She has developed drought planning guidance documents, contributed to state drought planning efforts, and designed one of the first "drought tournaments" in the United States. Courtney earned a BS in civil and environmental engineering from Lehigh University and an MS in environmental engineering from the University of Florida.

Rudolf Brázdil graduated in geography and mathematics from the J. E. Purkyně University in Brno. He is affiliated as a professor of physical geography at the Institute of Geography, Masaryk University in Brno, and as a scientist at the Global Change Research Institute, Czech Academy of Sciences in Brno. Rudolf participated in several national and international grant projects. His general area of interest is climate variability and climate change in central Europe, with particular focus on spatiotemporal variability of the climate in the instrumental period, climate reconstructions based on documentary data, and analysis of hydrometeorological extremes from documentary data and systematic meteorological and hydrological measurements.

Carmelo Cammalleri is an environmental engineer, currently a contract agent at the Joint Research Centre of the European Commission. He was educated at the Università degli Studi di Palermo, Italy, and he was formerly a researcher at the USDA Hydrology and Remote Sensing Laboratory in Beltsville, Maryland. Carmelo's research interests include the use of optical and thermal remote sensing data for spatially distributed estimation of hydrological variables, the modeling of land–atmosphere flux exchange processes, and the coupling of the two approaches through ensemble and data assimilation techniques, with particular focus on soil moisture and evapotranspiration for drought monitoring and water management.

Hugo Carrão is an applied statistician and environmental scientist with expertise in remote sensing and geographic information systems (GIS). He holds a degree in biophysical engineering from the University of Évora, Portugal, and a PhD in information management from the New University of Lisbon, Portugal. Since 2001, he has been working on the topics of land use and land cover accounting, disaster risk management, and climate change mitigation. Hugo was initially a research associate with the Portuguese Geographic Institute and later was a scientific project officer with the European Commission's Joint Research Centre. His research interests focus on the development and application of statistical indicators to support spatial planning activities and regional policies in Europe, Africa, and Latin America. This research includes the use of predictive modeling, machine learning, and data mining algorithms to estimate drought hazard and map the global distribution of drought risk from environmental and socioeconomic data.

Neisha Cave has been working as a research intern at the Caribbean Institute for Meteorology and Hydrology (CIMH) since 2016. Neisha's current duties include data rescuing and drought impacts reporting for the Caribbean. Prior to this she helped support the work of the early warning information systems across climate timescales (EWISACTS) program under the WMO Regional Climate Centre at CIMH. She recently graduated with an MS in financial and business economics from the University of West Indies, Cave Hill. In 2016, Neisha also completed a three-month internship at the East Caribbean Central Bank. Her expertise lies in risk analysis policy and statistical and research analysis accompanied with analytical and quantitative skills.

Francis H. S. Chiew has more than 20 years of experience in research, teaching, and consulting in hydroclimate and water resources, and in science leadership and project management. Francis joined CSIRO as a science leader in Canberra in 2006 after a 15-year academic career at the University of Melbourne. In the past ten years, Francis has led 30–40 hydrologists working on hydrological modeling, climate impact on water, streamflow forecasting, and integrated river basin management. Francis is highly regarded internationally in the areas of hydrological modeling and hydroclimate, where he has received various awards and published over 250 research papers. Francis is a member of several global water expert committees, including lead author of the IPCC AR5 Assessment Report. Francis is also active in converting research outcomes into modeling tools and guidelines for the water industry. More recently, Francis has led high-impact climate and water modeling initiatives and water resources assessment and prediction in Australia, and collaborative hydroclimate projects in South Asia, China, and South America.

Joanne Chong is the research director at the Institute for Sustainable Futures at the University of Technology, Sydney, Australia. Joanne leads interdisciplinary, applied research teams to investigate and develop solutions for improving water security and resilience in cities and regions globally. Joanne's expertise spans WASH, wetlands and catchment management, metropolitan water infrastructure and service planning, pricing and regulatory arrangements, and rapid drought response. Joanne was recently appointed as a Board Committee member of the Australian Water Partnership.

Daniel Connell works on governance issues relating to transboundary rivers in the Crawford School of Public Policy at the Australian National University. He has written extensively about Australia's Murray–Darling Basin, including *Basin Futures*, a book coedited with Quentin Grafton and published by ANU E Press. Since publishing *Water Politics in the Murray–Darling Basin* (Federation Press 2007), Daniel has been conducting a comparative study of the governance arrangements for rivers in multilayered governance systems, focusing in particular on Australia, South Africa, the United States, Mexico, the European Union (Spain), India, China, and Brazil. Themes include water reform, environmental justice, public participation, cultural change, institutional design, the distribution of costs and benefits across borders, water markets, and risk created by the interaction of different levels of government. His recent projects include a book, *Federal Rivers: Water Management in Multi-Layered Governance Systems*, based on a workshop conducted at Oxford University in April 2012 and published in 2014.

Heather Cooley is the director of the Water and Sustainability Program at the Pacific Institute. Heather conducts and oversees research on a range of issues related to sustainable water use and management, the connections between water and energy, and the impacts of climate change on water resources. She has served on several boards and committees, and authored numerous publications on water resource management. In 2009, she received the Outstanding Achievement Award from the US Environmental Protection Agency for her work on agricultural water efficiency.

Shelly-Ann Cox is a postdoctoral researcher at the Caribbean Institute for Meteorology and Hydrology. Some of her main tasks include coediting the monthly Caribbean Drought Bulletin and codeveloping the Climate Impacts Database for the Caribbean, which supports regional growth resilient to climate risks. Shelly-Ann also plays an integral role in the development of sectoral early warning information systems across climate timescales. She recently completed her PhD in natural resource management, with an emphasis on fisheries management, at the Centre for Resource Management and Environmental Studies. Shelly-Ann is passionate about her work and the opportunities presented to bring positive change to climate-sensitive sectors in the Caribbean.

Wade T. Crow received his PhD in 2001 from Princeton University and is currently a research physical scientist and project lead scientist at the USDA-ARS Hydrology and Remote Sensing Laboratory in Beltsville, Maryland, United States. His research focuses on the development of hydrologic and agricultural applications for remote sensing data and the implementation of appropriate data assimilation approaches to facilitate this goal, with a special emphasis on techniques that fuse information from various disparate remote sensing sources. Wade currently serves on the science teams for the NASA Precipitation Measurement Mission program, the NASA Hydros mission, the NASA SMAP mission, and the NASA AirMOSS mission. Since 2010, he has served as principal investigator on six separate NASA Earth Science projects. Wade recently completed a three-year term as editor of the American Meteorological Society's *Journal of Hydrometeorology*.

Susan M. Cuddy is an environmental technologist at CSIRO Land and Water, Canberra, Australia. Her more than 25 years of history in CSIRO has been focused on working with and within multidisciplinary teams and stakeholder groups to deliver integrated assessment tools to support NRM decision-making—through all software and project life cycle phases from initial elicitation of user requirements, development and implementation, monitoring and evaluation, reporting, and dissemination of results and methods. Susan's key research interests are in the fields of knowledge representation, science packaging for multiple audiences, and stakeholder participatory processes.

Erwin De Nys is program leader for sustainable development at the World Bank, based in Bogotá, Colombia. During 2011–15, he coordinated the World Bank's portfolio in Brazil related to investment projects and analytical work in the field of water resources management, irrigation management, and adaptation to climate change. Erwin holds a PhD in irrigation management from the University of Leuven–KU Leuven, Belgium, and master's degrees from the International Centre for Higher Education in Agricultural Sciences (Montpellier, France) and applied biological sciences from the University of Leuven–KU Leuven. Erwin was coleader of the Drought Preparedness and Climate Resilience technical collaboration program in Brazil described in Chapter 21 of this book.

Lucia De Stefano is associate professor at Complutense University of Madrid (Spain) and deputy director of the Water Observatory of the Botín Foundation, a Spanish think-tank working on water policy in Spain and Latin America. She has worked as a consultant for USAID, the World Bank, the University of Oxford, and Oregon State University. Lucia has been senior researcher at the Botín Foundation and postdoctoral researcher at Oregon State University, United States. Previously she worked as a policy officer for the World Wide Fund for Nature (WWF) and as a water management specialist in the private sector. A hydrogeologist by training, Lucia holds an advanced degree in geological sciences from University of Pavia (Italy) and a PhD in water policy evaluation from Complutense University of Madrid. Her main fields of interest are multilevel water planning, drought management, groundwater governance, transboundary waters, and the assessment of good governance attributes.

Mary Ann Dickinson is the president and CEO of the Alliance for Water Efficiency, a North American NGO focusing on the efficient and sustainable use of water. Headquartered in Chicago, the Alliance works with over 400 water utilities, water conservation professionals in business and industry, planners, regulators, and consumers. In 2014, the Alliance for Water Efficiency received the US Water Prize. Mary Ann has over 40 years of water resources experience. She has authored numerous publications on water conservation, land use planning, and natural resources management. She chairs the International Water Association's Efficient Urban Water Management Specialist Group.

Petr Dobrovolný graduated in physical geography from the J. E. Purkyně University in Brno. Since 2014, he has been affiliated as a professor of physical geography at the Institute of Geography, Masaryk University in Brno, and as a scientist at the Global Change Research Institute, Czech Academy of Sciences in Brno. Petr is especially interested in climate history and quantitative climate reconstructions based on documentary evidence, phenological observations, and tree ring widths. He has a long-term experience with the analysis of frequency and intensity of hydrometeorological extremes derived from documentary evidence and instrumental measurements in central Europe.

Nolan Doesken grew up in rural Illinois with a keen interest in weather. After receiving degrees from the University of Michigan and the University of Illinois, he accepted the position of assistant state climatologist at Colorado State University in 1977. In 2006, Nolan was appointed state climatologist. In that position, he has the responsibility of monitoring and tracking all aspects of Colorado's climate. Weather data collection, drought monitoring, research, and outreach have been key parts of Nolan's career. He has served on Colorado's Water Availability Task Force since it was established in 1981 and has been involved in developing, testing, and applying several drought indexes.

Nathan L. Engle is senior climate change specialist at the World Bank in Washington, DC. His work focuses on climate resilience and adaptation, climate finance, drought policy and planning, and water resources management. Nathan holds a PhD from the University of Michigan's School of Natural Resources and Environment, master's degrees from the University of Michigan in public policy (MPP) and natural resources and environment (MS), and a BS in earth sciences from the Pennsylvania State University. Nathan was coleader of the Drought Preparedness and Climate Resilience technical collaboration program in Brazil described in Chapter 21 of this book.

David Farrell has been the principal of the Caribbean Institute for Meteorology and Hydrology since 2006. He has a PhD and an MS in hydrogeology from the University of Manitoba, Canada, and received a BS (Hons.) in geophysics from the University of Western Ontario, Canada. David has over 20 years of experience working on a range of water resources, environmental, and geological hazard and risk assessment projects worldwide, including the United States, Canada, Morocco, Barbados, Dominica, and other Caribbean islands. He was formerly employed as a research scientist/senior research scientist in the Center for Nuclear Waste Regulatory Analyses at the Southwest Research Institute in San Antonio, Texas, from 1998 to 2006, where he managed several applied research and development projects related to the safe disposal and management of nuclear waste. David provides significant pre- and post-disaster support to the disaster management community in the region.

Taryn Finnessey works on climate change and natural hazard risk management for the State of Colorado. She coordinates Colorado's climate change efforts as well as manages the Colorado state drought mitigation and response planning and implementation efforts. She has overseen and provided technical expertise on the development of approaches for quantifying and considering uncertainties and vulnerability in water resource planning and management. Taryn currently cochairs the Governor's Water Availability Task Force. She holds a BA in earth and environmental science from Wesleyan University and an MA in global environmental policy from the American University.

Brian A. Fuchs is a faculty member and climatologist at the NDMC in the School of Natural Resources at the University of Nebraska–Lincoln. He received a BS in meteorology/climatology in 1997 and an MS in geosciences with an emphasis on climatology in 2000, both from the University of Nebraska. Brian joined the School of Natural Resources in May 2000, working as a climatologist for the High Plains Regional Climate Center. He started working with the NDMC in December 2005. Brian's work is focused on drought-related issues and research involving mitigation, risk assessment, monitoring, and impact assessment. Brian serves as one of the authors of the US Drought Monitor (http://www.droughtmonitor.unl.edu) and the North American Drought Monitor. He directs much of his effort at improving our understanding of drought impacts across a diverse group of sectors such as agriculture, energy, tourism, and transportation as well as the social and environmental impacts of drought.

Dustin E. Garrick is director of the Water Programme at the Smith School of Enterprise and the Environment, University of Oxford. His work focuses at the interface of water and the economy, specializing in the political economy of water allocation reform and water markets as responses to climate change, urbanization, and sustainable development challenges. Dustin's recent book, *Water Allocation in Rivers under Pressure*, assesses the evolution and performance of water markets and basin governance reforms in Australia and the United States. He currently leads a research observatory on decentralization and drought resilience in the Rio Grande/Rio Bravo Basin of Mexico and the United States, where he is developing indicators of fiscal decentralization and assessing new tools and incentives for adapting to drought-related shocks. Earlier this year, Dustin started a new project on water reallocation from agriculture to cities in rapidly growing economies, examining the implications of urban-rural linkages for sustainable development, inequality, and freshwater resilience. His work has included extensive field research in Australia, Mexico, and the United States and has been supported by the Australian Research Council, Canadian Research Council, Global Water Partnership, Fulbright Commission, OECD, and the World Bank.

Nicolas Gerber is senior researcher at the Center for Development Research (ZEF), University of Bonn, since 2007. He received his PhD in economics from the University of New South Wales, Australia. Over the last 15 years, Nicolas has worked broadly on the management of biodiversity, land and water resources, and their relations with poverty, food and nutrition security, and health, mostly in developing countries. He currently leads teams investigating the impacts of innovations and technological change in agriculture, from household- to system-level, and on the linkages between agriculture, water and health, and nutrition.

Christopher Hain is a research scientist at NASA's Marshall Space Flight Center in Huntsville, Alabama. He received his BS in meteorology from Millersville University in 2004 and his MS and PhD in atmospheric science from the University of Alabama in Huntsville in 2007 and 2009. Christopher's research interests include thermal infrared remote sensing with applications in surface energy balance modeling, soil moisture retrieval, hydrologic data assimilation, and drought monitoring. He has played a significant role in the development of the atmosphere land exchange inverse (ALEXI) model in ongoing collaboration with scientists at the USDA-ARS Hydrology and Remote Sensing Lab. ALEXI is currently used to monitor continental evapotranspiration, soil moisture, and drought. Christopher also actively works on finding synergistic relationships between soil moisture retrievals from thermal infrared and microwave methods, while showing the benefit of these two soil moisture methodologies in an EnKF dual data assimilation framework. He actively collaborates with a number of national and international scientists in the field of hydrology, land–surface interactions, and water resources.

Michael J. Hayes is a professor and applied climatologist within the Applied Climate and Spatial Sciences Mission Area in the School of Natural Resources at the University of Nebraska–Lincoln. In October 2016, he stepped down as the director for the NDMC to begin this role as a more traditional faculty member. Prior to this change, he began working at the NDMC in 1995 and he became the NDMC's director in 2007. Dr. Hayes' main interests continue to deal with drought risk management and climate- and water-related issues. Dr. Hayes received a BS degree in meteorology from the University of Wisconsin-Madison and his MS and PhD degrees in atmospheric sciences from the University of Missouri-Columbia.

Petr Hlavinka graduated from the Mendel University of Agriculture and Forestry in Brno. He is affiliated as a research assistant at the Department of Agrosystems and Bioclimatology at the same university and as a scientist at the Global Change Research Institute, Czech Academy of Sciences in Brno. Petr's research is focused on weather and drought impacts within field crops production using various methods, including field experiments, regional statistical data analysis, crop growth models, and remote sensing tools. His particular areas of interest are focused on conditions in central Europe, considering the current climate as well as aspects of expected climate change, including uncertainty assessment.

Mike Hobbins holds a bachelor's degree from the University of Leeds (1989) and an MS and PhD in hydrologic science and engineering from Colorado State University (2000 and 2004). He has since focused his research on evapotranspiration, evaporative demand, and drought, first at the Australian National University and now as a research scientist for NOAA and the University of Colorado in Boulder, Colorado. Mike's recent work supports drought early warning across the United States for the National Integrated Drought Information System and famine early warning across the globe for the Famine Early Warning Systems Network. This work includes the development and dissemination of reanalyses of evaporative demand, the development of the forecast reference evapotranspiration (FRET) product for daily and weekly evaporative demand forecasts across the United States, and the development of the evaporative demand drought index, and a drought monitoring and early warning product.

Jin Huang has been the chief of the Earth System Science and Modeling Division in NOAA's CPO since July 2016. She served as the director of NOAA's Climate Test Bed during 2011–2016 and a program manager at the NOAA/ CPO for the Climate Prediction Program for the Americas during 2001–2010. Jin led and supported a range of research activities, including the North American Multi-Model Ensemble system, climate predictability studies, climate process studies, and model development (e.g., land data assimilation system). As a research scientist at the NCEP Climate Prediction Center (CPC) in 1991–2000, she developed and implemented several operational climate forecast tools, including the optimal climate norm and the CPC soil moisture model, and drought monitoring and outlook products. Jin received her PhD degree in atmospheric science from the University of Illinois, Champaign-Urbana, in 1991.

Oscar F. Ibáñez is a professor at the Universidad Autónoma de Ciudad Juárez, Chihuahua; SNI level 1. He holds a PhD in environmental policy and politics from Colorado State University and an MS degree in environmental engineering from the University of Texas at El Paso. He also has a BS in civil engineering from the Universidad Autónoma de Chihuahua. Dr. Ibáñez is currently an advisor to the president of the Junta Central de Agua y Saneamiento in Chihuahua. He has 30 years of professional experience, including ten years as a consultant on geotechnical, environmental engineering, and public policy in Mexico, the United States, and Canada. Dr. Ibáñez has more than 20 years of experience as a university professor involved in binational environmental cooperation and water resources research and over 12 years of federal and municipal government experience in urban planning, environmental affairs, binational cooperation, and institutional coordination. He is the author of several articles, three book chapters, and four books on water, drought, environmental cooperation, energy, and sustainable development.

Anthony J. Jakeman is a professor in the Fenner School of Environment and Society and director of the Integrated Catchment Assessment and Management Centre at The Australian National University, Canberra, Australia. He has research interests in hydrological modeling and in integrated assessment and decision support of water resource issues. Anthony leads the integration program of Australia's National Centre for Groundwater Research and Training. In 2015, he was listed as a highly cited researcher by Thomson-Reuters and in 2016 was elected a fellow of the American Geophysical Union. Anthony is honorary editor-in-chief of the journal *Environmental Modelling and Software*.

Antonio Joyette has worked for 25 years in the field of meteorology with the St. Vincent and Grenadines Meteorological Service. His experience ranges from operational management to consultant researcher, in which he worked on national and regional projects. Antonio is presently a doctoral candidate and an associate lecturer in the postgraduate program at the Centre for Resources Management and Environmental Studies, University of the West Indies, Cave Hill Campus. He has written several publications and presented some of his research at regional and international forums, winning awards. His current research focuses on drought exposure in the eastern Caribbean. Antonio's research interests include small island vulnerability to mesoscale weather hazards, natural hazard risk reduction, and the impacts of climate change on small states.

Christophe Lavaysse is an atmospheric physicist and climatologist working on atmospheric processes and ensemble forecasts. He received an MS in physics and a PhD from the University of Grenoble, France. During Christophe's PhD and two years of postdoc at LATMOS (Paris, France), he studied the origins of precipitation variabilities in West Africa. He then worked at the LMD (Paris) to assess the uncertainties of downscaling extreme meteorological events using statistical and numerical methods. In 2009, Christophe studied surface/precipitation interactions at McGill University (Montreal, Quebec, Canada) and, in 2012, he was visiting scientist at NCAR (Boulder, Colorado, USA), quantifying the impacts of aerosols on precipitation. In 2013, Christophe joined the European Commission's Joint Research Centre, where he is in charge of the assessment and improvement of early forecasts of extreme meteorological events using Ensemble Prediction Systems.

Rebecca (Letcher) Kelly is an experienced researcher and modeler who has worked in the field of natural resources management for over a decade. Before forming is NRM, Rebecca was a fellow at the Australian National University. She has worked very actively with industry, government, NGOs, and the community on a broad range of environmental issues in Australia as well as overseas. In this capacity, Rebecca has developed substantial experience in community participation and workshop facilitation, decision support design, and integrated modeling. She has undergraduate degrees in science and economics with 1st Class Honors in mathematics and a PhD in resource and environmental management.

Doug Kluck is the Central Region Climate Services director for the NOAA. He has worked for NOAA since 1992 with the National Weather Service and the National Center for Environmental Information. In his capacity with NOAA, Doug has developed a regional network among federal, state, tribal, academics, and private interests for delivering risk management information around extreme climate events, such as drought and major flooding.

Antonio Rocha Magalhães is a Brazilian economist who served in different positions in the Brazilian and State of Ceará governments. He taught economics at the University of Ceará and public policy at the University of Texas, Austin. Antonio was secretary of planning of the State of Ceará, vice minister of planning of Brazil, and president of the Committee of Science and Technology of the UNCCD. He also worked for the World Bank in Brazil and currently coordinates research projects at the Center for Strategic Studies and Management (CGEE), in Brasilia, with a focus on water and droughts in Northeast Brazil. Antonio was a member of the Committee for the Preparation of the High-level Meeting on National Drought Policy (HMNDP). He has published extensively on climate and development issues.

Rodrigo Maia has a PhD in civil engineering from the Faculty of Engineering of the University of Porto (FEUP). He is an associate professor at FEUP and vice president of European Water Resources Association (EWRA). Rodrigo is a research project coordinator on water resources management with an emphasis on hydrological extreme events, namely, due to climate change effects, fluid mechanics and its applications to channel flows, hydraulic studies, and river rehabilitation. He has been involved in studies and research on international river basins, namely, the Iberian Peninsula. Rodrigo has been a research group coordinator in several national and international (EU) projects and was involved in the creation of a drought early warning management system for the Portuguese National Water Authority. He is the author/coauthor of six chapters in books, with more than 140 publications in refereed international periodicals and papers in conference proceedings.

Roche Mahon is a postdoctoral researcher at the Caribbean Institute for Meteorology and Hydrology. Roche is a social scientist with a background in disaster risk management and climate change adaptation. He is part of the regional team of climatologists, meteorologists, and hydrologists designing, developing, and delivering user-oriented climate information products and services to six climate-sensitive sectors in the Caribbean. A core focus of that work is establishing sectoral early warning information systems that explicitly integrate climate information into the decision-making contexts of the agriculture, water, health, tourism, energy, and disaster risk management sectors.

Annarita Mariotti is senior scientist and manager at NOAA's Office of Oceanic and Atmospheric Research, Climate Program Office. She is the director of the modeling, analysis, predictions, and projections program, a mission-driven research program engaging over 120 scientists annually to advance NOAA's capabilities to monitor, simulate, and predict earth system variability and change. In her tenure, Annarita pioneered the organization of the drought task force that over the years has made significant research contributions to the advancement of the national integrated drought information system. In her research career, she has focused on investigating water cycle–climate linkages in an interdisciplinary earth system framework considering the broad range of relevant processes and impacts. In particular, Annarita's research has advanced the understanding of water cycle variability and predictability in the European-Mediterranean region. Her work has resulted in over 30 research papers and many book chapters.

Daniel McEvoy's research is interdisciplinary, spanning the fields of climatology, hydrology, and meteorology, with a focus on drought monitoring and prediction, and the role of evaporative demand on drought dynamics. He holds a BS in environmental science from Plattsburgh State University and an MS and PhD in atmospheric science from the University of Nevada, Reno. Daniel is currently an assistant research professor of climatology at the Western Regional Climate Center. His work at the Western Regional Climate Center has created numerous advances and products related to drought monitoring. Daniel was a co-developer of the WestWide Drought Tracker, Climate Engine, and the Evaporative Demand Drought Index. His other areas of research interest include quality and uncertainty assessment of weather observations, climate modeling, and snow drought.

Chad McNutt has worked with federal, state, tribal, and local governments, as well as individuals, to develop drought early warning systems around the United States. Before joining NIDIS, he served as a John A. Knauss Marine Policy Fellow in Washington, DC, and as a policy advisor and executive director of NOAA's Office of the Assistant Secretary. While in the Office of the Assistant Secretary, Chad worked on climate science policy related to NOAA and the US Climate Change Science Program. He received a PhD in biology and biochemistry from the University of Houston and a BS degree from Texas A&M University.

Jiří Mikšovský graduated in physics from Charles University in Prague. He is an assistant professor at the Department of Atmospheric Physics, Faculty of Mathematics and Physics, Charles University, Prague, while also being a researcher at the Global Change Research Institute, Czech Academy of Sciences in Brno. Jiří specializes in the application of statistical methods to meteorological and climatic data, with particular focus on multivariable time series analysis and investigation of manifestations of nonlinearity in climatic measurements and simulations.

Alisher Mirzabaev is a senior researcher at the Center for Development Research (ZEF), University of Bonn. His current research areas include economics of land degradation, sustainable energy and energy transitions, and the water-energy-food security nexus. Before joining ZEF in 2009, he was an economist with the International Center for Agricultural Research in the Dry Areas (ICARDA). Dr. Mirzabaev holds a PhD in agricultural economics from the University of Bonn in Germany.

Blair E. Nancarrow is a director of a social science consulting partnership, Syme and Nancarrow Water. Now semiretired, she spent 21 years as a social scientist in CSIRO Land & Water Australia and three years as a visiting fellow at the Fenner School of Environment and Society at the Australian National University, Canberra. Blair's particular area of expertise has been in the design and implementation of large community-based research programs and experiments in the people/environment interface. She has a special interest in the development of methods and processes to ensure the meaningful incorporation of social science in triple-bottom-line analysis of environmental policy and developments. In her quest to integrate social science with the more technical sciences, she received a Biennial Medal for services to the Modelling and Simulation Society in Australia and New Zealand.

Jason Otkin has been involved in research at SSEC/CIMSS for more than 13 years. He has extensive experience interpreting drought signals in satellite remote sensing datasets and has developed new analysis and visualization tools to quickly identify areas experiencing flash drought conditions. Jason has authored or coauthored more than 40 peer-reviewed publications covering diverse topics, including assessing the response of soil moisture and evapotranspiration datasets to rapid drought onset and intensification, conducting focus group meetings to facilitate the use of drought early warning information by agricultural stakeholders, land surface flux modeling, and validating satellite-derived insolation datasets.

Theib Oweis is currently interim director of the Integrated Water and Land Management Program (IWLMP) at the International Center for Agricultural Research in the Dry Areas (ICARDA). Formerly, he was a Distinguished Guest Professor of water management in agriculture at the International Platform for Dryland research and Education (IPDRE) of Tottori University in Japan. Since 1991, Theib worked for ICARDA as director of research, scientist, and research manager. Earlier, he worked for the University of Jordan, Amman, as an assistant professor in irrigation and drainage, and in the 1970s he worked for Dar Alhandash Consultants (Shaer and Partners) as a field irrigation engineer in south Yemen. Theib received his BSc degree in agriculture from Aleppo University, Syria, and his MSc and PhD degrees in agricultural and irrigation engineering from Utah State University, Logan, Utah. He has over 40 years of experience in international research, development, human capacity building, and in the management of water for agriculture in dry areas. Theib has authored over 200 refereed journal publications, books/book chapters, and conference proceedings in the areas of water use efficiency, supplemental irrigation, water harvesting, water productivity, deficit irrigation, salinity, and the management of scarce water resources.

Mario López Pérez is an agronomy engineer from UAM Xochimilco and has an MS in pedology and soil survey from the University of Reading, United Kingdom. He has served for 34 years in the Federal Water Sector of Mexico, particularly in the National Water Commission. Mario is currently the hydrology coordinator at the Mexican Water Technology Institute. Among his professional experiences, he has headed joint projects with US institutions like EPA, DOI, IBWC, NOAA, USBR, and USGS, as well as WWF, WMO, IHP-UNESCO, and Guatemalan, Belizean, Colombian, Brazilian, and Peruvian water institutions. He has offered conferences, panels, and workshops in more than 15 countries regarding water planning, management, basin councils, drought management, climate change, and transboundary water management. Mario has authored publications in national and international congresses, scientific journals, and technical books, and he has been a member of the Mexican Hydraulic Association since 1980.

Christa D. Peters-Lidard is currently the deputy director for Hydrosphere, Biosphere, and Geophysics in the Earth Sciences Division at NASA's Goddard Space Flight Center. She was a physical scientist in the Hydrological Sciences Laboratory from 2001 to 2015 and laboratory chief from 2005 to 2012. Christa's research interests include land–atmosphere interactions, soil moisture measurement and modeling, and the application of high-performance computing and communications technologies in earth system modeling, for which her land information system team was awarded the 2005 NASA Software of the Year Award. She is currently the chief editor for the American Meteorological Society (AMS) *Journal of Hydrometeorology* and has served as an elected member of the AMS Council and Executive Committee. Christa is a member of Phi

Beta Kappa, and was awarded the COSPAR Zeldovich Medal in 2004 and the Arthur S. Flemming Award in 2007. She was elected as an AMS fellow in 2012.

Frederik Pischke is the senior program officer of the Global Water Partnership seconded to the World Meteorological Organization in Geneva, Switzerland, for the IDMP and the Associated Programme on Flood Management. His work includes developing the IDMP's regional efforts in the Horn of Africa, West Africa, Central and Eastern Europe, South Asia, and Latin America. Frederik has worked on natural resources management for the UN Secretariat and UN-Water in New York. He has also worked for the FAO in Rome and Latin America, the International Water Management Institute in Sri Lanka, the German Development Cooperation in Bolivia, and the Inter-American Development Bank at Mexico's National Water Commission in Mexico City. He has an MS in environmental sciences from the University of East Anglia.

Kelly T. Redmond was the deputy director and regional climatologist of the Western Regional Climate Center at the Desert Research Institute in Reno, Nevada. He completed a BS degree in physics from the Massachusetts Institute of Technology and MS and PhD degrees in meteorology from the University of Wisconsin in Madison. His research and professional interests spanned every facet of climate and climate behavior; climate's physical causes and behavior; studies of how climate interacts with other human and natural processes; and how such information is acquired, used, communicated, and perceived.

Ladislava Řezníčková graduated in physical geography from the Masaryk University in Brno. She is affiliated with the Institute of Geography, Masaryk University in Brno, and at the Global Change Research Institute, Czech Academy of Sciences in Brno. Ladislava participated in several national and international grant projects. Her scientific research is primarily oriented to historical climatology based on the study of weather records derived from documentary evidence in the pre-instrumental period and on the analysis of systematic meteorological measurements. Ladislava also focuses on the analysis of hydrometeorological extremes and spatiotemporal variability of the climate in the Czech lands.

Siegfried Schubert is a senior research scientist (emeritus) in NASA's Global Modeling and Assimilation Office. His research interests include climate variability and predictability, droughts, the hydrological cycle, extreme weather and climate events, and reanalysis. Siegfried has authored or coauthored over 140 research articles, serving as an editor of the *Journal of Climate* and helping to organize two special collections of the *Journal of Climate* on drought. He helped organize three international workshops to assess the current understanding of drought worldwide with the goal of developing a global drought information system. Siegfried also helped to coordinate and facilitate NOAA-funded drought research projects, including the development of a special collection of the *Journal of Hydrometeorology* on the topic of drought monitoring and prediction. He is a fellow of the American Meteorological Society (since 2013), and received the NASA Exceptional Service Medal in 2016.

Mannava V. K. Sivakumar obtained a BS (agriculture) from the Acharya NG Ranga Agricultural University in India, an MS (agronomy) from the Indian Agricultural Research Institute in India, and a PhD in agricultural climatology from Iowa State University. Mannava has over 40 years of international experience in the area of agroclimatology, having served in the International Crops Research Institute for the Semi-Arid Tropics for 20 years in India (1977–83) and in West Africa (1984–96). He also served in the WMO for 20 years as the chief of the Agricultural Meteorology Division (1996–2008), director of the Climate Prediction and Adaptation Branch (2008–2012), senior consultant (2012–2015), and acting secretary of the Intergovernmental Panel on Climate Change (2016). He has over 300 publications to his credit, including 54 books and 87 articles in various international journals. In recognition of his international contributions, Mannava was elected fellow of several international scientific societies, including the National Environmental Science Academy of India, the American Society of Agronomy, the Royal Academy of Overseas Sciences of Belgium, the Academy of Georgofili of Italy, and the Association for the Advancement of Biodiversity Science.

Robert Stefanski is currently chief of the Agricultural Meteorology Division of the WMO in Geneva, Switzerland. His duties include developing the scientific and technical aspects of this division and providing technical and scientific support to the WMO Commission for Agricultural Meteorology. Bob is also the head of the Technical Support Unit of the IDMP, which works with countries in developing drought early warning systems and national drought policies. From 2005 to 2010, he was the scientific officer in the same program at WMO. From 1991 to 2004, he was an agricultural meteorologist for the US Department of Agriculture, where he monitored worldwide weather and climate conditions and how they impacted global agricultural production. He has an MS degree in agricultural climatology from Iowa State University (1988) and a BS in agricultural meteorology from Purdue University (1986).

Petr Štěpánek graduated in physical geography at the Masaryk University in Brno. Later, he worked as head of the Department of Meteorology and Climatology at the Czech Hydrometeorological Institute, regional office Brno. Petr is affiliated as a scientist with the Global Change Research Institute, Czech Academy of Sciences. His scientific activities are homogenization and quality control of the climatological data, administration of climatic databases, and analysis of the data (both spatial and temporal), as well as climate variability. Petr is author of a software package used worldwide for homogenization and processing of climatological data (www.climahom. eu). He also devotes time to the processing and correction of regional climate model outputs and is an expert in statistical downscaling. Petr has participated in many national and international research projects.

Roger C. Stone has a lengthy career in climatological research and operational meteorology extending over 35 years. He received his PhD from the University of Queensland in 1992 and has since held positions of senior and principal research scientists in state and federal departments, director of three government research and development centers, and, more recently, director of the International Centre for Applied Climate Sciences at the University of Southern Queensland, Toowoomba. Roger has provided frequent science policy input to many international, national, and state organizations, and directly to rural industry. He served as a member of the Federal Rural Adjustment Scheme Advisory Council and has provided high-level advice to state premiers and associated departments on drought issues. Roger currently serves as Open Program chair and Management Group member of the WMO Commission for Agricultural Meteorology and an "expert team leader," and cochair of expert teams and advisory committees within the WMO Commission for Climatology. He has represented Australia at eight Commission for Climatology Congress Sessions and Commission for Agricultural Meteorology Sessions since 1996.

Mark D. Svoboda is an associate research professor and director of the NDMC in the School of Natural Resources at the University of Nebraska–Lincoln. A geographer and climatologist by training, he has been a member of the NDMC since it was founded in 1995. As director of the NDMC, Mark's duties include overseeing the center's staff and mission. The NDMC's mission is directed at the conduct of research and outreach activities associated with drought policy, planning, and monitoring/early warning. Mark cofounded the development of the US Drought Monitor (USDM) in 1999 and serves as one of the principal authors of both the weekly USDM and monthly North American Drought Monitor products. He is heavily involved with drought monitoring, assessment, and prediction committees at state, regional, and national levels. In addition, Mark has extensive experience working with the international drought, water, and climate community via project collaboration, advisory panels, and consultation, and through various hands-on training and outreach opportunities with over 50 countries and international organizations to date. He earned his BS in geography specializing in climatology; his MS in geography with a specialization in remote sensing, climatology, and GIS; and his PhD in natural resources with a human dimensions specialization, all from the University of Nebraska–Lincoln.

Tsegaye Tadesse is a research associate professor at the University of Nebraska–Lincoln's School of Natural Resources and a climatologist/ remote sensing expert at the National Drought Mitigation Center. His research interests and areas of expertise include drought monitoring, remote sensing/GIS, data mining, risk management, and climate variability and change, specifically focused on drought monitoring and prediction at global, regional, and local scales. Tsegaye's research interests also include the impacts of drought on food security. Tsegaye has extensive experience in developing drought monitoring and vegetation outlook tools. He has led/co-led several projects funded by NASA, NOAA, USDA, and USAID, among others. Tsegaye received his PhD in agro-meteorology from the University of Nebraska–Lincoln, his MS in space studies from the International Space University (Strasbourg, France), and his BS in physics from Addis Ababa University (Ethiopia).

Dennis Todey is the director of the US Department of Agriculture Midwest Climate Hub in Ames, Iowa. He works at the nexus of climate and agriculture issues in an eight-state region covering most of the US Corn Belt. Dennis was previously the state climatologist for South Dakota and associate professor at South Dakota State University. He has worked on regional collaboration of climate services in the Missouri River Basin and Midwest for over a decade in conjunction with NOAA, Regional Climate Centers, state climatologists, and state extension. Dennis was the president of the American Association of State Climatologists from 2010 to 2012.

Miroslav Trnka studied applied landscape ecology (Mendel University in Brno) and law and the law science (Masaryk University in Brno). Presently, he leads research in the Domain of Climate Analysis and Modelling at the Global Change Research Institute, Czech Academy of Sciences in Brno, and teaches graduate courses as professor of agronomy at the Mendel University in Brno. Using his experience gained at the University of Applied Life Sciences in Vienna and the National Drought Mitigation Center at the University of Nebraska–Lincoln, Miroslav has initiated and has been leading a research program focused on drought monitoring and forecasting for the Czech Republic and central Europe (www.drought.cz). He also leads climate change research projects that have resulted in a national climate change information system (www.czechadapt.cz). Miroslav has been trying to improve the current understanding of the drought phenomena through multidisciplinary collaboration using a wide range of experimental, laboratory, and modeling techniques. These efforts are intertwined with his research of climate change impacts and adaptation in the field of agronomy.

Adrian Trotman is the chief of Applied Meteorology and Climatology at the CIMH. Adrian has responsibility for climate-related issues, applications of meteorology to multiple economic and social sectors, data management and dissemination, and the provision of weather and climate information and advice to multiple economic sectors. He is a trained agrometeorologist with more than two decades of experience, and driven by his interest in agricultural drought and its impacts, he established the Caribbean Drought and Precipitation Monitoring Network in 2009. Adrian has managed, and continues to manage, other regional programs and projects that provide the impetus for CIMH to become a WMO Regional Climate Centre. He supports the work of WMO as a member of Expert Teams of the Commission for Agricultural Meteorology and the Commission for Climatology, and on other committees.

Andrea Turner is a research director at the Institute for Sustainable Futures at the University of Technology Sydney. With over 20 years' engineering and research experience in water, wastewater, storm water, and environmental assessment, and having worked in the United Kingdom, Hong Kong, and Australia, Andrea has worked with water service providers in most major cities in Australia and many regional centers. She has presented the findings of her research and provided workshops on end-use analysis, demand management, and IRP nationally and internationally. Andrea has also been involved in research associated with smart meters, water efficiency, distributed systems, and water recycling.

Cedric Van Meerbeeck is climatologist at the CIMH. He was previously a researcher in Paleoclimate Modeling at the VU University Amsterdam, The Netherlands. Cedric's current responsibilities include developing climate information products and services, coordinating applied climate research, and delivering training of professional meteorologists. He is a CIMH lecturer at the University of the West Indies, Cave Hill Campus, Barbados, in the BS Meteorology Programme. Cedric is especially credited for implementing regional climate early warning efforts through the Caribbean Climate Outlook Forum, which is setting a standard for small island developing states (SIDS) around the world.

Sergio M. Vicente-Serrano is a scientific researcher at the Department of Geoenvironmental Processes and Global Change of the Pyrenean Institute of Ecology (Spanish National Research Council). Sergio's main interest deals with different environmental topics related to global change such as changes in land cover, the influence of general atmospheric circulation on water resources availability, and, mainly, drought studies from different perspectives: drought indices, variability, trends, environmental consequences, and mapping. He has worked on national and European projects devoted to remote sensing, water resources management, and droughts, and published more than 150 papers in international journals in the fields of meteorology and atmospheric sciences, water resources, geosciences, and remote sensing. Sergio received a BS in geography from the University of Zaragoza, an MS in remote sensing from the Instituto de Estudios Espaciales de Cataluña, and a PhD in physical geography from the University of Zaragoza.

Amy L. Vickers is a nationally recognized water conservation and efficiency expert, engineer, and author of the award-winning *Handbook of Water Use and Conservation: Homes, Landscapes, Businesses, Industries, Farms* (WaterPlow Press). She is president of Amy Vickers & Associates, Inc., in Amherst, Massachusetts, USA, a consulting practice that works with US and overseas water utilities, agencies, and organizations on a wide range of water use and conservation projects. Amy is also the author of the first national water efficiency standards for plumbing fixtures that were adopted under the US Energy Policy Act of 1992—now saving 7 billion to 9 billion gallons daily. She holds an MS in engineering from Dartmouth College and a BA in philosophy from New York University. More: www.amyvickers.com.

Jürgen V. Vogt is a senior scientist at the European Commission's Joint Research Centre in Ispra, Italy. He received a diploma in rural survey from ITC Enschede, the Netherlands, and MS and PhD degrees in physical geography, climatology, and remote sensing from Trier University, Germany. Jürgen has been active in the implementation of the European Water Framework Directive and led the development of the CCM River and Catchment database for Europe. Currently, he is leading a research team on drought, including the development of the European Drought Observatory. In 2016, Jürgen initiated the development of a global drought observatory in support of the EU's civil protection and humanitarian aid services. His research interests include remote sensing and GIS applications in the fields of land management, natural disasters, and desertification. Jürgen's current focus is on monitoring, forecasting, and projecting droughts and their impacts and on the assessment of the societal and environmental vulnerability to extreme events.

Adam Vizina graduated from the Czech University of Life Sciences in Prague. He is affiliated as a scientist with the T. G. Masaryk Water Research Institute in Prague and with the Czech University of Life Sciences in Prague. Adam is a representative of the Czech Republic in FRIEND (Flow Regimes from International Experimental and Network Data–UNESCO). He also participated in several national and international research and commercial projects. Adam's activities are focused on water management, surface and subsurface hydrology, assessing the potential climate change impacts on catchment water balance, extreme hydrological events and hydrological processes in landscape, and evaluation of the effect of climate change adaptation measures.

Brian D. Wardlow is director of the Center for Advanced Land Management Information Technologies and associate professor in the School of Natural

Resources at the University of Nebraska–Lincoln, USA. His expertise is in the field of remote sensing as applied to drought monitoring, crop characterization, land surface phenology, land use/land cover mapping, and agricultural and natural resource applications. Brian has worked extensively in the development of remote sensing-based drought monitoring tools throughout the world involving methods to assess vegetation health, evapotranspiration, soil moisture, and groundwater. He coedited a book entitled *Remote Sensing of Drought: Innovative Monitoring Approaches* and currently cochairs the WMO's Task Team of the Use of Satellite Remote Sensing for Climate Monitoring. Brian was also a NASA Earth System Science graduate research fellow at the Kansas Applied Remote Sensing Program, University of Kansas, and a remote sensing specialist at the US Geological Survey Earth Resources Observation Science Center. He received a BS in geography from Northwest Missouri State University, an MA in geography from Kansas State University, and a PhD in geography from the University of Kansas.

Stuart White is the director of the Institute for Sustainable Futures at the University of Technology Sydney, and leads a team of researchers across a range of aspects of "creating change toward sustainable futures." Stuart is a member of the International Water Association Strategic Council and deputy chair of the Efficient Urban Water Management Specialist Group. He has led research, written, and spoken on integrated water supply demand planning widely in Australia and internationally over the last 25 years. In 2012, he received the Australian Museum Eureka Prize for Environmental Research.

Andy Wood is currently a project scientist at the National Center for Atmospheric Research in Boulder, Colorado. He worked as a research assistant professor at the University of Washington before becoming the lead scientist of the Seattle firm 3TIER, Inc., focusing on forecasting and assessment of hydropower, solar, and wind energy. Andy later worked as a hydrologist with two NOAA River Forecast Centers. Major accomplishments include developing statistical climate downscaling techniques and creating operational streamflow and drought monitoring and forecasting systems. He chaired the AMS Hydrology Committee from 2011 to 2013, and is currently an editor with the AMS *Journal of Hydrometeorology*. Andy recently served as a coleader of the NOAA MAPP Drought Task force, and currently coleads the international Hydrologic Ensemble Prediction Experiment. His recent work applies scientific advances in hydrologic, weather, and climate modeling, prediction, and projection to improve our management of water, energy, and terrestrial ecosystems.

Eric F. Wood holds the Susan Dod Brown Professorship in Civil and Environmental Engineering at Princeton University, where he has taught since 1976. His research area is hydroclimatology, with an emphasis on the modeling and analysis of the global water and energy cycles through land surface modeling, satellite remote sensing, and data analysis. Eric's foci include the monitoring and forecasting of drought, hydrologic impacts from climate change, and seasonal hydrological forecasting. He participates in WCRP's Global Energy and Water EXchange (GEWEX) activities to develop long-term Climate Data Records of the terrestrial surface heat flux data sets for climate studies. Eric has received a *Doctor Honoris Causa* from Gent University (Belgium) in 2011, the European Geosciences Union's Alfred Wegener Medal and John Dalton Medal, AMS's Jules G. Charney Award and Robert E. Horton Memorial Lectureship, and AGU's Hydrology Section's Robert E. Horton Award. He was elected to the U.S. National Academy of Engineering, is a fellow of the Royal Society of Canada, and a foreign fellow of the Australian Academy of Technological Sciences and Engineering (ATSE), the American Geophysical Union, and the American Meteorological Society.

Zdeněk Žalud graduated from the University of Agriculture in Brno. Currently, he is working as a head of the Institute for Agrosystems and Bioclimatology at Mendel University in Brno, where he is responsible for subjects dealing with agrometeorology and bioclimatology. Zdeněk is also an affiliated scientist at the Global Change Research Institute, Czech Academy of Sciences in Brno. He has participated in many national and international projects. His scientific work supports the scientific leadership of the university research plan oriented to biological and technological aspects of sustainability of controlled ecosystems and their adaptability to climate change. Zdeněk's areas of research involve crop modeling, climate change impacts on crop growth and development, and ecosystem services focused on meteorology and climatology aspects.

Sergio A. Zelaya-Bonilla is currently the senior land and water officer of FAO's Land and Water Division. Since the 1990s, he has been involved in advancing and implementing sustainable development policies, particularly related to the Rio Conventions at the regional and global levels. Sergio was vice minister of the environment of Honduras during 1996–1998, during the first stages of implementation of the Rio conventions and Agenda 21. Later on, he also actively participated in the Rio+20 and SDG process. From 2001 to 2015, Sergio worked in the Secretariat of the UNCCD in different capacities, first as a coordinator of the Latin America and the Caribbean Programme, then as coordinator of the unit on policies for advocacy on global and emerging issues, and as a special advisor to the executive secretary. His professional career also includes academic and consultancy experience with the National University of Honduras and with UNDP, ECLAC, and several civil society organizations. Sergio is originally from Honduras. His training was as an environmental economist (Vanderbilt University) and he received a master's degree in international trade (Duke University) and on economic development (National University of Honduras).

Section I

Overview

1

Drought as Hazard: Understanding the Natural and Social Context*

Donald A. Wilhite and Roger S. Pulwarty

CONTENTS

1.1 Introduction

Drought is an insidious natural hazard that results from a deficiency of precipitation from average or "normal" that, when extended over a season or longer, results in water supplies that are insufficient to meet the demands of human activities and the environment. Other factors, such as temperature, low humidity, and wind, can also contribute to the severity and duration of a drought episode. Temperature is an especially important additional variable because of its impact on atmospheric demand, a factor of increasing importance in a warming world. As observed by the UN Convention to Combat Desertification and others, desertification, agricultural demands, land degradation, and drought are contributing to a global water crisis.

Water-related crises, ranging from drought impacts on the most productive farmlands to access to safe drinking water, pose the most significant threats facing the planet over the next decade. The World Economic Forum

---

* This chapter is a significant revision of the original chapter from the first edition of *Drought and Water Crises*. We recognize the contributions of Margie Buchanan-Smith to the original chapter.

(2015) has concluded that, at the global level, demand is anticipated to exceed supply by 40 percent within 15 years. As noted in Wilhite and Pulwarty (2005), a crisis can be defined as an unstable or crucial time or state of affairs whose outcome will make a decisive difference for better or for worse. The word *crisis*, taken from the Greek *krisis*, literally means *decision*. Decision makers will be forced to make hard choices about allocations of water that will have significant impacts across the economy and the environment. The difficulty of these choices will only be exacerbated during drought, and even more so, if droughts are severe and sustained (World Economic Forum 2015).

Drought by itself is not a disaster. Whether it becomes a disaster depends on its impact on local people and the environment, and their level of resilience to an extended period of deficient precipitation. Adaptations to reduce vulnerability before drought events, such as improving water use efficiency, are similar to actions recommended or implemented as an event unfolds, differing from emergency response for other hazards. Early warning systems (hereafter EWS) in such contexts are needed not only for event onset, at which a threshold above some socially acceptable or safe level is exceeded, but also for intensification and duration, ranging temporally from a season to decades and spatially from a few hundred to hundreds of thousands of square kilometers (Pulwarty and Verdin 2013).

Drought resilience is enhanced significantly by preparedness plans and policies that emphasize vulnerability assessments for key sectors and the implementation of risk reduction measures that will mitigate future impacts associated with droughts. Therefore, the key to understanding and managing drought more effectively is an enhanced understanding of both its natural and social dimensions, including the decision-making arrangements that allow or hinder proactive responses. Integrated drought and water scarcity management approaches are increasingly recognizing the urgent need for multistakeholder platforms, at the country, community, and transboundary levels, for the implementation of joint strategies and the coordinated response and prevention of crises.

Drought is a normal part of climate, rather than a departure from normal climate (Glantz 2003). The latter view of drought has often led policy- and other decision makers to treat this complex phenomenon as a rare and random event. This perception has typically resulted in little effort targeted toward those individuals, population groups, economic sectors, regions, and ecosystems most at risk (Wilhite 2000; Sivakumar et al. 2014). Improved drought policies and preparedness plans that are proactive rather than reactive, and that aim at reducing risk rather than responding to crisis, are more cost-effective and can lead to more sustainable resource management and reduced interventions by government and aid agencies (Wilhite et al. 2000a, 2014; WMO and GWP 2014; also see Chapters 2, 3, 4, and 5).

The primary purpose of this chapter is to discuss drought in terms of both its natural characteristics and its human dimensions, as these shape the effectiveness of responses during water-related crises (Figure 1.1). As this diagram illustrates, all droughts begin with a deficiency of precipitation (i.e., meteorological drought) that can be aggravated by high temperatures and other factors, as noted previously. As illustrated in the canonical case shown in Figure 1.1, when this period of precipitation deficiency continues, it begins to affect agricultural production of biomass through reductions in soil moisture, leading to a loss of biomass production (i.e., agricultural drought). If drought persists for longer periods, the impacts become more and more complex with adaptation buffers (storage, aquifers, and efficiency practices) being depleted, resulting in increasing conflicts between water users from a multitude of sectors. For example, persistent drought conditions reduce streamflow, reservoir and lake levels, snowpack, and groundwater levels (i.e., hydrological drought), and result in significant impacts on hydroelectric power production, recreation and tourism, irrigated agriculture, ecosystems, and other sectors. Although all droughts originate from a deficiency of precipitation, the magnitude of impacts associated with these types of drought, as well as with socioeconomic and political drought, is largely the result of water and land management practices and policies. As a result, the impacts that occur with these other types of drought are less directly associated with

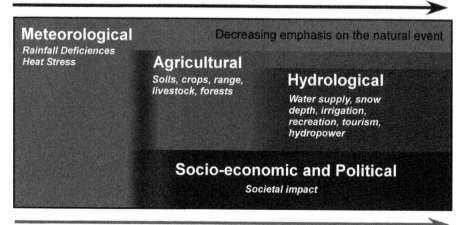

FIGURE 1.1
Natural and social dimensions of drought. (Courtesy of National Drought Mitigation Center, University of Nebraska, Lincoln.)

the physical event (i.e., a precipitation deficit) and more with how water and other natural resources are managed prior to and during a drought episode (see Sections 1.2 and 1.3 for more discussion on this topic). Drought risk reduction must focus heavily on changing management practices and policies at various levels.

This chapter provides readers with an overview of the concepts, characteristics, and impacts of drought, and a foundation for a more complete understanding of this complex hazard—how it affects people and society, and, conversely, how societal use and misuse of natural resources and government policies can exacerbate vulnerability to this natural hazard. The chapters in this volume promote a holistic and multidisciplinary approach to drought management—one that is focused on managing risk rather than the more typical approach of responding to an event after it has become a disaster (i.e., crisis management). This discussion is critical to an understanding of the material presented in subsequent sections of this volume (Parts II and III) as well as in the various case studies presented in Part IV.

We use the term *hazard* to describe the natural phenomenon of drought and the term *disaster* to describe significant negative human and environmental impacts that result after adjustment systems have been overwhelmed and external support is needed.

1.2 Drought as Hazard: Concepts, Definition, and Types

Drought differs from other natural hazards in several ways. First, drought is a slow-onset hazard, often referred to as a *creeping phenomenon* (Gillette 1950). Figure 1.2 further illustrates the hypothetical life cycle of a typical drought and the compounding aspects of its development and impacts. Because of the creeping nature of drought, its effects accumulate slowly, usually over several months or longer. Therefore, the onset and end of drought are difficult to determine, and scientists and policymakers often disagree on the bases (i.e., criteria) for declaring an end to drought. Tannehill (1947) notes:

> We may truthfully say that we scarcely know a drought when we see one. We welcome the first clear day after a rainy spell. Rainless days continue for some time and we are pleased to have a long spell of fine weather. It keeps on and we are a little worried. A few days more and we are really in trouble. The first rainless day in a spell of fine weather contributes as much to the drought as the last, but no one knows how serious it will be until the last dry day is gone and the rains have come again ... we are not sure about it until the crops have withered and died.

Should drought's end be signaled by a return to normal precipitation and, if so, over what period of time does normal or above-normal precipitation

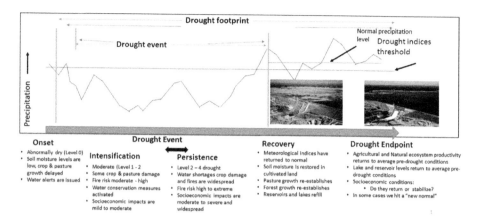

FIGURE 1.2

Increasing recognition of the multiple dimensions and timescales of drought. (Courtesy of World Meteorological Organization's Climatology and Agrometeorology Commissions.)

need to be sustained for the drought to be declared officially over? Is the drought event over only when precipitation deficits that accumulated during the drought event are erased? Do reservoirs and groundwater levels need to return to normal or average conditions? Impacts linger for a considerable time following the return of normal precipitation, so is drought terminated by meteorological or climatological factors, or by the diminishing negative impact on society and the environment?

Second, the absence of a precise and universally accepted definition of drought adds to the confusion about whether a drought exists and, if it does, its degree of severity. Realistically, definitions of drought must be region and application (or impact) specific (Wilhite and Glantz 1985). Definitions must be region specific because each climate regime has distinctive climatic characteristics (i.e., the characteristics of drought differ significantly between regions such as the North American Great Plains, Australia, eastern and southern Africa, western Europe, and northwestern India). Definitions need to be application specific because drought, like beauty, is defined by the beholder and how it affects his or her activity or enterprise. Thus, drought means something different for a water manager, a commodities producer, a hydroelectric power plant operator, a subsistence farmer, and a wildlife biologist. Even within sectors, such as agriculture, there are many different perspectives of drought because impacts may differ markedly for crop and livestock producers and agribusiness. For example, the impacts of drought on crop yield will differ for maize, wheat, soybeans, and sorghum because each crop is planted at a different time during the growing season and has different sensitivities to water and temperature stress at various growth stages. Management factors also play a significant role in crop yields. This is one of the reasons why numerous definitions of drought exist. For this reason, the search for a universal definition of drought is a rather pointless endeavor. Policymakers are at times frustrated by disagreements among scientists on whether a drought exists and its degree of severity. A policymaker is trying to determine if government should respond and, if so, through what types of response measures. The suite of responses employed is often based on those used for past drought events, with little or no consideration of whether these measures were actually effective. This book strives to change the paradigm for drought management from being reactive to proactive, the latter being an approach focused on risk reduction and, thus, reduced societal vulnerability.

Third, drought impacts are nonstructural and spread over larger geographical areas and temporal scales than are damages that result from other natural hazards such as floods, tropical storms, and earthquakes. These features of drought, combined with its creeping nature, make it particularly challenging to quantify and attribute specific impacts, and, therefore, more challenging to provide disaster relief in a timely and effective manner for drought than for other natural hazards.

These three characteristics of drought have hindered development of accurate, reliable, and timely forecasts; estimates of severity and impacts; and, ultimately, the formulation of drought preparedness plans and the implementation of appropriate risk reduction strategies or measures. Similarly, emergency managers, who have the assignment of responding to drought, struggle to deal with the large spatial coverage usually associated with drought.

Drought is a temporary aberration, unlike aridity, which is a permanent feature of the climate. Seasonal aridity (i.e., a well-defined dry season) also must be distinguished from drought. Considerable confusion exists among scientists and policymakers on the differentiation of these terms, especially in arid and semiarid regions. For example, Pessoa (1987) presented a map illustrating the frequency of drought in northeastern Brazil in his discussion of the impacts of and governmental response to drought. For a significant portion of the northeast region, he indicated that drought occurred 81–100 percent of the time. Much of this region is arid, and drought is a recurrent feature of its climate. However, drought is a temporary feature of the climate, so it cannot, by definition, occur 100 percent of the time. Similarly, researchers have defined a relative minimum during the Central American and Caribbean rainy season as a "midsummer drought" even though it occurs as part of the annual rainfall cycle each year (Magaña et al. 1999).

Nevertheless, it is important to identify trends over time and whether drought is becoming a more frequent and severe event. Today, concern exists that the threat of a warming climate may increase the frequency and/or severity of extreme climate events for some regions in the future (IPCC 2012; Melillo et al. 2014). As pressure on finite water supplies and other limited natural resources continues to build, more frequent and severe droughts are cause for concern in both water-short and water-surplus regions where tensions within (e.g., upstream vs. downstream) and between countries are growing. Anticipating and reducing the impacts of future drought events is paramount. It must be part of a sustainable development strategy, a theme developed later in this chapter and throughout this book.

Drought is a relative, rather than absolute, condition occurring in virtually all climate regimes. Our experience suggests scientists, policymakers, and the public associate drought primarily with arid, semiarid, and subhumid regions. For example, while drought has been traditionally associated with the southwestern United States and other parts of the western United States, the relatively humid Apalachicola-Chattahoochee-Flint Basin in the south-eastern region of the United States has been among the most contentious watersheds in the country, owing to the combination of drought and water extraction. In reality, drought occurs in most nations, in both dry and humid regions, and often on a yearly basis, especially in larger countries character-ized by multiple climate zones that result from different climatic controls. Drought characteristics and management will vary across each of these climatic zones. Drought is increasingly realized as having major impacts not only on agriculture but also on water supplies affecting health, energy, transportation, and recreation. This reality supports the need for a national strategy or policy that emphasizes drought risk reduction (Wilhite et al. 2014; WMO and GWP 2014; Chapters 2, 3, and 4).

1.3 Types of Drought

As previously stated, all types of drought originate from a deficiency of precipitation (Wilhite and Glantz 1985). When this deficiency spans an extended period of time (i.e., meteorological drought), its existence is defined initially in terms of the precipitation deficit, although temperature and other factors can also have a significant influence on its severity. Drought results from persistent large-scale disruptions in the global circulation pattern of the atmosphere (see Chapter 6). Exposure to drought varies spatially, and there is little, if anything, we can do to alter drought occurrence. However, the other common drought types (i.e., agricultural, hydrological, and socioeconomic) place greater emphasis on human or social aspects of drought, highlight-ing the interaction or interplay between the natural characteristics of the event and the human activities that depend on precipitation and water management to provide adequate supplies to meet societal and environ-mental demands (see Figure 1.1). For example, soils play an important role in determining how drought conditions will affect agricultural production because no direct relationship exists between precipitation and infiltration of precipitation into the soil. Infiltration rates vary according to antecedent moisture conditions, slope, soil type, and the intensity of the precipitation event. Soils also vary in their characteristics, with some soils having a high water-holding capacity and others a low water-holding capacity. Soils with a low water-holding capacity are more drought prone.

The characterization of hydrological drought is associated less with the precipitation deficiency because it is normally associated with the departure of surface and subsurface water supplies from some average condition at various points in time. Like agricultural drought, no direct relationship exists between precipitation amounts and the status of surface and subsurface water supplies in lakes, reservoirs, aquifers, and streams because these components of the hydrological system are used for multiple and competing purposes (e.g., irrigation, recreation, tourism, flood control, hydroelectric power production, domestic water supply, protection of endangered species, and environmental and ecosystem preservation). The use and management of surface and subsurface water supplies is a major factor that determines their availability when drought occurs. There is also considerable time lag between a deficiency of precipitation from average and when these deficiencies become evident in other components (e.g., reservoirs, groundwater, and streamflow) of the hydrologic system. Recovery of these components is also slow because of long recharge periods for surface and subsurface water supplies and how they are managed. In areas where the primary source of water is snowpack, such as in the western United States, the determination of drought severity is also complicated by infrastructures, institutional arrangements, and legal constraints. For example, reservoirs increase this region's resilience to drought because of their potential for storing large amounts of water as a buffer during dry years. However, the operating plans for these reservoirs try to accommodate the multiple, often conflicting, uses of the water (e.g., protection of fisheries, hydroelectric power production, recreation and tourism, irrigation) and the priorities set by governments when the funds were appropriated to construct the reservoir. The allocation of water between these various water use sectors is generally fixed and inflexible, making it difficult to manage a drought of unforeseen duration. Also, legal agreements between political jurisdictions (i.e., states and countries) concerning the amount of water to be delivered from one jurisdiction to another impose legal requirements on water managers to maintain flows at certain levels. During drought, conflicts heighten because of multiple values being advocated and because of limited available water. These shortages may result from poor water and land management practices that exacerbate the problem.

Socioeconomic drought differs markedly from the other types because it associates human activity with elements of meteorological, agricultural, and hydrological drought. It may result from factors affecting the supply of, or demand for, some commodity or economic good (e.g., water, forage, and hydroelectric power) that is dependent on precipitation. It may also result from the differential impact of drought on different groups within a population, depending on their access or entitlement to particular resources, such as land, and/or their access or entitlement to relief resources. Drought may fuel conflict between different groups as they compete for limited resources. A classic example in the Horn of Africa is the tension, which may become violent in drought years, between nomadic pastoralists in search of grazing and settled agriculturalists wishing to use the same land for cultivation. The Integrated Drought Management Program (discussed in Chapter 3) provides new efforts to inform drought management in this and other regions.

The interplay between drought and human activities raises a serious question with regard to attempts to define it in a meaningful way. The concept of socioeconomic drought is thus of primary concern to policymakers. It was previously stated that drought results from a deficiency of precipitation from expected or "normal" that is extended over a season or longer time period and is insufficient to meet the demands of human activities and the environment. Conceptually, this definition assumes that the demands of human activities are in balance or harmony with the availability of water supplies during periods of normal or average precipitation. However, if development demands exceed the supply of water available, then demand will exceed supply even in years of normal precipitation. This can result in human-induced drought or what is commonly referred to as water scarcity. In this situation, water supply for development is sustained only through mining of groundwater and/or the transfer of water into the region from other watersheds.

Drought severity is not only aggravated by other climatic factors, such as high temperatures, high winds, and low relative humidity, but also by the timing (i.e., principal season of occurrence, delays in the start of the rainy season, and occurrence of rains in relation to principal crop growth stages) and effectiveness of the rains (i.e., rainfall intensity and number of rainfall events). Thus, each drought event is unique in its climatic characteristics, spatial extent, impacts (i.e., no two droughts are identical), and likelihood of amelioration or demise. The area affected by drought is rarely static during the course of the event. As drought emerges and intensifies, its core area or epicenter shifts and its spatial extent expands and contracts. A comprehensive drought early warning system is critical for detecting emerging precipitation deficiencies and tracking these changes in spatial coverage, severity, and potential impacts, as explained below.

1.4 Characterizing Drought and Its Severity

In technical terms, droughts differ from one another in three essential characteristics: intensity, duration, and spatial coverage. Intensity refers to the degree of the precipitation shortfall and/or the severity of impacts associated with the shortfall. It is generally measured by the departure of some climatic parameter (e.g., precipitation), indicator (e.g., reservoir levels), or index (e.g., Standardized Precipitation Index or SPI) from normal and is closely linked to duration in the determination of impact. These tools for monitoring drought are discussed in detail in Chapters 7 and 8. Another distinguishing feature of drought is its duration. Droughts usually require a minimum of 2 to 3 months to become established but then can continue for months or years. The magnitude of drought impacts is associated with the timing of the onset of the precipitation shortage, its intensity, and the duration of the event. It should also be noted that quick onset or "flash" droughts occur, especially when precipitation deficiencies are associated with high temperature stress and internal atmospheric variability, such as that occurred during the 2012 drought in the United States.

Droughts also differ in terms of their spatial characteristics. The areas affected by severe drought evolve gradually, and regions of maximum intensity (i.e., epicenter) shift from season to season. In larger countries, such as Brazil, China, India, the United States, and Australia, drought rarely, if ever, affects the entire country. During the severe drought of the 1930s in the United States, for example, the area affected by severe and extreme drought reached 65 percent of the country in 1934. The 2012 drought in the United States was of a comparable spatial extent. These two droughts represent the maximum spatial extent of drought in the period from 1895 to 2016. The climatic diversity and size of countries such as the United States suggest that drought is likely to occur somewhere in the country each year.

From a planning perspective, the spatial characteristics of drought pose serious problems. For example, nations should determine the probability that drought may simultaneously affect all or several major crop-producing regions or river basins within their borders and develop contingencies for such an event. Likewise, it is important for governments to calculate the chances of a regional drought simultaneously affecting agricultural productivity and water supplies in their country and adjacent or nearby nations on which they may depend for food supplies or water transfers. A drought mitigation strategy that relies on the importation of food from neighboring countries or even distant markets may not be viable when regional-scale drought occurs. For example, the South African region experienced food insecurity from the simultaneous reduction of local corn production and the drop in rice imports from Southeast Asia, as a result of droughts associated with the 2015–2016 El Niño Southern Oscillation (ENSO) event.

1.5 Drought as Disaster: The Social and Political Context

As mentioned previously, drought, like all natural hazards, has both natural and social dimensions. The risk associated with drought for any region is a product of the region's exposure to the event (i.e., probability of occurrence at various severity levels), the vulnerability of society to the event, and its capacity to reduce or manage impacts. Vulnerability can be defined as "defenselessness, insecurity, exposure to risk, shocks and stress," and difficulty in coping with them (Chambers 1989). It is determined by both micro- and macro-level factors, and it is cross-sectoral—it is dependent on economic, social, cultural, and political factors. The Blaikie et al. (1994) and Wisner et al. (2004) disaster pressure or political ecology model represents well the interaction of hazard with drivers of vulnerability that are derived from local to global scales, as noted in Wilhite and Buchanan-Smith (2005). This model explores vulnerability in terms of three levels. First, there are root causes. These may be quite remote and are likely to relate to the underlying political and economic systems and structures. Second, there are dynamic pressures, which translate the effects of the root causes into particular forms of insecurity. These pressures might include rapid population growth, rapid urbanization, and epidemics. Finally, as a result, unsafe conditions are created; for instance, through people living in dangerous locations whether by choice or being forced to do so through displacement or job location and/ or the state or private agents such as insurers failing to provide adequate protection. Understanding people's vulnerability to drought is complex yet essential for designing drought preparedness, mitigation and response measures and drought risk reduction programs and policies.

Traditionally, the approach to understanding vulnerability has emphasized economic and social factors. This is most evident in the livelihoods frameworks that have underpinned much vulnerability assessment work. Some livelihood frameworks attempt to make sense of the complex ways in which individuals, households, and communities achieve and sustain their livelihoods and the likely impact of an external shock such as drought on both lives and livelihoods (Save the Children [UK] 2000; Young et al. 2001). Political factors and power relationships are sometimes underplayed in these frameworks. For example, institutionalized exploitation and discrimination between individuals, households, and groups are often overlooked. Yet these may be a key determinant of whether a particular ethnic or age group will have access to productive assets such as land and to relief resources. Similarly, many war-torn countries and fragile states are also drought prone. Understanding the dynamics and impact of the conflict—from national to local level—is critical to understanding the population's vulnerability to drought.

Understanding and measuring the vulnerability of a population or of particular groups within that population to drought is not an easy task. It requires an in-depth knowledge of the society and the relationships within that society. It is not a job for the novice. Instead, it benefits from long-term familiarity and collaborative networks, while retaining the ability to remain objective. Also, vulnerability is not static. Hence, no two droughts will have the same human impact. Ideally, a vulnerability assessment will capture dynamic trends and processes, not just a snapshot. And, the relationship is circular: high levels of vulnerability mean that a population is particularly at risk to the negative impact of drought. In turn, the impact of a prolonged drought may erode the asset base of that population, leaving them more vulnerable to future drought events in the absence of mitigating or preparedness measures.

Although we can do little, if anything, to alter drought occurrence, there are things we can do to reduce vulnerability. This is where government policy and the development of capacity come into play. For example, underlying vulnerability is reduced through development programs targeted to the most vulnerable, to strengthen their asset base. Governments can provide relief from the immediate impact of drought through livestock support and provision of subsidies or food. However, as discussed in subsequent chapters, relief programs should be consistent with a national drought policy directed at risk reduction.

1.6 The Challenge of Drought Early Warning

A drought early warning and information system (DEWIS) is designed to identify trends of key meteorological, hydrological, and social indicators to predict both the occurrence and the impact of a particular drought *and* to elicit appropriate mitigation and response measures (Buchanan-Smith and Davies 1995; WMO and GWP 2014). For most locations, the continuum from drought forecasting to early warning is still a linear process based on a "sender–receiver" model of risk communication. In the following discussion, the phrase *early warning information system* is used to describe an integrated process of risk assessment, communication, and decision support, of which an early warning is a central output. An early warning information system involves much more than development and dissemination of a forecast; it is the systematic collection and analysis of relevant information about, and coming from, areas of impending risk that (1) informs the development of strategic responses to anticipate crises and crisis evolution, (2) provides capabilities for generating problem-specific risk assessments and scenarios, and (3) effectively communicates options to critical actors for the purposes of decision-making, preparedness, and mitigation (Pulwarty and Verdin 2013).

The DEWIS when coupled with vulnerability assessments aimed at understanding and reducing risk becomes a powerful tool in risk reduction. The goal of a national drought policy is to define the overarching principles for the development of effective and timely programs that target risk reduction (see Chapter 4).

Numerous natural indicators of drought should be monitored routinely to determine drought onset, end, and spatial characteristics. Severity must also be evaluated continuously and at frequent time steps. Although droughts originate from a deficiency of precipitation, it is insufficient to rely only on this climate element to assess severity and resultant impacts. An effective DEWIS must integrate precipitation data with other data such as streamflow, snowpack, groundwater levels, reservoir and lake levels, and soil moisture in order to assess drought and water supply conditions (see Chapters 7, 8, and 9).

These physical indicators and climate indices must then be combined with socioeconomic indicators in order to predict human impact. Socioeconomic indicators include market data—for example, grain prices and the changing terms of trade between staple grains and livestock as an indicator of purchasing power in many rural communities—and other measures of coping strategies. Communities usually employ a sequence of strategies in response to drought. Early coping strategies rarely cause any lasting damage and are reversible. In many poor rural communities, examples of early coping strategies include the migration of household members to look for work, searching for wild foods, and selling nonproductive assets. If the impact of the drought intensifies, these early strategies become unviable and people are forced to adopt more damaging coping strategies, such as selling large numbers of livestock, choosing to go hungry or reduce nutritional sources of food in order to preserve some productive assets, and abandonment of traditional homelands. Once all options are exhausted, people are faced with destitution and resort to crisis strategies such as mass migration or displacement (Corbett 1988; Young et al. 2001). Monitoring these coping strategies provides a good indicator of the impact of drought on the local population, although by the time there is evidence of the later stages of coping, it is usually too late to launch a preventative response.

Effective DEWISs are an integral part of efforts worldwide to improve drought preparedness. Many DEWISs are, in fact, a subset of an early warning system with a broader remit—to warn of other natural disasters and sometimes also conflict and political instability. Timely and reliable data and information must be the cornerstone of effective drought policies and plans. Monitoring drought presents some unique challenges because of the hazard's distinctive characteristics, as noted previously.

An expert group meeting on early warning systems for drought preparedness, sponsored by the World Meteorological Organization (WMO) and others, examined the status, shortcomings, and needs of DEWISs and made recommendations on how these systems can help in achieving a greater level of drought preparedness (Wilhite et al. 2000b). This meeting was organized as part of WMO's contribution to the Conference of the Parties of the UN Convention to Combat Desertification (UNCCD). The proceedings of this meeting not only documented efforts in DEWISs in countries such as Brazil, China, Hungary, India, Nigeria, South Africa, and the United States, but also noted the activities of regional drought monitoring centers in eastern and southern Africa and efforts in West Asia and North Africa. Shortcomings of current DEWISs were noted in the following areas:

- *Data networks*—Inadequate station density, poor data quality of meteorological and hydrological networks, and lack of networks on all major climate and water supply indicators reduce the ability to represent the spatial pattern of these indicators accurately.

- *Data sharing*—Inadequate data sharing between government agencies and the high cost of data limit the application of data in drought preparedness, mitigation, and response. In 2017, this continues to be a serious problem in the majority of countries.

- *Early warning system products*—Data and information products are often too technical and detailed. They are not accessible to busy decision makers who, in turn, may not be trained in the application of this information to decision-making.

- *Drought forecasts*—Unreliable seasonal forecasts and the lack of specificity of information provided by forecasts limit the use of this information by farmers and others.

- *Drought monitoring tools*—Inadequate indices exist for detecting the early onset and end of drought, although the SPI was cited as an important new monitoring tool to detect the early emergence of drought. Significant advances in drought monitoring tools have been made since this meeting was held in 2000. These advances are discussed in much greater detail in Chapter 8, *Handbook on Drought Indices and Indicators*, published by the Integrated Drought Management Programme (IDMP) of the WMO and the Global Water Partnership (GWP), and republished in this volume with the permission of WMO and GWP (see Chapter 8; WMO and GWP 2016).

- *Integrated drought/climate monitoring*—Drought monitoring systems should be integrated and based on multiple physical *and* socioeconomic indicators to fully understand drought magnitude, spatial extent, and impacts. There has been considerable progress on this issue in some countries, as noted in numerous case studies included in this book.

- *Impact assessment methodology*—Lack of impact assessment methodology hinders impact estimates and the activation of mitigation and response programs. This continues to be a shortcoming in most countries.

- *Delivery systems*—Data and information on emerging drought conditions, seasonal forecasts, and other products are often not delivered to users in a timely manner.

- *Global early warning system*—No historical drought database exists and there is no global drought assessment product that is based on one or two key indicators, which could be helpful to international organizations, nongovernmental organizations (NGOs), and others. There has been considerable advancement in attempts to develop a global drought monitoring system in recent years (see Chapter 6 in this book; Heim et al. 2017).

As documented in this volume, efforts to address these specific concerns are now being undertaken through the UN International Drought Management Program, the US National Integrated Drought Information System, and PRONACOSE in Mexico, among others. As has now been well established, early warning alone is not enough to improve drought preparedness. Effective early warning depends on a multisectoral and interdisciplinary collaboration among all concerned actors at each stage in the warning process, from monitoring to response and evaluation, and informs longer-term planning beyond emergency responses. The key is whether decision makers listen to the warnings and act on them in time to protect livelihoods before lives are threatened and use the opportunities to create more proactive policies. The links between community-based approaches and the national and global EWSs remain relatively weak. There are many reasons why this is often the "missing link" and has been referred to as "the last mile." For example, risk-averse decisionmakers may be reluctant to respond to predictions, instead waiting for certainty and quantitative evidence. This invariably leads to a late response to hard evidence that the crisis already exists. Who "owns" the early warning information is also critical to how it is used. Does it come from a trusted source, or is it treated with suspicion? Ultimately, sufficient political will must exist to launch a timely response and hence to heed the early warnings (Buchanan-Smith and Davies 1995; Wilhite et al. 2014). How the political will to act is derived in noncrisis situations remains an area in need of focused policy sciences research.

1.7 The Three-Pillar Approach to Drought Risk Management

One of the major outcomes of the High-level Meeting on National Drought Policy (see Chapter 2) and subsequent activities such as the development of the IDMP and the conduct of a series of regional capacity-building workshops on national drought policy sponsored by WMO, the UN's Food and Agricultural Organization (FAO), UNCCD, UN-Water and the Convention on Biological Diversity (see Chapters 3 and 4 and other chapters in Part IV) has been the emergence of a three-pillar approach to drought risk management and policy. These pillars are illustrated in Chapter 3, Figure 3.1. The three pillars are monitoring and early warning, vulnerability and impact assessment, and mitigation, preparedness, and response. The concept of the three pillars is discussed in numerous chapters throughout this book and is promoted as the new model for drought risk management.

1.8 Summary and Conclusion

Drought occurs without the level of predictability that we ascribe to seasons (e.g., winter, summer, wet and dry seasons), yet it is a normal part of the climate experienced in virtually all regions. It should not be viewed as merely a physical phenomenon. Rather, drought is the result of the interplay between a natural event and the demand placed on a water supply by human-use systems. It becomes a disaster if it has a serious negative impact on people in the absence of adequate mitigating measures.

Since many definitions of drought exist, it is unrealistic to expect a universal definition to be derived. Drought can be grouped by type or disciplinary perspective as follows: meteorological, agricultural, hydrological, and socioeconomic. Each discipline incorporates different physical and biological factors in its definition. But above all, we are concerned with the impact of drought on people and the environment. Thus, definitions should incorporate both the physical aspects of drought (i.e., the intensity and duration of the event) and the impacts of the event on human activities and the environment, in order to be used for planning and operationally by decision makers. Definitions should also reflect the unique regional climatic characteristics. The three characteristics that differentiate one drought from another are intensity, duration, and spatial extent. The impacts of drought are thus diverse and fundamentally depend on the underlying vulnerability of the population. Vulnerability, in turn, is determined by a combination of social, economic, cultural, and political factors, at both micro- and macro levels. In many parts of the world, it appears that societal vulnerability to drought is escalating, and at a significant rate. Two additional elements have now been introduced: awareness that risks are changing and additional risks may arise, and the need for creating and communicating new knowledge about future conditions that is understood, trusted, and used (IPCC 2012; Pulwarty and Verdin 2013). Cognizant of these emerging elements, modern early warning information systems should provide the underpinning of a preparedness strategy aimed at risk reduction, which depends on adequate resources and collaborative networks and engages both the public and leadership. Understanding vulnerability is a critical first step in reducing drought risk, impacts, and the need for emergency response measures (i.e., the three-pillar approach).

It is imperative that increased emphasis be placed on mitigation, preparedness, and prediction and early warning if society has to reduce the social, economic, and environmental damages associated with drought. This will require interdisciplinary cooperation and a collaborative effort with policy-makers at all levels. This book provides concrete examples of how drought management can be enhanced through the development of research-based information and adoption of proactive actions that build institutional capacity directed at risk reduction. Numerous case studies are also presented that provide examples of how these proactive strategies are being crafted and applied in both developed and developing countries.

Going forward, there is considerable concern in both the scientific and policy communities about the linkages between drought and other key environmental and social issues. For example, the links between drought and climate change, water scarcity, national security, development, poverty, environmental degradation, food security, environmental refugees, and political stability are often cited in both the scientific and popular literature. Many of these linkages are discussed and explored in more detail for various settings in chapters included in this book.

References

Blaikie, P., T. Cannon, I. Davis, and B. Wisner. 1994. *At Risk: Natural Hazards, People's Vulnerability, and Disasters*. London: Routledge.

Buchanan-Smith, M., and S. Davies. 1995. *Famine Early Warning and Response—The Missing Link*. London: IT Publications.

Chambers, R. 1989. Vulnerability: How the poor cope. *IDS Bulletin* 20(2): 1–7.

Corbett, J. 1988. Famine and household coping strategies. *World Development* 16(9): 1092–1112.

Gillette, H. P. 1950. A creeping drought under way. *Water and Sewage Works* 104–105.

Glantz, M. H. 2003. *Climate Affairs: A Primer*. Washington, DC: Island Press.

Heim, R. R. Jr., M. J. Brewer, R. S. Pulwarty, D. A. Wilhite, M. J. Hayes, and M. V. K. Sivakumar. 2017. Drought early warning and information systems. In *Handbook of Drought and Water Scarcity: Principles of Drought and Water Scarcity*, eds. S. Eslamian and F. A. Eslamian, pp. 303–319. Boca Raton, FL: CRC Press.

IPCC. 2012. Managing the risks of extreme events and disasters to advance climate change adaptation. In *A Special Report of Working Groups I and II of the Intergovernmental Panel on Climate Change*, eds. C. B. Field, V. Barros, T. F. Stocker, et al. Cambridge: Cambridge University Press, pp. 594.

Magaña, V., J. A. Amador, and S. Medina. 1999. The mid-summer drought over Mexico and Central America. *Journal of Climate* 12: 1577–1588.

Melillo, J. M., T.C. Richmond, and G. W. Yohe, eds. 2014. *Climate Change Impacts in the United States: The Third National Climate Assessment*. Washington, DC: U.S. Global Change Research Program.

Pessoa, D. 1987. Drought in Northeast Brazil: Impact and government response. In *Planning for Drought: Toward a Reduction of Societal Vulnerability*, eds. D. A. Wilhite and W. E. Easterling, pp. 471–488. Boulder, CO: Westview Press.

Pulwarty, R., and J. Verdin. 2013. Crafting integrated early warning information systems: The case of drought. In *Measuring Vulnerability to Natural Hazards: Towards Disaster Resilient Societies*, ed. J. Birkmann, pp. 124–147. Tokyo: United Nations University Press.

Save the Children (UK). 2000. *The Household Economy Approach: A Resource Manual for Practitioners*. London: Save the Children.

Sivakumar, M. V. K., R. Stefanski, M. Bazza, S. Zelaya, D. A. Wilhite, and A. R. Magalhaes. 2014. High Level Meeting on National Drought Policy: Summary and major outcomes. *Weather and Climate Extremes* 3: 126–132.

Tannehill, I. R. 1947. *Drought: Its Causes and Effects*. Princeton, NJ: Princeton University.

Wilhite, D. A., ed. 2000. *Drought: A Global Assessment*. London: Routledge.

Wilhite, D. A., and M. Buchanan-Smith. 2005. Drought as hazard: Understanding the natural and social context. In *Drought and Water Crises: Science, Technology, and Management Issues*, ed. D. A. Wilhite, pp. 3–29. Boca Raton, FL: CRC Press.

Wilhite, D. A., and M. H. Glantz. 1985. Understanding the drought phenomenon: The role of definitions. *Water International* 10: 111–120.

Wilhite, D. A., M. J. Hayes, C. Knutson, and K. H. Smith. 2000a. Planning for drought: Moving from crisis to risk management. *Journal of the American Water Resources Association* 36: 697–710.

Wilhite, D. A., and R. S. Pulwarty. 2005. Drought and water crises: Lessons learned and the road ahead. In *Drought and Water Crises: Science, Technology, and Management Issues*, ed. D. A. Wilhite, 389–398. Boca Raton, FL: CRC Press.

Wilhite, D. A., M. V. K. Sivakumar, and R. Pulwarty. 2014. Managing drought risk in a changing climate: The role of national drought policy. *Weather and Climate Extremes* 3: 4–13. doi: 10.1016/j.wace.2014.01.002.

Wilhite, D. A., M. K. V. Sivakumar, and D. A. Wood, eds. 2000b. Early warning systems for drought preparedness and management. *Proceedings of an Expert Group Meeting*. AGM-2WMO/TD No. 1037. World Meteorological Organization, Geneva, Switzerland.

Wisner, B., P. Blaikie, T. Cannon, and I. Davis. 2004. *At Risk: Natural Hazards, People's Vulnerability, and Disasters*. 2nd ed. London: Routledge.

WMO and GWP. 2014. *National Drought Management Policy Guidelines: A Template for Action (D. A. Wilhite)*. Integrated Drought Management Programme (IDMP) Tools and Guidelines Series 1. WMO, Geneva, and GWP, Stockholm. http://www.droughtmanagement.info/literature/IDMP_NDMPG_en.pdf. Accessed March 2017.

WMO and GWP. 2016. *Handbook of Drought Indicators and Indices (M. Svoboda and B. Fuchs)*. Integrated Drought Management Programme (IDMP), Integrated Drought Management Tools and Guidelines Series 2. WMO, Geneva, and GWP, Stockholm. http://www.droughtmangement.info/literature/GWP_Handbook_of_Drought_Indicators_and_Indices_2016.pdf. Accessed March 2017.

World Economic Forum. 2015. *Global Risks 2015*. 10th ed. Geneva, Switzerland: World Economic Forum.

Young, H., S. Jaspars, R. Brown, J. Frize, and H. Khogali. 2001. *Food-security Assessments in Emergencies: A Livelihoods Approach*. HPN Network Paper No. 36. London: ODI.

Section II

Drought Risk Reduction: Shifting the Paradigm from Managing Disasters to Managing Risk

2

*The High-level Meeting on National Drought Policy: A Summary of Outcomes**

Mannava V. K. Sivakumar, Robert Stefanski,
Mohamed Bazza, Sergio A. Zelaya-Bonilla,
Donald A. Wilhite, and Antonio Rocha Magalhães

CONTENTS

2.1 Introduction

Drought is widely recognized as a slow, creeping phenomenon (Tannehill 1947) that occurs as a consequence of natural climatic variability and ranks first among all natural hazards according to Bryant (1991). In recent years,

* This chapter is an amended version of a paper that was originally published in a special issue of *Weather and Climate Extremes*. This paper is available online at: http://www.sciencedirect.com/science/article/pii/S2212094714000267.

concern has grown worldwide that droughts may be increasing in frequency and severity given the changing climatic conditions. Responses to droughts in most parts of the world are generally reactive, and this crisis management approach has been untimely, less effective, poorly coordinated, and disintegrated. Consequently, the economic, social, and environmental impacts of droughts have increased significantly worldwide, as water is integral to the production of goods and the provision of several services. The socioeconomic impacts of droughts may arise from the interaction between natural conditions and human factors, such as changes in land use and land cover, and water demand and water use. Excessive water withdrawals can exacerbate the impact of drought. Some direct impacts of drought are reduced crop, rangeland, and forest productivity; reduced water levels; increased fire hazard; reduced energy production; reduced opportunities and income for recreation and tourism; increased livestock and wildlife death rates; and damage to wildlife and fish habitat. A reduction in crop productivity usually results in less income for farmers, hunger, increased prices for food, unemployment, and migration.

The lessons learned from crisis management of droughts make it clear that future responses must be proactive. Despite the repeated occurrences of droughts throughout human history and their enormous impacts on different socioeconomic sectors, no concerted efforts have ever been made to initiate a dialogue on the formulation and adoption of national drought policies. Without a coordinated national drought policy (Sivakumar et al. 2011) that includes effective monitoring and early warning systems to deliver timely information to decision makers, effective impact assessment procedures, proactive risk management measures, preparedness plans aimed at increasing coping capacity, and effective emergency response programs directed at reducing the impacts of drought, nations will continue to respond to drought in a reactive, crisis management mode.

In order to address the issue of national drought policy, the World Meteorological Organization (WMO) Congress at its 16th session in Geneva in 2011 recommended the organization of a High-level Meeting on National Drought Policy (HMNDP). In parallel, the Conference of the Parties of the United Nations Convention to Combat Desertification and Drought (UNCCD) at its 10th session (held in 2011 in Changwon, Republic of Korea) welcomed the WMO recommendation. The member countries of the Food and Agriculture Organization (FAO) of the United Nations (UN) also requested the organization of the HMNDP for support in addressing drought issues since 2000. Accordingly, WMO, the Secretariat of UNCCD, and FAO, in collaboration with a number of UN agencies, international and regional organizations, and key national agencies, organized the HMNDP from 11 to March 15, 2013, in Geneva. The theme of the HMNDP was "Reducing Societal Vulnerability—Helping Society (Communities and Sectors)."

The HMNDP was sponsored by the African Development Bank (AfDB); the Ministry of National Integration (MI), Brazil; the Center for Strategic Studies and Management (CGEE), Brazil; the China Meteorological Administration (CMA); the OPEC Fund for International Development (OFID); the National Oceanic and Atmospheric Administration (NOAA); the Ministry of Foreign Affairs, Government of Norway; Saudi Arabia; the Swiss Agency for Development and Cooperation (SDC); and the United States Agency for International Development (USAID).

Four hundred and fourteen participants from 87 countries as well as representatives of international and regional organizations and UN agencies participated in the HMNDP.

2.2 Goals of the National Drought Policies

The objective of the HMNDP was to provide practical insight into useful, science-based actions to address the key drought issues being considered by governments and the private sector under the UNCCD and the various strategies to cope with drought. National governments must adopt policies that engender cooperation and coordination at all levels of government in order to increase their capacity to cope with extended periods of water scarcity in the event of a drought. The ultimate goal is to create more drought-resilient societies.

The goals of the national drought policies are:

- Proactive mitigation and planning measures, risk management, public outreach, and resource stewardship as key elements of effective national drought policy
- Greater collaboration to enhance the national/regional/global observation networks and information delivery systems to improve public understanding of, and preparedness for, drought
- Incorporation of comprehensive governmental and private insurance and financial strategies into drought preparedness plans
- Recognition of a safety net for emergency relief based on sound stewardship of natural resources and self-help at diverse governance levels
- Coordination of drought programs and response in an effective, efficient, and customer-oriented manner

2.3 Organization of HMNDP

The HMNDP was organized in two parts, a three-and-a-half-day scientific segment followed by a one-and-a-half-day high-level segment. The opening session of the scientific segment was chaired by His Excellency Mr. Nicholas

Tasunungurwa Goche, Honorable Minister of Transport, Communication, and Infrastructural Development, Zimbabwe, and Chair of the African Ministerial Conference on Meteorology (AMCOMET). The scientific segment of HMNDP addressed seven major themes relevant to the national drought policy: drought monitoring, early warning and information systems; drought prediction and predictability; drought vulnerability and impact assessment; enhancing drought preparedness and mitigation; planning for appropriate response and relief within the framework of national drought policy; and constructing a framework for national drought policy: the way forward. The scientific segment was organized in 15 sessions, including 7 plenary sessions, 2 roundtable discussion sessions, and 6 parallel sessions. Nineteen invited speakers made presentations on specific topics in these sessions and 28 experts from around the world served as discussants.

The high-level segment was addressed by heads of state and government, ministers, and heads and representatives of international organizations and sponsors.

2.4 Main Outcomes of the Scientific Segment of HMNDP

2.4.1 General

In the first general session of the scientific segment, setting the stage, Dr. Donald Wilhite, Professor of Applied Climate Science, University of Nebraska (USA), presented *Managing Drought Risk in a Changing Climate: The Role of National Drought Policy* (Wilhite et al. 2014). Following are the main recommendations from this session:

- It is important to develop national drought policies and preparedness plans that place emphasis on risk management rather than crisis management.
- There is a need to harmonize drought policies at regional levels with those at national to local levels, and vice versa.
- Several drought indicators recommended by WMO should be used in monitoring and forecasting of impending drought.
- The HMNDP should formulate networks/collaborations to enhance knowledge and information sharing to improve public understanding of and preparedness for drought.

2.4.2 Drought Monitoring, Early Warning, and Information Systems

To cover the topics of drought monitoring, early warning, and information systems, a plenary session was held in which Dr. Roger Pulwarty,

Director of the National Integrated Drought Information System (NIDIS) of the NOAA Climate Program Office, USA, presented *Information Systems in a Changing Climate: Early Warning and Drought Risk Management* (Pulwarty and Sivakumar 2014). Following the plenary session, comments were provided by four discussants from Brazil, Romania, the United States, and Kenya.

After the plenary session, a roundtable (implementing drought monitoring, early warning and information systems [DEWS]) and three parallel sessions (regional drought monitoring centers: progress and future plans, drought education at different levels, and advocacy to foster the outcomes of HMNDP) were held simultaneously.

Following are the main recommendations that were made in the five sessions on drought monitoring, early warning, and information systems:

- Establish scientifically sound, comprehensive, and integrated drought early warning systems (EWS); this will need additional research and development.
- Enable EWS to operate under data-rich as well as data-poor conditions.
- In data-poor conditions, explore using satellite-derived products, global modeling outcomes, and input from global initiatives to trigger action (e.g., using a fully developed global database management system [GDMS] in a way similar to using the European flood awareness system [EFAS]).
- Operate EWS at all times, not just during droughts.
- Prepare guidance material for developing the drought monitoring and early warning information systems. Key features should include integrated climate, surface and groundwater, and on-ground information from drought-impacted vulnerable sectors to provide decision makers (ranging from politicians, public servants, and nongovernmental organizations [NGOs] to communities and individuals) with comprehensive regional, national, district, and local information.
- Design and construct drought information, products, and services for end users by incorporating input from them, and deliver information using their preferred mode of receiving information (digital platforms including mobile phones, paper, face-to-face briefings, etc.).
- Educate end users to interpret information and demonstrate how they can use the information to trigger actions to reduce risks.
- Catalog three operating EWSs working in rich, medium, and poor data environments to illustrate what is possible.

2.4.3 Drought Prediction and Predictability

Major drought patterns are forced by major sea surface temperature (SST) patterns, and skillful drought predictability depends on skillful predictability of major SST patterns. Understanding physics of teleconnections between SST patterns and drought patterns is very important. Recent publications show that skillful prediction of decadal global-average temperature and North Atlantic SSTs is possible. Very encouraging preliminary results are emerging from the multiyear to decadal drought hindcasting using output from the World Climate Research Program (WCRP) coupled model inter-comparison project 5 (CMIP5). Tests are also underway using the hybrid dynamical-statistical prediction system for decadal climate and hydrome-teorology. Prediction of impacts and continuous interactions with stakehold-ers are vital for the success of drought policies guided by drought prediction and other information.

To cover the topic of drought prediction and predictability, a plenary session was held in which Dr. Vikram Mehta of the Center for Research on the Changing Earth System (USA) presented *Drought Prediction and Predictability—An Overview* (Mehta et al. 2014). This was followed by com-ments from three discussants from Kenya, Brazil, and the United States. The following recommendations were made in the session on drought prediction and predictability:

- Prediction of impacts and continuous interaction with stakeholders is vital for the success of drought policies guided by drought predic-tion and other information.
- Drought predictions cannot substitute for EWSs, but should be used to enhance existing drought monitoring and EWSs.
- A collaborative approach for research that takes into account user community needs should be promoted. Drought prediction needs many collaborative efforts from climate scientists and end users.
- The formation of collaborative platforms for scientists from devel-oping countries to work with leading agencies such as NOAA and WMO should be promoted to develop capacity and to ensure sus-tainability in forecasting and communication.
- Establishment of networks to enhance knowledge and information sharing should be promoted to improve public understanding of and preparedness for drought.

2.4.4 Drought Vulnerability, Impact Assessment, Drought Preparedness, and Mitigation

In the equation of risk, there are two factors, exposure to the hazard and vulnerability. Vulnerability is very context- and location-specific, takes into account socioeconomic and cultural aspects, and includes the coping capacity of the affected communities. Risk assessment involves the use of (1) drought risk models to account for drought losses and impacts; (2) ongo-ing monitoring of drought risk through observations (e.g., climate, remote sensing, food prices); and (3) assessment of drought impacts, number of households affected, and so forth.

The fragile agroecosystems of dry areas cover 41 percent of the earth's surface and are home to more than 2 billion inhabitants—and the majority of the world's poor. About 16 percent of the population live in chronic poverty, particularly in marginal rainfed areas. The challenges to coping with drought and enhancing food security in dry areas include inadequate agricultural policies for sustainable agricultural development and insufficient investment in agricultural research and development. We cannot prevent drought, but actions can be taken to better prepare to cope with drought, develop more resilient ecosystems and a better ability to recover from drought, and mitigate the impacts of droughts.

To address the theme of drought vulnerability and impact assessment, a plenary session was held in which Mr. John Harding from the United Nations Office for Disaster Risk Reduction (UNISDR) presented *Current Approaches to Drought Vulnerability and Impact Assessment*, which was followed by comments from four discussants from Germany, Kenya, Argentina, and Uzbekistan.

Another plenary session was held to cover the topic of drought preparedness and mitigation, in which Dr. Mahmoud Solh, Director General, International Center for Agricultural Research in the Dryland Areas (ICARDA), presented *Drought Preparedness and Drought Mitigation for Sustainable Agricultural Production* (Solh and van Ginkel 2014). This presentation was followed by comments from three discussants from India, Mexico, and the Russian Federation.

Following these two plenary sessions, a roundtable (vulnerability and impact assessment for risk reduction) and three parallel sessions (drought preparedness and mitigation strategies in different regions, drought impacts in key sectors and coping strategies, and strategy and recommendations for policy development) were held simultaneously.

The following recommendations were made on the issue of drought vulnerability and impact assessment:

- Pursue the efforts undertaken by WMO to promote standard indicators to measure drought throughout the world.
- Encourage countries to systematically collect data that will allow the assessment of drought impacts.
- Institutionalize the collection of disaster loss data that covers all hazards, including droughts.
- Facilitate comparison of drought vulnerability assessment among countries by the collection of a common minimum dataset.
- Factor climate change dimension in drought risk assessment and management policies.
- Account for context specificity by involving local communities in drought impact and vulnerability assessments.

- Conduct long-term monitoring to ensure reliability of vulnerability and impact assessments.
- Use not just top-down but also bottom-up approaches in designing adaptation strategies to allow inclusion of local knowledge and facilitate appropriation by the target communities.
- Go beyond economic cost-benefit considerations and include social and cultural dimensions in designing drought adaptation strategies.
- Use the inclusive wealth index (IWI), rather than gross domestic product (GDP) or income, for evaluation of success or failure.

The following recommendations were made on the issue of drought preparedness and management:

- Drought policies play a vital role in drought risk management and should be promoted.
- Policy processes should target institutional/interagency collaboration.
- Implementation of preparedness and mitigation strategies at the community and farm levels should be promoted.
- Ensure that technologies, measures, and practices adapted to drought conditions are freely available.
- Promote indigenous species/crops, plants, trees, etc.
- Consider both long- and medium-term measures for drought preparedness and mitigation.
- Link drought relief and drought plans at local and state levels.
- Ensure that information to meet users' needs is disseminated on accessible mediums.
- Promote efficient water management for irrigated, rainfed, and mixed systems.
- Emphasize water productivity optimization in lieu of yield maximization.
- Promote community approach in drought preparedness and mitigation.
- Ensure economic inclusion: youth programs are very important.
- Promote integrated approach to drought preparedness and mitigation.
- Determine most vulnerable zones and accessibility.
- Emphasize effective communication.
- Translate forecasts into a language/concept that users can understand.
- Focus on jobs and other long-term issues; drought management involves more than just providing food/water.
- Promote the development of safety nets and their implementation.

2.4.5 Planning for Appropriate Response and Relief within the Framework of National Drought Policy

There is a need to move from reactive to proactive approaches within the framework of national drought management policy. There is also a need to establish interlinkages between early warning, preparedness, and long-term resilience building. Appropriate approaches should consider the cross-sectoral and multidisciplinary nature of drought management, and strengthen collaborative decision-making. It is crucial to engage all stakeholders concerned, including private sectors, and to seek coordination of response measures at all levels.

To cover the issue of planning for appropriate response and relief within the framework of national drought policy, a plenary session was held in which Dr. Harvey Hill, Agriculture and Agri-Food Canada, presented *The Invitational Drought Tournament: Can It Support Drought Preparedness and Response?* (Hill et al. 2014). This presentation was followed by comments from three discussants from the United States, Italy, and Switzerland.

The following recommendations were made on the issue of planning for appropriate response and relief within the framework of national drought policy:

- Bridge the gaps between early warning and preparedness by utilizing traditional and newly developed tools to evaluate cross-sectoral impacts and the effects of relief measures.
- Enhance better understanding of drought phenomena and the associated risks and implications at all levels.
- Encourage immediate assistance (quick response) in a science-based and user-oriented manner.
- Promote the application of tools in support of proactive response, risk reduction, and long-term adaptation.
- The invitational drought tournament (IDT) approach could serve as a model to engage stakeholders in coordinated discussions and planning for drought events preparedness and response in the main sectors.
- IDT could serve as support for institutional preparedness and response to drought by providing frameworks within which to conduct assessments, identify strengths/gaps in preparedness and response, build upon assets, and address vulnerabilities.

2.4.6 Constructing a Framework for National Drought Policy: The Way Forward

To cover the theme of constructing a framework for national drought policy: the way forward, a plenary session was held in which Dr. Roger Stone,

University of Southern Queensland, Australia, presented *Constructing a Framework for National Drought Policy: The Way Forward—The Way Australia Developed and Implemented the National Drought Policy* (Stone 2014), followed by comments from three discussants from China, Mexico, and the African Development Bank.

Following the plenary session, in order to facilitate broad-based discussion among the participants on the subject of constructing a framework for national drought policy: the way forward, six breakout groups were established to cover the following six regions: Africa, Asia, North America and Caribbean, South America, Southwest Pacific, and Europe. The following recommendations were made:

- Understand the key climate drivers since the climate system links directly to farm cash income.
- Recognize the key value in the use of crop simulation modeling in planning preparedness for agricultural droughts.
- Government programs should help farmers to manage risk through appropriate decisions.
- Promote cooperation, consultation, communication, evidence-based policy and timing, and partnerships between several organizations—national and international, NGOs, private sector, and media.
- Establish national campaigns with the participation of national services, academic, research, and cultural organizations.
- Avoid duplication of efforts and resources.
- Create regional meteorological and support systems.
- Promote proactive response, especially for EWSs.
- Establish a system that allows integrated management of the different resources, especially in the least developed countries (LDCs).
- Emphasize dissemination of information to all users and in all languages.
- Evaluate the different activities that compete for water usage.
- Promote legislation dedicated to water resources usage and management.
- Improve wastewater treatment systems.

2.5 Main Outcomes of the High-level Segment of HMNDP

The high-level segment was addressed by heads of state and government, ministers, and heads and representatives of international organizations and sponsors.

His Excellency Mr. Brigi Rafini, Prime Minister of the Republic of Niger, chaired and addressed the opening of the high-level segment of the meeting, with supporting keynote addresses by Mr. Ban Ki-moon, Secretary General of the United Nations; His Excellency Mr. Jakaya Mrisho Kikwete, President of the United Republic of Tanzania; His Royal Highness Willem-Alexander, the Prince of Orange, Chair of the UN Secretary General's Advisory Board on Water and Sanitation (UNSGAB); and Dr. Bernard Lehmann, Director General, Swiss Federal Office of Agriculture. The Ministerial segment that was addressed by ministers from different parts of the world was chaired by Hon. Robert Sichinga, Minister of Agriculture and Livestock of the Republic of Zambia.

The high-level segment adopted the following meeting declaration encouraging all governments to develop and implement national drought policies (Sivakumar et al. 2014).

2.5.1 Final Declaration of HMNDP

(DECLARATION OPENING)

DO 1: We, the heads of state and government, ministers, heads of delegations and experts, attending the HMNDP in Geneva, March 11–15, 2013:

(PREAMBULAR PART)

Urgency of the problem

PP 1: Acknowledging that droughts are natural phenomena that have caused human suffering since the beginning of humanity, and are being aggravated as a result of climate change;

PP 2: Noting the interrelationships between drought, land degradation and desertification (DLDD), and the high impacts of DLDD in many countries, notably the developing and the least developed countries, and the tragic consequences of droughts, particularly in Africa;

PP 3: Acknowledging the role of the UN agencies, and in particular the United Nations Convention to Combat Desertification (UNCCD) in line with its mandate, provisions, and principles, in particular Parts II and III of the Convention, to assist in the combat against drought and desertification;

PP 4: Observing that drought has major implications in terms of the loss of human lives, food insecurity, degradation of natural resources, negative consequences on the environment's fauna and flora, poverty and social unrest and that there are increasingly immediate short-term and long-term economic losses in a number of economic sectors including, inter alia, agriculture, animal husbandry, fisheries, water supply, industry, energy production, and tourism.

PP 5: Concerned with the impacts of climate variability and change and the likely shift in the patterns of droughts and possible increase in the frequency, severity, and duration of droughts, thus further increasing the risk of social, economic and environmental losses;

PP 6: Underscoring that addressing climate change can contribute to reducing the aggravation of droughts and that it requires action, in accordance with the principles and provisions of the United Nations Framework Convention on Climate Change;

PP 7: Noting that desertification, land degradation and drought are global challenges that continue to pose serious challenges for the sustainable development of all countries, in particular the developing countries;

PP 8: Acknowledging that there are insufficient policies for appropriate drought management and proactive drought preparedness in many countries around the world and that there is need for enhancing international cooperation to support all countries, in particular developing countries in managing droughts and building resilience, and that countries continue to respond to droughts in a reactive, crisis management mode;

PP 9: Recognizing also the urgent needs for countries to manage droughts effectively and better cope with their environmental, economic, and social impacts;

PP 10: Recognizing that to better cope with droughts, countries need to understand the need for improved risk management strategies and develop preparedness plans to reduce drought risks.

Scientific progress in drought monitoring and early warning systems

PP 11: Recognizing that advances in drought monitoring and early warning and information systems, under government authority, and the use of local knowledge and traditional practices can contribute to enhanced societal resilience and more robust planning and investment decisions, including the reduction of consequences of drought impacts;

PP 12: Recognizing that scientific advances in seasonal to inter-annual and multi-decadal climate predictions offer an additional opportunity for the continued development of new tools and services to support improved management of droughts.

Need for vulnerability and impact assessment

PP 13: Noting the need for urgent intersectoral coordination of the assessment of drought vulnerability and drought management.

Need for rapid relief and response

PP 14: Noting the need to identify emergency measures that will reduce the impact of current droughts while reducing vulnerability to future occurrences, relief must be targeted to the affected communities and socioeconomic sectors and reach them in a timely fashion.

PP 15: Noting also the need to create synergies between drought relief measures and the preparedness, mitigation, and adaptation actions for long-term resilience.

Need for effective drought policies

PP 16: Recalling the commitment in the outcome document of the UN Conference on Sustainable Development (Rio +20) to significantly improve the implementation of Integrated Water Resources Management at all levels, as appropriate.

PP 17: Recalling that the UNCCD is pertinent to the promotion of sustainable development and that it calls for the establishment of effective policies to combat land degradation and desertification and mitigate the effects of droughts.

PP 18: Recalling also the call of the COP10 of UNCCD for an advocacy policy framework on drought for promoting the establishment of national drought management policies.

PP 19: Recalling the decision of governments to create the Global Framework for Climate Services (GFCS) to strengthen production, availability, delivery, and application of science-based climate prediction and services.

(OPERATIVE PART)

OP 1: Encourage all governments around the world to develop and implement national drought management policies, consistent with their national development laws, conditions, capabilities and objectives, guided, inter alia, by the following:

- Develop proactive drought impact mitigation, preventive and planning measures, risk management, fostering of science, appropriate technology and innovation, public outreach, and resource management as key elements of effective national drought policy.
- Promote greater collaboration to enhance the quality of local/national/regional/global observation networks and delivery systems.
- Improve public awareness of drought risk and preparedness for drought.
- Consider, where possible within the legal framework of each country, economic instruments, and financial strategies, including risk reduction, risk sharing, and risk transfer tools in drought management plans.
- Establish emergency relief plans based on sound management of natural resources and self-help at appropriate governance levels.
- Link drought management plans to local/national development policies.

OP 2: Urge the World Meteorological Organization, the UNCCD and the Food and Agriculture Organization of the United Nations (FAO), other related UN agencies, programs and treaties, as well as other concerned parties, to assist governments, especially the developing countries, in the development of national drought management policies and their implementation;

OP 3: Urge the developed countries to assist developing countries, especially the least developed countries, with the means of implementation toward the comprehensive development and implementation of national drought management policies in accordance with the principles and provisions of the UNCCD;

OP 4: Encourage the promotion of international cooperation, including north-south cooperation complemented by south-south cooperation, as appropriate, to foster drought policies in developing countries;

OP 5: Invite WMO, UNCCD, and FAO to update the draft versions of the science and policy documents taking into account the recommendations from the HMNDP and circulate them to all governments for their review prior to finalization, to assist governments in the development and implementation of the national drought management policies.

2.6 Summary and Conclusions

There is growing evidence that the frequency and extent of drought has increased as a result of global warming. Crisis management has typically characterized governmental response to drought. This approach has been ineffective, leading to untimely and poorly coordinated responses. Hence, the HMNDP was organized by WMO, UNCCD, and FAO, in collaboration with a number of UN agencies, international and regional organizations, and key national agencies. HMNDP provided practical insight into useful science-based actions to address the key drought issues being considered by governments and the various strategies to cope with drought. The HMNDP declaration, adopted unanimously by the participants in the meeting, encourages all governments around the world to develop and implement national drought management policies, consistent with their national development laws, conditions, capabilities, and objectives.

References

Bryant, E. A. 1991. *Natural Hazards*. Cambridge: Cambridge University Press.
Hill, H., M. Hadarits, R. Rieger, G. Strickert, E. G. D. Davies, and K. M. Strobbe. 2014. The invitational drought tournament: What is it and why is it a useful tool for drought preparedness and adaptation? *Weather and Climate Extremes* 3:107–116.

Mehta, V. M., H. Wang, K. Mendoza, and N. J. Rosenberg. 2014. Predictability and prediction of decadal hydrologic cycles: A case study in Southern Africa. *Weather and Climate Extremes* 3:47–53.

Pulwarty, R. S., and M. V. K. Sivakumar. 2014. Information systems in a changing climate: Early warnings and drought risk management. *Weather and Climate Extremes* 3:14–21.

Sivakumar, M. V. K., R. P. Motha, D. A. Wilhite, and J. J. Qu, eds. 2011. *Towards a Compendium on National Drought Policy*. Proceedings of an Expert Meeting on the Preparation of a Compendium on National Drought Policy, July 14–15, Washington, DC, AGM-12, WAOB-2011. Geneva, Switzerland: World Meteorological Organization, pp. 135.

Sivakumar, M. V. K., R. Stefanski, M. Bazza, S. Zelaya, D. A. Wilhite, and A. R. Magalhaes. 2014. High level meeting on national drought policy: Summary and major outcomes. *Weather and Climate Extremes* 3:126–132.

Solh, M., and M. van Ginkel. 2014. Drought preparedness and drought mitigation in the developing world's drylands. *Weather and Climate Extremes* 3:62–66.

Stone, R. C. 2014. Constructing a framework for national drought policy: The way forward—The way Australia developed and implemented the national drought policy. *Weather and Climate Extremes* 3:117–125.

Tannehill, I. R. 1947. *Drought, its Causes and Effects.* Princeton, NJ: Princeton University Press.

Wilhite, D. A., M. V. K. Sivakumar, and R. Pulwarty. 2014. Managing drought risk in a changing climate: The role of national drought policy. *Weather and Climate Extremes* 3:4–13.

3
Integrated Drought Management Initiatives[*]

Frederik Pischke and Robert Stefanski

CONTENTS

3.1 Introduction

This chapter provides an overview of the development of national drought management policies. It explores collaborative efforts that were started at the High-level Meeting on National Drought Policy (HMNDP) (see Chapter 2) and are implemented through the World Meteorological Organization (WMO) and Global Water Partnership's (GWP) integrated drought management programme (IDMP) and related initiatives, particularly the UN-Water Initiative on Capacity Development to Support National Drought Management Policies. Early outputs—for example, the *National Drought Management Policy Guidelines—A Template for Action* (see Chapter 4) and the *Handbook on Drought Indices and Indicators* (see Chapter 9)—provide an indication of how expert-reviewed guidance is brought together and used by drought practitioners for developing national drought management policies and applying drought indices/indicators. Regional examples from central and eastern Europe and the Horn of Africa highlight how these guidelines

---

[*] This chapter is a modified version of the following journal article: Pischke, F. and R. Stefanski, 2016. Drought management policies—From global collaboration to national action. *Water Policy* 18(6): 228–244.

and collaborative efforts and outputs are applied. The role of IDMP is to provide a framework and commensurate technical support to countries, but the actual development and implementation of national drought plans and policies still needs to be done by governmental ministries and national stakeholders. This chapter emphasizes how information from different sources is used to support countries to shift from only reacting to droughts when they occur to adopting proactive national drought policies that focus on improved collaboration and the mitigation of drought impacts through appropriate risk reduction measures.

Droughts have generally been addressed in a reactive manner, only responding after drought impacts have occurred. This reactive or crisis management approach is untimely, poorly coordinated, and disintegrated, and it provides negative incentives for adapting to a changing climate.

Despite recognition of the need to move away from crisis management to risk management, no concerted efforts have been made to initiate a dialogue on the formulation and adoption of national drought policies. In addition, since there are different impacts of drought across economic sectors and society as well as different providers of information and solutions, there must be a collaborative effort to effectively manage drought. For example, at the country level, the National Meteorological and Hydrological Services (NMHS) provide drought monitoring and early warning of the weather and climate variables, the water resource management agency provides information on reservoir levels, the Ministry of Agriculture provides information on crop yields and estimates production, the Ministry of the Environment provides data on environmental flows, and the Ministries of Planning and Finance are often key actors in the overall development planning. All of these institutions must collaborate to develop a coherent drought management policy for the country.

With the aim of addressing these issues, the HMNDP was organized by WMO, the Secretariat of the United Nations Convention to Combat Desertification (UNCCD), and the Food and Agriculture Organization of the United Nations (FAO), in collaboration with a number of UN agencies and international and regional organizations, in Geneva from March 11 to 15, 2013, as detailed in Chapter 2.

In its final declaration (WMO 2013), the HMNDP encouraged all governments to develop and implement national drought management policies guided by the following principles:

- Develop proactive drought impact mitigation, preventive and planning measures, risk management, fostering of science, appropriate technology and innovation, public outreach, and resource management as key elements of effective national drought policy.
- Promote greater collaboration to enhance the quality of local/national/regional/global observation networks and delivery systems.
- Improve public awareness of drought risk and preparedness for drought.

- Consider, where possible within the legal framework of each country, economic instruments and financial strategies, including risk reduction, risk sharing, and risk transfer tools in drought management plans.
- Establish emergency relief plans based on sound management of natural resources and self-help at appropriate governance levels.
- Link drought management plans to local/national development policies.

In addition, the policy document of the HMNDP (UNCCD et al. 2013) stated the essential elements of a national drought policy, namely:

- Promoting standard approaches to vulnerability and impact assessment
- Implementing effective drought monitoring, early warning, and information systems
- Enhancing preparedness and mitigation actions
- Implementing emergency response and relief measures that reinforce national drought management policy goals.

One of the successes of HMNDP is that it has helped focus the attention of international organizations and national governments on proactive policies.

The strong call for a framework in the form of a policy that combines different approaches that have been considered key in moving from a crisis management approach to a risk management approach led to the launch of the IDMP by WMO and GWP at the HMNDP in March 2013. With the objective of supporting stakeholders at all levels by providing policy and management guidance, and by sharing scientific information, knowledge, and best practices for an integrated approach to drought management, the IDMP aims:

- To shift the focus from reactive (crisis management) to proactive measures through drought mitigation, vulnerability reduction, and preparedness
- To integrate the vertical planning and decision-making processes at regional, national, and community levels into a multistakeholder approach including key sectors, especially agriculture and energy
- To promote the evolution of the drought knowledge base and to establish a mechanism for sharing knowledge and providing services to stakeholders across sectors at all levels
- To build capacity of various stakeholders at different levels

Based on the High-level Meeting on National Drought Policies (UNCCD et al. 2013), the IDMP and its partners have adopted three pillars of drought

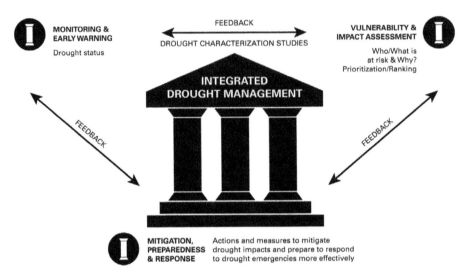

FIGURE 3.1
The three pillars of integrated drought management.

management (Figure 3.1), which have been advanced by Wilhite (WMO and GWP 2014): (1) drought monitoring and early warning systems; (2) vulnerability and impact assessment; and (3) drought preparedness, mitigation, and response.

The pillars have been reflected in many different initiatives, including the UN-Water Initiative on Capacity Development to Support National Drought Management Policies and the Windhoek Declaration of the African Drought Conference (UNCCD 2016), because they represent a common way of structuring the work toward an integrated approach to drought management.

3.2 Early Outputs of Global Collaboration—The Approach of the Integrated Drought Management Programme

The strength of the initiatives that have been formed following the HMNDP is that they provide a common framework, to which previously disparate efforts can contribute. The efforts that the IDMP and the UN-Water initiative were carried out in partnership and have leveraged the activities of its partners to determine the status and needs of countries and move forward collectively to support addressing these needs. To date, more than 30 organizations have agreed to support and provide input to the goals of the IDMP. The IDMP also uses the network of NMHS and related institutions affiliated with WMO (the United Nations specialized agency for weather, climate and water) and the regional and country water partnerships of the GWP as the multistakeholder platform to bring together actors from government, civil society, the private sector, and academia working on water resources management, agriculture, and energy. In addition, the IDMP liaises with related initiatives that are not formally part of IDMP but which are contributing to WMO and GWP.

Against the background of the HMNDP, the IDMP developed guidance for national drought policy development and implementation. Based on one of the tools that has been instrumental for the development of drought preparedness plans in the United States (Wilhite 1991; Wilhite et al. 2005), Wilhite adapted the 10-step planning process within the framework of the IDMP. These guidelines (WMO and GWP 2014) focus on a national policy context and draw on experiences from different countries. The purpose of these guidelines is to provide countries with a template that they can use and modify for their own purposes. Countries should not blindly use the 10-step process. The process should be modified by local experiences and context. For example, countries in central and eastern Europe have adapted the guidelines to the context of the European Union (EU) Water Framework Directive (see Section 3.4.1), and Mexico has also modified the 10-step process (see Chapter 19) to its national context (WMO and GWP 2014). The guidelines are further elaborated in Chapter 4 and the complete publication is available at the IDMP website (http:www.droughtmanagement.info) in the six official languages of the United Nations.

Another early accomplishment of the IDMP is the publication of the "Handbook of Drought Indicators and Indices" (see Chapter 9), which provides options for identifying the severity, location, duration, onset, and cessation of such conditions (WMO and GWP 2016). The purpose of this handbook is to provide to drought practitioners some of the most commonly used drought indicators/indices that are being used across drought-prone regions, with the goal of further advancing monitoring, early warning, and information delivery systems in support of risk-based drought management policies and preparedness plans. The handbook is a reference book with details of more than 50 drought indices and indicators, including information on their ease of use, origins, characteristics, input parameters, applications, strengths, weaknesses, resources (including access to software code), and references. Information derived from indicators and indices is useful in planning and designing applications (such as risk assessment, drought early warning systems, and decision support tools for managing risks), provided that the climate regime and drought climatology is known for the location. The handbook is further described in Chapter 9.

With the aim of improving understanding of the benefits of action and costs of inaction on drought mitigation and preparedness, the IDMP has been developing a work stream on this issue with a literature review as a first output (WMO and GWP 2017). The literature review is presented in Chapter 5.

The IDMP uses information from its partners to keep track of the status of national drought policies and activities from around the world. One example of using an output from a partner organization is the use of a WMO survey to assess the status of the agricultural meteorological products and services provided by the NMHS of the world. In the most recent survey (2010–2014), the NMHS were requested to list current drought indices in use in their country and whether the country has a national drought policy or plan. This is not an all-inclusive list; out of the 52 countries that responded, 17 indicated that they have some sort of national drought policy or plan. The results of this survey are only a starting point. The IDMP uses various sources of information such as this survey to keep track of the status of national drought policies around the world. The work of the IDMP and its partners is then to liaise with these countries to see if these policies or plans are actually implemented and if their objectives are fulfilled.

In order to support countries in developing and implementing drought management actions, an integrated drought management help desk is in the process of being established. The help desk approach draws on the sister program of the IDMP, the WMO/GWP Associated Programme for Flood Management (APFM), which established the Integrated Flood Management Help Desk in 2009 to provide support for the implementation of the principles of integrated flood management. The Integrated Drought Management Help Desk will consist of a "Find" section (providing existing knowledge resources), an "Ask" section (offering a point of contact to expertise), and a "Connect" facility (providing an overview and a connection to ongoing initiatives). The resources that have been developed and the regional initiatives of the IDMP will populate the help desk. It should be stressed that these help desks are available to any government agency, national institution, or individual. Requests for assistance can be made via email, correspondence, or phone call. Depending on the type of request, the expected assistance would include pointing out relevant resource material, providing detailed information on a procedure, advisory services, and, in some cases, the development of training or a country visit.

3.3 UN-Water Initiative on Capacity Development to Support National Drought Management Policies

The IDMP liaises with similar initiatives to assist its work and mandate. A related activity that also originated from the HMNDP was the aforementioned UN-Water Initiative on Capacity Development to Support National Drought Management Policies. This was a collaborative initiative of several UN-Water organizations: WMO, UNCCD, FAO, the Convention on Biological Diversity (CBD), and the UN-Water Decade Programme on Capacity Development (UNW-DPC). Since WMO and other organizations were partners in both initiatives, complementarities and synergies with the IDMP were realized.

The UN-Water initiative was initially funded by UNW-DPC, and the other organizations also provided funds to enable the organization of six regional workshops (Eastern European, July 2013, Romania; Latin America and the Caribbean, December 2013, Brazil; Asia-Pacific, May 2014, Vietnam; eastern and southern Africa, August 2014, Ethiopia; near east and north Africa, November 2014, Egypt; and west and central Africa, May 2015, Ghana). These regional workshops were organized for participants from drought-prone countries, with a focus on developing countries and transition economies (UNW-DPC 2015a).

The overarching goals of the UN-Water initiative were to enhance capacities of key government stakeholders dealing with drought issues in their countries, and ensure effective coordination within all levels of governments in order to develop more drought-resilient societies by reducing the risk associated with the incidence of drought.

Based on these goals, the key targets of the workshops were to:

- Improve awareness of drought issues and countries' needs to establish strategies for national drought management policies based on the principles of risk reduction
- Equip key government stakeholders concerned with drought with tools and strategies to support decision-making and for risk assessments of vulnerable sectors, population groups, and regions
- Advance national drought management policies by taking into account long-term benefits of risk-based and proactive approaches that address drought and water scarcity problems at large and move beyond short-term planning, which addresses drought as a crisis
- Promote collaboration between the various sectors (agriculture, water resources, meteorological/climatological, ecological, and urban) at country and regional levels

Each partner organization was responsible for one of the main thematic sessions across the workshops. Although these workshops were not a direct IDMP activity, the 10-step process (WMO and GWP 2014) described above and in more detail in Chapter 4 was used during workshops. The UN-Water initiative has officially ended but the organizations are discussing ways to assist countries individually with the development of national drought policies, with the integrated drought management help desk, described above, as a potential vehicle to deliver sustained support to the national process where needed. There is currently discussion on how the lessons and experience of the UN-Water initiative can be further developed by the IDMP.

During the course of these workshops, several challenges were identified by the various participants. One of the main challenges was the availability

of relevant data. Many countries concluded that data on drought characterization in countries were scarce. Data issues that needed to be resolved
included developing a country-level database on past drought incidences
and impacts; promoting exchange and integration of data needed for
drought monitoring; developing assessment tools and approaches to quantify drought impacts; and increasing the density of rain gauges and sensors
or stations for drought-related parameters such as stream flow, soil moisture,
and reservoir levels. Many countries reported that there was no consistent
methodology for assessing drought impacts or archiving this information in
a database. Other significant challenges included the lack of political will,
which can hinder progress on national drought management policies, and
the lack of funding, which limits developing and implementing national
drought policies.

The participants also listed several next steps that will be needed at the
national level to successfully implement national drought policies. These
steps include continuing improvement of human and institutional capacities,
improving understanding of the economics of drought, raising awareness
of the ineffectiveness of the current approach to drought management, and
strengthening cooperation at all levels.

3.4 Regional and National Application of Outputs

Although it is too early to see the impacts of implementing this approach,
some progress has been made: (1) in central and eastern Europe, 20 national
consultations were conducted in 10 countries over 2 years, and on the basis of
these consultations, guidance has been developed and drought management
plans in the regional policy framework have been elaborated; (2) several
regional projects have been established in eastern Africa, West Africa, South
Asia, and Central America; and (3) the three publications previously mentioned (WMO and GWP 2014, 2016, 2017).

3.4.1 Central and Eastern Europe

The longest-established regional program, the IDMP Central and Eastern
Europe (IDMP CEE), managed by GWP Central and Eastern Europe (GWP
CEE), brings together more than 40 partners from the region. It has, since
2013, provided practical advice on how droughts can be managed, with the
goal of increasing the capacity and ability of countries in central and eastern
Europe to adapt to climate variability and change by enhancing resilience to
drought. Outputs are a compendium of good practices, support for a drought
information exchange platform, demonstration projects testing innovative

solutions for better resilience to drought, and capacity-building training and workshops at the national and regional level. The actions of the IDMP CEE also focused on the status and implementation of drought management plans. A first analysis (GWP CEE and Falutova 2014) showed that the majority of countries in the region had not produced drought management plans, and key elements of the drought management plans—namely, indicators and thresholds establishing different drought stages, measures to be taken in each drought stage, and the organizational framework for drought management—had not been implemented.

Based on the realization that drought management plans in the policy framework of central and eastern Europe were lacking or insufficient, the IDMP CEE set out to support countries in developing drought management plans, with the guidance of the 10-step process for drought management policies produced by WMO and GWP (2014). Part of this process included two rounds of consultations in the 10 participating countries: Bulgaria, Czech Republic, Hungary, Lithuania, Moldova, Poland, Romania, Slovakia, Slovenia, and Ukraine. The consultations involved key actors involved in drought management in each country, including government ministries and competent authorities, hydrometeorological services, and universities, as well as affected stakeholders, including farmers, energy utilities, and fisheries. The first round of consultation focused on analyzing the current state of drought policy in the individual countries. Draft guidelines were developed for the preparation of drought management plans within the context of the EU Water Framework Directive river basin management plans, as the overarching policy context of the region. The second round of national consultations aimed to gather national experiences and information relevant to drought planning and further develop the draft guidelines from the first round of consultations.

This process of using the input from the global level with the 10-step process and adapting it to the regional context through the 20 national consultations led to the publication of guidelines for the preparation of drought management plans in the context of the EU Water Framework Directive (GWP CEE 2015) and the definition of seven steps that are specifically tailored for the EU Water Framework Directive, the regional policy context:

1. Develop a drought policy and establish a drought committee

2. Define objectives of drought risk-based management policy

3. Make an inventory of data for drought management plan development

4. Produce/update a drought management plan

5. Publicize the drought management plan for public involvement

6. Develop scientific and research programs

7. Develop educational programs

The steps are forming the basis for a third round of consultations to develop drought management plans. The publication has also been translated into a majority of the national languages in the countries participating in its development.

It was found that adapting the step-by-step planning process proposed by the *National Drought Management Policy Guidelines* (WMO and GWP 2014) in the context of the EU Water Framework Directive by linking this planning process to river basin management plans has synergistic effects in achieving environmental objectives of the EU Directive.

3.4.2 Horn of Africa

The IDMP also established regional programs in 2015 to support the practical application of its principles at the regional and national level in the Horn of Africa and West Africa. Both programs are aiming to close the gap of current efforts and provide an impetus to existing drought management initiatives in these regions. These regional initiatives use the institutional capability of GWP through the GWP country water partnerships to bring together the key actors not only from the water community but also from the agriculture and energy communities. The IDMP regional initiatives thus liaise with existing institutions and activities to further promote integrated drought management.

The Greater Horn of Africa is at high risk to extreme climate events such as droughts and floods. Before the HMNDP and the establishment of the IDMP, this region had developed an innovative way to address these issues at a regional level. In 1989, 24 countries across eastern and southern Africa established the Drought Monitoring Centre, with headquarters in Nairobi. In 2003, the Eastern African Regional Intergovernmental Authority on Development (IGAD) adopted this center as a specialized IGAD institution. In 2007, the name of this institution was changed to the IGAD Climate Prediction and Application Centre (ICPAC) in order to better reflect its mandates, mission, and objectives within the IGAD system (ICPAC 2016; UNW-DPC 2015b).

This center is responsible for the 11 countries in the Greater Horn of Africa (Burundi, Djibouti, Eritrea, Ethiopia, Kenya, Rwanda, Somalia, South Sudan, Sudan, Tanzania, and Uganda) and works closely with the NMHS of member countries as well as regional and international centers for data and information exchange. Its main objectives are to provide timely early warning information on climate change and to support sector-specific applications for the mitigation of poverty and the management of environment and sustainable development in relation to the impact of climate variability; to improve the technical capacity of producers and users of climatic information; to develop an improved, proactive, timely, broad-based system of information/product dissemination and feedback; and to expand the climate knowledge base and applications within the subregion in order to facilitate informed decision-making on climate risk-related issues (ICPAC 2016).

One of the outputs of the IDMP Horn of Africa (IDMP HOA) regional project was the publication of the "Assessment of Drought Resilience Frameworks on the Horn of Africa" (GWP EA 2015). This publication provides an overview of drought policy and institutional frameworks for Djibouti, Ethiopia, Kenya, Somalia, South Sudan, Sudan, and Uganda.

After the severe regional drought in 2010/2011, the Summit of Heads of State and Government of the IGAD and the East African Community (EAC) met in Nairobi in September 2011. This summit decided to address the effects of recurring droughts on vulnerable communities in the IGAD region, calling for an increased commitment by affected countries and development partners to support investments in sustainable development, especially in arid and semiarid areas. According to the assessment (GWP EA 2015), Ethiopia, Kenya, and Uganda have national policies on disaster risk management while South Sudan is in the process of developing similar policies. Djibouti, Somalia, and Sudan either have policies that focus more on emergency responses or they do not yet have disaster risk management policies. The assessment concluded that even the existing policies in many countries of the Horn of Africa are not yet comprehensive enough to fully address integrated drought management.

Most countries in the region have a government institution responsible for leading and coordinating the implementation of disaster risk management, but the structure of the arrangements varies among the countries. For example, Kenya has established the National Drought Management Authority (NDMA) (GoK 2012). South Sudan has the Ministry of Environment, which includes the Ministry of Humanitarian Affairs and Disaster Management. Ethiopia has established the Disaster Risk Management and Food Security Sector, led by the Minister of State under the Ministry of Agriculture. Uganda and Somalia have high-level coordination under their respective prime ministers' offices.

IGAD developed the IGAD Drought Disaster Resilience and Sustainability Initiative (IDDRSI), which member states can use to prevent, mitigate, and adapt to the adverse impacts of drought (IGAD 2013b). The approach developed and recommended by IDDRSI combines relief with development interventions in dealing with drought and related emergencies in the region through the IGAD Secretariat. The IDMP HOA regional project is working closely with IDDRSI to assist Eritrea, Ethiopia, Kenya, Sudan, and Uganda. Djibouti, Somalia, and South Sudan will be supported by the IDMP HOA for some IDDRSI components.

The IDDRSI developed the IGAD Regional Programming Paper (RPP), which provides the framework for the operationalizing of drought-related actions at both country and regional levels (IGAD 2013a). With IDDRSI in place, member states have developed country programming papers (CPPs) (IGAD 2012), which can serve as the planning, coordination, and resource mobilization tools for projects and investments needed to help end drought emergencies. The CPPs have been able to identify the root causes of vulnerability to drought, areas of intervention, and investments and establish adequate national coordinating mechanisms to implement drought resilience programs (IGAD 2012). For example, the Ethiopia CPP identified the following challenges: recurrence of drought, population growth versus shrinking resources, low levels of infrastructure, low implementation capacity, violent conflicts, and climate change (Government of Ethiopia 2012).

Although the region has been taking some positive measures toward building drought resilience, the assessment noted the following major gaps that still need to be addressed through promotion of integrated drought management (GWP EA 2015):

1. Limitations in human and institutional capacities that are needed to coordinate and implement drought risk management and resilience-building initiatives

2. Inadequate policy and legislative frameworks for disaster risk management and particularly for drought risk management

3. Lack of information on water and other natural resources in the arid and semiarid areas of the countries

4. Weak market, communication, and transport infrastructure in areas vulnerable to drought

5. Weak early warning systems to inform vulnerable communities of weather and disasters, and alerts for effective preparedness and response

6. Low level of educational information coupled with a strong adherence to traditional ways of keeping large herds of livestock by pastoralist communities

7. Limitation of resources to finance drought risk management and resilience-building initiatives

8. Inadequate participatory infrastructure in drought management programs

9. Continued reactive crisis management approach to drought management, including an overreliance on relief aid

The IDMP HOA regional project has aligned various interventions to address these gaps based on prioritization by the countries in the region. The IDMP HOA is implementing a capacity development program. The support is tailored for countries to develop and revise their drought plans and develop practices that increase drought resilience.

These gaps highlight the main challenge in the region, which is to reverse the growing human vulnerability to environmental hazards, such as droughts and man-made disturbances such as conflicts and economic crises (GWP EA 2015). Other challenges include the threat to pastoral and agricultural production systems due to rapid population growth, migration, environmental degradation, land reallocation, fragmentation of rangelands, decreasing mobility for herds, and growing competition in using scarce pasture and water resources. Since land access and water rights are not sufficiently regulated, conflicts can arise between different competing stakeholders, especially among cross-border communities. The combination of these adverse factors is accelerating environmental degradation and therefore exacerbating the vulnerability of societies. When droughts occur, the whole agro-pastoral production system can collapse, with disastrous consequences for the affected populations. Large financial resources are then needed for humanitarian aid and even more to recover the production systems and livelihoods of the drought-affected communities.

In the past, efforts were more concentrated in managing the drought disaster and related humanitarian emergencies. The new approach will focus on the underlying causes of the need for humanitarian aid and will approach disaster management through proactive and preventive solutions.

The following opportunities and steps taken were identified for the promotion and implementation of an integrated approach to drought management in the Horn of Africa (GWP EA 2015):

- Existence of IGAD to establish regional and international mechanisms for cooperation to address regional drought issues
- Existence of the IDDRSI framework, including the adopted CPPs for drought resilience and sustainable development
- Availability of political will and commitment to drought risk reduction by governments in the region
- Existing national implementing and coordination structures and institutions for drought management
- Availability of relevant national policies, strategies, and initiatives on drought management in the HOA countries
- Availability of institutions with experience and well-developed frameworks in implementing programs and projects that can provide examples of good practices
- Interest of development partners, IGAD member countries, and the private sector to support national and regional initiatives to enhance drought resilience

The IDMP HOA assessment of drought resilience status shows that there is commitment by governments to combat drought for sustainable national growth and development. IDDRSI, the IGAD initiative to strengthen food security and drought resilience in the region, has also resulted in national programs with specific institutional frameworks for their implementation. In addition, development partners are ready to support actions that are aimed at strengthening drought resilience instead of reactive emergency and relief operations.

Based on the assessments, the following priority areas were recommended for building drought resilience in the HOA region:

1. Demonstrating innovative drought resilience cases
2. Developing capacity for drought management and resilience building
3. Promoting partnerships for integrated drought management
4. Facilitating regional cooperation/collaboration for drought management in the HOA region
5. Facilitating policy development for integrated drought management
6. Mainstreaming drought mitigation and adaptation strategies in relevant government sector ministries and agencies
7. Strengthening early warning systems

Innovative drought resilience case studies would include using lessons learned and best practices that use an approach of integrated water resources management from countries.

3.5 Conclusions

The HMNDP in 2013 put the spotlight on what needs to be done and provided several key elements essential for developing national drought policies. It has resulted in a concerted effort through various initiatives to move forward. However, much remains to be done in terms of developing fully operational drought management policies that fulfill the multiple objectives of being proactive, mitigating impacts, improving public awareness, employing appropriate economic instruments, and linking to local and national development frameworks. The three pillars of integrated drought management—drought monitoring and early warning systems; vulnerability and impact assessments; and drought preparedness, mitigation, and response—provide a guide to structure and focus the work that needs to be done.

To support countries in putting these recommendations into practice, the IDMP uses information from its partners to keep track of the status of national drought policies and activities from around the world, working in partnership to close the gaps to move from crisis to risk management and foster horizontal integration among different sectors and actors, and vertical exchange—learning from the local through to the global and vice versa. The IDMP does not try to coordinate all drought activities around the world but rather aims to synthesize and apply existing knowledge and approaches to integrated drought management in collaboration with many international and regional organizations. The IDMP and its many partners provide a framework and commensurate technical support to countries, but the actual development and implementation of national drought plans and policies needs to be done by the governmental ministries and national stakeholders. The IDMP has already used these partnerships to create several early outputs in assisting countries to develop more proactive drought policies and plans with practical guidance and applications tailored to regional and national circumstances.

References

GoK (Government of Kenya). 2012. *Programming Framework to End Drought Emergencies in the Horn of Africa—Ending Drought Emergencies in Kenya: Country Programme Paper.* Ministry of State for Development of Northern Kenya and other Arid Lands, Government of Kenya, Nairobi, Kenya.

Government of Ethiopia. 2012. *Ethiopia Country Programming Paper to End Drought Emergencies in the Horn of Africa.* Ministry of Agriculture, Addis Ababa, Ethiopia.

GWP CEE (Global Water Partnership Central and Eastern Europe). 2015. *Guidelines for the Preparation of Drought Management Plans. Development and Implementation in the Context of the EU Water Framework Directive.* Global Water Partnership Central and Eastern Europe, Bratislava, Slovakia.

GWP CEE (Global Water Partnership Central and Eastern Europe) and E. Falutova. 2014. *Report on Review of the Current Status of Implementation of the Drought Management Plans and Measures.* Global Water Partnership Central and Eastern Europe, Bratislava, Slovakia.

GWP EA (Global Water Partnership Eastern Africa). 2015. *Assessment of Drought Resilience Frameworks in the Horn of Africa.* Integrated Drought Management Program in the Horn of Africa (IDMP HOA). Global Water Partnership Eastern Africa, Entebbe, Uganda.

ICPAC. 2016. *IGAD Climate Prediction & Applications Centre (ICPAC): Profile.* http://www.icpac.net/?page_id=4.

IGAD. 2012. *IDDRSI Country Programming Papers.* http://resilience.igad.int/index.php/programs-projects/national.

IGAD. 2013a. *The IGAD Drought Disaster Resilience and Sustainability Initiative (IDDRSI).* Regional Programming Paper. Final Draft. Intergovernmental Authority on Development (IGAD), Djibouti.

IGAD. 2013b. *Drought Disaster Resilience and Sustainability Initiative (IDDRSI): The IDDRSI Strategy.* Intergovernmental Authority on Development (IGAD), Djibouti.

UNCCD. 2016. *Windhoek Declaration for Enhancing Resilience to Drought in Africa.* http://www.unccd.int/Documents/Windhoek%20Declaration%20Final%20Adopted%20by%20the%20ADC%20of%2015-19%20August%202016.pdf.

UNCCD, FAO, and WMO. 2013. *High Level Meeting on National Drought Policy (HMNDP) Policy Document: National Drought Management Policy.* http://www.droughtmanagement.info/literature/WMO_HMNDP_policy_document_2012.pdf.

UNW-DPC (UN-Water Decade Programme on Capacity Development). 2015a. *Capacity Development to Support National Drought Management Policies.* http://www.ais.unwater.org/ais/pluginfile.php/516/course/section/168/NDMP-Synthesis.pdf.

UNW-DPC (UN-Water Decade Programme on Capacity Development). 2015b. *Proceedings of the Regional Workshops on Capacity Development to Support National Drought Management Policies for Eastern and Southern Africa and the Near East and North Africa Regions.* http://www.ais.unwater.org/ais/pluginfile.php/516/course/section/168/proceedings-no-14_WEB.pdf.

Wilhite, D. A. 1991. Drought planning: A process for state government. *Water Resources Bulletin* 27:29–38.

Wilhite, D. A., M. J. Hayes, and C. L. Knutson. 2005. Drought preparedness planning: Building institutional capacity. In *Drought and Water Crises: Science, Technology and Management Issues*, ed. D. A. Wilhite, 93–136. Boca Raton, FL: CRC Press.

WMO (World Meteorological Organization). 2013. *High Level Meeting on National Drought Policy (HMNDP) Final Declaration.* World Meteorological Organization, Geneva, Switzerland.

WMO (World Meteorological Organization) and GWP (Global Water Partnership). 2014. *National Drought Management Policy Guidelines: A Template for Action* (D. Wilhite). Integrated Drought Management Programme (IDMP) Tools and Guidelines Series 1. World Meteorological Organization, Geneva, Switzerland.

WMO (World Meteorological Organization) and GWP (Global Water Partnership). 2016. *Handbook of Drought Indicators and Indices* (M. Svoboda and B. A. Fuchs, eds.). Integrated Drought Management Tools and Guidelines Series 2. Integrated Drought Management Programme (IDMP), Geneva, Switzerland.

WMO (World Meteorological Organization) and GWP (Global Water Partnership). 2017. *Benefits of Action and Costs of Inaction: Drought Mitigation and Preparedness—A Literature Review* (N. Gerber and A. Mirzabaev, eds.). Integrated Drought Management Programme (IDMP) Working Paper 1. WMO, Geneva, Switzerland.

4

National Drought Management Policy Guidelines: A Template for Action*

Donald A. Wilhite

CONTENTS

---

* This chapter is included in this book with the permission of the World Meteorological Organization (WMO) and Global Water Partnership (GWP) (2014) *National Drought Management Policy Guidelines: A Template for Action* (D. A. Wilhite). *Integrated Drought Management Programme (IDMP) Tools and Guidelines Series 1.* WMO, Geneva, Switzerland and GWP, Stockholm, Sweden. The full guidelines report is available at http://www.droughtmanagement.info/literature/IDMP_NDMPG_en.pdf.

4.1 Introduction

The implementation of a drought policy based on the philosophy of risk reduction can alter a nation's approach to drought management by reducing the associated impacts (risk). This was the idea that motivated the World Meteorological Organization (WMO), the Secretariat of the United Nations Convention to Combat Desertification (UNCCD), and the Food and Agriculture Organization of the United Nations (FAO), in collaboration with a number of UN agencies, international and regional organizations, and key national agencies, to organize the High-level Meeting on National Drought Policy (HMNDP), which was held March 11–15, 2013, in Geneva, Switzerland. The theme of the HMNDP was "Reducing Societal Vulnerability—Helping Society (Communities and Sectors)" (see Chapter 2).

The spiraling impacts of drought on a growing number of sectors is cause for significant concern. No longer is drought primarily associated with the loss or reduction of agricultural production. Today, the occurrence of drought is also associated with significant impacts in the energy, transportation, health, recreation/tourism, and other sectors. Equally important is the direct impact of water shortages on water, energy, and food security. With the current and projected increases in the incidence of drought frequency, severity, and duration as a result of climate change, the time to move forward with a paradigm shift from crisis to risk management is now. This approach is directed at improving the resilience or coping capacity of nations to drought.

The outcomes and recommendations emanating from the HMNDP are drawing increased attention from governments, international and regional organizations, and nongovernmental organizations. One of the specific outcomes of the HMNDP was the launch of the Integrated Drought Management

Programme (IDMP) by WMO and the Global Water Partnership (GWP). The IDMP is addressing these concerns with a number of partners with the objective of supporting stakeholders at all levels by providing them with policy and management guidance through globally coordinated generation of scientific information and sharing best practices and knowledge for integrated drought management. The IDMP especially seeks to support regions and countries to develop more proactive drought policies and better predictive mechanisms, and these guidelines are a contribution to this end.

During the opening session of the High-level Meeting on National Drought Policy in March 2013, the Secretary General of the WMO, Michel Jarraud, stated:

> *In many parts of the world, the approach to droughts is generally reactive and tends to focus on crisis management. Both at the national and regional scale, responses are known to be often untimely, poorly coordinated and lacking the necessary integration. As a result, the economic, social and environmental impacts of droughts have increased significantly in many regions of the world. We simply cannot afford to continue in a piecemeal mode, driven by crisis rather than prevention. We have the knowledge, we have the experience and we can reduce the impacts of droughts. What we need now is a policy framework and action on the ground for all countries that suffer from droughts. Without coordinated national drought policies, nations will continue to respond to drought in a reactive way. What we need are monitoring and early warning systems to deliver timely information to decision makers. We must also have effective impact assessment procedures, proactive risk management measures, preparedness plans to increase coping capabilities and effective emergency response programmes to reduce the impact of drought.*

In 2013, the Secretary General of the United Nations, Ban Ki-moon, stated:

> *Over the past quarter-century, the world has become more drought-prone, and droughts are projected to become more widespread, intense and frequent as a result of climate change. The long-term impacts of prolonged drought on ecosystems are profound, accelerating land degradation and desertification. The consequences include impoverishment and the risk of local conflict over water resources and productive land. Droughts are hard to avert, but their effects can be mitigated. Because they rarely observe national borders they demand a collective response. The price of preparedness is minimal compared to the cost of disaster relief. Let us therefore shift from managing crises to preparing for droughts and building resilience by fully implementing the outcomes of the High-level Meeting on National Drought Policy held in Geneva last March.*
>   *(The complete statement from Ban Ki-moon is available at: http://www. un.org/sg/statements/?nid=6911)*

4.2 Drought Policy and Preparedness: Setting the Stage

Drought is a complex natural hazard, and the impacts associated with it are the result of numerous climatic factors and a wide range of societal factors that define the level of societal resilience. Population growth and redistribution and changing consumption and production patterns are two of the factors that define the vulnerability of a region, economic sector, or population group. Many other factors, such as poverty and rural vulnerability, weak or ineffective governance, changes in land use, environmental degradation, environmental awareness and regulations, and outdated or ineffective government policies, also contribute to changing vulnerability.

Although the development of drought policies and preparedness plans can be a challenging undertaking, the outcome of this process can significantly increase societal resilience to these climatic shocks. One of the primary goals of the guidelines presented in this document is to provide a template in order to make the development of national drought policies and associated preparedness plans at the subnational level less daunting.

Simply stated, a national drought policy should establish a clear set of principles or operating guidelines to govern the management of drought and its impacts. The overriding principle of drought policy should be an emphasis on risk management through the application of preparedness and mitigation* measures (HMNDP 2013). This policy should be directed toward reducing risk by developing better awareness and understanding of the drought hazard and the underlying causes of societal vulnerability, along with developing a greater understanding of how being proactive and adopting a wide range of preparedness measures can increase societal resilience. Risk management can be promoted by:

- Encouraging the improvement and application of seasonal and shorter-term forecasts
- Developing integrated monitoring and drought early warning systems and associated information delivery systems
- Developing preparedness plans at various levels of government
- Adopting mitigation actions and programs
- Creating a safety net of emergency response programs that ensure timely and targeted relief
- Providing an organizational structure that enhances coordination within and between levels of government and with stakeholders.

---

* In the natural hazards field, mitigation measures are commonly defined as actions taken in advance of the hazard event (e.g., drought) to lessen impacts when the next drought occurs. In contrast, mitigation in the context of climate change is focused on reducing greenhouse gas (GHG) emissions and thereby mitigating or limiting future temperature increases.

The policy should be consistent and equitable for all regions, population groups, and economic sectors and consistent with the goals of sustainable development.

As vulnerability to and the incidence of drought has increased globally, greater attention has been directed to reducing risks associated with its occurrence through improved planning to improve operational capabilities (e.g., climate and water supply monitoring, and building institutional capacity) and mitigation measures that are aimed at reducing drought impacts. This change in emphasis is long overdue. Mitigating the effects of drought requires the use of all components of the cycle of disaster management (Figure 4.1), rather than only the crisis management portion of this cycle. Typically, when drought occurs, governments and donors have followed with impact assessment, response, recovery, and reconstruction activities to return the region or locality to a pre-disaster state. Historically, little attention has been given to preparedness, mitigation, or prediction/early warning actions (i.e., risk management) and the development of risk-based national drought management policies that could avoid or reduce future impacts and lessen the need for government and donor interventions in the future. Crisis management only addresses the symptoms of drought, as they manifest themselves in the impacts that occur as a direct or indirect consequence of drought. Risk management, on the other hand, is focused on identifying where vulnerabilities exist (particular sectors, regions, communities, or

FIGURE 4.1
Cycle of disaster management. (Courtesy of National Drought Mitigation Center, University of Nebraska–Lincoln.)

population groups) and addresses these risks through systematically implementing mitigation and adaptation measures that will lessen the risk associated with future drought events. Since societies have emphasized crisis
management in past attempts at drought management, countries have generally moved from one drought event to another with little, if any, reduction
in risk. In addition, in many drought-prone regions, another drought event
is likely to occur before the region fully recovers from the last event. If the
frequency of drought increases in the future, as projected for many regions,
there will be less recovery time between these events.

Progress on drought preparedness and policy development has been slow
for a number of reasons. It is certainly related to the slow-onset characteristics of drought and the lack of a universal definition. Drought shares with climate change the distinction of being a creeping phenomenon—the challenge
being getting people to recognize changes that occur slowly or incrementally
over a long period of time. These characteristics of drought make early warning, impact assessment, and response difficult for scientists, natural resource
managers, and policy makers. The lack of a universal definition often leads
to confusion and inaction on the part of decision makers since scientists may
disagree on the existence and severity of drought conditions (i.e., the onset
and recovery time differences between meteorological, agricultural, and
hydrological drought). Severity is also difficult to characterize since it is best
evaluated on the basis of multiple indicators and indices, rather than on the
basis of a single variable. The impacts of drought are also largely nonstructural and spatially pervasive. These features make it difficult to assess the
effects of drought and to respond in a timely and effective manner. Drought
impacts are not as visual as the impacts of other natural hazards, making it
difficult for the media to communicate the significance of the event and its
impacts to the public. Public sentiment to respond is often lacking in comparison to other natural hazards that result in loss of life and property.

Associated with the crisis management approach is the lack of recognition
that drought is a normal part of the climate. Climate change and associated
projected changes in climate variability will likely increase the frequency
and severity of drought and other extreme climatic events. In the case of
drought, the duration of these events may also increase. Therefore, it is
imperative for all drought-prone nations to adopt a drought management
approach that is aimed at risk reduction. This approach will increase resilience to future episodes of drought.

It is important to note that each occurrence of drought provides a window
of opportunity to move toward a more proactive risk management policy.
Immediately following a severe drought episode, policy makers, resource
managers, and all affected sectors are aware of the impacts that have
occurred, and at this time the causal factors associated with these impacts
(i.e., the roots of the vulnerability) are more easily recognized. Any deficiencies in the government's response or that of donor organizations could also
be more easily identified. There is no better time to approach policy makers

with the concept of developing a national drought policy and preparedness plan aimed at increasing societal resilience.

To provide guidance on the preparation of national drought policies and planning techniques, it is important to define the key components of drought policy, its objectives, and steps in the implementation process. An important component of national drought policy is increased attention to drought preparedness in order to build institutional capacity to deal more effectively with this pervasive natural hazard. The lessons learned by a few countries that have been experimenting with this approach will be helpful in identifying pathways to achieve more drought-resilient societies. For this reason, several case studies are included in this document. It is a living document, which will be revised with experiences gained from further case studies.

A constraint to drought preparedness has been the dearth of methodologies available to policy makers and planners to guide them through the planning process. Drought differs in its physical characteristics between climate regimes, and impacts are locally defined by unique economic, social, and environmental characteristics. A methodology developed by Wilhite (1991) and revised to incorporate greater emphasis on risk management (Wilhite et al. 2000, 2005) has provided a set of generic steps that can be adapted to any level of government (i.e., national to subnational) or geographical setting for the development of a drought preparedness plan.

The IDMP, an initiative of the WMO and the GWP, recognizes the urgent need to provide nations with guidelines for the development of national drought management policies. To achieve this goal, the drought preparedness planning methodology referred to above has been modified to define a generic process by which governments can develop a national drought policy and drought preparedness plans at the national and subnational level that support the principles of that policy. This process is described below with the aim of providing a template that governments or organizations can adapt to their needs to reduce societal vulnerability to drought, thus creating greater resilience for future droughts across all sectors. A national drought policy can be a standalone policy or a subset of a natural disaster risk reduction, sustainable development, integrated water resources, or climate change adaptation plan that may already exist.

4.3 Drought Policy: Characteristics and the Way Forward

As a beginning point in the discussion of drought policy, it is important to identify the various types of drought policies that are available and have been employed for drought management. The first and most common approach followed by both developing and developed nations is post-impact government (or nongovernment) interventions. These interventions are normally relief measures in the form of emergency assistance programs aimed at providing money or other specific types of assistance (e.g., livestock feed, water, and food) to the victims (or those experiencing the most severe impacts) of the drought. This reactive approach, characterized by the hydro-illogical cycle (Figure 4.2), is seriously flawed from the perspective of vulnerability reduction since the recipients of this assistance are not expected to change behaviors or resource management practices as a condition of the assistance. Brazil, a country that has typically followed the crisis management approach, is currently reevaluating this approach and strongly considering the development of a national drought policy that is focused on risk reduction (see Chapter 21).

Although drought assistance provided through emergency response interventions may address a short-term need, it may in the longer term actually decrease the coping capacity of individuals and communities by fostering greater reliance on these interventions rather than increasing self-reliance. For example, livestock producers that do not maintain adequate on-farm storage of feed for livestock as a drought management strategy will be the first to experience the impacts of extended precipitation shortfalls, and they will be the first to turn to the government or other organizations for assistance

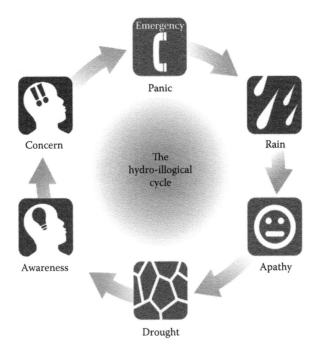

FIGURE 4.2
The hydro-illogical cycle. (Courtesy of National Drought Mitigation Center, University of Nebraska–Lincoln.)

to maintain herds until the drought is over and forage supplies return to adequate levels. Likewise, urban communities that have not augmented water supply capabilities in response to population growth or maintained or updated delivery systems may turn to government for assistance during periods of drought-induced water shortages. The shortages that result are the product of poor planning rather than a direct impact of drought. This reliance on the government for relief is contrary to the philosophy of encouraging risk preparedness through an investment by producers, water managers, and others to improve their drought-coping capacity. Government assistance or incentives that encourage these investments would be a philosophical change in how governments respond and would promote a change in the expectations of livestock producers as to the role of government in these response efforts. The more traditional approach of providing relief is also flawed in terms of the timing of assistance being provided. It often takes weeks or months for assistance to be received, at times well beyond the window of when the relief would be of greatest value in addressing the impacts of drought. In addition, those livestock producers who previously employed appropriate risk reduction techniques are likely ineligible for assistance since the impacts they experienced were reduced and therefore do not meet the eligibility requirements. This approach rewards those that have not adopted appropriate resource management practices.

COMMUNITY-BASED RESILIENCE ANALYSIS (COBRA) IN KENYA AND UGANDA

The Drylands Development Centre of the United Nations Development Programme (UNDP) has demonstrated through Community Based Resilience Analysis (CoBRA) in Kenya and Uganda the existence of resilient households, which have been able to sustain their life and livelihoods without humanitarian aid even in the hardest hit areas. Consultations with these households showed that they are resilient to any hazard because of their strong asset base and diversified risk management options. One of the primary reasons for this higher level of resilience in all four arid and semiarid assessment areas in Kenya and Uganda was education, not at elementary but higher (secondary or tertiary) levels, which provided them with the knowledge needed to cope with any type of hazard. A higher level of education provided more income-generating opportunities, leading to better access to different goods and services.

Although at times there is a need to provide emergency response to various sectors (i.e., post-impact assessment interventions), it is critically important for the purpose of moving toward a more proactive risk management approach that the two drought policy approaches described below become the cornerstone of the policy process.

The second type of drought policy approach is the development and implementation of policies and preparedness plans, which include organizational frameworks and operational arrangements developed in advance of drought and maintained between drought episodes by government or other entities. This approach attempts to create greater institutional capacity focused on improved coordination and collaboration within and between levels of government; stakeholders in the primary impact sectors; and the plethora of private organizations with a vested interest in drought management (i.e., communities, natural resource or irrigation districts or managers, utilities, agribusiness, farmers' organizations, and others).

The third type of policy approach emphasizes the development of pre-impact government programs or measures that are intended to reduce

DROUGHT MITIGATION

As previously noted, mitigation in the context of natural hazards is different from mitigation in the context of climate change, where the focus is on reducing greenhouse gas (GHG) emissions. Mitigation in the context of natural hazards refers to actions taken in advance of drought to reduce impacts in the future.

Drought mitigation measures are numerous, but they may be more confusing to the general public in comparison to mitigation measures for earthquakes, floods, and other natural hazards where the impacts are largely structural. Impacts associated with drought are generally nonstructural, and thus the impacts are less visible, more difficult to assess (e.g., reductions in crop yield), and do not require reconstruction as part of the recovery process. Drought mitigation measures would include establishing comprehensive early warning and delivery systems, improved seasonal forecasts, increased emphasis on water conservation (demand reduction), increased or augmented water supplies through greater utilization of groundwater resources, water reutilization and recycling, construction of reservoirs, interconnecting water supplies between neighboring communities, drought preparedness planning to build greater institutional capacity, and awareness-building and education.

In some cases, such water resource augmentation measures are best developed jointly with a neighboring state (or country), or at least such measures should be coordinated if they might have an impact on the other riparian state (or downstream use in general). Insurance programs, currently available in many countries, would also fall into this category of policy types.

vulnerability and impacts. This approach could be considered a subset of the second approach listed above. In the natural hazards field, these types of programs or measures are commonly referred to as mitigation measures.

4.4 National Drought Management Policy: A Process

The challenges that nations face in the development of a risk-based national drought management policy are complex. The process requires political will at the highest level possible and a coordinated approach within and between levels of government and with the diversity of stakeholders that must be engaged in the policy development process. A national drought policy could be a standalone policy. Alternatively, it could contribute to or be a part of a national policy for disaster risk reduction with holistic and multihazard approaches that is centered on the principles of risk management (UNISDR 2009).*

The policy should provide a framework for shifting the paradigm from one traditionally focused on reactive crisis management to one that is focused on a proactive risk-based approach that is intended to increase the coping capacity of the country and thus create greater resilience to future episodes of drought.

The formulation of a national drought policy, while providing the framework for a paradigm shift, is only the first step in vulnerability reduction. The development of a national drought policy must be intrinsically linked to the development and implementation of preparedness and mitigation plans at the subnational level. These plans will be the instruments through which a national drought policy is executed.

The 10 steps below provide an outline of the process for policy and preparedness planning. The process is intended to be a generic template or road map—in other words, applying this methodology requires adapting it to the current institutional capacity, political infrastructure, and technical capacity within the country concerned. It has been modified from a 10-step drought planning process or methodology developed in the United States for application at the state level. Currently, 47 of the 50 US states have developed

---

* To this end, the *Hyogo Framework for Action 2005–2015: Building the Resilience of Nations and Communities to Disasters*, adopted by member states in 2005, gives strategic directions to cover all phases of disaster risk reduction, from policy and legislation development to institutional frameworks, multihazard risk identification, people-centered early warning systems, knowledge and innovation to build a culture of resilience, reduction of underlying risk factors, and strengthening disaster preparedness. Consultations on the implementation of the Hyogo Framework for Action and its successor are underway. This process intends to culminate at the Third World Conference on Disaster Risk Reduction agreed on by the UN General Assembly for March 14–18, 2015, in Sendai, Japan.

drought plans, and the majority of these states have followed these guide-lines in the preparation or revision of drought plans.* This drought planning methodology has also been followed in other countries in the development of national drought strategies. For example, Morocco applied it beginning in 2000 as part of a process to develop a national drought strategy. Their strategy has continued to evolve over the past decade.

The process, originally developed in the early 1990s, has been revised numerous times, placing greater emphasis on mitigation planning with each revision. Now, it has been modified once again to reflect an emphasis on developing a national drought management policy, including the develop-ment of drought preparedness plans at the subnational level that support the goals of a national policy.

The 10 steps in the drought policy and preparedness process are:

Step 1: *Appoint* a national drought management policy commission

Step 2: *State* or *define* the goals and objectives of a risk-based national drought management policy

Step 3: *Seek* stakeholder participation; *define* and *resolve* conflicts between key water use sectors, considering also transboundary implications

Step 4: *Inventory* data and financial resources available and *identify* groups at risk

Step 5: *Prepare/write* the key tenets of the national drought manage-ment policy and preparedness plans, which would include the fol-lowing elements: monitoring; early warning and prediction; risk and impact assessment; and mitigation and response

Step 6: *Identify* research needs and *fill* institutional gaps

Step 7: *Integrate* science and policy aspects of drought management

Step 8: *Publicize* the national drought management policy and pre-paredness plans and *build* public awareness and consensus

Step 9: *Develop* educational programs for all age and stakeholder groups

Step 10: *Evaluate* and *revise* national drought management policy and supporting preparedness plans

4.4.1 Step 1: Appoint a National Drought Commission

The process for creating a national drought management policy should begin with the establishment of a national commission to oversee and facili-tate policy development. Given the complexities of drought as a hazard, and

---

* Drought planning resources by state. Available at http://drought.unl.edu/Planning/ PlanningInfobyState.aspx

the cross-cutting nature of managing all aspects of monitoring, early warning, impact assessment, response, mitigation, and planning, it is critical to coordinate and integrate the activities of the many agencies/ministries of government at all levels; the private sector, including key stakeholder groups; and civil society. To ensure a coordinated process, the president/prime minister or other key political leader must take the lead in establishing a national drought policy commission. Otherwise, it may not garner the full support and participation of all relevant parties.

The purpose of the commission is twofold. First, the commission will supervise and coordinate the policy development process. This includes bringing together all of the necessary resources of the national government and integrating these resources from the various ministries and levels of government in order to develop the policy and supporting preparedness plans. By pooling the government's resources, this initial phase will likely require only minimal new resources coupled with a redirection of existing resources (e.g., financial, data, and human) in support of the process. Second, once the policy is developed, the commission will be the authority responsible for the implementation of the policy at all levels of government. The principles of this policy will be the basis for the development and implementation of preparedness or mitigation-based plans at the subnational level. In addition, the commission will be tasked with the activation of the various elements of the policy during times of drought. The commission will coordinate actions and implement mitigation and response programs or will delegate this action to governments at the subnational level. They will also initiate policy recommendations to the political leader and/or the appropriate legislature body and implement specific recommendations within the authority of the commission and the ministries represented.

The commission should reflect the multidisciplinary nature of drought and its impacts, and it should include all appropriate national government ministries. It is also appropriate to consider the inclusion of key drought experts from universities to serve either in an advisory capacity to the commission or as an official member of the body. A representative from the president's office should also be included in order to facilitate communication as well as an awareness of drought impacts, status, and actions.

It may also be appropriate to consider the inclusion of representatives from key sectors, professional associations, and environmental and public interest groups. If members of these groups are not included, an alternative would be the creation of a citizen's advisory committee composed of these representatives in order for these groups to have a voice in the policy development process and in the identification and implementation of appropriate response and mitigation actions. Having said that, representatives of these groups will also be involved in the development process for the drought preparedness plans at the state/provincial level, so their inclusion on the commission or as a separate citizen's advisory committee may be redundant.

It is also important for the commission to include a public information specialist as an expert on communication strategies. This person can formulate effective communication messages to all media. It is imperative for the commission to communicate with the media with a single voice so the message to the public is clear and concise. Because of the scientific, regional, and sectoral complexities of drought, the severity of drought and related impacts, and the wide-ranging response and mitigation programs/actions that may be involved, the public can be easily confused when information is forthcoming from multiple release points.

Given the wide range of stakeholder groups that will be involved in policy development, implementation, and activation, a public participation practitioner should be engaged. This person would be an observer or ex-officio member of the commission and regularly attend commission meetings. This person would also assist in the orchestration of many aspects of the policy development process in order to solicit input from the multitude of stakeholder groups that will be engaged. This person can also ensure that all groups, both well-funded and disadvantaged stakeholder or interest groups, are included in the process.

The composition of the membership of national drought commissions that have been engaged in the policy development process in specific countries may provide useful insights. For example, in Mexico, a national drought program was announced by President Enrique Peña Nieto on January 10, 2013. The goals of this program are early warning and early action to identify preventive actions leading to timely decisions to prevent and/or mitigate the effects of drought (see Chapter 19 in this book).

4.4.2 Step 2: State or Define the Goals and Objectives of a Risk-Based National Drought Management Policy

Drought is a normal part of climate, but there is considerable evidence and growing concern that the frequency, severity, and duration of droughts are increasing in many parts of the world—or will increase in the future—as a result of anthropogenic climate change. The HMNDP, held in March 2013, was organized largely in response to this concern, as well as the ineffectiveness of the traditional crisis management approach or response to the occurrence of drought. It provided a forum and launched the IDMP.

The essential elements of a national drought management policy, as identified through the HMNDP, are:

- Developing proactive mitigation and planning measures, risk management approaches, and public outreach and resource stewardship.
- Enhancing collaboration between national, regional, and global observation networks and developing information delivery systems that improve public understanding of, and preparedness for, drought.

- Creating comprehensive governmental and private insurance and financial strategies.
- Recognizing the need for a safety net of emergency relief based on sound stewardship of natural resources and self-help at diverse governance levels.
- Coordinating drought programs and response efforts in an effective, efficient, and customer-oriented manner.

Following the formation of the commission, its first official action should be to establish specific and achievable goals for the national drought policy and a timeline for implementing the various aspects of the policy, as well as a timeline for achieving the goals. Several guiding principles should be considered as the commission formulates a strategy to move from crisis management to a drought risk reduction approach. First, assistance measures, if employed, should not discourage agricultural producers, municipalities, and other sectors or groups from the adoption of appropriate and efficient management practices that help to alleviate the effects of drought (i.e., assistance measures should reinforce the goal of increasing resilience or coping capacity to drought events). Those assistance measures employed should help to build self-reliance to future drought episodes. Second, assistance should be provided in an equitable (i.e., to those most affected), consistent, and predictable manner to all without regard to economic circumstances, sector, or geographic region. It is important to emphasize that the assistance provided is not counterproductive or a disincentive for self-reliance. Third, the protection of the natural and agricultural resource base is paramount, so any assistance or mitigation measures adopted must not run counter to the goals and objectives of the national drought policy and long-term sustainable development goals.

As the commission begins its work, it is important to inventory all emergency response and mitigation programs that are available through the various ministries at the national level. It is also important to assess the effectiveness of these programs and past disbursement of funds through these programs. A similar exercise should be implemented at the state or provincial level in association with the development of drought preparedness and mitigation plans.

To provide guidance in the preparation of national drought policies and planning techniques, it is important to define the key components of a drought policy, its objectives, and steps in the implementation process. Commission members, supporting experts, and stakeholders should consider many questions as they define the goals of the policy:

- What is the purpose and role of government in drought mitigation and response efforts?
- What is the scope of the policy?

- What are the country's most vulnerable economic and social sectors and regions?
- Historically, what have been the most notable impacts of drought?
- Historically, what has been the government's response to drought and what has been its level of effectiveness?
- What is the role of the policy in addressing and resolving conflict between water users and other vulnerable groups during periods of shortage?
- What current trends (e.g., climate, drought incidence, land and water use, and population growth) may increase vulnerability and conflicts in the future?
- What resources (human and financial) is the government able to commit to the planning process?
- What other human and financial resources are available to the government (e.g., climate change adaptation funds)?
- What are the legal and social implications of the plan at various jurisdictional levels, including those extending beyond the state borders?
- What principal environmental concerns are exacerbated by drought?

A generic statement of purpose for the drought policy and preparedness plans is to reduce the impacts of drought by identifying principal activities, groups, or regions most at risk and developing mitigation actions and programs that reduce these vulnerabilities. The policy should be directed at providing government with an effective and systematic means of assessing drought conditions, developing mitigation actions and programs to reduce risk in advance of drought, and developing response options that minimize economic stress, environmental losses, and social hardships during drought.

UNITED STATES DROUGHT MANAGEMENT, POLICY, AND PREPAREDNESS

Drought is a normal part of the climate for virtually all portions of the United States; it is a recurring, inevitable feature of climate that results in serious economic, environmental, and social impacts. In 1995, the Federal Emergency Management Agency (FEMA) estimated average annual losses because of drought in the United States to be US$6 billion–US$8 billion, more than for any other natural hazard. The recent 2012 drought resulted in impacts estimated at US$35 billion–US$70 billion. Yet the United States has, historically, been ill-prepared for the recurrence of severe drought and responds, like most nations, in a reactive, crisis management approach, focusing on responding to the symptoms (impacts) of drought through a wide assortment of emergency response or relief programs. These programs can best be characterized as too little and too late. More importantly, drought relief does little if anything to reduce the vulnerability of the affected area to future drought events. Today, the nation has a better understanding of the pathway needed for improving drought management, which will require a new paradigm, one that encourages preparedness and mitigation through the application of the principles of risk management.

Since the early 1980s, a growing number of states have developed drought plans. To date, 47 of the 50 states have developed such plans, and of these, 11 are more proactive, stressing the importance of mitigation in the preparedness process. The majority of states have relied upon the 10-step drought planning process as a guide in the plan preparation process, either by directly applying the process or by replicating the plans of other states that have followed this 10-step process.

The most significant progress in drought preparedness at the state level has occurred since the mid-1990s and, especially, since 2000. In these more recent years, there has been a stronger emphasis on mitigation. This progress can be attributed largely to several key factors. First, a series of significant droughts have affected nearly all portions of the country since 1996 and, in many cases, for five to seven consecutive years. These droughts have raised the awareness of drought within the science and policy communities, as well as with the public. The US Drought Monitor Map, a weekly product produced since 1999 through a partnership between the National Drought Mitigation Center (NDMC) at the University of Nebraska, the National Oceanic and Atmospheric Administration (NOAA), and the US Department of Agriculture, has helped to raise awareness of drought conditions and impacts across the nation. It is highly regarded by both federal and state government as an excellent integrated approach to characterize the severity of drought and its spatial dimensions across the nation. The US Drought Monitor Map is not only effectively used at the federal level but also by states for drought assessment and as a trigger for drought response and mitigation programs. Second, the spiraling impacts of drought and the increasing number of key sectors affected, as well as the conflicts between sectors, has elevated the importance of drought preparedness within the policy community at all levels. Third, the creation of the NDMC at the University of Nebraska in 1995 has resulted in increased attention on issues of drought monitoring, impact assessment, mitigation, and preparedness. Many states have benefited from the existence of this expertise to guide the drought planning process. This is especially noticeable through the trend in the number of states developing or revising plans with a substantial emphasis on mitigation. As states have moved along the continuum from response to mitigation planning, there is an increasing need for better and timelier information on drought status and early warning, including improved seasonal forecasts and the delivery of that information to decision makers and other users. It is also important for these users or stakeholders to be involved in the development of products or decision support tools to ensure that their concerns and needs are being met.

Although the United States has not developed a national drought policy, there has been considerable pressure from states for the federal government to move toward a risk-based national drought policy. This pressure has been quite effective, leading to the introduction of legislation in the US Congress to improve preparedness and early warning. The National Drought Policy Act of 1998 created the National Drought Policy Commission (NDPC), charged with making recommendations to the Congress on future approaches to drought management. The final report of the Commission was submitted to Congress in 2000 and included a recommendation that the United States move forward with the development of a national drought policy based on the principles of risk management (NDPC 2000). The National Drought Preparedness Act, largely embodying the most significant recommendations from the NDPC, was introduced in Congress in 2001, and then reintroduced in 2003 and 2005. Although this bill did not pass and become law, it did generate another bill, the National Integrated Drought Information System (NIDIS) Act, which passed Congress in 2006 and was signed by the president later that year. This system (NIDIS) has been implemented by NOAA with partners from other federal agencies, state and regional organizations, and universities. NIDIS was recently reauthorized for a period of 5 years by Congress.

Largely in response to the severe drought of 2012 in the United States, which at its peak affected 65 percent of the contiguous states, the Obama Administration authorized the creation of a National Drought Resilience Partnership through an Executive Order in November 2013. This partnership includes seven federal agencies with the goal of assisting communities to better prepare for and reduce the impact of drought events on communities, families, and businesses. This action by the president has the potential to continue moving the United States on a path toward a risk-based national drought policy as part of the Obama Administration's Climate Change Action Plan.

4.4.3 Step 3: Seek Stakeholder Participation; Define and Resolve Conflicts between Key Water Use Sectors, Considering also Transboundary Implications

As noted in Step 1, a public participation specialist is an important contributor in the policy development process because of the complexities of drought as it intersects with society's social, economic, and environmental sectors, and the dependence of these sectors on access to adequate supplies of water in support of diverse livelihoods. As drought conditions intensify, competition for scarce water resources increases and conflicts often arise. These conflicts cannot be addressed during a crisis and thus it is imperative for potential conflicts to be addressed during nondrought periods when tension between these groups is minimal. As a part of the policy development process, it is essential to identify all citizen groups (i.e., stakeholders), including the private sector, that have a stake in the process and their interests. These groups must be involved early and continuously for fair representation to ensure an effective drought policy development process at the national and subnational levels. In the case of transboundary rivers, international obligations under agreements that the state is a party to should also be taken into account. Discussing concerns early in the process gives participants a chance to develop an understanding of one another's various viewpoints, needs, and concerns, leading to collaborative solutions. Although the level of involvement of these groups will vary notably from country to country and even within countries, the power of public interest groups in policy making is considerable in many settings. In fact, these groups are likely to impede progress in the policy development process if they are not included in the process. The commission should also protect the interests of stakeholders who may lack the financial resources to serve as their own advocates. One way to facilitate public participation is to establish a citizen's advisory council (as noted in Step 1) as a permanent feature of the commission's organizational structure in order to keep information flowing and address/resolve conflicts between stakeholders.

A national drought policy development process must be multilevel and multidimensional in its approach, as noted in the example of Mexico (see Chapter 19). In the case of Mexico, 26 district basin plans are being developed in concert with the national drought program initiative. Thus, the goals of basin plans should mirror or reflect national policy goals. State or provincial governments need to consider if district or regional advisory councils should be established and what their composition might be. These councils could bring stakeholder groups together to discuss their water use issues and problems and seek collaborative solutions in advance of the next drought.

4.4.4 Step 4: Inventory Data and Financial Resources Available and Identify Groups at Risk

An inventory of natural, biological, human, and financial resources, including the identification of constraints that may impede the policy development, may need to be initiated by the commission. In many cases, much information already exists about natural and biological resources through various provincial and national agencies/ministries. It is important to determine the vulnerability of these resources to periods of water shortage that result from drought. The most obvious *natural* resource of importance is water (i.e., location, accessibility, quantity, and quality), but a clear understanding of other natural resources such as climate and soils is also important. *Biological/ecological resources* refer to the quantity and quality of grasslands/rangelands, forests, wildlife, wetlands, and so forth. *Human resources* include the labor needed to develop water resources, lay pipeline, haul water and livestock feed, process and respond to citizen complaints, provide technical assistance, provide counseling, and direct citizens to available services.

It is also imperative to identify constraints to the policy development process and to the activation of the various elements of the policy and preparedness plans as drought conditions develop. These constraints may be physical, financial, legal, or political. The costs associated with policy development must be weighed against the losses that will likely result if no plan is in place (i.e., the cost of inaction). As stated previously, the goal of a national drought policy is to reduce the risk associated with drought and its economic, social, and environmental impacts. Legal constraints can include water rights, existing public trust laws, requirements for public water suppliers, transboundary agreements (e.g., specifying that a certain volume or share of river flow across the border has to be guaranteed), and liability issues.

The transition from crisis to risk management is difficult because, historically, little has been done to understand and address the risks associated with drought. To solve this problem, areas of high risk should be identified, as should actions that can be taken before a drought occurs to reduce those risks. Risk is defined by both the exposure of a location to the drought hazard and the vulnerability of that location to periods of drought-induced water shortages (Blaikie et al. 1994). Drought is a natural event; it is important to define the exposure (i.e., frequency of drought of various intensities and durations) of various parts of the country, province, or watershed to the drought hazard. Some areas are likely to be more at risk than others because of greater exposure to the hazard, which inhibits or shortens the recovery time between successive droughts. As a result of current and projected changes in climate and the frequency of occurrence of extreme climatic events, such as droughts, it is important to assess historical as well as projected future exposure to droughts. Vulnerability, on the other hand, is affected by social factors such as population growth and migration trends, urbanization, changes in land use, government policies, water use trends, diversity of economic base, and cultural composition. The commission can address these issues early in the policy development process, but the more detailed work associated with this risk or vulnerability process will need to be directed to specific working groups at the state or provincial level as they embark on the process of drought preparedness planning. These groups will have more precise local knowledge and will be better able to garner input from local stakeholder groups.

4.4.5 Step 5: Prepare/Write the Key Tenets of a National Drought Management Policy and Preparedness Plans, Including the Following Elements: Monitoring, Early Warning, and Prediction; Risk and Impact Assessment; and Mitigation and Response

Drought preparedness/mitigation plans, as stated earlier, are the instruments through which a national drought policy is carried out. It is essential for these plans to reflect the principles of the national drought policy, which is centered on the concept of risk reduction. What is defined below is the creation of institutional capacity that should be replicated within each state or province within a country, with formal communication and reporting links to a national drought commission.

At the outset, it is important to point out that preparedness planning can take two forms. The first form, response planning, is directed toward the creation of a plan that is activated only during drought events and usually for the purpose of responding to impacts. This type of planning is reactive, and the responses that are forthcoming, whether from national or state government or donor organizations, are intended to address specific impacts on sectors, population groups, and communities and, therefore, reflect the key areas of societal vulnerability. In essence, responding to impacts through emergency measures addresses only the symptoms of drought (impacts), and these responses are usually untimely, poorly coordinated, and, often, poorly targeted to those most affected. As noted earlier, this largely reactive approach actually leads to an increase in societal vulnerability since the recipients of drought relief or assistance programs become dependent on government and other programs through the assistance provided to survive the crisis. This approach discourages the development of self-reliance and implementation of improved resource management practices that will reduce risk in the longer term. Stated another way, why should the potential recipients of emergency assistance institute more proactive mitigation measures if government or others are likely to bail them out of a crisis situation? Emergency measures are appropriate in some cases, particularly with regard to providing humanitarian assistance, but they need to be used sparingly and be compatible with the longer-term goals of a national drought policy that is focused on improving resilience to future events.

The second form of preparedness planning is mitigation planning. With this approach, the vulnerabilities to drought are identified as part of the planning process through the analysis of both historical and more recent impacts of droughts. These impacts represent those sectors, regions, and population groups that are most at risk. The planning process then can focus on identifying actions and governmental or nongovernmental authorities that can assist in providing the necessary resources to reduce the vulnerability. In support of a risk-based national drought policy, mitigation planning is the best choice if risk reduction is the goal of the planning process. The discussion below shows how states/provinces might go about creating a plan that emphasizes mitigation.

Each drought task force at the subnational level should identify the specific objectives that support the goals of the plan. The objectives that should be considered include the following:

- Collect and analyze drought-related information in a timely and systematic manner.
- Establish criteria for declaring drought emergencies and triggering various mitigation and response activities.
- Provide an organizational structure and delivery system that ensures information flow between and within levels of government and to decision makers at all levels.
- Define the duties and responsibilities of all agencies or ministries with respect to drought.
- Maintain a current inventory of government programs used in assessing and responding to drought emergencies and in mitigating impacts in the longer term, if available.
- Identify drought-prone areas of the state and vulnerable economic sectors, individuals, or environments.
- Identify mitigation actions that can be taken to address vulnerabilities and reduce drought impacts.
- Provide a mechanism to ensure timely and accurate assessment of drought's impacts on agriculture, industry, municipalities, wildlife, tourism and recreation, health, and other areas.
- Keep the public informed of current conditions and response actions by providing accurate and timely information to media in print and electronic form (e.g., via TV, radio, and the Internet).
- Establish and pursue a strategy to remove obstacles to the equitable allocation of water during shortages and establish requirements or provide incentives to encourage water conservation.
- Establish a set of procedures to continually evaluate and exercise the plan and periodically revise the plan so it will remain responsive to local needs and reinforce national drought policy.

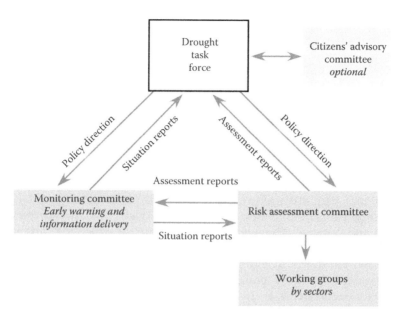

FIGURE 4.3
Drought preparedness and mitigation plan organizational structure. (Courtesy of National Drought Mitigation Center, University of Nebraska–Lincoln.)

The development of a drought preparedness plan that emphasizes mitigation begins with the establishment of a series of committees to oversee development of institutional capacity necessary for the plan as well as its implementation and application during times of drought when the various elements of the plan are activated. At the heart of the mitigation plan is the formation of a drought task force at the subnational level (e.g., state or provincial, and community) that mirrors to a large extent the makeup of the national drought commission (i.e., representatives from multiple agencies/ministries and key stakeholder groups). The organizational structure for the drought plan (Figure 4.3) reflects the three primary elements of the plan: monitoring, early warning, and information delivery; risk and impact assessment; and mitigation, preparedness, and response. It is recommended that a committee be established to focus on the first two of these requirements; the drought task force can, in most instances, carry out the mitigation and response functions since these are heavily policy oriented.

These committees will have their own tasks and goals, but well-established communication and information flow between committees and the task force is a necessity to ensure effective planning.

4.4.5.1 Monitoring, Early Warning, and Information Delivery Committee

A reliable assessment of water availability and its outlook for the near and long term is valuable information in both dry and wet periods.

During drought, the value of this information increases markedly. A monitoring committee should be a part of each state or provincial committee since it is important to interpret local conditions and impacts and communicate this information to the NDPC and its representative from the national meteorological service. In some instances, a monitoring committee may be set up for certain regions with similar climatic conditions and exposure to drought, rather than for each state or province. However, the makeup of this committee should include representatives from all agencies with responsibilities for monitoring climate and water supply. It is recommended that data and information on each of the applicable indicators (e.g., precipitation, temperature, evapotranspiration, seasonal climate forecasts, soil moisture, streamflow, groundwater levels, reservoir and lake levels, and snowpack) be considered in the committee's evaluation of the water situation and outlook. The agencies responsible for collecting, analyzing, and disseminating data and information will vary considerably from country to country and province to province. Also, the data included in systematic assessments of water availability and future outlooks will need to be adjusted for each setting to include those variables of greatest importance for local drought monitoring.

The monitoring committee should meet regularly, especially in advance of the peak demand season and/or beginning of the rainy season(s). Following each meeting, reports should be prepared and disseminated to the provincial level drought task force, the NDPC, and the media. The chairperson of the monitoring committee should be a permanent member of the provincial drought task forces. In many countries, this person would be the representative from the national meteorological service. If conditions warrant, the task force leadership should brief the provincial governor or appropriate government official about the contents of the report, including any recommendations for specific actions. Public dissemination of information should be screened by a public information specialist to avoid confusing or conflicting reports on the current status of conditions.

The primary objectives of the monitoring committee are to:

- Adopt a workable definition of drought that could be used to phase in and phase out levels of state and national mitigation actions and emergency measures associated with drought conditions. It may be necessary to adopt more than one definition of drought in identifying impacts in various economic, social, and environmental sectors, since no single definition of drought applies in all cases.

  The committee will need to consider appropriate indicators (e.g., precipitation, temperature, soil moisture and streamflow) and indices as integral to the water supply assessment process. Many indices are available, and the strengths and weaknesses of each index should be carefully considered (see Chapter 8). The trend is to rely on multiple drought indices to trigger mitigation and response actions, which are calibrated to various intensities of drought and/or impacts. The current thought is that no single index of drought is adequate to measure the complex interrelationships between the various elements of the hydrological cycle and impacts.

It is helpful to establish a sequence of descriptive terms for drought and water supply alert levels, such as *advisory, alert, emergency,* and *rationing* (as opposed to more generic terms such as *phase 1* and *phase 2,* or sensational terms such as *disaster*). It would be helpful to review the terminology used by other entities (i.e., local utilities, irrigation districts, and river basin authorities) and choose terms that are consistent so as not to confuse the public with different terms in areas where there may be authorities with overlapping regional responsibilities. Consistency of terminology between state preparedness plans is essential. These alert levels should be defined in discussions with both the risk assessment committee and the provincial task force.

In considering emergency measures such as rationing, it is important to remember that the impacts of drought may vary significantly from one area to the next, depending on the sources and uses of water and the degree of planning previously implemented. For example, some cities may have expanded their water supply capacity while other adjacent communities may have an inadequate water supply capacity during periods of drought. Imposing general emergency measures on people or communities without regard for their existing vulnerability may result in political repercussions and loss of credibility.

A related consideration is that some municipal water systems may be out of date or in poor operating condition, so that even moderate drought strains a community's ability to supply customers with water. Identifying inadequate (i.e., vulnerable) water supply systems and putting in place programs to upgrade those systems should be part of a long-term drought mitigation strategy.

- Establish drought management areas (i.e., subdivide the province or region into more conveniently sized districts by political boundaries, shared hydrological characteristics, climatological characteristics, or other means such as drought probability or risk). These subdivisions may be useful in drought management since they may allow drought stages and mitigation and response options to be regionalized as the severity of drought changes over time.

- Develop a drought monitoring system. The quality of meteorological and hydrological networks is highly variable from country to country and region to region within countries (e.g., number of stations, length of record, and amount of missing data). Responsibility for collecting, analyzing, and disseminating data is divided between many government authorities. The monitoring committee's challenge is to coordinate and integrate the analysis so decision makers and the public receive early warning of emerging drought conditions.

Considerable experience has been gained in recent years with automated weather data networks that provide rapid access to climate data. These networks can be invaluable in monitoring emerging and ongoing drought conditions. The experiences of regions with comprehensive automated meteorological and hydrological networks should be investigated and lessons learned should be applied, where appropriate. It is essential that automated weather networks be established and networked in order to retrieve the data in a timely manner.

- Inventory data quantity and quality from current observation networks. Many networks monitor key elements of the hydrologic system. Most of these networks are operated by national or provincial agencies, but other networks may also exist and could provide critical information for a portion of a province or region. Meteorological data are important but represent only one part of a comprehensive monitoring system. These other physical indicators (soil moisture, streamflow, reservoir and groundwater levels, etc.) must be monitored to reflect impacts of drought on agriculture, households, industry, energy production, transportation, recreation and tourism, and other water use sectors.

It is also imperative to establish a network of observers to gather impact information from all of the key sectors affected by drought and to create an archive of this information. Both quantitative and qualitative information is important. The value of this information is twofold. First, this information is of pronounced importance in assisting researchers and managers to identify the linkages or correlations between thresholds of various drought indices and indicators and the emergence of specific impacts. It is those correlations between indices/indicators and impacts that can be used to trigger a wide range of mitigation actions as key components of the preparedness plan, which is based on the principles of risk reduction. Second, the establishment of an archive of drought impacts will illustrate the trend in impacts over time on specific sectors. This information is critically important to policy makers who must demonstrate how those investments in mitigation measures up front are paying off in the longer term through vulnerability reduction, as measured by reduced impacts and government expenditures on drought assistance.

- Determine the data needs of primary users for information and decision support tools. Developing new or modifying existing data collection systems is most effective when the people who will be using the data are consulted early and often to determine their specific needs or preferences and the timing for critical decision points. Soliciting input on expected new products/decision support tools or obtaining feedback on existing products is critical to ensuring that products meet the needs of primary users and, therefore, will be used in decision- making. Training on how to use or apply products in routine decision-making is also essential.

- Develop and/or modify current data and information delivery systems. People need to be warned of drought as soon as it is detected, but often they are not. Information must reach people in time for them to use it in making decisions. In establishing information channels, the monitoring committee needs to consider when people need what kinds of information. Knowledge of these decision points will make a difference as to whether the information provided is used or ignored.

4.4.5.2 Risk Assessment Committee

Risk is the product of exposure to the drought hazard (i.e., probability of occurrence) and societal vulnerability, represented by a combination of economic, environmental, and social factors. Therefore, in order to reduce vulnerability to drought, it is essential to identify the most significant impacts and assess their underlying causes. Drought impacts cut across many sectors and across normal divisions of government authority.

Membership of the risk assessment committee should include representatives or technical experts from economic sectors, social groups, and ecosystems most at risk from drought. The committee's chairperson should be a member of the drought task force to ensure seamless reporting. Experience has demonstrated that the most effective approach to follow in determining vulnerability to and impacts of drought is to create a series of working groups under the aegis of the risk assessment committee. The responsibility of the committee and working groups is to assess sectors, population groups, communities, and ecosystems most at risk and identify appropriate and reasonable mitigation measures to address these risks.

Working groups would be composed of technical specialists representing those areas referred to above. The chair of each working group, as a member of the risk assessment committee, would report directly to the committee. Following this model, the responsibility of the risk assessment committee is to direct the activities of each of the working groups. These working groups will then make recommendations to the drought task force on mitigation actions to consider for inclusion in the mitigation plan. Mitigation actions are identified in advance and implemented in order to reduce the impacts of drought when it occurs. Some of these actions represent programs that are long term in nature while others may be actions that are activated when drought occurs. The activation of these measures at appropriate times is determined by the triggers (i.e., indicators and indices) identified by the monitoring committee in association with the risk assessment committee in relation to the key impacts (i.e., vulnerabilities) associated with drought.

The number of working groups that are set up under the risk assessment committee will vary considerably between provinces, states, or river basins, reflecting the principal impact sectors of importance to the region and their respective vulnerabilities to drought because of differences in the exposure to drought (frequency and severity) and the most important economic, social, and environmental sectors. More complex economies and societies will require a larger number of working groups to reflect these sectors. It is common for the working groups to focus on some combination of the following sectors: agriculture, recreation and tourism, industry, commerce, drinking water supplies, energy, environment and ecosystem health, wildfire protection, and health.

To assist in the drought preparedness and mitigation process, a methodology is proposed to identify and rank (prioritize) drought impacts through an examination of the underlying environmental, economic, and social causes of these impacts, followed by the selection of actions that will address these underlying causes. What makes this methodology different and more helpful than previous methodologies is that it addresses the causes behind drought impacts. Previously, responses to drought have been reactive in nature and focused on addressing a specific impact, which is a symptom of the vulnerability that exists. Understanding why specific impacts occur provides the opportunity to lessen these impacts in the future by addressing these vulnerabilities through the identification and adoption of specific mitigation actions. Other vulnerability or risk assessment methodologies exist, and nations are encouraged to evaluate these for application in their specific setting (Iglesias et al. 2009; Sonmez et al. 2005; Wilhelmi and Wilhite 2002).

The methodology proposed here is divided into six specific tasks. Once the risk assessment committee establishes the working groups, each of these groups would follow this methodology in the risk assessment process.

Task 1. Assemble the team

It is essential to bring together the right people and supply them with adequate data to make fair, efficient, and informed decisions pertaining to drought risk. Members of this group should be technically trained in the specific topical areas covered by each working group. Also important is the need to include public input and consideration when dealing with the issues of appropriateness, urgency, equity, and cultural awareness in drought risk analysis. Public participation could be warranted at every step, but time and money may limit their involvement to key stages in the risk analysis and planning process (public review vs. public participation). The amount of public involvement is at the discretion of the drought task force and other members of the planning team. The advantage of publicly discussing questions and options is that the procedures used in making any decision will be better understood, and it will also demonstrate a commitment to participatory management. At a minimum, decisions and reasoning should be openly documented to build public trust and understanding.

The choice of specific actions to deal with the underlying causes of the drought impacts will depend on the economic resources available and related social values. Typical concerns are associated with cost and technical feasibility, effectiveness, equity, and cultural perspectives. This process has the potential to lead to the identification of effective and appropriate drought risk reduction activities that will reduce long-term drought impacts, rather than *ad hoc* responses or untested mitigation actions that may not effectively reduce the impact of future droughts.

Task 2. Drought impact assessment

Impact assessment examines the consequences of a given event or change. For example, drought is typically associated with a number of outcomes that result from the shortage of water, either directly or indirectly. Drought impact assessments begin by identifying direct consequences of the drought, such as reduced crop yields, livestock losses, and reduced reservoir levels. These direct outcomes can then be traced to secondary consequences (often social effects), such as the forced sale of household assets, food security, reduced energy production, dislocation, or physical and emotional stress. This initial assessment identifies drought impacts but does not identify the underlying reasons for these impacts.

The impacts from drought can be classified as economic, environmental, or social, even though many impacts may span more than one sector. A detailed checklist of impacts that could affect a region or location is provided in the IDMP publication referred to on the first page of this chapter. This list should be expanded to include other impacts that may be important for the region. Recent drought impacts, especially if they are associated with severe to extreme drought, should be weighted more heavily than the impacts of historical drought (in most cases), since they better reflect current vulnerabilities, which is the purpose of this exercise. Attention should also be given to specific impacts that are expected to emerge or increase in magnitude because of new vulnerabilities resulting from recent or projected societal changes or changes in drought incidence.

It is appropriate at this point to classify the types of impacts according to the severity of drought, noting that in the future, droughts of lesser magnitude may produce more serious impacts as vulnerability increases. Hopefully, interventions taken now will reduce these vulnerabilities in the future. It is also important to identify the "drought of record" for each region. Droughts differ from one another according to intensity, duration, and spatial extent. Thus, there may be several droughts of record, depending on the criteria emphasized (i.e., most severe drought of a season or 1-year duration vs. most severe multiyear droughts). These analyses would yield a range of impacts related to the severity of drought. In addition, by highlighting past, current, and potential impacts, trends may become evident that will also be useful for planning purposes. These impacts highlight sectors, populations, or activities that are vulnerable to drought, and when evaluated with the probability of drought occurrence, they help identify varying levels of drought risk.

Task 3. Ranking impacts

After each working group has completed the checklist referred to in Task 2, the unchecked impacts can be omitted from further consideration. This new list will contain the relevant drought impacts for each location or activity. From this list, impacts should be ranked/prioritized by working group members. To be effective and equitable, the ranking should take into consideration concerns such as cost of mitigation actions, the areal extent of the impact, trends over time, public opinion, and fairness. Be aware that social and environmental impacts are often difficult to quantify. It is recommended that each working group complete a preliminary ranking of impacts. The drought task force and other work groups can participate in a plenary discussion of these rankings following the initial ranking iterations. It is recommended that a matrix be constructed (see an example in Table 4.1) to help rank or prioritize impacts. From this list of prioritized impacts, each working group should decide which impacts should be addressed and which can be deferred to a later time or stage in the planning process.

Task 4. Vulnerability assessment

Vulnerability assessment provides a framework for identifying the social, economic, and environmental causes of drought impacts. It bridges the gap between impact assessment and policy formulation by directing policy attention to underlying causes of vulnerability rather than to its result, the negative impacts, which follow triggering events such as drought. For example, the direct impact of precipitation deficiencies may be a reduction of crop yields. The underlying cause of this vulnerability, however, may be that some farmers did not use drought-resistant seeds or other management

TABLE 4.1

Drought Impact Decision Matrix

Impacts	Cost	Equally Distributed?	Growing?	Public Priority	Equitable Recovery?	Impact Rank

Source: FAO and NDMC, *The Near East Drought Planning Manual: Guidelines for Drought Mitigation and Preparedness Planning*, FAO, Rome, 2008.

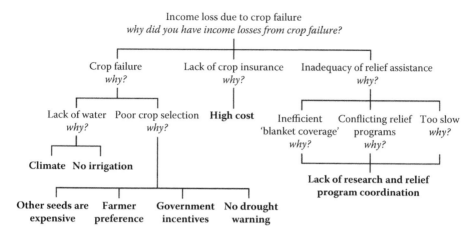

FIGURE 4.4

An example of a simplified agricultural impact tree diagram. (From FAO and NDMC, *The Near East Drought Planning Manual: Guidelines for Drought Mitigation and Preparedness Planning*, FAO, Rome, 2008.)

practices, because of concerns about their effectiveness or high cost, or some commitment to cultural beliefs. Another example might be associated with the vulnerability of a community's water supply. The vulnerability of their water supply system might be largely the result of the lack of expansion of the system to keep pace with population growth, aging infrastructure, or both. The solution to vulnerability reduction would be the development of new supply sources and/or the replacement of infrastructure. Therefore, for each of the identified impacts from Table 4.1, the members of the working group should ask why these impacts occurred. It is important to realize that a combination of factors might produce a given impact. It might be beneficial to visualize these causal relationships in some form of a tree diagram. Two examples are shown in Figures 4.4 and 4.5. Figure 4.4 demonstrates a typical agricultural example and Figure 4.5 a potential urban scenario. Depending on the level of analysis, this process can quickly become somewhat complicated. This is why it is necessary to have each working group composed of the appropriate mix of people with technical expertise.

The tree diagrams illustrate the complexity of understanding drought impacts. The two examples provided are not meant to be comprehensive or represent an actual scenario. Basically, their main purpose is to demonstrate that impacts must be examined from several perspectives to expose their true underlying causes. For this assessment, the lowest causes—the items in boldface on the tree diagrams—will be referred to as basal causes. These basal causes are the items that have the potential to be acted on to reduce the associated impact. Of course, some of these impact causes should not or cannot be acted on for a wide variety of reasons (discussed in Task 5).

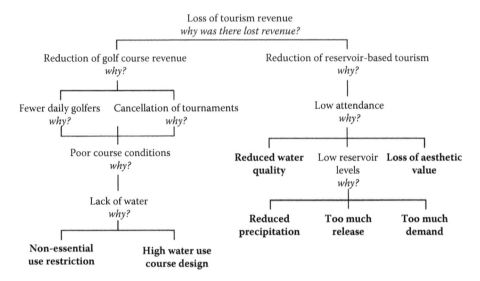

FIGURE 4.5
An example of a simplified urban impact tree diagram. (Courtesy of National Drought Mitigation Center, University of Nebraska–Lincoln.)

Task 5. Action identification

Mitigation is defined as actions taken in advance or in the early stages of drought that reduce the impacts of the event. Once drought impact priorities have been set and the corresponding underlying causes of vulnerability have been exposed, actions can be identified that are appropriate for reducing drought risk. The matrix lists the impact as well as the described basal causes of the impact. From this point, the working group should investigate what actions could be taken to address each of these basal causes. The following sequence of questions may be helpful in identifying potential actions:

- Can the basal cause be mitigated (can it be modified before a drought)? If yes, then how?
- Can the basal cause be responded to (can it be modified during or after a drought)? If so, then how?
- Is there some basal cause, or aspect of the basal cause, that cannot be modified and must be accepted as a drought-related risk for this activity or area?

As will be discussed (in Task 6), not all mitigation actions are appropriate in all cases. Many of the actions are more in the realm of short-term emergency response or crisis management, rather than long-term mitigation or risk management. Emergency response is an important component of drought planning, but should only be one part of a more comprehensive mitigation strategy.

Task 6. Developing the To-Do list

After the impacts, causes, and relevant potential actions have been identified, the next step is to determine the sequence of actions to take as part of the risk reduction planning exercise. This selection should be based on such concerns as feasibility, effectiveness, cost, and equity. Additionally, it will be important to review the impact tree diagrams when considering which groups of actions need to be considered together. For example, if you wanted to reduce crop losses by promoting the planting of a more drought-resistant crop, it would not be effective to educate farmers on the benefits of the new crop if markets do not currently exist or there are government incentives for continuing to grow the current crop. Government policies may often be out of sync with vulnerability reduction actions.

In choosing the appropriate actions, it might be helpful to ask some of the following questions:

- What are the cost/benefit ratios for the actions identified?
- Which actions are considered to be feasible and appropriate by the general public?
- Which actions are sensitive to the local environment (i.e., sustainable practices)?
- Are actions addressing the right combination of causes to adequately reduce the relevant impact?
- Are actions addressing short-term and long-term solutions?
- Which actions would equitably represent the needs of affected individuals and groups?

This process has the potential to lead to the identification of effective and appropriate drought risk reduction activities that will reduce future drought impacts.

Completion of risk analysis

Following Task 6, the risk analysis is completed at this point in the planning process. Remember, this is a planning process, so it will be necessary to periodically reevaluate drought risk and the various mitigation actions identified. Step 10 in the mitigation planning process is associated with evaluating, testing, and revising the drought plan. Following a severe drought episode would be an appropriate time to revisit mitigation actions to evaluate their effectiveness in association with an analysis of lessons learned.

4.4.5.3 Mitigation and Response Committee

It is recommended that mitigation and response actions be under the purview of the drought task force. The task force, working in cooperation with the monitoring and risk assessment committees, has the knowledge and experience to understand drought mitigation techniques, risk analysis (economic, environmental, and social aspects), and drought-related decision-making processes. The task force, as originally defined, is composed of senior policy makers from various government agencies and, possibly, key stakeholder groups. Therefore, they are in an excellent position to recommend and/or implement mitigation actions, request assistance through various national programs, or make policy recommendations to a legislative body or political leader.

As a part of the drought planning process, the NDPC should inventory all assistance programs available from national sources to mitigate or respond to drought events. Each provincial drought task force should review this inventory of programs available from governmental and nongovernmental authorities for completeness and provide feedback to the commission for the improvement of these programs to address short-term emergency situations as well as long-term mitigation programs that may be useful in addressing risk reduction. In some cases, additional programs might be available from the provinces or states that have supplemented programs available at the national level. Assistance should be defined in a very broad way to include all forms of technical, mitigation, and relief programs available. As stated previously, the national drought commission should undertake a similar exercise with national programs and evaluate their effectiveness in responding to and mitigating the effects of previous droughts.

4.4.5.4 Writing the Mitigation Plan

With input from each of the committees and working groups and the assistance of professional writing specialists, the drought task force will draft the drought mitigation plan. After completion of a working draft, it is recommended that public meetings or hearings be held at several locations to explain the purpose, scope, and operational characteristics of the plan and how it will function in relation to the objectives of the national drought policy. Discussion must also be presented on the specific mitigation actions and response measures recommended in the plan. A public information specialist for the drought task force can facilitate planning for the hearings and also prepare news stories announcing the meetings and providing an overview of the plan.

After the draft plan has been vetted at the state or provincial level, it should be submitted to the national drought commission for review to determine if the plan meets the requirements mandated by the commission. Although each state-level plan will contain different elements and procedures, the basic structure should conform to policy standards provided to the states at the outset of the planning process by the national drought commission.

4.4.6 Step 6: Identify Research Needs and Fill Institutional Gaps

The NDPC should identify specific research needs that would contribute to a better understanding of drought, its impacts, mitigation alternatives, and needed policy instruments, leading to a reduction of risk. These needs are likely to originate from the state-level drought task forces that are implemented to develop mitigation plans. It will be the task of the commission to collate these needs into a set of priorities for future action and funding priorities.

Many examples of potential research needs could be mentioned. First, improving understanding of how climate change may affect the incidence of drought events and their severity, particularly at a regional scale, would provide critical information that could facilitate the risk reduction measure. As the science of climate change improves and the resolution of computer models increases, this information will be invaluable to policy makers, managers, and other decision makers. Also critically important are improved early warning techniques and delivery systems, improved understanding of the linkages between indicators and indices and impacts to provide key decision points or thresholds for implementing mitigation actions, and the development of decision support tools for managers.

It will also become apparent during the policy development and preparedness planning process that institutional gaps exist that will hamper the policy and planning process. For example, serious gaps in monitoring station networks may exist, or existing meteorological, hydrological, and ecological networks may need to be automated and networked so that data can be retrieved in a timely manner in support of an early warning system. Archiving the impacts of drought is also a critical component of the process to help identify and quantify losses and discern trends in impact reduction. It is expected that Step 6 will be carried out concurrently with Steps 4 and 5 of the policy and plan development process.

4.4.7 Step 7: Integrate Science and Policy Aspects of Drought Management

An essential part of the policy and planning process is integrating the science and policy aspects of drought management. Policy makers' understanding of the scientific issues and technical constraints involved in addressing problems associated with drought is often limited. Likewise, scientists and managers may have a poor understanding of existing policy constraints for responding to the impacts of drought. In many cases, communication and understanding between the science and policy communities must be enhanced if the planning process is to be successful. This is a critical step in the development of a national drought policy. Members of the NDPC have a good understanding of the policy development process and the political and financial constraints associated with proposed changes in public policy. They are also aware of the difficulties inherent in a change in the paradigm for the recipients of drought emergency assistance to a new approach focused on drought risk reduction. However, those persons at the state or community level that are embedded in the preparedness planning process are less aware of these constraints but have an excellent understanding of drought management actions, local conditions, and the key sectors affected and their operational needs. Linking the policy process with critical needs requires an excellent communication conduit from state-based drought task forces and the commission.

In essence, this communication conduit is necessary to distinguish what is feasible from what is desirable for a broad range of science and policy options. Integration of science and policy during the planning process will also be useful in setting research priorities and synthesizing current understanding. The drought task force should consider a wide range of options for drought risk reduction and evaluate the pros and cons of each in terms of their feasibility and potential outcomes.

4.4.8 Step 8: Publicize the National Drought Management Policy and Preparedness Plans and Build Public Awareness and Consensus

If there has been good communication with the public throughout the process of establishing a drought policy and plan, there may already be an improved awareness of goals of the drought policy, the rationale for policy implementation, and the drought planning process by the time the policy is ready to be implemented. The public information specialists that are engaged in this process at the commission level and at the state level are vital in this regard. Throughout the policy and planning development process, it is imperative for local and national media to be used effectively in the dissemination of information about the process. Themes to emphasize in writing news stories during the drought policy and planning process could include:

- How the drought policy and plan is expected to reduce impacts of drought in both the short and long term. Stories can focus on the social dimensions of drought, such as how it affects local economies and individual families; environmental consequences, such as reduced wildlife habitat; human health; and the impacts on the regional and national economy and the development process.
- Behavioral changes that will be required to reduce drought impacts, various aspects of state drought preparedness plans, new policies associated with water allocations, and water management during the various stages of drought severity.

In subsequent years, it may be useful to do "drought policy and planning refresher" news releases at the beginning of the most drought-sensitive season, letting people know the current status of water supplies and projections regarding water availability. News releases can also focus on the various aspects of the drought policy and plan. Success stories regarding the application of the plan in various sectors or communities will help to reinforce the goals of the mitigation plan and the national policy. It may be useful to refresh people's memories ahead of time on circumstances that would lead to water use restrictions. The timing of these news releases would be associated with regular meetings of the monitoring committee at the local and national levels, pinpointing regions and/or sectors of particular concern.

During drought, the commission and state drought task forces should work with public information professionals to keep the public well informed of the current status of water supplies, whether conditions are approaching trigger points that will lead to requests for voluntary or mandatory use restrictions, and how victims of drought can access information and assistance. Websites should be created and updated on a regular basis so the public and managers can get information directly from the task force without having to rely on mass media. Products or dissemination strategies and tools need to be available that effectively communicate information to the user community.

4.4.9 Step 9: Develop Education Programs for All Age and Population Groups

A broad-based education program focused on all age groups is necessary to raise awareness of the new strategy for drought management, the importance of preparedness and risk reduction, short- and long-term water supply issues, and other crucial prerequisites for public acceptance and implementation of drought policy and preparedness goals. This education program will help ensure that people know how to manage drought when it occurs and that drought preparedness will not lose ground during nondrought years. It would be useful to tailor information to the needs of specific groups (e.g., elementary and secondary education, small business, industry, water managers, agricultural producers, homeowners, and utilities). The drought task force in each state or province and participating agencies should consider developing presentations and educational materials for events such as a water awareness week, community observations of Earth Day, and other events focused on environmental awareness; relevant trade shows; specialized workshops; and other gatherings that focus on natural resource stewardship or management.

4.4.10 Step 10: Evaluate and Revise National Drought Management Policy and Supporting Preparedness Plans

The tenets of a national drought policy and each of the preparedness or mitigation plans that serve as the implementation instruments of the policy require periodic evaluation and revision in order to incorporate new technologies, lessons learned from recent drought events, changes in vulnerability, and so forth. The final step in the policy development and preparedness process is to create a detailed set of procedures to ensure an adequate evaluation of the successes and failures of the policy and the preparedness plans at all levels. Oversight of the evaluation process would be provided by the NDPC, but the specific actions taken and outcomes exercised in the drought-affected states or provinces would need to have the active involvement of those specific drought task forces. The policy and preparedness process must be dynamic; otherwise, the policies and plans will quickly become outdated. Periodic testing, evaluation, and updating of the drought policy are needed to keep the plan responsive to the needs of the country, states, and key sectors. To maximize the effectiveness of the system, two modes of evaluation must be in place: ongoing and post-drought.

4.4.10.1 Ongoing Evaluation

An ongoing or operational evaluation keeps track of how societal changes such as new technology, new research, new laws, and changes in political leadership may affect drought risk and the operational aspects of the drought policy and supporting preparedness plans. The risk associated with drought in various sectors (economic, social, and environmental) should be evaluated frequently while the overall drought policy and preparedness plans may be evaluated less often. An evaluation under simulated drought conditions (i.e., computer-based drought exercise) is recommended before the drought policy and state-level plans are implemented and periodically thereafter. It is important to remember that the drought policy and preparedness planning process is dynamic, not a discrete event.

Another important aspect of the evaluation process and the concept of drought exercises is linked to changes in government personnel, which, in most settings, occurs frequently. If the goals and elements of the national drought policy are not reviewed periodically and the responsibilities of all agencies revisited, whether at the national or state level, governmental authorities will not be fully aware of their roles and responsibilities when drought recurs. Developing and maintaining institutional memory is an important aspect of the drought policy and preparedness process.

4.4.10.2 Post-Drought Evaluation

A post-drought evaluation or audit documents and analyses the assessment and response actions of government, nongovernmental organizations, and others, and provides for a mechanism for implementing recommendations for improving the system. Without post-drought evaluations of both the drought policy and the preparedness plans at the local level, it is difficult to learn from past successes and mistakes, as institutional memory fades.

Post-drought evaluations should include an analysis of the climatic, social, and environmental aspects of the drought (i.e., its economic, social, and environmental consequences); the extent to which predrought planning was useful in mitigating impacts, in facilitating relief or assistance to stricken areas, and in post-drought recovery; and any other weaknesses or problems caused or not covered by the policy and the state-based plans. Attention must also be directed to situations in which drought-coping mechanisms worked and where societies exhibited resilience; evaluations should not focus only on those situations in which coping mechanisms failed. Evaluations of previous responses to severe drought are also a good planning aid, if they have been done. These evaluations establish a baseline for later comparisons so trends in resiliency can be documented.

To ensure an unbiased appraisal, governments may wish to place the responsibility for evaluating the effectiveness of the drought policy and each of the preparedness plans in the hands of nongovernmental organizations such as universities and/or specialized research institutes.

4.5 Summary and Conclusion

For the most part, previous responses to drought in all parts of the world have been reactive, reflecting what is commonly referred to as the crisis management approach. This approach has been ineffective (i.e., assistance poorly targeted to specific impacts or population groups), poorly coordinated, and untimely; more importantly, it has done little to reduce the risks associated with drought. In fact, the economic, social, and environmental impacts of drought have increased significantly in recent decades. A similar trend exists for all natural hazards.

The intent of the policy development and planning process described in this report is to provide a set of generic steps or guidelines that nations can use to develop the overarching principles of a national drought policy aimed at risk reduction. This policy would be implemented at the subnational (i.e., provincial or state) level through the development and implementation of drought preparedness plans that follow the framework or principles of the national drought policy. These plans are the instruments for implementing a national drought policy based on the principles of risk reduction. Following these guidelines, a nation can significantly change the way they prepare for and respond to drought by placing greater emphasis on proactively addressing the risks associated with drought through the adoption of appropriate mitigation actions. The guidelines presented here are generic in order to enable governments to choose those steps and components that are most applicable to their situation. The risk assessment methodology embedded in this process is designed to guide governments through the process of evaluating and prioritizing impacts and identifying mitigation actions and tools that can be used to reduce the impacts of future drought episodes. Both the policy development process and the planning process must be viewed as ongoing, continuously evaluating the nation's changing exposure and vulnerabilities and how governments and stakeholders can work in partnership to lessen risk.

References

Blaikie, P., T. Cannon, I. Davis, and B. Wisner. 1994. *At Risk: Natural Hazards, People's Vulnerability, and Disasters.* London: Routledge.
FAO and NDMC. 2008. *The Near East Drought Planning Manual: Guidelines for Drought Mitigation and Preparedness Planning.* Rome: FAO.
HMNDP. 2013. *Final Declaration from the High-level Meeting on National Drought Policy.* http://hmndp.org. Accessed June 2017.
Iglesias, A., M. Moneo, and S. Quiroga. 2009. Methods for evaluating social vulnerability to drought. In *Coping with Drought Risk in Agriculture and Water Supply Systems,* ed. A. Iglesias, L. Garrotte, A. Cancelliere, F. Cubillo, and D. A. Wilhite, 153–159. Advances in Natural and Technological Hazards Research 26. New York: Springer.
NDPC (National Drought Policy Commission). 2000. *Preparing for Drought in the 21st Century.* Washington, DC: U.S. Department of Agriculture.
Sonmez, F. K., A. U. Komuscu, A. Erkan, and E. Turgu. 2005. An analysis of spatial and temporal dimensions of drought vulnerability in Turkey using the standardized precipitation index. *Natural Hazards* 35:243–264.
UNISDR. 2009. *Drought Risk Reduction Framework and Practices: Contributing to the Implementation of the Hyogo Framework for Action.* Geneva, Switzerland: United Nations International Strategy for Disaster Reduction.
Wilhelmi, O. V., and D. A. Wilhite. 2002. Assessing vulnerability to agricultural drought: A Nebraska case study. *Natural Hazards* 25:37–58.
Wilhite, D. A. 1991. Drought planning: A process for state government. *Water Resources Bulletin* 27(1):29–38.
Wilhite, D. A., M. J. Hayes, and C. L. Knutson. 2005. Drought preparedness planning: Building institutional capacity. In *Drought and Water Crises: Science, Technology, and Management Issues,* ed. D. A. Wilhite, 93–136. Boca Raton, FL: CRC Press.
Wilhite, D. A., M. J. Hayes, C. Knutson, and K. H. Smith. 2000. Planning for drought: Moving from crisis to risk management. *Journal of the American Water Resources Association* 36:697–710.

5

Benefits of Action and Costs of Inaction: Drought Mitigation and Preparedness—A Literature Review*

Nicolas Gerber and Alisher Mirzabaev

CONTENTS

5.1 Introduction

Droughts are major natural hazards and have wide-reaching economic, social, and environmental impacts. Their complex, slow, and creeping nature; the difficulty of determining their onsets and endings; their site-dependence; and the diffuse nature of their damage (Below et al. 2007) makes the task of comprehensively and accurately determining the cost of

---

* This chapter is reprinted in this book with the permission of the Integrated Drought Management Programme of the World Meteorological Organization and the Global Water Partnership. It was originally published in: N. Gerber and A. Mirzabaev. 2017. World Meteorological Organization (WMO) and Global Water Partnership (GWP). Benefits of action and costs of inaction: Drought mitigation and preparedness—a literature review. Integrated Drought Management Programme (IDMP) Working Paper 1, WMO, Geneva, Switzerland; GWP, Stockholm, Sweden.

droughts a highly challenging one. These difficulties are compounded by a lack of data on droughts and their impacts (Changnon 2003), especially in low-income countries.

Droughts are the most detrimental of all the natural disasters (Bruce 1994; Obasi 1994; Cook et al. 2007; Mishra and Singh 2010). Globally, about one-fifth of the damage caused by natural hazards can be attributed to droughts (Wilhite 2000), and the cost of droughts is estimated to be around USD 80 billion per year (Carolwicz 1996). In the United States—one of the few countries having relatively good data availability—the annual losses attributed to droughts were estimated to be around USD 6–8 billion in the early 1990s (Wilhite 2000, citing FEMA 1995). In the European Union, the damage caused by droughts is estimated to be around EUR 7.5 billion per year (CEC 2007; EC 2007). However, these estimates are likely to be quite conservative, since they often fail to take all the impacts into account. Indirect drought impacts in particular are seldom captured appropriately or systematically by drought monitoring and reporting systems. For example, in addition to affecting the quantity of water, droughts have negative effects on the quality of water systems. These effects include increased salinity, enhanced stratification leading to algal production and toxic cyanobacterial blooms, higher turbidity, and deoxygenation (Webster et al. 1996; Mosley 2015). The costs of these water quality impacts are yet to be quantified adequately.

Importantly, droughts may also have far-reaching social and economic impacts—for example, by leading to conflict and civil unrest (Johnstone and Mazo 2011; von Uexkull 2014; Linke et al. 2015), migration (Gray and Mueller 2012), gender disparities (Fisher and Carr 2015), reduced hydro-energy generation (Shadman et al. 2016), food security and famine (IFRC 2006), poverty (Pandey et al. 2007), and negative short- and long-term health effects (Hoddinott and Kinsey 2001; Ebi and Bowen 2015; Lohmann and Lechtenfeld 2015). Conway (2008) indicates that between 1993 and 2003, drought-induced famines affected 11 million people in Africa. According to the World Meteorological Organization (WMO), droughts may have caused 280,000 human deaths between 1991 and 2000 globally (Logar and van den Bergh 2011). Other indirect impacts are mentioned in national post-disaster needs assessments supported by the Global Facility for Disaster Reduction and Recovery/World Bank and other technical and donor agencies, and extend to social (e.g., access to education) and environmental (e.g., loss of ecosystem services) issues (see for instance the reports from Kenya 2012, Djibouti 2011, and Uganda 2010–2011, available at https://www.gfdrr.org/post-disaster-needs-assessments). However, there is relatively little literature on the economic costs of such indirect impacts. Furthermore, indirect costs may increase to a greater extent than direct costs in the future because of increasing frequency and severity of droughts under climate change, and these will be particularly challenging to model (Jenkins 2011).

The difficulty of assessing the costs and impacts of droughts is complicated by the challenge of how to define drought. Drought is a temporary climatic feature, unlike aridity, which is a permanent characteristic of a climate (Wilhite 1992). Drought has numerous definitions, which may be mutually incompatible. Ideally, the definition should be set specifically for each location, taking into account the characteristics of that location (Wilhite and Glantz 1985).

Drought is a natural hazard, so its occurrence in any location and during a given time period could be evaluated by attaching probabilities depending on the biophysical and climatic characteristics of that location (Wilhite 2000). However, drought impacts are strongly modulated by the socioeconomic characteristics of affected areas, such as their vulnerability and resilience to drought, as well as their level of drought preparedness. The role of socioeconomic factors in determining drought impacts is complex and relations are not linear; for example, a higher level of socioeconomic development and water services infrastructure can mitigate or exacerbate the impacts of drought.

In a risk-based approach to drought (described in this study as drought risk management), we refer to mitigation of the risk of incurring negative impacts from drought events, rather than reducing the probability of occurrence of drought events. In this sense, vulnerability to drought is the susceptibility to be negatively affected by drought (Adger 2006), with the opposite being resilience, that is, the ability to cope successfully with drought and overcome its impacts. Vulnerability and resilience to drought are affected by actions taken to mitigate drought impacts and increase drought preparedness (Wilhite et al. 2014). These both reflect the degree of adaptive capacity of a community (Engle 2013). Drought preparedness involves actions undertaken before drought occurs and that will improve operational and institutional response to drought (Kampragou et al. 2011).

On the other hand, drought impact mitigation actions include a variety of activities carried out before drought occurs that will minimize the impacts of drought on people, the economy, and the environment. Wilhite et al. (2005) classified actions for drought preparedness in a 10-step process. This has been further refined for national drought management policies by WMO and GWP (2014). Based on the High-level Meeting on National Drought Policy (WMO et al. 2013), the integrated drought management programme (IDMP) and its partners have adopted three pillars of drought management: (1) drought monitoring and early warning systems; (2) vulnerability and impact assessments; and (3) drought preparedness, mitigation, and response.

The difficulty of accurately assessing the costs of droughts presents substantial challenges for the analysis of the costs and benefits of investments made and policy actions taken against droughts. At the same time, droughts are not weather or climatic anomalies, but a recurrent and normal feature of almost any climate (Kogan 1997), even in comparatively water-rich countries (Kampragou et al. 2011). NCDC (2002) indicates that about 10 percent of the territory of the United States is affected by drought at any given time. Between 2000 and 2006, 15 percent of the European Union's land area was affected by drought

(Kampragou et al. 2011), more than double the annual average for 1976–1990 (EC 2007). Droughts have occurred in different locations across Vietnam in 40 out of the past 50 years (Lohmann and Lechtenfeld 2015). Gan et al. (2016) provide an extensive review of climate change and variability in drought-prone areas of Africa and predicts critical negative impacts on a wide variety of drought-related indicators. Given the scale of the issue and the likely drought trends under climate change, it is essential to have a well-defined strategy for mitigating the impacts of drought and enhancing drought preparedness.

However, the default course of action used by many countries is to respond to the impacts of droughts once they have occurred, through drought relief (i.e., crisis management), rather than proactively improving resilience through appropriate risk management strategies (Wilhite 1996). Crisis management approaches usually fail to reduce future vulnerability to drought. On the contrary, by providing drought relief to activities that are vulnerable to drought, they may in fact incentivize their perpetuation. As a result, continued vulnerability makes crisis management costlier to society than ex ante investments that mitigate drought risks by building resilience. Moreover, since we currently lack comprehensive assessments of the full social and environmental costs of droughts, the ultimate costs of continued vulnerability are likely to be higher than estimated at present. Furthermore, climate change is expected to increase the frequency and severity of droughts (Stahl and Demuth 1999; Andreadis and Lettenmaier 2006; Bates et al. 2008). The changing climate is also likely to expand the geographical extent of drought-prone areas (Mishra and Singh 2009; IPCC 2014), making crisis management approaches even less affordable than they are today. This begs the question: if proactive risk management is socially optimal compared with reactive crisis management, why is the shift from crisis management to risk management happening so slowly?

This review seeks to shed light on responses to this question by evaluating current relevant literature. More specifically, we seek to summarize the key literature on the costs and benefits, and pros and cons, of reactive public crisis management versus ex ante government policies for drought risk management directed toward investment in mitigation actions and drought preparedness that reduce the impacts of future droughts. We also identify the obstacles and opportunities facing the transition from crisis management to risk management, presenting country experiences from around the world. In this regard, the findings highlight that many drought risk management actions and investments have substantial cobenefits and positive social returns even without droughts. Hence, they can be promoted widely as low- or no-regret strategies for sustainable development and building resilience to a variety of environmental, economic, and social shocks. Finally, this review discusses the major existing research and knowledge gaps in current drought-related literature and policy actions.

Selection of literature for this review was based on searches in Google Scholar and ScienceDirect platforms using the word *drought* in combination with other key words such as *vulnerability, resilience, early warning and monitoring, impacts, risk management,* and *crisis management*. IDMP partners and participants in the IDMP Expert Group Meeting on this topic in September 2016 (see Acknowledgments) also provided key references. Moreover, citations in key documents were followed to identify additional relevant publications. This review did not cover every aspect of the drought literature in detail, but focused on publications of most relevance to the specific research question mentioned above. Although peer-reviewed papers, institutional publications, and unpublished sources were included, we gave peer-reviewed papers a higher preference in shaping the conclusions of the review, while institutional publications served as valuable background material and sources of further reading.

5.2 Benefits of Action versus Costs of Inaction: Concepts and Methodologies

This review was developed and guided by the conceptual framework depicted in Figure 5.1. Drought events lead to numerous economic, social, and environmental costs of a magnitude modulated by social and household vulnerability and resilience to drought. When a drought occurs, bearing its

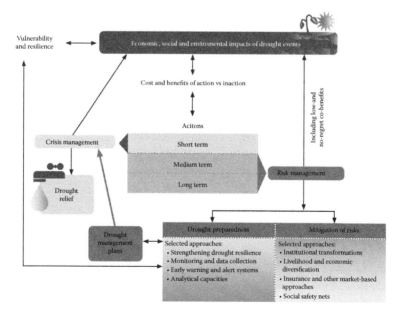

FIGURE 5.1
Conceptual framework. (Reproduced from WMO and GWP 2017. With permission.)

costs while taking no action could increase the overall cost of damage due to the drought, representing the cost of inaction, as compared to taking ex ante and ex post actions against drought. The costs of action against droughts can be classified into three categories: (1) preparedness costs; (2) drought risk mitigation costs; and (3) drought relief costs. If drought relief costs make up the costs of crisis management, drought preparedness costs and the costs of proactive mitigation of drought risks make up the costs of risk management (Figure 5.2). Risk management also leads to the preparation of drought management plans, which identify a set of ex ante and ex post actions against drought and its impacts.

The assumption made in this review is that the costs of action are usually lower than the costs of inaction, and the returns from investing in ex ante risk management actions are higher than those of investing in ex post crisis management, as indicated in Figure 5.2. Actions involving drought preparedness and drought risk mitigation lower the eventual drought relief costs, in addition to helping to mitigate the costs of inaction. For example, the US Federal Emergency Management Agency (FEMA) estimated that the United States would save at least USD 2 on future disaster costs from every USD 1 spent on drought risk mitigation (Logar and van den Bergh 2013). The facility to respond to drought events before and after they occur—amounting to "adaptive capacity," according to Engle (2013)—and reduce their economic and social costs depends on a number of factors, which are context-specific. Engle (2013) identifies a number of these factors for the United States, among which he crucially lists regulated flexibility, that is, balancing the trade-offs between state regulations and structural preparedness, and the capacity for adaptive capacity at the local community level (notably for community water suppliers).

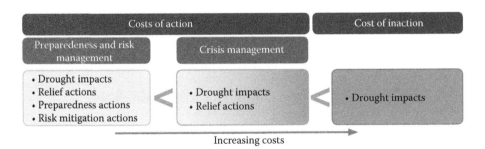

FIGURE 5.2
Summary of costs of drought under different action scenarios (reproduced from WMO and GWP 2017). Note: The figure suggests that the costs of droughts due to inaction are higher than the costs of addressing the impacts of droughts through crisis management approaches (using the inequality sign "<"). In their turn, the costs of actions against droughts using crisis management approaches are expected to be higher than those of using risk management approaches.

5.2.1 Methodologies for Drought Impact Assessment

The site- and time-specific nature of droughts have led to multiple and diverging methods of assessing their impacts. Methodologies vary across scales (from intra-household or crop-specific to economy-wide impacts) and causal channels (direct or indirect; see Birthal et al. 2015). The choice begins with selection of drought indicators (Bachmair et al. 2016). Specifically, econometric models are used to estimate the impact of droughts on crop losses (e.g., Quiroga and Iglesias 2009; Birthal et al. 2015; Bastos 2016), and sometimes also economy-wide, regional-level or basin-level drought costs (Gil et al. 2013; Kirby et al. 2014; Sadoff et al. 2015).

On the other hand, partial equilibrium, computable general equilibrium, and input–output models are used to evaluate the sectoral or economy-wide costs of droughts (Booker et al. 2005; Horridge et al. 2005; Rose and Liao 2005; Berrittella et al. 2007; Dudu and Chumi 2008; Pérez y Pérez and Barreiro-hurlé 2009; Dono and Mazzapicchio 2010; Peck and Adams 2010; Pauw et al. 2011) or of specific policy responses to drought, for example, water restrictions (González 2011). All these papers offer great insights on the methodologies and their improvements in the application of models to assess the costs of droughts or water scarcity. Pérez y Pérez and Barreiro-hurlé (2009) estimated that the direct drought costs in agriculture amounted to EUR 482 million in the Ebro river basin in Spain during 2005. At the same time, indirect costs in the energy sector amounted to EUR 377 million, indicating the substantial scale of indirect costs. Gil et al. (2011) used a combination of econometric and modeling approaches for ex ante assessment of potential drought impacts in Spain. Jenkins (2013) used an input–output model to show the importance of indirect drought costs in projections to 2050. Finally, Santos et al. (2014) used a mixed approach of input–output analysis and event decision trees to evaluate three risk management strategies: reducing the level of water supply disruption, managing water consumption, and prioritizing water use.

Naturally, all these valuation techniques are associated with some difficulties in their implementation or drawbacks in their results. More crucially, though, there is a need for mutually compatible methodologies that allow comparison of drought costs and impacts between sites and across time, or even across various types of natural hazard assessment (Meyer et al. 2013). This would help to target international and national drought mitigation investments or, more generally, investments in the mitigation of all natural hazards. It would also enable a more accurate understanding of vulnerabilities to droughts and impact pathways of droughts. At the same time, such methodologies should account adequately for intrinsic differences in the ways droughts occur in different biophysical settings.

The estimates of drought costs need to include both direct (e.g., reduced crop productivity) and indirect (e.g., increased food insecurity and poverty) impacts of droughts, immediate costs and long-term costs, and losses in both market-priced and nonmarketed ecosystem services (Ding et al. 2011).

Meyer et al. (2013) provide a complete review and classification of the costs of natural hazards, some of which overlap. Thus double counting, as in the case of assessing ecosystem services, must be avoided (Balmford et al. 2008). Indeed, Banerjee et al. (2013) claim that an ecosystem services approach for estimating the economic losses associated with droughts could be used for this purpose. Ecosystem services-based approaches could indeed be useful for including nonmarket impacts of droughts by applying such valuation techniques as avoided and replacement costs methods, contingent valuation, benefit transfer, and other ecosystem services valuation approaches (Nkonya et al. 2011). Using the ecosystem services approach, Banerjee et al. (2013) estimated the costs of the millennium drought of 1999–2011 in the South Australian Murray–Darling Basin to be about USD 810 million.

Building a pool of case studies that evaluate the costs of action versus inaction against droughts using consistent and mutually comparable methodological approaches could provide the basis for a more rigorous understanding of drought costs, impact pathways, vulnerabilities, and costs and benefits of various crisis and risk management approaches against droughts. This would ultimately lead to better informed policy and institutional action (Ding et al. 2011; Wilhite et al. 2014). Without more accurate estimations of the costs of inaction, it is obviously difficult to compare these with the costs and benefits of action against droughts (Changnon 2003).

5.2.2 Global and Local Drought Costs Evaluations

Meanwhile, existing evaluations of drought costs, although highly valuable, remain partial and are often contradictory. Table 5.1 provides some wide-ranging quantifications of drought impacts from the literature. For agriculture, a critical factor affecting the costs of droughts is the possibility to substitute surface water with groundwater resources. Use of groundwater is associated with additional pumping costs, due partly to falling groundwater levels (Howitt et al. 2014, 2015), but the future costs of such groundwater substitution seem to be unknown. In another example, the severe drought occurring in Spain and Portugal in 2005 reduced total European cereal production by 10 percent (UNEP 2006). EEA (2010) indicates that the average annual costs of droughts in the European Union doubled between 1976 and 1990, and 1991–2006, reaching EUR 6.2 billion after 2006, although it is not clear if this doubling was due to increased frequency and severity of droughts or due to the increased area of the European Union caused by new countries joining.

Many countries in Africa, especially in the Sahel region, have long been prone to severe droughts causing massive socioeconomic costs (Mishra and Singh 2010), but quantifications are generally more difficult to find for all developing countries. Uganda lost on average USD 237 million annually to droughts during the last decade (Taylor et al. 2015). Sadoff et al. (2015) found that droughts were likely to reduce gross domestic product (GDP) in Malawi

TABLE 5.1

Selected Examples of the Costs of Droughts

Drought Costs per Annum (USD Billion)	Period	Geographical Unit	Source
0.75	1900–2004	Global	Below et al. (2007)
6.0–8.0	Early 1990s	USA	FEMA (1995)
40.0	1988	USA	Riebsame et al. (1991)
2.2	2014	California	Howitt et al. (2014)
2.7	2015	California	Howitt et al. (2015)
2.5	2006	Australia	Wong et al. (2009)
6.2	2001–2006	European Union	EEA (2010)

by 20 percent and in Brazil by 7 percent. According to Sadoff et al. (2015), the countries that are most vulnerable to GDP losses due to droughts are located in eastern and southern Africa, South America, and South and Southeast Asia. Indeed, the World Bank reports that the frequency of droughts has been increasing in India (World Bank 2003). The magnitude of drought costs also seems to be increasing over time in India (World Bank 2003) and Morocco (MADRPM 2000), due mainly to the increasing value of drought-vulnerable assets. Another issue with these assessments is that they do not really capture costs in the sense of drought costs due to inaction, but implicitly cover the mitigating effects of various measures of either relief or risk management. For comparability and consistency, all assessments of the costs of droughts should be clear about which categories of costs they cover—from the broad categories described in Figure 5.2 or as described in Meyer et al. (2013).

Comprehensive evaluations of the costs of action versus inaction need to be informed by drought risk assessments. These would include analyses of drought hazards, vulnerability to drought, and drought risk management plans (Hayes et al. 2004). Analyses of drought hazards are important because proper risk assessments are impossible without knowledge of historical drought patterns and evolving probabilities of drought occurrence and magnitudes under climate change (Mishra and Singh 2010). This requires weather and drought monitoring networks with sufficient coverage, as well as sufficient human capacity to analyze and transform this information into drought preparedness and risk mitigation action (Pozzi et al. 2013; Wu et al. 2015). However, the operational forecasting of drought onset, its severity, and potential impacts several months in advance has not been broadly possible so far, especially in developing countries (Enenkel et al. 2015). Hallegatte (2012) indicates that the development of hydrometeorological capacities and early warning systems in developing countries to levels similar to those in developed countries would yield annual benefits of between USD 4 and 36 billion, with benefit–cost ratios between 4 and 35 (Pulwarty and Sivakumar 2014). Peck and Adams (2010), citing the case of the Vale Oregon Irrigation District in the United States, demonstrated that longer lead time weather forecasts are essential to enable appropriate responses to droughts. For example, if agricultural producers lack the knowledge that a second drought will shortly follow the first, they may mistakenly increase their future drought costs by expanding their earlier, vulnerable activities as a way to recoup their past losses. In this regard, in addition to physical meteorological infrastructures, wider innovative applications of available information and communication technologies, such as remotely sensed satellite data, have been instrumental in tracking vegetation cover change over long periods of time and with wide geographical coverage (Le et al. 2016). Similarly, mobile phone networks could help trace rainfall patterns with increased time and scale resolutions, especially in contexts where it could be time-consuming and costly to build physical weather monitoring infrastructure (Dinku et al. 2008; Hossain and Huffman 2008; Yin et al. 2008; Zinevich et al. 2008).

Although, as stated above, the literature on the impacts of droughts is fairly extensive, there is a lack of studies comparing the costs of inaction versus action. For example, Salami et al. (2009) traced the economy-wide effects of the 1999–2000 drought in Iran and found the total costs to be equal to 4.4 percent of the country's GDP. The same study also found that applying water-saving technologies to increase water-use productivity by 10 percent would reduce losses due to drought by 17.5 percent or USD 282 million. Furthermore, changing cropping patterns to suit the drought conditions allowed losses to be reduced by USD 597 million. Taylor et al. (2015) evaluated the viability of government drought risk mitigation strategies through increasing water-use efficiency, implementing integrated water resource management, and improving water infrastructures in Uganda. The results indicated that the rate of return could be more than 10 percent. Harou et al. (2010) used the case of California to show that mitigation action such as water markets could substantially reduce the costs of drought impacts, while Wheeler et al. (2014) showed how such markets have worked for Australia's Murray River Basin. Most of these examples of drought costs are linked to agriculture, yet droughts also have impacts in urban areas (Box 5.1).

5.3 Action against Drought: Risk Management versus Crisis Management

Drought risk management includes the following elements: drought preparedness, mitigation of drought risks, and forecasting and early warning of droughts. Drought risk assessments serve as the basis for drought preparedness and drought risk mitigation (Hayes et al. 2004). These feed into drought management plans and identify specific ex ante and ex post actions (Alexander 2002).

BOX 5.1 DROUGHT IMPACTS AND
RESPONSES IN URBAN AREAS

Although agriculture continues to be the major water user globally, the impacts and costs of droughts can be extensive in urban areas. In addition to specific industries (e.g., food and beverages), this also puts the service sector (e.g., tourism) at risk and could spark social tensions. The urban costs of droughts will continue to grow in the future because of climate change and expanding urbanization, and are magnified by relatively higher levels of returns from urban compared with agricultural water use. Therefore, drought preparedness and mitigation efforts in urban areas are important.

Several ways to increase urban drought resilience have been suggested. For example, reducing the overall costs of droughts may involve water transfers from low-value agricultural uses to higher-value urban uses during drought periods. Similarly, drought costs could be substantially reduced in the urban areas of northern California by purchasing water from lower-value agricultural uses.

Drought preparedness and mitigation plans in urban areas include increasing water conservation through appropriate policies and infrastructures (see Chapter 13). Water conservation measures could include nonmarket and market mechanisms. Nonmarket mechanisms usually involve water conservation education and explicit restrictions on specific water uses, while market-based mechanisms involve increasing water prices during droughts. Nonmarket mechanisms may be associated with significant transaction costs to enforce compliance, as well as loss of revenues to water utilities. Increasing the price of water during drought periods, on the other hand, may pose challenges in terms of social equity in water access. Beyond their immediate short-term impacts, droughts may also have longer-term indirect impacts on urban economies and livelihoods. For example, water conservation measures and higher water pricing may encourage a transition to more water-efficient home appliances (e.g., washing machines, dishwashers, showerheads, and toilets).

Sources: Moncur, J.E., Water Resour. Res., 23(3), 393–398, 1987; Michelsen, A.M, and Young, R.A., *Am. J. Agric. Econ.*, 75(4), 1010–1020, 1993; Dixon, L., et al., *Drought Management Policies and Economic Effects in Urban Areas of California, 1987–1992, Rand Corporation, Santa Monica, CA, 1996*; Rosegrant, M.W., et al., *Annu. Rev. Environ. Resour.*, 34(1), 205, 2009; Harou, J.J., et al., *Water Resour. Res.*, 46(5), 2010; Saurí, D., *Annu. Rev. Environ. Resour.*, 38, 227–248, 2013; Güneralp, B., et al., *Global Environ. Change, 31, 217–225, 2015*.

Drought risk management activities concern reducing vulnerability to droughts and are conducted at various scales. The micro-level actions involving households, communities, and individual businesses are often underappreciated but, arguably, are the most important elements of drought risk mitigation. For example:

- More secure land tenure and better access to electricity and agricultural extension were found to facilitate the adoption of drought risk mitigation practices among agricultural households in Bangladesh (Alam 2015). Similarly, Kusunose and Lybbert (2014) found that access to secure land tenure, markets, and credit played a major role in helping farmers cope with droughts in Morocco.

- Holden and Shiferaw (2004) found that improved access to credit helped farming households in Ethiopia to cope better with drought impacts since they no longer needed to divest their productive assets. Moreover, since many rural households in Ethiopia tend to channel their savings into livestock, which may be wiped out during droughts, developing access to financial services and alternative savings mechanisms could also help to mitigate drought risk.

- Land use change and modification of cropping patterns are frequently cited as ways to build resilience against droughts (Lei et al. 2014 in China; Deressa et al. 2009 in Ethiopia; Huntjens et al. 2010 in Europe; Willaume et al. 2014 in France).

- Dono and Mazzapicchio (2010) showed that agricultural producers in Italy's Cuga hydrographic basin could minimize the impact of future droughts by tapping into groundwater resources.

- Another frequently used drought risk mitigation strategy is to diversify livelihoods by adopting off-farm activities (Sun and Yang 2012 in China; Kochar 1999 in India; Kinsey et al. 1998 in Zimbabwe), and divesting of livestock assets (Kinsey et al. 1998 in Zimbabwe; Reardon and Taylor 1996 in Burkina Faso).

- Finally, UNDP (2014) found that a strong asset base and diversified risk management options were among the key characteristics of drought-resilient households in Kenya and Uganda. These aspects were due primarily to the households having better education and greater knowledge of coping actions against various hazards. This allowed them to diversify their income sources.

At the macro level, activities contributing to the mitigation of drought risks mostly involve institutional and policy measures. Booker et al. (2005) found that the establishment of interregional water markets could reduce drought costs by 20–30 percent in the US Rio Grande basin. Other examples include the development of an early warning system (Pulwarty and Sivakumar 2014), drought preparedness plans, increased water supply by investing in water infrastructure (Zilberman et al. 2011), demand reduction (e.g., water conservation programs) (Taylor et al. 2015), and crop insurance.

Although drought insurance is an effective and proactive measure, the development of formal drought insurance mechanisms is hindered in many developing countries by a number of obstacles, including high transaction costs, asymmetric information, and adverse selection (OECD 2016). At the same time, the covariate nature of droughts decreases the effectiveness of traditional community and social network-based informal risk sharing (Kusunose and Lybbert 2014). On the other hand, insurance can actually discourage ex ante drought mitigation behavior. However, this depends on the type of insurance used. In general, two types of insurance are used to ensure against drought damage in agriculture. Indemnity-based insurance protects against predefined losses, while index-based insurance protects against predefined risk events such as droughts (Barnett et al. 2008; GlobalAgRisk 2012).

Specifically, under indemnity-based insurance, crop producers are compensated for their drought-induced losses after a formal assessment of the extent, often compared with their preexisting productivity levels (Meherette 2009). As a result, the transaction costs of indemnity-based insurance schemes are high and they are more suitable for large-scale farming operations.

Index-based insurance schemes use shortfalls in rainfall, temperature, or soil moisture (without formal on-farm assessments of the extent of the damage) to trigger payouts to insured farmers. With significantly lower transaction costs, index-based insurance could be more suitable for smallholder farmers (Barnett et al. 2008; Meherette 2009). Index-based insurance, however, requires a well-functioning and relatively dense infrastructure of weather monitoring stations. Presently, the lack of such infrastructure presents a barrier to the wider rollout of index-based insurance schemes in many developing countries. Under index-based insurance, insurance payouts are not linked to actual damage, but to deviations in weather parameters. Insured farmers, therefore, would continue to have an incentive to take measures to limit the extent of their losses due to droughts. Moreover, index-based approaches allow for insurance against the indirect costs of droughts. For example, agro processors could take out index-based insurance, while they may find traditional indemnity-based insurance is not applicable to them within the context of droughts (GlobalAgRisk 2012).

A limitation of index-based insurance lies in appropriate identification of risk event thresholds that trigger payments, that is, minimizing the so-called basis risk, when the realized weather parameters in the area covered by the same index could be very heterogeneous (Barnett and Mahul 2007). If the threshold is too high, it may not cover some of the losses. If it is too low, the longer-term viability of the insurance scheme may be jeopardized. Identification of optimal payment trigger thresholds also requires the availability of sufficient past data to construct the index. Naturally, depending on the context, a blend of both index- and indemnity-based insurance approaches could be used.

Beyond local and national levels, international coordination of drought risk mitigation and drought responses are equally important in transboundary river basins (Cooley et al. 2009). Inadequate management of transboundary water systems during droughts could magnify both direct and indirect costs of droughts, especially in downstream countries. The existing transboundary agreements on water allocation may need to be reviewed for their flexibility to respond adequately to the increasing frequency of hydrological droughts under climate change (Fischhendler 2004). For example, whether the transboundary water allocation schemes are based on predefined minimum flow deliveries from upstream to downstream countries or on percentage quotas could have substantially different impacts during droughts (Hamner and Wolf 1998). Regional drought risk mitigation efforts would include increasing the flexibility of transboundary water allocation regimes in response to droughts (McCaffrey 2003). This includes the operation of large-scale water reservoirs, which could have considerable impacts on upstream–downstream water flow regimes (López-Moreno et al. 2009). Transboundary water management institutions could play a vital role in coordinating such responses to droughts (Cooley et al. 2009), and efforts are needed to promote the development of national and transboundary drought preparedness plans, assuring they are consistent in cases when they are interdependent.

As it is not possible or economically efficient to eliminate drought vulnerability completely, droughts will continue to affect society to some extent. It is, therefore, important to identify more efficient drought responses. Crisis management measures may include impact assessments, response, and reconstruction, involving such tools as drought relief funds, low-interest loans, transportation subsidies for livestock and livestock feed, provision of food, water transport, and drilling wells for irrigation and public water supplies (Wilhite 2000). Several studies identify ways to improve the efficiency of drought response measures. For example, pooling resources at the regional level in sub-Saharan Africa was found to be an effective strategy to hasten drought relief and reduce its costs (Clarke and Hill 2013), although this may not reduce future drought vulnerability. Experiences from Ethiopia showed that employment generation schemes could be effective in terms of immediate aid and strengthening local resilience against future droughts. These schemes paid drought-affected populations to work in drought mitigation activities (e.g., building terraces and check dams) rather than giving direct food relief (IFRC 2003).

Since it is difficult to evaluate the costs of droughts, it is even more challenging to compare the costs and benefits of proactive risk management versus reactive crisis management. Lack of comprehensive data on drought costs also makes it difficult to assess the effectiveness of mitigation investments (FEMA 1997). Moreover, because of the limited number of historical mitigation investments, any ex ante assessments of the rate of return from future mitigation actions will depend on modeling assumptions, which may not always prove to be consistent with the actual performance of the investments. However, once mitigation investments are made, governments and donors will want to know the returns from their investments. This should lead to additional impact assessments being conducted and will identify more efficient drought risk mitigation options (Changnon 2003). Most of the relevant past studies investigated the impact of adopting very specific drought mitigation options, where data were available and the uncertainty of assumptions could be reduced—for example, the impact of water-saving technologies (Ward 2014) or policies such as water trading (Booker et al. 2005; Ward et al. 2006). There is a need for additional such case studies. While it is plausible that drought risk management approaches are more efficient than crisis management measures, this review found a lack of rigorous empirical evidence to support this argument.

5.4 From Crisis Management to Risk Management: Obstacles and Opportunities

5.4.1 Drivers of Ex Ante and Ex Post Action against Drought

Over the past few decades, we have experienced an increasing frequency and severity of droughts (Changnon et al. 2000) associated with rising economic and social costs (Downing and Bakker 2000). We have also seen an increased perception of the greater efficiency of risk management strategies (Wilhite 2005), and their lower burden on public budgets compared with frequent drought relief actions. These trends are leading to shifts from drought crisis management to risk management in many countries, including Australia, India, the United States, and the countries of the European Union (EC 2008; Birthal et al. 2015). Among these factors, the escalation of drought relief costs and the increasing burden on government budgets seem to have played a major role in promoting risk management strategies in the United States (Changnon 2003), Australia (Stone 2014), and probably additional countries embarking on this transition path. Box 5.2 illustrates that even with the best dispositions toward risk management, governments are sometimes locked in crisis management strategies, especially during particularly long and acute drought episodes.

Nonetheless, path dependence and lack of information on the costs and benefits of risk management and crisis management actions are the leading causes of the persistence of crisis management approaches in many countries. When there is a lack of information on the costs and benefits of mitigation actions, governments are often reluctant to make costly investments in

BOX 5.2: DROUGHT IN BRAZIL: IMPACTS, COSTS, AND POLICY RESPONSES

Droughts in Brazil, especially in the northeast, are expected to increase in frequency and intensity as a result of global climate change. Drought and climate change combined with existing pressure on freshwater availability and quality are likely to lead to new and increased water management challenges. These have been recognized by the Brazilian water community, including resource managers and users, researchers, and policy makers.

The country has several semiarid regions, particularly in the northeast, where droughts are frequent events (see also Chapter 21). Parts of this region experience high rainfall variability, with the rainy season in February to May accounting for about 70 percent of annual rainfall. The country in general and the region in particular thus have a long history of institutional drought management. This dates back to the first reservoir built in 1886, followed by the creation of agencies to address drought throughout the twentieth century. Some of these are still in place in revised forms. The country also established a water code as early as 1930. According to the Brazilian Constitution, "water is a limited natural resource and an inalienable public good that belongs either to the federal or state government."

Yet, the recent multiyear drought event of 2010–2013 has been particularly severe. Precipitation during the rainy season of 2012 was classed as "dry" to "extremely dry" for most of the northeast, reaching only about 50 percent of the historical average for the season. The lack of water availability affected crops, livestock, and industries, as well as drinking water levels. Hence, despite its history of water management institutions, Brazil is struggling to cope with new, prolonged, and extreme drought events.

In the wake of these events, Brazil has reverted to emergency relief and response actions. These are listed in Bastos (2016) and include various measures aimed at mitigating the negative impacts on communities and farmers as a direct consequence of the lack of water (water truck deliveries, cisterns) or as an indirect consequence of reduced agricultural production (emergency credit lines, debt negotiation—the costliest measure). Additionally, infrastructure development such as well drilling or new dams has been included under the growth development plan. These measures have come with high costs; as of 2014, USD 4.5 billion had been allocated to emergency relief and infrastructure development. These costs are in addition to the estimated 13 percent loss in gross real value of agricultural output over the period 2010–2014.

The magnitude of these costs demonstrates the difficulty of implementing predrought plans and actions to cope with the economic impacts of droughts. This is true in Brazil, a country with a history of drought management, infrastructure, available indicators, and scientific knowledge and expertise in meteorological, climatological, and hydrological monitoring and forecasting. The gaps and opportunities for drought preparedness and policy in Brazil, identified in Gutiérrez et al. (2014), can help to improve the situation in the country and in similar emerging economies. These point largely toward more and better integration between monitoring and forecasting communities, as well as with state and municipal decision-making bodies, the keeping of national archives to determine vulnerabilities to and impacts of drought (and other disasters), and vulnerability assessments conducted in the context of climate change. Many of these gaps are of an organizational nature, pointing to the need for documentation of droughts and their impacts. Others point to the need for analysis of vulnerability to drought. Together, such action should ensure faster and better mitigation and response to drought in the future.

Sources: World Bank, *Water Resources Planning and Adaptation to Climate Variability and Climate Change in Selected River Basins in Northeast Brazil: Final Report on a Non-Lending Technical Assistance Program (P123869)*, World Bank, Washington, DC, 2013; Gutiérrez, A.P.A., et al., *Weather Clim. Extremes*, 3, 95–106, 2014; Bastos, P., *Drought Impacts and Cost Analysis for Northeast Brazil, in Drought in Brazil: Proactive Management and Policy*, eds. E. De Nys, N. L. Engle, and A. R. Magalhães, pp. 119–142, CRC Press, Boca Raton, FL, 2016.

mitigation (Ding et al. 2011). Moreover, under various uncertainties and with a shortage of empirical evidence on the greater efficiency of drought risk mitigation actions, it may be economically rational to respond to droughts only after shocks (Zilberman et al. 2011). Economic theory shows that under conditions of uncertainty, actors will delay irreversible investments until their net benefits exceed a positive critical value (McDonald and Siegel 1986). Meanwhile, Zilberman et al. (2011) indicate that major changes in institutions and technological adoptions are likely to happen ex post as a response to droughts. For example, the drought of 1987–1991 in California led to wide adoption of water conservation technologies (sprinkler irrigation), fallowing of land, lining of canals for reducing water loss, and the introduction of water trading, although these measures had been recommended for a long time before the occurrence of the drought (Zilberman et al. 2011).

Jaffee and Russell (2013) suggest that ex ante actions are not always preferable to ex post actions when individuals attach varying subjective probabilities to drought hazards, which then shape their investment decisions. In such contexts, they suggest to maximize social welfare it may be better to provide disaster relief rather than ex ante actions. Moreover, ex ante adjustments to droughts could increase resilience in the case of droughts, but could also simultaneously lead to choices that have lower returns during nondrought periods (Kusunose and Lybbert 2014). However, this analysis needs to compare ex ante and ex post interventions on farmers' production and investment decisions, and varying impacts of droughts on them (OECD 2016).

Drought preparedness plans need to include various trajectories of change that occur after they are implemented. For example, in the Segura river basin in Spain, drought preparedness plans imposing water supply restrictions from surface water led to the overexploitation of groundwater, which was not covered by the plan. This led to higher drought risks than would have occurred without the plan (Gómez and Perez-Blanco 2012). Therefore, drought preparedness plans, like other action plans, need to be evaluated and improved continuously to suit the evolving context and encompass learning from past mistakes (WMO and GWP 2014).

Although ex post actions seem to happen more often, there are economic reasons for ex ante actions. Drought is a business risk and agricultural producers will try to avoid its costs. Thus, while they have incentives to undertake mitigation actions, they face obstacles in the form of lack of knowledge about drought occurrences (early warning systems) and their impacts (extension and advisory services), and lack of funds (access to credit) (OECD 2016).

Similarly, numerous studies show that human and social systems evolve continuously to adapt to the changing environment. Biazin and Sterk (2013) showed that pastoral households in Ethiopia were shifting to more resilient mixed farming systems as a response to drought and that their earlier coping option involving migrating to alternative pastures was no longer feasible. Households in many drought-affected areas continuously apply risk management strategies as a normal part of their livelihood behavior. Such risk management strategies are often applied in response to past drought shocks with a view to minimize the impacts of future drought events, that is, households learn from their past experiences.

In the context of public goods, where experience plays a reduced role in fostering proactive behavior, the lack of visibility of the impacts of drought risk management versus drought response measures is critical. However, risk management strategies could be more efficient and forward-looking if they were supported by scientific data on climate, drought, and drought risk mitigation measures, with enabling ex ante government policies. Birthal et al. (2015) indicate that although agricultural households carry out coping actions after droughts, which could serve as risk management strategies by reducing their vulnerability to future droughts, they may rarely be able to recover fully the loss of their productive assets due to the impact of the past drought. Indeed, drought relief in many developing countries is not as comprehensive as it might be in some developed countries, or is simply nonexistent, so that affected households are left to their own means. On one hand, this may accelerate transitions to risk management approaches at the microeconomic level, but on the other, if governments do not need to save on drought relief costs (because they are small or none, or are borne by outside donors), there will be no urgency to make the transition at the macro level.

5.4.2 Cobenefits of Drought Risk Management Strategies

In addition to mitigating drought risks, risk management strategies have a major appealing characteristic in that they have substantial socioeconomic cobenefits. Many drought risk management actions build resilience against droughts and additional socioeconomic and environmental shocks. Thus, a number of approaches to risk management against droughts are low- or no-regret options (Figure 5.3). Therefore, their application makes sense as a precautionary measure to prevent the negative impacts of many direct and, especially, indirect costs of droughts about which we have little knowledge. Figure 5.3 highlights that the benefits of adopting risk management approaches include reducing drought costs and lowering drought relief costs as well as having substantial socioeconomic cobenefits.

For example, as elaborated earlier, more secure land tenure, better access to electricity and agricultural extension, access to credit, diverse livelihood options (including off-farm activities), and higher education levels were

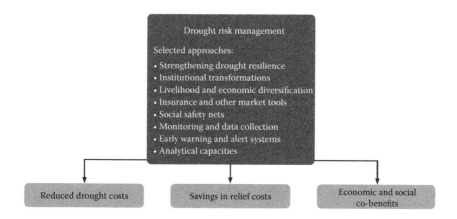

FIGURE 5.3
Approaches to drought risk management and benefits. (Reproduced from WMO and GWP 2017.)

associated with stronger resilience against drought impacts (Holden and Shiferaw 2004; Sun and Yang 2012; UNDP 2014; WMO and GWP 2014; Alam 2015). At the same time, these factors substantially increase adaptive capacities against climate change (Deressa et al. 2009), help address land degradation (Nkonya et al. 2016), facilitate poverty reduction (Khandker 1998), improve household food security (Babatunde and Qaim 2010), and promote broader sustainable development.

Another example—the adoption of improved irrigation techniques or alternative water sources (Hettiarachchi and Ardakanian 2016)—could have positive impacts on agricultural income and sustainable water and land use during normal conditions as well as during times of drought. For example, the adoption of conservation agriculture practices in Kazakhstan, which included zero tillage and mulching, had the effect of reducing soil erosion and fuel use for land preparation as well as helping people cope better with the effects of the 2010 drought (Kienzler et al. 2012). This was because conservation agriculture practices allowed better retention of available soil moisture, thus reducing losses in crop productivity compared with previous droughts. While the adoption of conservation agriculture was driven primarily by the desire to save on fuel costs, it eventually served as a drought risk management strategy (Kienzler et al. 2012).

As a result, investments in drought risk management strategies and actions that have significant cobenefits can serve as "low-hanging fruit" in drought risk mitigation—that is, they are the easiest to implement initially. Although literature exists on the links between poverty reduction/food security and such factors as income diversification, land tenure security, and access to extension and credit, there is a need for more studies incorporating the cobenefits of promoting these and other similar drought risk mitigation factors as part of drought risk management approaches. Ideally, such studies would include quantification of the contributions of these factors to reducing drought costs and the extent of their cobenefits.

It should be noted that drought risk management strategies, such as household options for proactive increases in resilience to drought events, are not without trade-offs and that their impact can be highly case-specific. For instance, UNDP (2014) provides a number of examples where such strategies can have negative effects economically and socially at the level of the household and beyond. Examples include early marriages to boost the asset base through dowries, or disinvestment in education in favor of immediate employment in low-skill jobs. In specific agroclimatic systems, income specialization in livestock activities can prove to be a more drought-resilient strategy than income diversification. Similarly, gender- and age-differentiated impact assessments might lead to interesting insights on the distributional impacts of drought events and drought risk management strategies. This could ultimately point to different cost–benefit ratios and recommendations for action tailored to population target groups.

5.5 Conclusions and Next Steps

This review shows that although significant progress has been made over the past decade in understanding droughts and their impacts, as well as the merits of risk management approaches compared with traditional crisis management approaches, important research and policy gaps remain. There is a need for mutually compatible methodologies to comprehensively assess drought costs and impacts. Presently, many available estimates of drought costs are partial and difficult to compare. The problem is compounded by the lack of data on droughts and their impacts. Moreover, there is relatively little knowledge available on the costs of indirect and longer-term drought impacts.

Potential next steps include the following:

- Case studies should evaluate the costs of action versus inaction against droughts using consistent and mutually comparable methodological approaches. This should allow better understanding of the drought costs, impact pathways, vulnerabilities, costs and benefits of various crisis and risk management approaches against droughts, and the cobenefits of risk management approaches, which will ultimately lead to better informed policy and institutional actions on droughts.

- Comprehensive evaluations of the costs of action versus inaction against droughts need to be informed by drought risk assessments. They require weather and drought monitoring networks with sufficient coverage, as well as adequate human capacity to analyze and transform this information into drought preparedness and mitigation actions.

- When the previous two points are fulfilled, a clearer picture of the cost–benefit ratio of actions before drought (drought preparedness) versus the cost–benefit ratio of reactive actions (crisis management) can emerge. This is required to guide policy and investments for building drought resilience.

- Since it is not possible or economically efficient to eliminate vulnerability to droughts, they will continue to affect society to some extent. Therefore, more efficient drought responses also need to be identified.

- To have impact, research and development partners need to demonstrate to governments that it will be unaffordable to continue with drought relief in the future. It is already putting a huge burden on budgets, thus requiring a shift to risk management approaches in both the discourse and through specific funded actions. A low-hanging fruit in this regard would be to choose mitigating actions that have immediate cobenefits beyond drought risk management, and that would be beneficial with or without droughts. There is a need for more research to identify such socioeconomic cobenefits of drought risk management strategies and approaches, and for more evidence-based advocacy on this issue.

References

Adger, W. N. 2006. Vulnerability. *Global Environmental Change* 16(3):268–281.

Alam, K. 2015. Farmers' adaptation to water scarcity in drought-prone environments: A case study of Rajshahi District, Bangladesh. *Agricultural Water Management* 148:196–206.

Alexander, D. 2002. From civil defense to civil protection and back again. *Disaster Prevention and Management: An International Journal* 11(3):209–213.

Andreadis, K. M., and D. P. Lettenmaier. 2006. Trends in 20th century drought over the continental United States. *Geophysical Research Letters* 33(10).

Babatunde, R. O., and M. Qaim. 2010. Impact of off-farm income on food security and nutrition in Nigeria. *Food Policy* 35(4):303–311.

Bachmair, S., K. Stahl, K. Collins, et al. 2016. Drought indicators revisited: The need for a wider consideration of environment and society. *WIREs Water* 3:516–536. doi: 10.1002/wat2.1154.

Balmford, A., A. Rodrigues, M. Walpole, et al. 2008. *Review on the economics of biodiversity loss: Scoping the science.* European Commission, Cambridge, UK.

Banerjee, O., R. Bark, J. Connor, and N. D. Crossman. 2013. An ecosystem services approach to estimating economic losses associated with drought. *Ecological Economics* 91:19–27.

Barnett, B. J., C. B. Barrett, and J. R. Skees. 2008. Poverty traps and index-based risk transfer products. *World Development* 36(10):1766–1785.

Barnett, B. J., and O. Mahul. 2007. Weather index insurance for agriculture and rural areas in lower-income countries. *American Journal of Agricultural Economics* 89(5):1241–1247.

Bastos, P. 2016. Drought impacts and cost analysis for Northeast Brazil. In *Drought in Brazil: Proactive management and policy*, eds. E. De Nys, N. L. Engle, and A. R. Magalhães, pp. 119–142. CRC Press, Boca Raton, FL.

Bates, B., Z. W. Kundzewicz, S. Wu, and J. Palutikof. 2008. *Climate change and water: Technical paper of the Intergovernmental Panel on Climate Change.* IPCC Secretariat, Geneva, Switzerland.

Below, R., E. Grover-Kopec, and M. Dilley. 2007. Documenting drought-related disasters: A global reassessment. *Journal of Environment and Development* 16(3):328–344.

Berrittella, M., A. Y. Hoekstra, K. Rehdanz, R. Roson, and R. S. J. Tol. 2007. The economic impact of restricted water supply: A computable general equilibrium analysis. *Water Research* 41:1799–1813.

Biazin, B., and G. Sterk. 2013. Drought vulnerability drives land-use and land cover changes in the Rift Valley dry lands of Ethiopia. *Agriculture, Ecosystems and Environment* 164:100–113.

Birthal, P. S., D. S. Negi, M. T. Khan, and S. Agarwal. 2015. Is Indian agriculture becoming resilient to droughts? Evidence from rice production systems. *Food Policy* 56:1–12.

Booker, J. F., A. M. Michelsen, and F. A. Ward. 2005. Economic impact of alternative policy responses to prolonged and severe drought in the Rio Grande Basin. *Water Resources Research* 41(2).

Bruce, J. P. 1994. Natural disaster reduction and global change. *Bulletin of the American Meteorological Society* 75(10):1831–1835.

Carolwicz, M. 1996. Natural hazards need not lead to natural disasters. *EOS* 77(16):149–153.

CEC. 2007. *Impact assessment. Accompanying document to the communication from the Commission to the European Parliament and the Council, COM (2007) 414, SEC (2007) 993*. Commission of the European Communities, Brussels, Belgium.

Changnon, S. A. 2003. Measures of economic impacts of weather extremes: Getting better but far from what is needed—A call for action. *Bulletin of the American Meteorological Society* 84(9):1231–1235.

Changnon, S. A., R. A. Pielke, Jr., D. Changnon, R. T. Sylves, and R. Pulwarty. 2000. Human factors explain the increased losses from weather and climate extremes. *Bulletin of the American Meteorological Society* 81(3):437–442.

Clarke, D. J., and R. V. Hill. 2013. *Cost-benefit analysis of the African risk capacity facility*. IFPRI Discussion Paper 01292. International Food Policy Research Institute, Washington, DC.

Conway, G. 2008. *The science of climate change in Africa: Impacts and adaptation*. Department for International Development (DFID), London, UK.

Cook, E. R., R. Seager, M. A. Cane, and D. W. Stahle. 2007. North American drought: Reconstructions, causes, and consequences. *Earth Science Reviews* 81(1):93–134.

Cooley, H., J. H. Christian-Smith, P. H. Gleick, L. Allen, and M. Cohen. 2009. *Understanding and reducing the risks of climate change for transboundary waters*. Pacific Institute, Oakland, CA.

Deressa, T. T., R. M. Hassan, C. Ringler, T. Alemu, and M. Yesuf. 2009. Determinants of farmers' choice of adaptation methods to climate change in the Nile Basin of Ethiopia. *Global Environmental Change* 19(2):248–255.

Ding, Y., M. J. Hayes, and M. Widhalm. 2011. Measuring economic impacts of drought: A review and discussion. *Disaster Prevention and Management: An International Journal* 20(4):434–446.

Dinku, T., S. Chidzambwa, P. Ceccato, S. J. Connor, and C. F. Ropelewski. 2008. Validation of high-resolution satellite rainfall products over complex terrain. *International Journal of Remote Sensing* 29(14):4097–4110.

Dixon, L., N. Y. Moore, and E. M. Pint. 1996. *Drought management policies and economic effects in urban areas of California, 1987–1992*. Rand Corporation, Santa Monica, CA.

Dono, G., and G. Mazzapicchio. 2010. Uncertain water supply in an irrigated Mediterranean area: An analysis of the possible economic impact of climate change on the farm sector. *Agricultural Systems* 103(6):361–370. doi:10.1061/(ASCE)HE.1943-5584.0000169.

Downing, T. E., and K. Bakker. 2000. Drought discourse and vulnerability. In *Drought, a global assessment*, ed. D. A. Wilhite, Vol. 2, pp. 213–230. Routledge, London.

Dudu, H., and S. Chumi. 2008. *Economics of irrigation water management: A literature survey with focus on partial and general equilibrium models*. World Bank Policy Research Working Paper Series. World Bank, Washington, DC.

Ebi, K. L., and K. Bowen. 2015. Extreme events as sources of health vulnerability: Drought as an example. *Weather and Climate Extremes* 11:95–102.

EC. 2007. *Water scarcity and droughts: In-depth assessment.* Second Interim Report. European Commission. http://ec.europa.eu/environment/water/quantity/pdf/comm_droughts/2nd_int_report.pdf. Accessed June 21, 2017.

EC. 2008. *Drought management plan report.* Technical Report 2008–023. Office for Official Publications of the European Commission (EC), Luxembourg.

EEA. 2010. *Mapping the impacts of natural hazards and technological accidents in Europe: An overview of the last decade.* European Environment Agency (EEA) Technical report No 13/2000. Publications Office of the European Union, Luxembourg.

Enenkel, M., L. See, R. Bonifacio, et al. 2015. Drought and food security—Improving decision-support via new technologies and innovative collaboration. *Global Food Security* 4:51–55.

Engle, N. L. 2013. The role of drought preparedness in building and mobilizing adaptive capacity in states and their community water systems. *Climatic Change* 118(2):291–306.

FEMA. 1995. *National mitigation strategy.* Federal Emergency Management Agency, Washington, DC.

FEMA. 1997. *Multi-hazard identification and risk assessment: A cornerstone of the national mitigation strategy.* Federal Emergency Management Agency, Washington, DC.

Fischhendler, I. 2004. Legal and institutional adaptation to climate uncertainty: A study of international rivers. *Water Policy* 6(4):281–302.

Fisher, M., and E. R. Carr. 2015. The influence of gendered roles and responsibilities on the adoption of technologies that mitigate drought risk: The case of drought-tolerant maize seed in eastern Uganda. *Global Environmental Change* 35:82–92.

Gan, T. Y., M. Ito, S. Hülsmann, et al. 2016. Possible climate change/variability and human impacts, vulnerability of drought-prone regions, water resources and capacity building for Africa. *Hydrological Sciences Journal* 61(7):1209–1226. doi:10.1080/02626667.2015.1057143.

Gil, M., A. Garrido, and A. Gómez-Ramos. 2011. Economic analysis of drought risk: An application for irrigated agriculture in Spain. *Agricultural Water Management* 98(5):823–833.

Gil, M., A. Garrido, and N. Hernández-Mora. 2013. Direct and indirect economic impacts of drought in the agri-food sector in the Ebro River basin (Spain). *Natural Hazards and Earth System Sciences* 3:2679–2694.

GlobalAgRisk. 2012. *Comparison of indemnity and index insurances.* Technical Note 4. Deutsche Gesellschaft für Internationale Zusammenarbeit (GIZ) GmbH, Lima, Peru.

Gómez, C. M. G., and C. D. Perez-Blanco. 2012. Do drought management plans reduce drought risk? A risk assessment model for a Mediterranean river basin. *Ecological Economics* 76:42–48.

González, J. F. 2011. Assessing the macroeconomic impact of water supply restrictions through an input–output analysis. *Water Resources Management* 25(9):2335–2347.

Gray, C., and V. Mueller. 2012. Drought and population mobility in rural Ethiopia. *World Development* 40(1):134–145.

Güneralp, B., I. Güneralp, and Y. Liu. 2015. Changing global patterns of urban exposure to flood and drought hazards. *Global Environmental Change* 31:217–225.

Gutiérrez, A. P. A., N. L. Engle, E. De Nys, C. Molejón, and E. S. Martins. 2014. Drought preparedness in Brazil. *Weather and Climate Extremes* 3:95–106.

Hallegatte, S. 2012. *A cost effective solution to reduce disaster losses in developing countries: Hydro-meteorological services, early warning, and evacuation.* World Bank Policy Research Working Paper (6058). World Bank, Washington, DC.

Hamner, J., and A. Wolf. 1998. *Colorado Natural Resources, Energy, & Environmental Law Review, Uni, Colorado, Boulder.* http://heinonline.org/HOL/Page?handle=hein. journals/colenvlp9&div=43&g_sent=1&collection=journals. Accessed June 21, 2017.

Harou, J. J., J. Medellín-Azuara, T. Zhu, et al. 2010. Economic consequences of optimized water management for a prolonged, severe drought in California. *Water Resources Research* 46(5).

Hayes, M. J., O. V. Wilhelmi, and C. L. Knutson. 2004. Reducing drought risk: Bridging theory and practice. *Natural Hazards Review* 5(2):106–113.

Hettiarachchi, H., and R. Ardakanian. 2016. *Safe use of wastewater in agriculture: Good practice examples.* United Nations University Institute for Integrated Management of Material Fluxes and of Resources (UNU-FLORES), Dresden, Germany.

Hoddinott, J., and B. Kinsey. 2001. Child growth in the time of drought. *Oxford Bulletin of Economics and Statistics* 63(4):409–436.

Holden, S., and B. Shiferaw. 2004. Land degradation, drought and food security in a less-favoured area in the Ethiopian highlands: A bio-economic model with market imperfections. *Agricultural Economics* 30(1):31–49.

Horridge, M., J. Madden, and G. Wittwer. 2005. The impact of the 2002–2003 drought on Australia. *Journal of Policy Modeling* 27(3):285–308.

Hossain, F., and G. J. Huffman. 2008. Investigating error metrics for satellite rainfall data at hydrologically relevant scales. *Journal of Hydrometeorology* 9(3):563–575.

Howitt, R., J. Medellín-Azuara, D. MacEwan, J. Lund, and D. Sumner. 2014. *Economic analysis of the 2014 drought for California agriculture.* Center for Watershed Sciences, University of California, Davis, CA.

Howitt, R., J. Medellín-Azuara, D. MacEwan, J. Lund, and D. Sumner. 2015. *Economic analysis of the 2015 drought for California agriculture.* Center for Watershed Sciences, University of California, Davis, CA.

Huntjens, P., C. Pahl-Wostl, and J. Grin. 2010. Climate change adaptation in European river basins. *Regional Environmental Change* 10(4):263–284.

IFRC. 2003. *Ethiopian droughts: Reducing the risk to livelihoods through cash transfers.* International Federation of Red Cross and Red Crescent Societies, Geneva, Switzerland.

IFRC. 2006. *Eastern Africa: Regional drought response.* DREF Bulletin No. MDR64001. International Federation of Red Cross and Red Crescent Societies, Geneva, Switzerland.

IPCC. 2014. *Climate change 2014: Impacts, adaptation, and vulnerability. Part A: Global and sectoral aspects.* Contribution of Working Group II to the Fifth Assessment Report of the Intergovernmental Panel on Climate Change. Cambridge University Press, Cambridge.

Jaffee, D., and T. Russell. 2013. The welfare economics of catastrophe losses and insurance. *The Geneva Papers on Risk and Insurance—Issues and Practice* 38(3):469–494.

Jenkins, K. 2013. Indirect economic losses of drought under future projections of climate change: A case study for Spain. *Natural Hazards* 69(3):1967–1986.

Jenkins, K. L. 2011. Modelling the economic and social consequences of drought under future projections of climate change. PhD dissertation, Darwin College, University of Cambridge.

Johnstone, S., and J. Mazo. 2011. Global warming and the Arab Spring. *Survival* 53(2):11–17.

Kampragou, E., S. Apostolaki, E. Manoli, J. Froebrich, and D. Assimacopoulos. 2011. Towards the harmonization of water-related policies for managing drought risks across the EU. *Environmental Science and Policy* 14(7):815–824.

Khandker, S. R. 1998. *Fighting poverty with microcredit: Experience in Bangladesh.* Oxford University Press, New York.

Kienzler, K. M., J. P. A. Lamers, A. McDonald, et al. 2012. Conservation agriculture in Central Asia—What do we know and where do we go from here? *Field Crops Research* 132:95–105.

Kinsey, B., K. Burger, and J. W. Gunning. 1998. Coping with drought in Zimbabwe: Survey evidence on responses of rural households to risk. *World Development* 26(1):89–110.

Kirby, M., R. Bark, J. Connor, M. E. Qureshi, and S. Keyworth. 2014. Sustainable irrigation: How did irrigated agriculture in Australia's Murray–Darling Basin adapt in the millennium drought? *Agricultural Water Management* 145:154–162.

Kochar, A. 1999. Smoothing consumption by smoothing income: Hours-of-work responses to idiosyncratic agricultural shocks in rural India. *Review of Economics and Statistics* 81(1):50–61.

Kogan, F. N. 1997. Global drought watch from space. *Bulletin of the American Meteorological Society* 78(4):621–636.

Kusunose, Y., and T. J. Lybbert. 2014. Coping with drought by adjusting land tenancy contracts: A model and evidence from rural Morocco. *World Development* 61:114–126.

Le, Q. B., E. Nkonya, and A. Mirzabaev. 2016. Biomass productivity-based mapping of global land degradation hotspots. In *Economics of land degradation and improvement—A global assessment for sustainable development*, eds. E. Nkonya, A. Mirzabaev, and J. von Braun, pp. 55–84. Springer International Publishing, Heidelberg.

Lei, Y., Y. Yue, Y. Yin, and Z. Sheng. 2014. How adjustments in land use patterns contribute to drought risk adaptation in a changing climate—A case study in China. *Land Use Policy* 36:577–584.

Linke, A. M., J. O'Loughlin, J. T. McCabe, J. Tir, and F. D. Witmer. 2015. Rainfall variability and violence in rural Kenya: Investigating the effects of drought and the role of local institutions with survey data. *Global Environmental Change* 34:35–47.

Logar, I., and J. C. J. M. van den Bergh. 2011. *Methods for assessment of the costs of droughts.* CONHAZ Rep. WP05. Available from: http://mp.mountaintrip.eu/uploads/media/workpackagereport/CONHAZ_REPORT_WP05_1_FINAL.pdf. Accessed June 21, 2017.

Logar, I., and J. C. J. M. van den Bergh. 2013. Methods to assess costs of drought damages and policies for drought mitigation and adaptation: Review and recommendations. *Water Resources Management* 27(6):1707–1720.

Lohmann, S., and T. Lechtenfeld. 2015. The effect of drought on health outcomes and health expenditures in rural Vietnam. *World Development* 72:432–448.

López-Moreno, J. I., S. M. Vicente-Serrano, S. Beguería, J. M. García-Ruiz, M. M. Portela, and A. B. Almeida. 2009. Dam effects on droughts magnitude and duration in a transboundary basin: The Lower River Tagus, Spain and Portugal. *Water Resources Research* 45(2).

MADRPM. 2000. *Programme de sécuritisation de la production végétale: Rapport de synthèse campagne 1999–2000*. Technical report, Ministère de l'Agriculture, du Développement Rural et des Pêches Maritimes, Rabat, Morocco.

McCaffrey, S. C. 2003. The need for flexibility in freshwater treaty regimes. *Natural Resources Forum* 27(2):156–162.

McDonald, R., and D. Siegel. 1986. The value of waiting to invest. *Quarterly Journal of Economics* 101:707–728.

Meherette, E. 2009. *Innovations in insuring the poor. Providing weather index and indemnity insurance in Ethiopia. Vision 2020 for food, agriculture, and the environment.* Focus 17, Brief 8. International Food Policy Research Institute, Washington, DC.

Meyer, V., N. Becker, V. Markantonis, et al. 2013. Review article: Assessing the costs of natural hazards—State of the art and knowledge gaps. *Natural Hazards and Earth System Sciences* 13(5):1351–1373.

Michelsen, A. M, and R. A. Young. 1993. Optioning agricultural water rights for urban water supplies during drought. *American Journal of Agricultural Economics* 75(4):1010–1020.

Mishra, A. K., and V. P. Singh. 2009. Analysis of drought severity–area–frequency curves using a general circulation model and scenario uncertainty. *Journal of Geophysical Research: Atmospheres* 114(D6).

Mishra, A. K., and V. P. Singh. 2010. A review of drought concepts. *Journal of Hydrology* 391(1):202–216.

Moncur, J. E. 1987. Urban water pricing and drought management. *Water Resources Research* 23(3):393–398.

Mosley, L. M. 2015. Drought impacts on the water quality of freshwater systems; review and integration. *Earth Science Reviews* 140:203–214.

NCDC. 2002. *US National Percent Area severely to extremely dry and severely to extremely wet*. US National Climatic Data Center. http://www.ncdc.noaa.gov/oa/climate/research/2002/may/uspctarea-wetdry.txt (accessed November 25, 2016).

Nkonya, E., N. Gerber, P. Baumgartner, et al. 2011. *The economics of desertification, land degradation, and drought toward an integrated global assessment. Discussion Papers on Development Policy No. 150.* Center for Development Research, University of Bonn, Bonn, Germany.

Nkonya, E., A. Mirzabaev, and J. von Braun. 2016. *Economics of land degradation and improvement—A global assessment for sustainable development.* Springer International Publishing, Cham, Switzerland. doi:10.1007/978-3-319-19168-3.

Obasi, G. O. P. 1994. WMO's role in the international decade for natural disaster reduction. *Bulletin of the American Meteorological Society* 75(9):1655–1661.

OECD. 2016. *Mitigating droughts and floods in agriculture: Policy lessons and approaches, OECD studies on water*. Organisation for Economic Cooperation and Development, Paris, France.

Pandey, S., H. Bhandari, and B. Hardy. 2007. *Economic costs of drought and rice farmers' coping mechanisms: A cross-country comparative analysis from Asia.* International Rice Research Institute, Manila, the Philippines.

Pauw, K., J. Thurlow, M. Bachu, and D. E. Van Seventer. 2011. The economic costs of extreme weather events: A hydrometeorological CGE analysis for Malawi. *Environment and Development Economics* 16(02):177–198.

Peck, D. E., and R. M. Adams. 2010. Farm-level impacts of prolonged drought: Is a multiyear event more than the sum of its parts? *Australian Journal of Agricultural and Resource Economics* 54(1):43–60.

Pérez y Pérez, L., and J. Barreiro-Hurlé, J. 2009. Assessing the socio-economic impacts of drought in the Ebro River Basin. *Spanish Journal of Agricultural Research* 7(2):269–280.

Pozzi, W., J. Sheffield, R. Stefanski, et al. 2013. Toward global drought early warning capability: Expanding international cooperation for the development of a framework for monitoring and forecasting. *Bulletin of the American Meteorological Society* 94(6):776–785.

Pulwarty, R. S., and M. V. Sivakumar. 2014. Information systems in a changing climate: Early warnings and drought risk management. *Weather and Climate Extremes* 3:14–21.

Quiroga, S., and A. Iglesias. 2009. A comparison of the climate risks of cereal, citrus, grapevine and olive production in Spain. *Agricultural Systems* 101:91–100.

Reardon, T., and J. E. Taylor. 1996. Agroclimatic shock, income inequality, and poverty: Evidence from Burkina Faso. *World Development* 24(5):901–914.

Riebsame, W. E., S. A. Changnon, and T. Karl. 1991. *Drought and natural resources management in the United States: Impacts and implications of the 1987–1989 drought.* Westview Press, Boulder, CO.

Rose, A., and S. Y. Liao. 2005. Modeling regional economic resilience to disasters: A computable general equilibrium analysis of water service disruptions. *Journal of Regional Science* 45(1):75–112.

Rosegrant, M. W., C. Ringler, and T. Zhu. 2009. Water for agriculture: Maintaining food security under growing scarcity. *Annual Review of Environment and Resources* 34(1):205.

Sadoff, C. W., J. W. Hall, D. Grey, et al. 2015. *Securing water, sustaining growth: Report of the Global Water Partnership/Organisation for Economic Cooperation and Development Task Force on Water Security and Sustainable Growth.* University of Oxford, Oxford, UK.

Salami, H., N. Shahnooshi, and K. J. Thomson. 2009. The economic impacts of drought on the economy of Iran: An integration of linear programming and macroeconometric modelling approaches. *Ecological Economics* 68(4): 1032–1039.

Santos, J. R., S. T. Pagsuyoin, L. C. Herrera, R. R. Tan, and D. Y. Krista. 2014. Analysis of drought risk management strategies using dynamic inoperability input–output modeling and event tree analysis. *Environment Systems and Decisions* 34(4):492–506.

Saurí, D. 2013. Water conservation: Theory and evidence in urban areas of the developed world. *Annual Review of Environment and Resources* 38:227–248.

Shadman, F., S. Sadeghipour, M. Moghavvemi, and R. Saidur. 2016. Drought and energy security in key ASEAN countries. *Renewable and Sustainable Energy Reviews* 53:50–58.

Stahl, K., and S. Demuth. 1999. Linking streamflow drought to the occurrence of atmospheric circulation patterns. *Hydrological Sciences Journal* 44(3):467–482.

Stone, R. C. 2014. Constructing a framework for national drought policy: The way forward—The way Australia developed and implemented the national drought policy. *Weather and Climate Extremes* 3:117–125.

Sun, C., and S. Yang. 2012. Persistent severe drought in southern China during winter–spring 2011: Large-scale circulation patterns and possible impacting factors. *Journal of Geophysical Research: Atmospheres* 117(D10).

Taylor, T., A. Markandya, P. Droogers, and A. Rugumayo. 2015. Economic assessment of the impacts of climate change in Uganda. National Level Assessment: Water Sector report. Climate Change Department, Ministry of Water and Environment, Uganda. Climate & Development Knowledge Network (CDKN), London, UK.

UNDP. 2014. *Understanding community resilience: Findings from Community-Based Resilience Analysis (CoBRA) assessments.* United Nations Development Programme. http://www.undp.org/content/undp/en/home/librarypage/environment-energy/sustainable_land_management/CoBRA/CoBRA_assessment.html (accessed November 25, 2016).

UNEP. 2006. *Geo Year Book 2006: An overview of our changing environment.* United Nations Environment Programme, Nairobi, Kenya.

von Uexkull, N. 2014. Sustained drought, vulnerability and civil conflict in Sub-Saharan Africa. *Political Geography* 43:16–26.

Ward, F. A. 2014. Economic impacts on irrigated agriculture of water conservation programs in drought. *Journal of Hydrology* 508:114–127.

Ward, F. A., J. F. Booker, and A. M. Michelsen. 2006. Integrated economic, hydrologic, and institutional analysis of policy responses to mitigate drought impacts in Rio Grande Basin. *Journal of Water Resources Planning and Management* 132(6):488–502.

Webster, K. E., T. K. Kratz, C. J. Bowser, J. J. Magnuson, and W. J. Rose. 1996. The influence of landscape position on lake chemical responses to drought in northern Wisconsin, USA. *Limnology and Oceanography* 41:977–984.

Wheeler, S. A., A. Loch, and J. Edwards. 2014. The role of water markets in helping irrigators adapt to water scarcity in the Murray–Darling Basin, Australia. In *Applied Studies in Climate Adaptation*, eds. J. P. Palutikof, S. L. Boulter, J. Barnett, and D. Rissik, pp. 166–174. Wiley-Blackwell, Hoboken, NJ.

Wilhite, D. A. 1992. *Preparing for drought: A guidebook for developing countries.* Climate Unit, United Nations Environment Programme, Nairobi, Kenya.

Wilhite, D. A. 1996. A methodology for drought preparedness. *Natural Hazards* 13(3):229–252.

Wilhite, D. A. 2000. *Drought: A global assessment. Natural Hazards and Disasters Series*, Vol. 1. Routledge, London.

Wilhite, D. A. 2005. *Drought and water crises: Science, technology and management issues.* CRC Press, Boca Raton, FL.

Wilhite, D. A., and M. H. Glantz. 1985. Understanding the drought phenomenon: The role of definitions. *Water International* 10:111–120.

Wilhite, D. A., M. J. Hayes, and C. L. Knutson. 2005. Drought preparedness planning: Building institutional capacity. In *Drought and water crises: Science, technology, and management issues*, ed. D. A. Wilhite, pp. 93–135. CRC Press, Boca Raton, FL.

Wilhite, D. A., M. V. Sivakumar, and R. Pulwarty. 2014. Managing drought risk in a changing climate: The role of national drought policy. *Weather and Climate Extremes* 3:4–13.

Willaume, M., A. Rollin, and M. Casagrande. 2014. Farmers in southwestern France think that their arable cropping systems are already adapted to face climate change. *Regional Environmental Change* 14(1):333–345.

WMO and GWP. 2014. *National drought management policy guidelines: A template for action*, ed. D. A. Wilhite. Integrated Drought Management Programme (IDMP) Tools and Guidelines Series 1. World Meteorological Organization, Geneva, Switzerland.

WMO and GWP. 2017. *Benefits of action and costs of inaction: Drought mitigation and preparedness – a literature review*, ed. N. Gerber and A. Mirzabaev. Integrated Drought Management Programme (IDMP) Working Paper 1. WMO, Geneva, Switzerland and GWP, Stockholm, Sweden.

WMO, UNCCD, and FAO. 2013. High Level Meeting on National Drought Policy, Geneva, 11–15 March 2013. Policy Document: National Drought Management Policy. World Meteorological Organization, Geneva, Switzerland.

Wong, G., M. F. Lambert, M. Leonard, and A. V. Metcalfe. 2009. Drought analysis using Trivariate Copulas conditional on climatic states. *Journal of Hydrological Engineering* 15(2):129–141.

World Bank. 2003. *Report on financing rapid onset natural disaster losses in India: A risk management approach. Report No. 26844-IN*. World Bank, Washington, DC.

World Bank. 2013. *Water resources planning and adaptation to climate variability and climate change in selected river basins in Northeast Brazil: Final report on a non-lending technical assistance program (P123869)*. World Bank, Washington, DC.

Wu, J., L. Zhou, X. Mo, H. Zhou, J. Zhang, and R. Jia. 2015. Drought monitoring and analysis in China based on the Integrated Surface Drought Index (ISDI). *International Journal of Applied Earth Observation and Geoinformation* 41:23–33.

Yin, Z.Y., X. Zhang, X. Liu, M. Colella, and X. Chen. 2008. An assessment of the biases of satellite rainfall estimates over the Tibetan Plateau and correction methods based on topographic analysis. *Journal of Hydrometeorology* 9(3):301–326.

Zilberman, D., A. Dinar, N. MacDougall, M. Khanna, C. Brown, and F. Castillo. 2011. Individual and institutional responses to the drought: The case of California agriculture. *Journal of Contemporary Water Research and Education* 121(1):3.

Zinevich, A., P. Alpert, and H. Messer. 2008. Estimation of rainfall fields using commercial microwave communication networks of variable density. *Advances in Water Resources* 31(11):1470–1480.

Section III

Advances in Tools for Drought Prediction, Early Warning, Decision Support, and Management

6

Research to Advance Drought Monitoring and Prediction Capabilities

Jin Huang, Mark D. Svoboda, Andy Wood,
Siegfried Schubert, Christa D. Peters-Lidard, Eric Wood,
Roger Pulwarty, Annarita Mariotti, and Dan Barrie

CONTENTS

6.1 Introduction

Droughts have significant economic and societal impacts. The National Integrated Drought Information System (NIDIS) works to prepare people, communities, businesses, and governments to mitigate the impacts of drought through preparation, improved monitoring and prediction, and building information system networks that extend from the local to the federal level. A critical component in building NIDIS's drought information system is research to (1) advance the scientific understanding of the physical mechanisms that lead to the onset, maintenance, and recovery of drought; (2) improve drought prediction skill; (3) improve current drought monitoring capabilities; and (4) improve drought information systems by incorporating the latest advances in monitoring and prediction, objective metrics relevant to various societal sectors, and advanced information delivery platforms.

In partnership with NIDIS, the National Oceanic and Atmospheric Administration (NOAA) Climate Program Office modeling, analysis, predictions, and projections (MAPP) program established a drought task force (DTF) to address the above research questions. The DTF leverages and contributes to drought research in NOAA research labs and operational centers and across the federal government as part of the U.S. Global Change Research Program and international research programs. This chapter offers an overview of the state of science and practice in Monitoring, Forecasting, and Understanding Droughts. It highlights the research advances in these areas, and the remaining challenges and opportunities. More technical details can be found in the MAPP/DTF *Research to Capability Assessment Report* (Huang et al. 2016), a special collection of scientific papers organized by the MAPP/DTF (Schubert et al. 2015), and in particular a synthesis paper by Wood et al. (2015).

6.2 Research to Improve Drought Monitoring

The US Drought Monitor (USDM) (Svoboda et al. 2002), which has been providing a weekly assessment of drought conditions throughout the United States since 1999, represents the nation's current state-of-the-science operational monitoring capability. A detailed description of the USDM is included in Chapter 7. This section focuses on overarching research efforts aimed at improving drought monitoring capabilities. The goals of drought monitoring research are to develop increasingly accurate, reliable, comprehensive, and high-resolution characterizations of the geophysical variables sensitive to drought through science-based methods, data, and understanding. The following subsections describe research efforts in the context of (1) real-time operational utilization of land surface models (LSMs) that quantitatively and reproducibly depict surface hydrometeorological conditions using operational, real-time meteorological input data, and long-term retrospective hydro-climate system datasets and (2) observational surface analyses based on satellite remote sensing retrievals of drought-relevant parameters. LSM-based drought prediction capabilities will be discussed in Section 6.3.

6.2.1 Land Surface Modeling and Indices

The goal of the North American land data assimilation system (NLDAS) is to construct quality-controlled, spatially, and temporally consistent LSM datasets derived from the best available observations and model output. The NLDAS project commenced in 1999 and has been steadily enhanced primarily through NOAA and NASA research programs (Mitchell et al. 2004). Hosted at the NOAA National Centers for Environmental Prediction (NCEP) Environmental Modeling Center (EMC), NLDAS runs four LSMs at an hourly time-step over the continental United States (CONUS) at a spatial resolution of 0.125 degree resolution (approximately 12 kilometers). The observed meteorological forcing inputs (e.g., precipitation, temperature, humidity, wind speed, and radiation) and land surface model outputs (e.g., soil moisture, snow-water equivalent, evapotranspiration, and river discharge) represent a central thrust of advances in objective drought monitoring and the core component of an effort to advance drought early warning systems. For example, the USDM products currently make use of NOAA's Climate Prediction Center (CPC) soil moisture analysis (Huang et al. 1996), but NLDAS modeling efforts surpass the CPC product in physical realism. Thus the NLDAS data products can now support a finer resolution and higher quality version of the USDM.

Numerous drought products and innovations have emerged from the NLDAS efforts described already, among which are newly derived indices and new objective strategies for integrating indices and multiple sources of information. Examples of real-time systems that apply modern LSMs for drought quantification and prediction include the University of Washington's experimental surface water monitor (Wood 2008); Princeton University's African flood and drought monitor (Sheffield et al. 2014), and the global integrated drought monitoring and prediction system (GIDMaPS). These systems have supported the development of numerous new model-based indices, such as the NCEP objective blended NLDAS drought index (Xia et al. 2014), the multivariate drought severity index (Hao and AghaKouchak 2013), and the standardized runoff index (Shukla and Wood 2008), among others. Compared to the existing USDM, which broadly integrates drought factors, the LSM-based drought indices tend to depict specific drought variables (e.g., soil moisture alone, or a combination of soil moisture and snow). Further, the LSM-based systems can assimilate remotely sensed observations of soil moisture, snowpack, and terrestrial water storage to further improve the holistic assessment of drought conditions (e.g., Houborg et al. 2012; Kumar et al. 2014). These LSM-driven indices also provide objective, quantitative, and reproducible retrospective drought analyses, in contrast to the interpretive approaches behind the USDM. The long and consistent retrospective drought analyses allow for scientific assessment of drought trends and variability and provide drought monitoring information, whereas the current USDM cannot generate this type of information for the United States or around the globe.

6.2.2 Remotely Sensed Observational Analyses

In addition to LSM-derived drought-related analyses, research supported by various agencies (including NOAA) has led to the development of new strategies for using satellite data to monitor droughts (and floods), which can provide an assessment of drought characteristics independent of LSM analyses. Like most current LSMs, the NLDAS models do not include a dynamical vegetation component, and therefore do not capture the reduction in evaporation that can arise from vegetation changes caused by drought (e.g., crop damage or delay).

A key success in this area has been the expansion of near-real time satellite-based analyses that are relevant to drought, particularly those describing vegetation and evapotranspiration. For example, the evaporative stress index (ESI) (Otkin et al. 2013) provides a thermal infrared satellite-based index to estimate evapotranspiration deficits, and may provide complementary information to the NLDAS systems. In addition, these products add to the information resources that can be utilized for characterizing current droughts as part of the USDM. For example, rapid-onset droughts are typically driven by warm air temperatures and low humidity and clear skies, and often with high winds that enhance evaporation and dry soils. The remotely sensed ESI captures these phenomena and can provide an early warning of drought impacts on agricultural systems in some cases, whereas an integrated multivariate drought monitoring system may be slower to depict rapid changes because of inherent lags in some of the component analyses.

6.3 Advancing Drought Prediction Capabilities

The overarching goals of drought prediction research have been to improve our understanding of physical mechanisms of drought, sources of predictability, and the nature and magnitude of unpredictable variability (i.e., noise). Goals also include improving operational drought prediction skill through the full utilization of sources of predictability and the development of improved models, and the observations and data assimilation systems needed to initialize and validate the models. Specifically, research seeks to better understand the physical mechanisms and advance the ability to predict various aspects of drought, including its onset, duration, severity, and recovery. To facilitate progress toward these objectives, the MAPP DTF developed a research framework and the drought capability assessment protocol (Wood et al. 2015), which proposes performance metrics, test cases, and verification datasets to guide individual researchers in testing and evaluating their methods and ideas against the operational or state-of-the-art capabilities. The framework, as originally developed, focuses on the analysis of four major historical drought events over North America to standardize evaluations over particular reference periods. To provide a more general evaluation of prediction skills, the DTF has also embraced the North American multimodel ensemble (NMME) seasonal prediction protocol for evaluation of capabilities over a standard 30-year (1981–2010) period (Kirtman et al. 2014).

Section 6.3.1 describes the current drought prediction capability and pre-diction research advances, and Section 6.3.2 highlights key research results related to drought mechanisms and predictability.

6.3.1 Current Operational and Experimental Prediction Capabilities

The current US seasonal drought outlook (SDO) produced by the CPC relies on forecaster expertise to combine climate and weather forecasts (such as the official CPC temperature and precipitation outlooks, long-lead forecasts from the NCEP's climate forecast system [CFS], and short-term forecasts from NCEP's global forecast system [GFS] and the European Centre for Medium-Range Weather Forecasts) and initial drought conditions depicted by the USDM. This forecast consolidation process produces a map of projected changes in drought severity category from the current USDM. Figures 6.1 and 6.2 compare SDO forecasts for two events (the 2012 drought in the Upper Great Plains and the 2011 Tex-Mex drought, respectively) with a verifying USDM map, showing markedly different performance: the 2012 prediction failed to indicate the upcoming Midwest drought development, whereas the 2011 prediction correctly foresaw the persisting Tex-Mex drought conditions.

The newly developed seasonal NMME climate forecast system affords the potential to analyze major US droughts as well as drought predictions more broadly across a number of coupled global models, which will further our understanding of operational drought prediction capabilities and also sup-port the diagnostic evaluation of forecast uncertainties. The development of NMME was led by NOAA Climate Test Bed (CTB) in 2011 with support from the NOAA Climate Program Office MAPP program, as well as the National Science Foundation, US Department of Energy, and NASA programs and NOAA/National Weather Service. The NMME was transitioned to NCEP operations in 2015. NMME leverages considerable research and development activities that support coupled model prediction systems, and which are car-ried out at universities and various research labs and centers throughout North America. Public access to the NMME hindcast and real-time forecasts provides the community with a great opportunity for research to improve operational drought prediction capabilities. The analyses of the 30-year NMME hindcast showed that, in general, the NMME improves seasonal forecast skill because of an increased size of forecast ensembles and the diversity of models.

The southeastern US precipitation forecast skill of the NMME system typi-cally equals or surpasses that of individual models throughout most seasons and lead times (Kirtman et al. 2014). NMME skill can vary seasonally; for example, the Southeast shows more skill in winter seasons versus summer seasons and NMME is generally able to predict winter season variability based on the 30-year hindcast assessment. During the 2006/07 US drought in the Southeast, the NMME showed moderate precipitation forecast skill at short leads during more extreme seasonal phases of this drought, but a lack

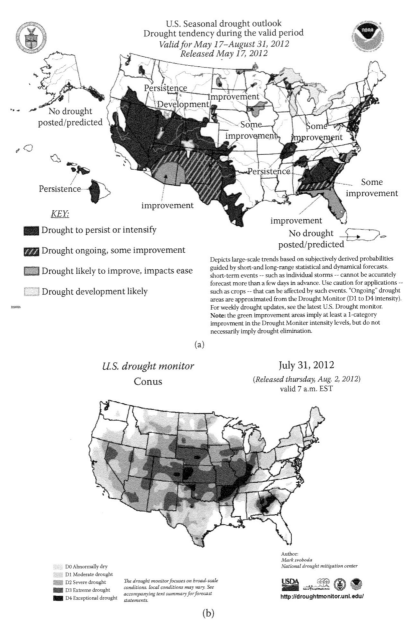

(a)

(b)

FIGURE 6.1
The seasonal drought outlook (a) and US Drought Monitor (b) for summer 2012. The outlook depicts the expected evolution of drought over a 3.5-month period, in this case forecasting from an initial condition of May 17 through the end of August. The outlook communicates where drought is expected to develop, persist, intensify, or improve. The drought monitor provides a categorical view of current drought conditions in the United States by classifying drought into four categories with an additional category depicting drier-than-normal conditions. The significant drought in the central US was not anticipated in the outlook.

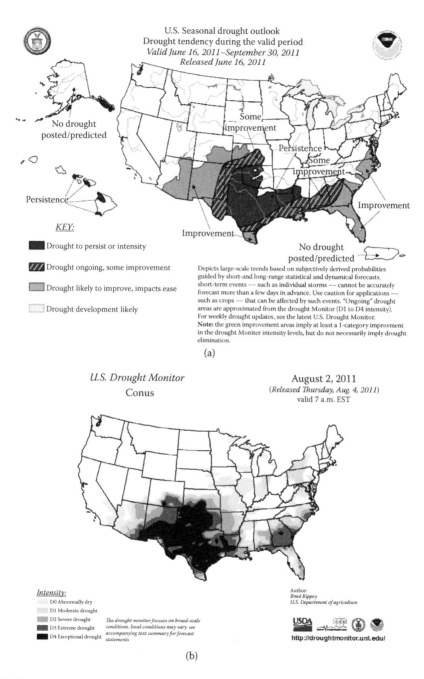

FIGURE 6.2
(a) shows the U.S. seasonal outlook for the period June 16 to September 30, 2011. The lower map (b) shows the U.S. Drought Monitor for August 2, 2012, illustrating the actual pattern of drought conditions in 2011.

of skill at long leads, particularly during the driest phase of the drought. The availability of NMME climate forecasts has also led to the implementation of drought-relevant forecast products derived directly from NMME output, such as the NMME-based Standard Precipitation Index (SPI), which is computed specifically for drought prediction applications at CPC.

Another notable research effort of the last decade has been the development of seasonal hydrological forecast systems linking climate model precipitation and temperature forecasts to uncoupled LSM simulations. A number of previous and current quasi-operational GCM-LSM efforts (e.g., Sheffield et al. 2014; Wood 2008; Wood et al. 2005) have demonstrated skillful soil moisture, snowpack, and runoff predictions at lead times of 1–6 months arising from the persistence land surface moisture anomalies, and the potential benefits of incorporating climate information. Leveraging the frameworks and advances from this line of work, a recent system linking CFSv2 climate forecasts and the VIC LSM (Yuan et al. 2013) was transitioned to NCEP operations through NOAA CTB support, enhancing the suite of continuous and real-time drought information products available to NIDIS drought portals. Yuan et al. (2013) demonstrated that the system, which downscales the current CFSv2 climate forecast to drive the VIC model, yielded better seasonal hydrologic and climate forecasts than running Variable Infiltration Capacity (VIC) using climatological meteorological forcings, although streamflow forecasts using the CFSv2 precipitation as input to the VIC LSM yield limited skill beyond 1 month. El Niño Southern Oscillation (ENSO) conditioning was also found to marginally enhance the CFSv2-VIC forecast skill, confirming the feasibility of skillful soil moisture forecasts out to 6 months in some seasons and locations. Such offline LSM systems, driven by climate model output, have a strong heritage from the interagency NLDAS initiative and are now seen as an important template for a seamless hydrologic drought monitoring and forecasting capability.

6.3.2 Drought Mechanisms and Predictability

DTF research projects have had a long-term focus on improving our understanding of various hydrological and coupled processes (land, ocean, and atmosphere) and how these contribute to the development of drought, and specifically in determining the potential to predict drought. In particular, there has been considerable focus on the more recent 2010–12 period of intense droughts over the United States to explore sources of predictability that can contribute to forecast skill and to learn about predictability limits (in the case of 2012). The key findings are:

- Substantial progress has been made in our understanding and quantification of the role of SSTs in producing drought over North America. Seager et al. (2014) found that La Niña conditions in the tropical Pacific initiated the 2010/11 Tex-Mex drought. By comparing the roles of SSTs in the 2011 and 2012 US droughts, Wang et al. (2014) found that other oceans (Indian and Atlantic) can play an important role in either enhancing or suppressing the role of the Pacific.

- We now have a better appreciation of the role of internal atmospheric variability in producing some of the most extreme droughts, limiting the predictability of such events (e.g., the 2012 upper Great Plains drought) on seasonal and longer time scales.

- Hoerling et al. (2014) also indicated that the 2012 upper Great Plains drought event might be linked to a regime shift toward a warmer and drier summer in the Great Plains as part of natural decadal variability.

- Research has improved our understanding of the role of land surface processes/feedbacks during drought and the potential benefits of higher-resolution precipitation information for streamflow forecasts. Koster et al. (2014) demonstrated that high-resolution precipitation forecasts will only be effective in improving (large-scale) stream-flow forecasts in areas with limited evaporation from land surfaces. Dirmeyer et al. (2014) found that changes in local and remote surface evaporation sources of moisture supplying precipitation over land are more of a factor during droughts than in wet periods over much of the globe.

6.4 Toward Future Progress

The investments in drought-related science, technology, and information systems over the past decade through the efforts of NOAA/NIDIS and others have clearly enhanced and expanded the quality and range of drought products, the number of people engaged in drought-related activities, and our understanding of drought as a phenomenon in the United States. This section discusses remaining challenges and opportunities for future progress in drought monitoring, predictability, predictions, and understanding.

6.4.1 Drought Monitoring

In drought monitoring research, one major success of the last decade is the development of the LSM-based NLDAS and its application to operational drought monitoring. Another success is the expanded use of remotely sensed data in drought monitoring efforts, especially data describing vegetation and evapotranspiration. The USDM authors have historically used research products in a somewhat subjective manner, but there are now many more objective geophysical analyses of different facets of drought (e.g., precipitation, soil moisture, and runoff) available for the official USDM product to take into consideration. A significant challenge is to objectively integrate these myriad inputs into the USDM in a reproducible fashion, especially without undermining consistency with current popular USDM products. There is also a critical need to create a quantitative and structured pathway for developing and testing new monitoring-related research products in the operational USDM process, including benchmarking them against current operational versions.

6.4.2 Drought Prediction

The development of the NMME seasonal climate forecast system is a significant success in demonstrating the potential for forging a collaboration of operational and research groups focusing on both the generation of forecasts and their analysis. Key issues that remain to be resolved are how to optimally combine the multiple model ensembles based on their hindcast skill or conditioned on the phase of teleconnection patterns for major climate variability patterns such as ENSO, the Pacific Decadal Oscillation, or the North Atlantic Oscillation. Promising new research is emerging in the area of hybrid dynamical/statistical approaches to climate prediction—for example, the finding that a skill-weighted statistical combination of dynamical model outputs improves upon simple ensemble averages (e.g., Wanders and Wood 2016).

Such advances have arisen since the operational community recognized the value of reforecasting (or "hindcasting"; Hamill et al. 2005), which enables forecast model post-processing and weighting, and multimodel efforts. New multimodel hindcast archives are becoming available and being tested for both medium range (1–15 days) and subseasonal to seasonal (S2S) predictions, complementing the NMME effort (e.g., as part of the SubX project). However, it is essential to improve the individual dynamical prediction model systems that contribute to systems such as the NMME via continued investments in a number of critical areas: model development; data assimilation; observational networks for model validation, assimilation, and initialization; and high performance computing infrastructure that can allow experimentation at high resolution; and the generation of more sophisticated models and larger ensembles. Ultimately, enhancing the understanding and modeling of predictable phenomena and processes is foundational to making improved predictions.

Important goals for enhancing US drought management capabilities include achieving seamless systems for monitoring and forecasting of drought, and advancing our ability to quantify uncertainties in monitoring and forecasting products. Toward the first objective, NOAA can build on success with the NLDAS and LSM-based hydrologic monitoring and prediction systems, in which the approaches for monitoring and predicting drought-related variables are consistent and integrated. To pursue the second goal, NOAA is working to link the four NLDAS LSMs and the NMME suite of seasonal climate forecasting models within an operational drought information system.

The overall system will provide broader estimates of uncertainty in both current land surface moisture states and future climate forcings, and enhance our probabilistic drought monitoring and prediction capabilities. Although the initial LSM ensemble in this system is small, limiting the depiction of land modeling uncertainty, the framework can be extended to leverage unified modeling concepts that can allow for more comprehensive and deliberate LSM uncertainty quantification (e.g., Clark et al. 2015).

6.4.3 Understanding Drought Predictability

Despite good progress, there are still important limitations on our understanding and ability to predict various aspects of drought, including onset, duration, severity, and recovery. A critical need is to improve precipitation forecasts beyond a 1-month lead, to provide skill beyond lead times at which initial atmospheric and land conditions dominate forecast skill. Improving the prediction of the full life cycle of droughts requires a better understanding of how predictable water and energy signals propagate through the ocean–atmosphere–land system. This in turn should shed light on the necessary model improvements for advancing drought predictions as well as the fundamental predictability limitations imposed on our ability to produce skillful forecasts of the various facets of drought, including precipitation, temperature, soil moisture, snow, and runoff.

Key challenges regarding the potentially predictable signal beyond a 1-month lead time involve isolating the information content of the SST signal (spatially and temporally) to identify what aspects of the SST drives the atmospheric response over North America. Land initialization is another key source of predictability, which is widely accepted as useful at one- to 2-month lead times, but is potentially important for forecasts of longer lead time. Incorporation of LSMs that represent groundwater into systems like NLDAS should provide avenues to explore the potential for longer lead time land initialization impacts. In addition to improving land–atmosphere coupling in climate models, there are uncertainties about the sensitivities to the land models, including how the skill of lead times varies with LSM, and how the skill in both soil moisture and streamflow depends on the model physics. Recent advances in the development of ultra-high-resolution global climate models offer new capabilities for addressing these challenges.

References

Clark, M. P., B. Nijssen, J. D. Lundquist, et al. 2015. A unified approach to hydrologic modeling: Part 1. Model structure. *Water Resources Research* 51(4):2498–2514. doi:10.1002/2015WR017198.

Dirmeyer, P. A., J. Wei, M. G. Bosilovich, and D. M. Mocko. 2014. Comparing evaporative sources of terrestrial precipitation and their extremes in MERRA using relative entropy. *Journal of Hydrometeorology* 15:102–116. doi:10.1175/JHM-D-13-053.1.

Hamill, T. M., J. S. Whitaker, and S. L. Mullen. 2005. Reforecasts, an important dataset for improving weather predictions. *Bulletin of the American Meteorological Society* 87:33–46.

Hao, Z., and A. AghaKouchak. 2013. Multivariate standardized drought index: A parametric multi-index model. *Advances in Water Resources* 57:12–18.

Hoerling, M., J. Eischeid, A. Kumar, et al. 2014. Causes and predictability of the 2012 Great Plains drought. *Bulletin of the American Meteorological Society* 95:269–282. doi:10.1175/BAMS-D-13-00055.1.

Houborg, R., M. Rodell, B. Li, R. Reichle, and B. Zaitchik. 2012. Drought indicators based on model assimilated GRACE terrestrial water storage observations. *Water Resources Research* 48:W07525. doi:10.1029/2011WR011291.

Huang, J., M. Svoboda, A. Wood, et al. 2016. *Research to Advance National Drought Monitoring and Prediction Capabilities. NOAA Drought Task Force report.* http://cpo.noaa.gov/sites/cpo/MAPP/pdf/rtc_report.pdf. Accessed March 1, 2016.

Huang, J., H. M. van den Dool, and K. G. Georgakakos. 1996. Analysis of model-calculated soil moisture over the United States (1931–1993) and applications to long-range temperature forecasts. *Journal of Climate* 9:1350–1362. doi:10.1175/1520-0442(1996)009,1350:AOMCSM.2.0.CO;2.

Kirtman, B. P., D. Min, J. M. Infanti, et al. 2014. The North American Multimodel Ensemble: Phase-1 seasonal-to-interannual prediction; Phase-2 toward developing intraseasonal prediction. *Bulletin of the American Meteorological Society* 95:585–601. doi:10.1175/BAMS-D-12-00050.1.

Koster, R. D., G. K. Walker, S. P. P. Mahanama, and R. H. Reichle. 2014. Soil moisture initialization error and subgrid variability of precipitation in seasonal streamflow forecasting. *Journal of Hydrometeorology* 15:69–88. doi:10.1175/JHM-D-13-050.1.

Kumar, S. V., C. D. Peters-Lidard, D. M. Mocko, et al. 2014. Assimilation of passive microwave-based soil moisture and snow depth retrievals for drought estimation. *Journal of Hydrometeorology* 15(6):2446–2469. doi:10.1175/JHM-D-13-0132.1.

Mitchell, K. E., D. Lohmann, P. R. Houser, et al. 2004. The multi-institution North American Land Data Assimilation System (NLDAS): Utilizing multiple GCIP products and partners in a continental distributed hydrological modeling system. *Journal of Geophysical Research* 109:D07S90. doi:10.1029/2003JD003823.

Otkin, J., M. Anderson, C. Hain, I. Mladenova, J. Basara, and M. Svoboda. 2013. Examining rapid onset drought development using the thermal infrared-based evaporative stress index. *Journal of Hydrometeorology* 14:1057–1074. doi:10.1175/JHM-D-12-0144.1.

Schubert, S., M. Kingtse, and A. Mariotti, eds. 2015. Advancing drought monitoring and prediction. *Journal of Hydrometeorology.* http://journals.ametsoc.org/topic/drought_monitor. Accessed January 12, 2016.

Seager, R., L. Goddard, J. Nakamura, N. Henderson, and D. E. Lee. 2014. Dynamical causes of the 2010/11 Texas–northern Mexico drought. *Journal of Hydrometeorology* 15:39–68. doi:10.1175/JHM-D-13-024.1.

Sheffield, J., E. F. Wood, N. Chaney, et al. 2014. A drought monitoring and forecasting system for sub-Sahara African water resources and food security. *Bulletin of the American Meteorological Society* 95:861–882. doi:10.1175/BAMS-D-12-00124.1.

Shukla, S., and A. W. Wood. 2008. Use of a standardized runoff index for characterizing hydrologic drought. *Geophysical Research Letters* 35:L02405. doi:10.1029/2007GL032487.

Svoboda, M., D. LeComte, M. Hayes, et al. 2002. The drought monitor. *Bulletin of the American Meteorological Society* 83(8):1181–1190.

Wanders, N., and E. F. Wood. 2016. Improved sub-seasonal meteorological forecast skill of extremes using a weighted multi-model ensemble simulation. *Environmental Research Letters* 11:940007. doi:10.1088/1748-9326/11/9/094007.

Wang, H., S. Schubert, R. Koster, Y.-G. Ham, and M. Suarez. 2014. On the role of SST forcing in the 2011 and 2012 extreme U.S. heat and drought: A study in contrasts. *Journal of Hydrometeorology* 15:1255–1273. doi:10.1175/JHM-D-13-069.1.

Wood, A. W. 2008. *The University of Washington Surface Water Monitor: An experimental platform for national hydrologic prediction.* Proceedings of the American Meteorological Society Annual Meeting, New Orleans, LA. http://ams.confex.com/ams/pdfpapers/134844.pdf. Accessed May 1, 2008.

Wood, A. W., A. Kumar, and D. P. Lettenmaier. 2005. A retrospective assessment of NCEP climate model-based ensemble hydrologic forecasting in the western U.S. *Journal of Geophysical Research* 110: D04105.

Wood, E. F., S. D. Schubert, A. W. Wood, et al. 2015. Prospects for advancing drought understanding, monitoring, and prediction. *In Advancing Drought Monitoring and Prediction, a special collection of the Journal of Hydrometeorology* 16(4): 1636–1657. doi:10.1175/JHM-D-14-0164.1.

Xia, Y., M. B. Ek, D. Mocko, et al. 2014. Uncertainties, correlations, and optimal blends of drought indices from the NLDAS multiple land surface model ensemble. *Journal of Hydrometeorology* 15:1636–1650. doi:10.1175/JHM-D-13-058.1.

Yuan, X., E. F. Wood, J. K. Roundy, and M. Pan. 2013. CFSv2-based seasonal hydroclimatic forecasts over the conterminous United States. *Journal of Climate* 26:4828–4847. doi:10.1175/JCLI-D-12-00683.1.

7

Drought Monitoring and Early Warning: Twenty-First Century Advancements and Challenges

Michael J. Hayes, Mark Svoboda, and Kelly T. Redmond

CONTENTS

7.1 Introduction: The Importance of Drought Early Warning

Water crises were identified as the top global risk facing society over the next 10 years according to a recent survey of 750 of the world's leading economists (World Economic Forum 2016). As a normal natural hazard in most climates, drought will compound these crises and play a fundamental direct or indirect role in water stress issues occurring around the world, particularly given that droughts are expected to increase in frequency and intensity as a consequence of climate change (Glotter and Elliott 2016). Water stress issues are prevalent across the United States. For example, increasing growth and development continue to strain water supplies not only for the major metropolitan areas of the arid West but also for metropolitan areas in the relatively humid eastern

United States. Issues surrounding shared water resources across international boundaries, such as the Colorado and Rio Grande river basins between the United States and Mexico, and the Great Lakes and Columbia River basins between the United States and Canada, will also continue to grow. Therefore, because of serious drought impacts on water resource-related issues, planning for and responding effectively to future droughts is critically important within the United States and around the world.

Drought early warning is a key component within a drought risk management approach, which can help planners and decision makers break the hydro-illogical cycle (see Chapter 4). While drought monitoring involves the continuous assessment of the natural indicators of drought severity and spatial extent, drought early warning refers to the use of that information to produce an appropriate and timely response (Hayes et al. 2012). A drought early warning system, which combines both assessment and decision-maker response, provides decision makers accurate early warning information to implement effective drought policies and response and recovery programs. The components of a drought early warning system vary and can be adapted for any region. Generally, these components include an operational drought monitoring network, access to timely data, "value-added" analyses, synthesis, and dissemination of data that can then be used and integrated into decision support tools, communication strategies, and educational efforts (Hayes et al. 2012). Decision makers also benefit from short- and long-term drought forecasting tools that allow them to anticipate and respond to a drought event with better understanding and confidence, and these forecasting tools should be incorporated into a drought early warning system.

One constraint to effective drought early warning has always been the lack of a universally accepted definition for drought. Scientists and decision makers must accept that the search for a single definition of drought is a hopeless exercise. Drought definitions must be specific to the region, application, or impact. Drought must be characterized by many different climate and water supply indicators, and an effective early warning system must build on these indicators (see also Chapter 10). Impacts are complex and vary regionally, and at a variety of timescales. Drought monitoring indicators, ideally, should be tied directly to triggers that assist decision makers with timely and effective responses both before and during drought events.

Recent widespread and severe droughts resulting in serious economic, social, and environmental impacts in many countries highlight the need for continual improvement in drought early warning systems. In the United States, these droughts have fostered development of improved drought monitoring data, decision support tools, and collaborations between scientists. This chapter discusses some of these new developments, as well as the current status of drought forecasting in the United States. The chapter also provides an opportunity to recognize one of the true heroes of drought early warning, author Kelly Redmond. Dr. Redmond passed away on

November 3, 2016, following a very distinguished climatology career. On the topic of drought early warning, Dr. Redmond provided tremendous leadership, guidance, and wisdom. He was known for his ability to verbalize complex issues in simple and witty terms, and one of his favorite quotes directly applies to drought early warning: "An ounce of observation is worth a pound of forecasts" (Redmond 2014). As the quote suggests, Dr. Redmond was an emphatic proponent of the value of drought monitoring systems within the context of early warning.

7.2 Recent Advancements

Tremendous progress in drought early warning has taken place since the first edition of *Drought and Water Crises: Science, Technology, and Management Issues* was released in 2005. At that time, the US drought monitor (USDM) product was relatively new and was just beginning to be used as a decision-making tool. The famine early warning systems network (FEWS NET) is another example of an early warning system targeted at addressing food security issues for specific locations around the world, and drought was, and still is, an important component within this system. More drought indicators and indices were being developed, but relatively few were available for people to use or access. As illustrated in Chapter 8 of this book, there are now more than 50 indicators and indices available for use by decision makers. This section of the chapter reviews some of the recent advancements in drought monitoring and early warning that have taken place since that first edition.

7.2.1 US Drought Monitor

One of the tools highlighted in the first edition of this book was the USDM (Svoboda et al. 2002). The USDM product has now been produced and released every week since it became operational as a weekly assessment of drought conditions in August 1999 (http://drought.unl.edu/dm). The consistency and reliability of the product has led to it becoming the "state-of-the-science" for drought monitoring in the United States; it is a major tool for decision-making by resource managers and policy makers, a communication tool for the media, and a resource for teachers at all levels of educational instruction.

Several of the fundamental characteristics originating when the USDM process began still apply today and have likely helped make the process as successful as it is. Authorship of the map rotates between the National Drought Mitigation Center (NDMC) at the University of Nebraska–Lincoln, the US Department of Agriculture (USDA), and the National Oceanic and Atmospheric Administration (NOAA). Within NOAA, authors specifically come from the Climate Prediction Center, National Centers for Environmental Information, and Western Regional Climate Center. In addition, the USDM incorporates information from approximately 420 scientists and local experts around the country. This number was close to 150 in 2005, highlighting the growth in collaboration and awareness that has taken place since then. The USDM continues to seek corroborative monitoring and impact data, and information from this group of participants in order to provide added confidence in the initial assessments gained from purely quantitative information describing the physical environment. This kind of "ground truth" is important, and it increases broad-based credibility and trust in the product.

The USDM is not a forecast; rather, it was designed to be a comprehensive drought assessment that reflects the current drought situation (i.e., snapshot) across the country. Because multiple physical conditions may be present at one time and no preferred scale exists for assessing drought, the USDM also depends on, incorporates, and weights human expertise and judgment in the assessment of the associated impacts.

A key strength of the USDM product is that it is based on multiple indicators. One indicator is not adequate to represent the complex characteristics of drought across a region. Therefore, it is important for a product like the USDM to use a variety of quantitative and qualitative indicators. The key indicators used in creating the weekly USDM map include streamflow, measures of recent precipitation, drought indices, remotely sensed products, and modeled soil moisture. Many other ancillary indicators are also used, depending on the region and the season. For example, in the western United States, indicators such as snow water content, reservoir information, and water supply indices are important for evaluating the current and future availability of water. These indicators inherently incorporate the effects of hydrological lag and relationships across space and time between climate and the surface or groundwater system.

The USDM defines four categories of drought severity based on increasing intensity (D1–D4), with a fifth category (D0) indicating abnormally dry areas (possible emerging drought conditions or an area that is recovering from drought but may still be seeing lingering impacts). The drought categories represented by this scale are moderate (D1), severe (D2), extreme (D3), and exceptional (D4). Another of its strengths is that the five categories are based on a percentile approach, where D0 is approximately equal to the 30th percentile; D1, the 20th; D2, the 10th; D3, the 5th; and D4, the 2nd (Svoboda et al. 2002).

Recent improvements to the USDM have focused on providing value-added products and tools for assistance in decision-making. For example, USDM maps and weekly statistics for various regions, states, tribal reservations, and river basins are now available. A user can incorporate census data and make an approximate estimate of the number of people affected by various drought categories for each of these regions. Change maps (i.e., a comparison to the prior week) and animations (i.e., multiple weeks) are also available to the public, as well as all of the shape files related to the map each week, thus providing an opportunity for researchers and decision makers to tailor the information to their needs. Improvements like these continue to be made as the data availability and technology evolve.

As a tool for decision-making, the USDM provides a great example of how science can motivate policymaking. The USDM was first formally incorporated into the 2008 US Farm Bill for several livestock-related drought relief programs. The 2014 Farm Bill expanded the USDM's use for agricultural drought relief programs, and it is also a trigger to assist in fast-tracking USDA Secretarial Drought Disaster designations. Other federal agencies that use the USDM for decisions include the Internal Review Service, the National Weather Service, the Environmental Protection Agency, the Centers for Disease Control and Prevention, and the Bureau of Land Management. Multiple states and regional or local organizations also use the USDM for triggering various activities or as an information source.

Building upon the USDM experience in the United States, other nations have either experimented with or adapted the USDM process for drought early warning in their countries. Brazil, Mexico, and the Czech Republic also have operational USDM-like tools for drought early warning. The monthly North American Drought Monitor (NADM) continues to be produced by drought scientists in Canada, Mexico, and the United States (http://www.ncdc.noaa.gov/oa/climate/monitoring/drought/nadm/). The development of the NADM represented an important step in a cooperative, multinational effort to improve monitoring and assessment of climate extremes throughout the continent (Lawrimore et al. 2002).

7.2.2 NIDIS Regional Drought Early Warning Systems

When the US Congress passed the NIDIS Act in 2006 (Public Law 118-36), the goal of the national integrated drought information system (NIDIS) was to enable the nation to move toward a more proactive drought risk management approach. The act had three main objectives related to drought early warning: (1) provide effective drought early warning systems reflective of local, regional, and state differences; (2) coordinate and integrate, as practicable, federal research in support of a drought early warning system; and (3) build upon existing forecasting and assessment programs and partnerships. To accomplish these objectives, NIDIS has been tasked with coordinating efforts to improve drought early warning. NIDIS is led by NOAA, but its governance structure and various working groups include representatives of other federal, state, tribal, local, and regional agencies, as well as representatives of academic and private entities.

In order to establish drought early warning across the United States, NIDIS has developed a network of regional drought early warning systems

(RDEWS) to focus on regional issues, sectors, and stakeholders most affected by droughts in that region. Descriptions and a map of the existing RDEWS are highlighted in Chapter 15, which provides more details about the Missouri River Basin RDEWS.

7.2.3 Drought Indicators and Indices

One aspect of the substantial progress in drought monitoring in recent decades has been the ability to measure drought severity through a variety of drought indicators and indices. As highlighted in Chapter 8 and the World Meteorological Organization/Global Water Partnership (2016) *Indicator Handbook*, approaches to drought monitoring can consist of (1) a single indicator or index, (2) an approach incorporating multiple indicators or indices, or (3) an approach that uses a composite of indicators or indices. A drought indicator is a variable or parameter used to measure and track changes in various components of the hydrological cycle (e.g., precipitation, temperature, streamflow, and soil moisture), derived primarily from point-based, *in situ* observations. A drought index, however, is a calculated representation of a condition, with the Palmer Drought Severity Index and the Standardized Precipitation Index being two more commonly used drought indices. A composite indicator combines multiple indicators and indices, and the USDM is a great example of a composite indicator.

Indicators and indices are often reflective of a particular disciplinary perspective such as agricultural, hydrological, or ecological conditions. Composite indicators, however, often cover multiple disciplinary perspectives. As described in Chapter 8, no one indicator or index is going to describe everything related to drought. Therefore, decision makers might have to look for the appropriate option or options available to provide the most relevant information. Thus, the WMO and GWP (2016) guide is a great starting point for decision makers establishing and maintaining early warning systems. One of the advantages of a composite drought indicator is that it potentially simplifies the options for decision makers, who can often be confused by the variety of indicators and indices—and their corresponding characteristics—that are available (Hayes et al. 2012; Mizzell 2008).

7.2.4 Remote Sensing

Remote sensing applications offer unique opportunities for augmenting and/or improving drought monitoring efforts that complement the traditionally used climatological and hydrological drought indicators and indices. Satellite-derived remote sensing information is particularly useful in assisting with drought monitoring over larger spatial scales. Satellites provide synoptic, repeat coverage of spatially continuous information in a consistent, systematic, and objective manner (Hayes et al. 2012). This information can supplement or simulate data from regions by utilizing existing observation networks having abundant ground-based data, or where ground-based observational networks and monitoring data are sparse.

During the past decade, new satellite-based instruments and major advancements in computing, analyses, and modeling techniques have resulted in the rapid development of many remote sensing tools and products with drought monitoring applications (Hayes et al. 2012). These new tools and products, described in more detail in Chapter 10 of this book, cover a suite of environmental variables that are useful in drought monitoring, including vegetation health, precipitation, evapotranspiration, soil moisture, terrestrial water resources, and snow cover.

The specific advantages satellite remote sensing can provide within a drought early warning system, as described by Hayes et al. (2012, 5), include:

1. Provide information at spatial scales required for local-scale drought monitoring and decision-making, that cannot be adequately supported from information derived from traditional, point-based data sources (e.g., single area-based value over administrative geographic unit or spatially interpolated climate index grids).

2. Fill in informational gaps on drought conditions for locations between *in situ* observations and in areas that lack (or have very sparse) ground-based observational networks.

3. Enable earlier drought detection in comparison to traditional climatic indices.

4. Collectively provide a suite of tools and data sets geared to meet the observational needs (e.g., spatial scale, update frequency, and data type) for a broad range of decision support activities related to drought.

7.2.5 Drought Forecasting

The science of drought forecasting was described as being in its infancy in the chapter on drought monitoring in the 2005 edition of *Drought and Water Crises: Science, Technology, and Management Issues*. Since then, the need for decision makers to have accurate drought forecast information and tools to determine future conditions remains, and may be as important as the assessments of the current conditions that come from drought monitoring. To forecast drought, it is important to know something about the causes of drought. Drought is usually established by persisting high pressure that results in dryness because of subsidence of air, more sunshine and evaporation, and the deflection of precipitation-bearing storms. This is usually part of a persistent large-scale disruption in the global circulation pattern. Scientists have continued to look for local or distant influences that might create such atmospheric-blocking patterns.

Recent progress on drought prediction or outlooks, described in more detail in Chapter 6 of this book, now connects phenomena including SST anomalies (e.g., the El Niño southern oscillation, Pacific decadal oscillation, and Atlantic multidecadal oscillation), global-scale atmospheric changes (e.g., planetary waves, Hadley cell, and Walker circulations), and regional forcing and land feedbacks (e.g., changes in soil moisture, snow, dust, vegetation, and low level jets) with the potential development of drought conditions. There has also been more research on how droughts might end, such as through events known now as atmospheric rivers.

Multiple efforts are taking place to improve the understanding of drought outlooks, as well as the communication of outlooks with various stakeholder groups. NIDIS has been able to support research within the National Centers for Environmental Prediction's climate test beds and the modeling, analysis, predictions, and projections (MAPP) program established by NOAA's Climate Program Office beginning in 2011 (Chapter 6). NIDIS has also either funded or supported multiple regional climate forums around the country that provide stakeholders the opportunity to interact with the drought monitoring and drought forecasting communities.

7.3 Challenges

Even given the progress on drought early warning described, multiple challenges remain, especially related to how decision makers can use the monitoring and early warning information. For example, regardless of whether a region is data rich or data sparse, users often request that the early warning data and corresponding information be provided at improved temporal and spatial scales. Therefore, the improved resolution for decision-making is user-defined and is a major challenge given the various spatial scales of need. For example, agricultural producers may demand field level information and products while water supply managers may be more interested in basin-scale data and products. In addition, the information delivered to users is often too technical or complex so that its use by decision makers and the public is limited. Another challenge is that data and information sharing is often poor within and between government agencies and ministries, as well as between countries and regions. This section addresses several additional issues that remain challenges within drought monitoring and early warning efforts.

7.3.1 Drought Impacts

Impact assessment is one component within drought early warning systems that is frequently limited in scope or often forgotten. It is a key component because understanding impacts connects the monitoring of drought severity with decision-making and appropriate drought-related responses, including planning. Impacts also provide important clues for understanding vulnerabilities critical for targeting drought mitigation strategies designed to reduce future impacts. Monitoring impacts has been a challenge and is frequently overlooked because, typically, there have been no standard methods for reporting, distributing, or archiving drought impact data. Thus, the availability of quality impact data required for effective drought risk management often does not exist. Ultimately, a comprehensive drought early warning and information system should include drought impact collection as one of its key components and activities as a way of establishing an impact baseline.

In 2005, the NDMC started an operational tool called the drought impact reporter (DIR) (http://droughtreporter.unl.edu). This tool continues to build a drought impact database, or archive, that includes an interactive, Web-based mapping tool designed to display impact information across the United States gathered from a variety of sources such as media, government agencies, and the public. At present, the archive contains more than 42,000 reports and 21,000 impacts. This near real-time information within the DIR helps decision makers (i.e., policymakers and resource managers) identify and quantify the occurrence, severity, and types of impacts in order to help them understand the connection between drought severity and drought impacts (e.g., risk and vulnerability). Their management actions can be more efficient and timely if they can anticipate the impacts that may need to be addressed as drought severity increases and dissipates during a drought event. For example, impacts will often linger well beyond the time when climate indicators have returned to normal, and if the impact information is being collected, officials can be aware of these impacts.

The DIR highlights several additional challenges regarding drought impact collection. One of these challenges has been identifying and validating sources of information, whether this information comes from media reports or from user input information collected from the public. These are unique sources that, particularly in the case of public reports, may need additional verification or review. Another challenge relates to the value of qualitative versus quantitative information. Most drought impact information is qualitative by nature. Qualitative information is valuable, but quantitative information is encouraged because it provides an opportunity to compare impacts with current drought severity levels, other locations, or past and future impacts (see also Chapter 5 on the costs of action vs. inaction in association with drought preparedness). Also related is the challenge of what to do with positive impacts associated with droughts and how these positive impacts become categorized in a database. It is important to recognize that there can be winners and losers when it comes to the impacts of drought, and these can vary by region and season. Just as it is with droughts, timing is a critical component in determining the impacts that may occur for a particular region.

7.3.2 Drought Triggers

The chapter by Steinemann, Hayes, and Cavalcanti in the first edition of *Drought and Water Crises: Science, Technology, and Management Issues* covered the topic of drought triggers well, and that particular chapter remains one of the best descriptions of triggers available. As defined in Chapter 8, triggers are "specific values of an indicator or index that initiate and/or terminate" responses or management actions by decision makers based upon existing guidelines or plans. Triggers remain a challenge within drought early warning information because, as already described, the link between drought severity and impact levels remains difficult to quantify. As more information about drought severity and impacts becomes available, as well as the linkage between the two, the development of triggers will become less challenging. Triggers must be tailored to the local context, and as local characteristics change, triggers must also be adaptable to adjust to these changing vulnerabilities, for better or worse, as mitigation actions can ultimately help reduce risk to future droughts. Decision makers are slowly adopting more triggers into management actions. The USDA, for example, uses the USDM product as a trigger for multiple drought disaster relief programs for agricultural producers.

7.3.3 Connecting Drought Early Warning with Drought Risk Management

One of the challenges that officials struggle with is the concept of connecting drought early warning with the other aspects of drought risk management (as illustrated in Figure 4.1, Chapter 4). Oftentimes, these efforts of drought monitoring and drought management take place in isolation from one another. In and of itself, information from a drought early warning system generally provides limited benefits. The key is to integrate the early warning information with risk management. If integration is successful, a feedback loop becomes established involving the drought early warning and the risk management strategies. As better drought management occurs, it drives the need for improved drought early warning information at higher spatial and temporal scales. Similarly, improved drought early warning encourages more effective drought management and the incorporation of drought early warning information into management actions (Hayes et al. 2012). Again, the USDM provides a great example of this looped evolution in drought early warning and risk management in the United States. Improvements in the USDM product have led to shifts in national agricultural policies, leading to additional advancements in the available drought monitoring tools and information utilized in supporting implementation of these policies at a local scale.

7.3.4 Climate Change and Drought Impacts

Most expectations are that climate change will, in general, increase the frequency and severity of droughts worldwide (Kundzewicz et al. 2007; Meehl et al. 2007). However, the specific effects of climate change on regional and local droughts provide another challenge for drought early warning systems for several reasons (Hayes et al. 2011). First, Milly et al. (2008) highlighted how the past climate may not represent the best analog for the future. Second, recent climate trends do not necessarily reflect future projections. Third, future drought projections are going to reflect, in part, the projections being made for both temperature and precipitation (see Chapter 11). While temperature projections are more uniform and understood, the projections of precipitation are not as uniform and have higher uncertainty on both spatial and temporal scales.

Drought early warning systems will ultimately have to be able to account for how the local and regional characteristics of the hydrological cycle will be affected by climate change. This will have major implications for sectors susceptible to drought, including agriculture and water supply and management. Both rain-fed and irrigated agriculture could see drought impacts on production, particularly because of increased water deficits during summer growing months, even if they actually receive more precipitation than they do at present. The reason for this, of course, would be due to increased temperature, increased evapotranspiration, and possibly more days in between precipitation events. Projected reductions in general runoff, and in the runoff generated by snowpack and glaciers, would reduce water availability for the agricultural sector in areas where these reductions occur, resulting in greater vulnerability to drought impacts on agriculture (Backlund et al. 2008; Kundzewicz et al. 2007; Meehl et al. 2007). The overall global impact on agriculture is likely to be extremely variable and dependent on factors such as the local environmental and socioeconomic conditions (Eitzinger et al. 2009).

7.4 Conclusion

As drought early warning information systems evolve around the world, the demand for consistent, high-quality observations, datasets, decision tools, and value-added products and information in support of applications across a range of spatial scales (i.e., local, national, regional, and global) will continue to increase. To meet this demand, traditional climate data, in combination with new technologies such as remote sensing tools, should provide a more complete and accurate depiction of current drought conditions for decision makers. In the United States, the USDM has been a great catalyst for improving drought monitoring strategies, incorporating drought impact information, and connecting early warning with drought risk management. Although many challenges remain, the pace of progress points to continued optimism that drought early warning information systems will continue to improve well into the twenty-first century.

References

Backlund, P., A. Janetos, D. S. Schimel, et al. 2008. Executive summary. *The Effects of Climate Change on Agriculture, Land Resources, Water Resources, and Biodiversity.* Report by the U.S. Climate Change Science Program and Subcommittee on Global Change Research, Washington, DC. https://www.usda.gov/oce/climate_change/SAP4_3/CCSPFinalReport.pdf (accessed February 6, 2017).

Eitzinger, J., G. Kubu, and S. Thaler. 2009. Climate change impacts and adaptation options for agriculture in complex terrain and small scale agricultural systems. Results for case studies in Austria. Extended Abstracts of the International Symposium, University of Natural Resources and Applied Life Sciences (BOKU), Vienna, 22–23 June. *BOKU-Met Report* 17:9–12.

Glotter, M., and J. Elliott. 2016. Simulating US agriculture in a modern Dust Bowl drought. *Nature Plants* 3:1–6. doi:10.1038/nplants.2016.193.

Hayes, M. J., M. D. Svoboda, B. D. Wardlow, M. C. Anderson, and F. Kogan. 2012. Drought monitoring: Historical and current perspectives. In *Remote Sensing of Drought: Innovative Monitoring Approaches*, eds. B. Wardlow, M. Anderson, and J. Verdin, 1–19. Boca Raton, FL: Taylor and Francis.

Hayes, M., D. Wilhite, M. Svoboda, and M. Trnka. 2011. Investigating the connections between climate change, drought, and agricultural production. In *Handbook on Climate Change and Agriculture*, eds. R. Mendelsohn and A. Dinar, pp. 73–86. Cheltenham, United Kingdom: Edward Elgar Publishing Ltd.

Kundzewicz, Z. W., L. J. Mata, N. W. Arnell, et al. 2007. Freshwater resources and their management. In *Climate Change 2007: Impacts, Adaptation and Vulnerability*, eds. M. L. Parry, O. F. Canziani, J. P. Palutikof, P. J. van der Linden, and C. E. Hanson, pp. 173–210. Contribution of Working Group II to the Fourth Assessment Report of the Intergovernmental Panel on Climate Change. Cambridge: Cambridge University Press.

Lawrimore, J., R. R. Heim, Jr., M. Svoboda, V. Swail, and P. J. Englehart. 2002. Beginning a new era of drought monitoring across North America. *Bulletin of the American Meteorological Society* 83:1191–1192.

Meehl, G. A., T. F. Stocker, W. D. Collins, et al. 2007. Global climate projections. In *Climate Change 2007: The Physical Science Basis*, eds. S. Solomon, D. Qin, M. Manning, et al. Contribution of Working Group I to the Fourth Assessment Report of the Intergovernmental Panel on Climate Change, pp. 747–846. Cambridge: Cambridge University Press.

Milly, P. C. D., J. Betancourt, M. Falkenmark, et al. 2008. Stationarity is dead: Whither water management? *Science* 319(5863):573–574. doi: 10.1126/science.1151915.

Mizzell, H. 2008. Improving drought detection in the Carolinas: Evaluation of local, state, and federal drought indicators. PhD diss., Department of Geography-Climatology, University of South Carolina.

Redmond, K. 2014. A perspective on accelerating change. Tyndall Lecture, American Geophysical Union (AGU) Meeting, San Francisco, CA, December 13, 2014. https://www.youtube.com/watch?v=SjSwPMvBsCM.

Svoboda, M., D. Le Comte, M. Hayes, et al. 2002. The drought monitor. *Bulletin of the American Meteorological Society* 83:1181–1189.

World Economic Forum. 2016. *The Global Risks Report 2016*, 11th Edition, Geneva. http://www3.weforum.org/docs/GRR/WEF_GRR16.pdf (accessed January 30, 2017).

World Meteorological Organization (WMO) and Global Water Partnership (GWP). 2016. *Handbook of Drought Indicators and Indices* (M. Svoboda and B.A. Fuchs eds.), Integrated Drought Management Programme, Integrated Drought Management Tools and Guidelines Series 2., WMO-No. 1173. Geneva, Switzerland: WMO.

8

Handbook of Drought Indicators and Indices*

Mark D. Svoboda and Brian A. Fuchs

CONTENTS

8.1 Introduction

Why is it important to monitor droughts? Droughts are a normal part of the climate, and they can occur in any climate regime around the world, even deserts and rainforests. Droughts are one of the more costly natural hazards on a year-to-year basis; their impacts are significant and widespread, affecting many economic sectors and people at any one time. The hazard footprints of (areas affected by) droughts are typically larger than those for other hazards, which are usually constrained to floodplains, coastal regions, storm tracks,

---

* Published with permission from the World Meteorological Organization (WMO) and Global Water Partnership (GWP). M. Svoboda and B.A. Fuchs. (2016). *Handbook of Drought Indicators and Indices*. Integrated Drought Management Programme (IDMP), Integrated Drought Management Tools and Guidelines Series 2. WMO: Geneva, Switzerland; GWP: Stockholm, Sweden.

or fault zones. Perhaps no other hazard lends itself quite so well to monitoring, because the slow onset of droughts allows time to observe changes in precipitation, temperature, and the overall status of surface water and groundwater supplies in a region. Drought indicators, or indices, are often used to help track droughts and these tools can vary depending on the region and the season.

Like other hazards, droughts can be characterized in terms of their severity, location, duration, and timing. Droughts can arise from a range of hydrometeorological processes that suppress precipitation and/or limit surface water or groundwater availability, creating conditions that are significantly drier than normal or otherwise limiting moisture availability to a potentially damaging extent. The indicators and indices discussed in this *Handbook of Drought Indicators and Indices* provide options for identifying the severity, location, duration onset, and cessation of such conditions. It is important to note that the impacts of droughts can be as varied as the causes of droughts. Droughts can adversely affect agriculture and food security, hydropower generation and industry, human and animal health, livelihood security, personal security (e.g., women walking long distances to fetch water) and access to education (e.g., girls not attending school because of increased time spent on fetching water). Such impacts depend on the socioeconomic contexts in which droughts occur in terms of who, or what, are exposed to the droughts and the specific vulnerabilities of the exposed entities. Therefore, the type of impacts relevant in a particular drought monitoring and early warning context is often a crucial consideration in determining the selection of drought indicators.

A drought impact is an observable loss or change at a specific time because of drought. Drought risk management involves hazards, exposure, vulnerability and impact assessment, a drought early warning system (DEWS) (monitoring and forecasting, see Box 8.1), and preparedness and mitigation (WMO et al. 2013). It is important that drought indicators or indices accurately reflect and represent the impacts being experienced during droughts. As droughts evolve, the impacts can vary by region and by season.

BOX 8.1   DROUGHT EARLY WARNING SYSTEMS

Drought early warning systems typically aim to track, assess, and deliver relevant information concerning climatic, hydrologic, and water supply conditions and trends. Ideally, they have both a monitoring (including impacts) component and a forecasting component. The objective is to provide timely information in advance of, or during, the early onset of drought to prompt action (via threshold triggers) within a drought risk management plan as a means of reducing potential impacts. A diligent, integrated approach is vital for monitoring such a slow-onset hazard.

Monitoring different aspects of the hydrologic cycle may require a variety of indicators and indices. It is desirable to align these and their depiction with the impacts of emerging conditions on the ground and management decisions being taken by different individuals, groups, and organizations. Although a DEWS is ultimately concerned with impacts, drought impact assessment is a large gap in many DEWSs used around the globe at this time. Assessment of impacts is complicated, as socioeconomic factors other than the physical nature of droughts influence the levels and types of impacts related to drought exposure and vulnerability.

Understanding how droughts affect people, communities, businesses, or economic sectors is key to taking steps toward mitigating the impacts of future droughts.

Following publication of the *Intergovernmental Panel on Climate Change* report on extreme events (IPCC 2012), the issue of quantifying loss and damage from extreme climate events such as droughts has become important for policy implementation, especially with regard to the United Nations Framework Convention on Climate Change agenda. In addition, due to the magnitude of associated disaster losses, improved drought monitoring and management will be fundamental to implementing the Sendai Framework for Disaster Risk Reduction 2015–2030 and the sustainable development goals. Effective and accurate monitoring of hydrometeorological indicators is a key input to risk identification, to DEWSs, and for managing sector impacts. In light of this, the 17th World Meteorological Congress, held in June 2015, adopted Resolution 9: *Identifiers for Cataloguing Extreme Weather, Water and Climate Events*. This initiated a process of standardizing weather, water, climate, space weather and other related environmental hazards and risk information, and prioritized the development of identifiers for cataloging extreme weather, water, and climate events. This handbook will make an important contribution to these efforts.

The purpose of this handbook is to cover some of the most commonly used drought indicators/indices that are being applied across drought-prone regions, with the goal of advancing monitoring, early warning, and information delivery systems in support of risk-based drought management policies and preparedness plans. These concepts and indicators/indices are outlined in what is considered to be a living document that will evolve and integrate new indicators and indices as they become known and are applied in the future. The handbook is aimed at those who want to generate indicators and indices themselves, as well as for those who simply want to obtain and use products that are generated elsewhere. It is intended for use by general drought practitioners (e.g., meteorological/hydrological services and ministries, resource managers, and other decision makers at various levels) and aims to serve as a starting point, showing which indicators/indices are available and being put into practice around the world. In addition, the handbook has been designed with drought risk management processes in mind. However, this publication does not aim to recommend a "best" set of indicators and indices. The choice of indicators/indices is based on the specific characteristics of droughts most closely associated with the impacts of concern to the stakeholders.

This handbook does not attempt to address the full complexities of impacts and the entire range of socioeconomic drought indicators and indices. The indicators and indices included describe the hydrometeorological characteristics of droughts and do not cover socioeconomic and environmental factors such as those that may be needed to assess and anticipate drought-related impacts and outcomes. The handbook is intended as a reference, providing an overview and guide to other sources of information. The Integrated Drought Management Programme (IDMP) is establishing a complementary help desk on integrated drought management.

8.2 Definitions: Indicators versus Indices

It is important to define what is meant by drought indicators and indices.

Indicators are variables or parameters used to describe drought conditions. Examples include precipitation, temperature, streamflow, groundwater and reservoir levels, soil moisture, and snowpack.

Indices are typically computed numerical representations of drought severity, assessed using climatic or hydrometeorological inputs including the indicators listed above. They aim to measure the qualitative state of droughts on the landscape for a given time period. Indices are technically indicators as well. Monitoring the climate at various timescales allows identification of short-term wet periods within long-term droughts or short-term dry spells within long-term wet periods. Indices can simplify complex relationships and provide useful communication tools for diverse audiences and users, including the public. Indices are used to provide quantitative assessment of the severity, location, timing, and duration of drought events. Severity refers to the departure from normal of an index. A threshold for severity may be set to determine when a drought has begun, when it ends, and the geographic area affected. Location refers to the geographic area experiencing drought conditions. The timing and duration are determined by the approximate dates of onset and cessation. The interaction of the hazard event and the exposed elements (people, agricultural areas, reservoirs, and water supplies), and the vulnerabilities of these elements to droughts, determines the impacts. Vulnerabilities may have been exacerbated by previous droughts, which, for example, might have triggered the sale of productive assets to meet immediate needs. The timing of droughts may be as significant as their severity in determining impacts and outcomes. A short, relatively low severity, intraseason drought, if it occurs during the moisture sensitive period of a stable crop, can have a more devastating impact on crop yield than a longer, more severe drought occurring at a less critical time during the agricultural cycle. Thus, drought indices—in combination with additional information on exposed assets and their vulnerability characteristics—are essential for tracking and anticipating drought-related impacts and outcomes. Indices may also play another critical role, depending on the index, in that they can provide a historical reference for planners or decision makers. This provides users with a probability of occurrence, or recurrence, of droughts of varying severities. Importantly, however, climate change will begin to alter historical patterns.

Information derived from indicators and indices is useful in planning and designing applications (such as risk assessment, DEWSs and decision support tools for managing risks in drought- affected sectors), provided that the climate regime and drought climatology is known for the location. In addition, various indictors and indices can be used to validate modeled, assimilated, or remotely sensed indicators of drought.

8.3 Approaches for Monitoring Drought and Guiding Early Warning and Assessment

There are three main methods for monitoring drought and guiding early warning and assessment:

1. Using a single indicator or index
2. Using multiple indicators or indices
3. Using composite or hybrid indicators

In the past, decision makers and scientists employed one indicator or index because that was the only measurement available to them, or they had only limited time in which to acquire data and compute derivative indices or other deliverables. Over the past 20 years or so, there has been strong global interest and growth in the development of new indices based on various indicators that are suitable for different applications and scales, both spatial and temporal. These new tools have given decision makers and policymakers more choices, but, until recently, they have still lacked a clear-cut method to synthesize results into a simple message that can be relayed to the public. The advent of geographic information systems and increasing computing and display capabilities has increased the capacity to overlay, map, and compare various indicators or indices. For a more detailed discussion on mapping drought indices and indicators, see the *Standardized Precipitation Index User Guide* (WMO 2012).

Confusion can arise when trying to determine which indicators or indices to use, especially if they are linked to a comprehensive drought plan and used as a trigger for drought management actions. It takes time and a system of trial and error to determine the best fit for any given location, area, basin, or region. In the past decade or so, a new type of composite (sometimes referred to as hybrid) indicator has emerged as a means to merge different indicators and indices, either weighted or not, or in a modeled fashion. The idea is to use the strengths of a variety of inputs, yet maintain a single, simple source of information for decision makers, policymakers, or the public. Given that drought severity is best evaluated on the basis of multiple indicators associated with water availability for a given area or region, the composite or hybrid approach allows an increased number of elements to be incorporated into the assessment process.

While this handbook does not aim to state exactly which indicators or indices to integrate or apply in terms of drought management guidance, it is important to note the role of indices and indicators in a DEWS within an overall drought risk management strategy. They provide useful triggers to help direct decision makers and policymakers toward proactive risk management.

Triggers are specific values of an indicator or index that initiate and/or terminate each level of a drought plan and associated mitigation and emergency management responses. In other words, they trigger action and allow for accountability as to who is doing what and when they need to do it. This should ultimately tie in with a comprehensive drought management plan or policy (WMO and GWP 2014). It is essential to have a complete list of triggers for indicators or indices, which should also be aligned with an action plan to guide a coordinated set of actions by individual agencies or ministries. Without this alignment, there is likely to be considerable delay in action at the onset of drought in an area or region.

8.4 Selecting Indicators and Indices

Just as there is no "one-size-fits-all" definition of drought, there is no single index or indicator that can account for and be applied to all types of droughts, climate regimes, and sectors affected by droughts. This handbook is not intended to be prescriptive by telling readers which indices and indicators are best to use and when; in fact, many factors feed in to determining which indicator, index or trigger (or combination thereof) is the best to use for a particular need or application. The following questions may help users to decide which indicators and indices are most appropriate for their current situation:

- Do the indicators/indices allow for timely detection of drought in order to trigger appropriate communication and coordination of drought response or mitigation actions?

- Are the indicators/indices sensitive to climate, space, and time in order to determine drought onset and termination?

- Are the indicators/indices and various severity levels responsive and reflective of the impacts occurring on the ground for a given location or region?

- Are the chosen indicators, indices and triggers the same, or different, for going into and coming out of drought? It is critical to account for both situations.

- Are composite (hybrid) indicators being used in order to take many factors and inputs into account?

- Are the data and resultant indices/indicators available and stable? In other words, is there a long period of record for the data source that can give planners and decision makers a strong historical and statistical marker?

- Are the indicators/indices easy to implement? Do the users have the resources (time and human) to dedicate to efforts and will they be maintained diligently when not in a drought situation? This can be better justified if such a system is set up for monitoring all aspects of the hydrologic or climatic cycles, not just droughts.

The simplest indicator/index to use is typically one that is already being produced operationally and freely available, but this does not necessarily mean that it is the best or most applicable.

Ultimately, the choice has to be determined by users at the regional, national, or local levels. The preferred and recommended approach is for users to take a multiple or composite/hybrid indicator/index approach as part of a DEWS within the context of a comprehensive drought mitigation plan. Ideally, this requires thorough analyses and a research approach to determine which indicators work best in particular climate regimes, regions, basins, and locations. Research is also required to determine which seasons the indicators are most relevant to, representing impacts occurring on the ground. Once identified, the indicators/indices can be recommended or implemented in a DEWS as potential triggers tied to emergency response or mitigation actions within a drought plan.

8.5 Summary of Indicators and Indices

As already stated, no single indicator or index can be used to determine appropriate actions for all types of droughts given the number and variety of sectors affected. The preferred approach is to use different thresholds with different combinations of inputs. Ideally, this will involve prior study to determine which indicators/indices are best suited to the timing, area, and type of climate and drought. This takes time because it requires a trial-and-error approach. Decision-making based on quantitative index-based values is essential to the appropriate and accurate assessment of drought severity and as input into an operational DEWS or comprehensive drought plan.

The indicators and indices listed in Table 8.1 have been drawn from IDMP and partner literature, and online searches. They are categorized by type and ease of use, and grouped into the following classifications: (a) meteorology, (b) soil moisture, (c) hydrology, (d) remote sensing, and (e) composite or modeled. Although listed by "ease of use," it is possible that any, all, or none of the indicators may be suitable for a particular application, based on user knowledge, needs, data availability, and computer resources available to implement them. The resource needs increase from green to yellow to red, as outlined below. Again, the simplest index/indicator is not necessarily the best one to use.

The ease of use classification uses a traffic-light approach for each indicator/index as follows:

- Green: Indices are considered to be green if one or more of the following criteria apply:
  - A code or program to run the index is readily and freely available.
  - Daily data are not required.
  - Missing data are allowed for.
  - Output of the index is already being produced operationally and is available online.

NOTE: While a green ease of use classification may imply that the indicator/index may be the easiest to obtain or use, it does not mean it is the best for any given region or locality. The decision as to which indicators/indices to use has to be determined by the user and depends on the given application(s).

- Yellow: Indices are considered to be yellow if one or more of the following criteria apply:
  - Multiple variables or inputs are needed for calculations.
  - A code or program to run the index is not available in a public domain.
  - Only a single input or variable may be needed, but no code is available.
  - The complexity of the calculations needed to produce the index is minimal.

TABLE 8.1

Indicators and Indices Listed in this Handbook

	Page	Ease of Use	Input Parameters	Additional Information
Meteorology				
Aridity Anomaly Index (AAI)	11	Green	P, T, PET, ET	Operationally available for India
Deciles	11	Green	P	Easy to calculate; examples from Australia are useful
Keetch–Byram Drought Index (KBDI)	12	Green	P, T	Calculations are based upon the climate of the area of interest
Percent of Normal Precipitation	12	Green	P	Simple calculations
Standardized Precipitation Index (SPI)	13	Green	P	Highlighted by the World Meteorological Organization as a starting point for meteorological drought monitoring
Weighted Anomaly Standardized Precipitation (WASP)	15	Green	P, T	Uses gridded data for monitoring drought in tropical regions
Aridity Index (AI)	15	Yellow	P, T	Can also be used in climate classifications
China Z Index (CZI)	16	Yellow	P	Intended to improve upon SPI data
Crop Moisture Index (CMI)	16	Yellow	P, T	Weekly values are required
Drought Area Index (DAI)	17	Yellow	P	Gives an indication of monsoon season performance
Drought Reconnaissance Index (DRI)	17	Yellow	P, T	Monthly temperature and precipitation are required
Effective Drought Index (EDI)	18	Yellow	P	Program available through direct contact with originator
Hydro-thermal Coeffcient of Selyaninov (HTC)	19	Yellow	P, T	Easy calculations and several examples in the Russian Federation
NOAA Drought Index (NDI)	19	Yellow	P	Best used in agricultural applications
Palmer Drought Severity Index (PDSI)	20	Yellow	P, T, AWC	Not green due to complexity of calculations and the need for serially complete data
Palmer Z Index	20	Yellow	P, T, AWC	One of the many outputs of PDSI calculations
Rainfall Anomaly Index (RAI)	21	Yellow	P	Serially complete data required

(Continued)

TABLE 8.1 *(Continued)*

Indicators and Indices Listed in this Handbook

	Page	Ease of Use	Input Parameters	Additional Information
Self-Calibrated Palmer Drought Severity Index (sc-PDSI)	22	Yellow	P, T, AWC	Not green due to complexity of calculations and serially complete data required
Standardized Anomaly Index (SAI)	22	Yellow	P	Point data used to describe regional conditions
Standardized Precipitation Evapotranspiration Index (SPEI)	23	Yellow	P, T	Serially complete data required; output similar to SPI but with a temperature component
Agricultural Reference Index for Drought (ARID)	23	Red	P, T, Mod	Produced in southeastern United States of America and not tested widely outside the region
Crop-specific Drought Index (CSDI)	24	Red	P, T, Td, W, Rad, AWC, Mod, CD	Quality data of many variables needed, making it challenging to use
Reclamation Drought Index (RDI)	25	Red	P, T, S, RD, SF	Similar to the Surface Water Supply Index, but contains a temperature component
Soil moisture				
Soil Moisture Anomaly (SMA)	25	Yellow	P, T, AWC	Intended to improve upon the water balance of PDSI
Evapotranspiration Deficit Index (ETDI)	26	Red	Mod	Complex calculations with multiple inputs required
Soil Moisture Deficit Index (SMDI)	26	Red	Mod	Weekly calculations at different soil depths; complicated to calculate
Soil Water Storage (SWS)	27	Red	AWC, RD, ST, SWD	Owing to variations in both soil and crop types, interpolation over large areas is challenging
Hydrology				
Palmer Hydrological Drought Severity Index (PHDI)	27	Yellow	P, T, AWC	Serially complete data required
Standardized Reservoir Supply Index (SRSI)	28	Yellow	RD	Similar calculations to SPI using reservoir data
Standardized Streamflow Index (SSFI)	29	Yellow	SF	Uses the SPI program along with streamflow data

(Continued)

TABLE 8.1 *(Continued)*

Indicators and Indices Listed in this Handbook

	Page	Ease of Use	Input Parameters	Additional Information
Standardized Water-level Index (SWI)	29	Yellow	GW	Similar calculations to SPI, but using groundwater or well-level data instead of precipitation
Streamflow Drought Index (SDI)	30	Yellow	SF	Similar calculations to SPI, but using streamflow data instead of precipitation
Surface Water Supply Index (SWSI)	30	Yellow	P, RD, SF, S	Many methodologies and derivative products are available, but comparisons between basins are subject to the method chosen
Aggregate Dryness Index (ADI)	31	Red	P, ET, SF, RD, AWC, S	No code, but mathematics explained in the literature
Standardized Snowmelt and Rain Index (SMRI)	32	Red	P, T, SF, Mod	Can be used with or without snowpack information
Remote sensing				
Enhanced Vegetation Index (EVI)	32	Green	Sat	Does not separate drought stress from other stress
Evaporative Stress Index (ESI)	33	Green	Sat, PET	Does not have a long history as an operational product
Normalized Difference Vegetation Index (NDVI)	33	Green	Sat	Calculated for most locations
Temperature Condition Index (TCI)	34	Green	Sat	Usually found along with NDVI calculations
Vegetation Condition Index (VCI)	34	Green	Sat	Usually found along with NDVI calculations
Vegetation Drought Response Index (VegDRI)	35	Green	Sat, P, T, AWC, LC, ER	Takes into account many variables to separate drought stress from other vegetation stress
Vegetation Health Index (VHI)	35	Green	Sat	One of the first attempts to monitor drought using remotely sensed data
Water Requirement Satisfaction Index (WRSI and Geo-spatial WRSI)	36	Green	Sat, Mod, CC	Operational for many locations
Normalized Difference Water Index (NDWI) and Land Surface Water Index (LSWI)	37	Green	Sat	Produced operationally using Moderate Resolution Imaging Spectroradiometer data

(Continued)

TABLE 8.1 *(Continued)*

Indicators and Indices Listed in this Handbook

	Page	Ease of Use	Input Parameters	Additional Information
Soil Adjusted Vegetation Index (SAVI)	37	Red	Sat	Not produced operationally
Composite or modelled				
Combined Drought Indicator (CDI)	38	Green	Mod, P, Sat	Uses both surface and remotely sensed data
Global Integrated Drought Monitoring and Prediction System (GIDMaPS)	38	Green	Multiple, Mod	An operational product with global output for three drought indices: Standardized Soil Moisture Index, SPI and Multivariate Standardized Drought Index
Global Land Data Assimilation System (GLDAS)	39	Green	Multiple, Mod, Sat	Useful in data-poor regions due to global extent
Multivariate Standardized Drought Index (MSDI)	40	Green	Multiple, Mod	Available but interpretation is needed
United States Drought Monitor (USDM)	41	Green	Multiple	Available but interpretation is needed

Note: Indicators and indices are sorted by 'ease of use' and then alphabetically within each 'ease of use' category.

AWC, available water content; CC, crop coeffcient; CD, crop data; ER, ecoregion; ET, evapotranspiration; GW, groundwater; LC, land cover; Mod, modelled; Multiple, multiple indicators used; P, precipitation; PET, potential evapotranspiration; Rad, solar radiation; RD, reservoir; S, snowpack; Sat, satellite; SF, streamflow; ST, soil type; SWD, soil water deficit; T, temperature; Td, dewpoint temperature; W, wind data.

- Red: Indices are considered to be red if one or more of the following criteria apply:
  - A code would need to be developed to calculate the index based upon a methodology given in the literature
  - The index or derivative products are not readily available
  - The index is an obscure index, and is not widely used, but may be applicable
  - The index contains modeled input or is part of the calculations

8.6 Index and Indicator Resources

There are several sources of information on the many indices and indicators being applied today around the world. Some of the more common indices are documented and explained by the National Drought Mitigation Center (NDMC) at the University of Nebraska–Lincoln, United States, which maintains a dedicated drought indices resource section, http://drought.unl.edu/Planning/Monitoring/HandbookofDroughtIndices.aspx.

The World Meteorological Organization (WMO)/NDMC *Inter-Regional Workshop on Indices and Early Warning Systems for Drought* was held in 2009 at the University of Nebraska–Lincoln. One of the outcomes was to endorse the Standardized Precipitation Index (SPI) via the Lincoln Declaration on Drought Indices as the standard for determining the existence of meteorological drought (Hayes et al. 2011). WMO has developed a user guide to SPI—see http://www.droughtmanagement.info/literature/WMO_standardized_precipitation_index_user_guide_en_2012.pdf.

As a follow-up, WMO and the United Nations Office for Disaster Risk Reduction in collaboration with the Segura Hydrographic Confederation and Spain's Agencia Estatal de Meteorología (State Meteorological Agency) organized an expert group meeting on agricultural drought indices in Murcia, Spain, in 2010 (Sivakumar et al. 2011). A group of scientists from around the world represented WMO regions and reviewed 34 indices used for assessing drought impacts on agriculture, highlighting their strengths and weaknesses. The proceedings, *Agricultural Drought Indices: Proceedings of an Expert Meeting*, are documented in the form of 17 papers and can be found at http://www.wamis.org/agm/pubs/agm11/agm11.pdf.

See also the references listed at the end of this chapter, for example, Heim (2002), Keyantash and Dracup (2002), and Zargar et al. (2011), which review drought indices in use, both today and in the past.

For additional help with the selection, interpretation, and application of indicators and indices, contact IDMP at http://www.droughtmanagement.info/ or by e-mail at idmp@wmo.int.

8.7 Indicators and Indices

8.7.1 Meteorology

Index name: Aridity Anomaly Index (AAI)

Ease of use: Green

Origins: Developed in India by the India Meteorological Department

Characteristics: A real-time drought index in which water balance is considered. The Aridity Index (AI) is computed for weekly or 2-weekly periods. For each period, the actual aridity for the period is compared to the normal aridity for that period. Negative values indicate a surplus of moisture, while positive values indicate moisture stress.

Input parameters: Actual evapotranspiration and calculated potential evapotranspiration, which require temperature, wind, and solar radiation values

Applications: Impacts of drought in agriculture, especially in the tropics where defined wet and dry seasons are part of the climate regime. Both winter and summer cropping seasons can be assessed using this method.

Strengths: Specific to agriculture, calculations are simple, and descriptions of drought (mild, moderate, or severe) are based on departure from normal. Responds quickly with a weekly time step.

Weaknesses: Not applicable to long-term or multiseasonal events.

Resource: http://imdpune.gov.in/hydrology/methodology.html

Reference: http://www.wamis.org/agm/gamp/GAMP_Chap06.pdf

Index name: Deciles

Ease of use: Green

Origins: A simple mathematical approach described by Gibbs and Maher in 1967 through their work with the Australian Bureau of Meteorology

Characteristics: Using the entire period of record of precipitation data for a location, the frequency and distribution of precipitation are ranked. The first decile is composed of the rainfall amounts in which the lowest 10 percent of the values are not exceeded, and the fifth decile is the median. A wet scale is also available. Daily, weekly, monthly, seasonal, and annual values can all be considered in the methodology, as it is flexible when current data are compared to the historical record for any given period.

Input parameters: Precipitation only; the timescale considered is flexible

Applications: With the ability to look at different timescales and time steps, deciles can be used in meteorological, agricultural, and hydrological drought situations.

Strengths: With a single variable being considered, the methodology is simple and flexible for many situations. Using clearly defined thresholds, the current data are put into a historical context and drought status can be recognized. Useful in both wet and dry situations.

Weaknesses: As with other indicators that use only precipitation, the impacts of temperatures and other variables are not considered during the development of drought. A long record period provides the best results because many wet and dry periods will be included in the distribution.

Resources: There is no specific software code for deciles, and several online tools can provide output. Thus, it is important to clarify the underlying methodology, as there are a number of statistical approaches to calculate deciles from meteorological data; http://drinc.ewra.net/.

Reference: Gibbs and Maher (1967).

Index name: Keetch–Byram Drought Index (KBDI)

Ease of use: Green

Origins: Part of work done in the late 1960s by Keetch and Byram of the United States Department of Agriculture's Forest Service Division. It is mainly a fire index.

Characteristics: Developed to identify drought in the early stages using a uniform method specific to the climate of the region. It is the net effect of evapotranspiration and precipitation in producing a moisture deficiency in the upper layers of the soil and also gives an indication of how much precipitation is needed for saturation of the soil and eliminating drought stress.

Input parameters: Daily maximum temperature and daily precipitation. Tables are computed to relate KBDI to various precipitation regimes based upon the local climate.

Applications: Intended as a method of monitoring fire danger due to drought, KBDI was found to be useful in agricultural contexts because the measure of soil moisture was directly related to drought stress on crops.

Strengths: Expresses moisture deficiency for an area and can be scaled to indicate the characteristics of each particular location. Calculations are simple and the method is easy to use.

Weaknesses: Assumes a limit of available moisture and the necessity of certain climatic conditions for drought to develop, which may or may not be true for every location.

Resources: The method and calculation are available and well described in the literature. Many maps are available online for various locations, http://www.wfas.net/index.php/keetch-byram-index-moisture--drought-49.

Reference: Keetch and Byram (1968).

Index name: Percent of Normal Precipitation

Ease of use: Green

Origins: The percentage of any quantity is a simple statistical formulation. The exact origin or first use is not known in describing precipitation anomalies.

Characteristics: A simple calculation that can be used to compare any time period for any location. It can be computed on daily, weekly, monthly, seasonal, and annual timescales, which will suit many user needs. It is calculated by dividing actual precipitation by normal precipitation for the time being considered, and multiplying by 100.

Input parameters: Precipitation values suitable for the timescale being calculated. It is ideal to have at least 30 years' worth of data for calculation of the normal period.

Applications: Can be used for identifying and monitoring various impacts of droughts.

Strengths: A popular method that is quick and easy to calculate with basic mathematics.

Weaknesses: Establishing the normal for an area is a calculation that some users could confuse with mean or average precipitation. It is hard to compare different climate regimes with each other, especially those with defined wet and dry seasons.

Reference: Hayes (2006).

Index name: Standardized Precipitation Index (SPI)

Ease of use: Green

Origins: The result of research and work done in 1992 at Colorado State University, United States, by McKee et al. The outcome of their work was first presented at the 8th Conference on Applied Climatology, held in January 1993. The basis of the index is that it builds upon the relationships of drought to frequency, duration, and timescales.

In 2009, WMO recommended SPI as the main meteorological drought index that countries should use to monitor and follow drought conditions (Hayes et al. 2011). By identifying SPI as an index for broad use, WMO provided direction for countries trying to establish a level of drought early warning.

Characteristics: Uses historical precipitation records for any location to develop a probability of precipitation that can be computed at any number of timescales, from 1 to 48 months or longer. As with other climatic indicators, the time series of data used to calculate SPI does not need to be of a specific length. Guttman (1998, 1999) noted that if additional data are present in a long time series, the results of the probability distribution will be more robust because more samples of extreme wet and extreme dry events are included. SPI can be calculated on as little as 20 years' worth of data, but ideally the time series should have a minimum of 30 years of data, even when missing data are accounted for.

SPI has an intensity scale in which both positive and negative values are calculated, which correlate directly to wet and dry events. For drought, there is great interest in the "tails" of the precipitation distribution, and especially in the extreme dry events, which are the events considered to be rare based upon the climate of the region being investigated.

Drought events are indicated when the results of SPI, for whichever timescale is being investigated, become continuously negative and reach a value of −1. The drought event is considered to be ongoing until SPI reaches a value of 0. McKee et al. (1993) stated that drought begins at an SPI of −1 or less, but there is no standard in place, as some researchers will choose a threshold that is less than 0, but not quite −1, while others will initially classify drought at values less than −1.

Owing to the utility and flexibility of SPI, it can be calculated with data missing from the period of record for a location. Ideally, the time series should be as complete as possible, but SPI calculations will provide a null value if there are insufficient data to calculate a value, and SPI will begin calculating output again as data become available. SPI is typically calculated for timescales of up to 24 months, and the flexibility of the index allows for multiple applications addressing events that affect agriculture, water resources, and other sectors.

Input parameters: Precipitation. Most users apply SPI using monthly datasets, but computer programs have the flexibility to produce results when using daily and weekly values. The methodology of SPI does not change based upon using daily, weekly, or monthly data.

Applications: The ability of SPI to be calculated at various timescales allows for multiple applications. Depending on the drought impact in question, SPI values for 3 months or less might be useful for basic drought monitoring, values for 6 months or less for monitoring agricultural impacts and values for 12 months or longer for hydrological impacts. SPI can also be calculated on gridded precipitation datasets, which allows for a wider scope of users than those just working with station-based data.

Strengths: Using precipitation data only is the greatest strength of SPI, as it makes it very easy to use and calculate. SPI is applicable in all climate regimes, and SPI values for very different climates can be compared. The ability of SPI to be computed for short periods of record that contain missing data is also valuable for those regions that may be data poor, or lacking long-term, cohesive datasets. The program used to calculate SPI is easy to use and readily available. NDMC provides a program for use on personal computers that has been distributed to more than 200 countries around the world. The ability to be calculated over multiple timescales also allows SPI to have a wide breadth of application. Many articles relating to SPI are available in the science literature, giving novice users a multitude of resources to rely on for assistance.

Weaknesses: With precipitation as the only input, SPI is deficient when accounting for the temperature component, which is important to the overall water balance and water use of a region. This drawback can make it more difficult to compare events of similar SPI values but different temperature scenarios. The flexibility of SPI to be calculated for short periods of record, or on data that contain many missing values, can also lead to misuse of the output, as the program will provide output for whatever input is provided. SPI assumes a prior distribution, which may not be appropriate in all environments, particularly when examining short-duration events or entry into, or exit out of, drought. There are many versions of SPI available, implemented within various computing software packages other than that found in the source code distributed by NDMC. It is important to check the integrity of these algorithms and the consistency of output with the published versions.

Resource: The SPI program can be run on Windows-based personal computers: http://drought.unl.edu/MonitoringTools/DownloadableSPIProgram.aspx.

References: Guttman (1998, 1999); Hayes et al. (2011); McKee et al. (1993); World Meteorological Organization (2012); Wu et al. (2005).

Index name: Weighted Anomaly Standardized Precipitation Index (WASP)

Ease of use: Green

Origins: Developed by Lyon to monitor precipitation in the tropical regions within 30° latitude of the equator.

Characteristics: Uses gridded monthly precipitation data on a 0.5° × 0.5° resolution, and is based on 12-month overlapping sums of weighted, standardized monthly precipitation anomalies.

Input parameters: Monthly precipitation and annual precipitation values

Applications: Used mainly in wet tropical regions to monitor developing drought, taking into account the defined wet and dry periods in the climate regime. Can be used to monitor droughts that affect agriculture and other sectors.

Strengths: Using precipitation as a single input allows for simpler computations.

Weaknesses: Does not work so well in desert regions. Gridded precipitation data may be a challenge to obtain in an operational capacity.

Resources: The methods and calculations are provided and explained in the literature, http://iridl.ldeo.columbia.edu/maproom/Global/Precipitation/WASP_Indices.html.

Reference: Lyon (2004).

Index name: Aridity Index (AI)

Ease of use: Yellow

Origins: Developed from work done by De Martonne in 1925; aridity is defined as the ratio of precipitation to mean temperature.

Characteristics: Can be used to classify the climates of various regions, because the ratio of precipitation to temperature provides a method for determining an area's climate regime.

Monthly calculation of AI can be used to determine the onset of drought, as the index takes into account temperature impacts as well as precipitation.

Input parameters: Monthly mean temperature and precipitation. For climate classification, annual values are used.

Applications: Mainly used to determine the development of drought over shorter timescales, which is helpful for identifying and monitoring agricultural and meteorological impacts.

Strengths: Easy to compute with just two inputs. Flexible in that various time steps can be analyzed.

Weaknesses: Does not take into account carry-over of dryness from year to year. May be slow to react in certain climates.

References: Baltas (2007); De Martonne (1925).

Index name: China Z Index (CZI)

Ease of use: Yellow

Origins: Developed in China, CZI builds on the ease of calculation provided by SPI and improves on it by making the calculations even easier for the user. A statistical Z-score can be used to identify and monitor drought periods. The index was first used and developed in 1995 by the National Climate Centre of China.

Characteristics: CZI is similar to SPI because precipitation is used to determine wet and dry periods, assuming that the precipitation obeys a Pearson type III distribution. It uses monthly time steps from 1 to 72 months, giving it the ability to identify droughts of various durations.

Input parameters: Monthly precipitation.

Applications: Similar to SPI, in which both wet and dry events can be monitored over multiple timescales.

Strengths: Simple calculations, which can be computed for several time steps. Can be used for both wet and dry events. Allows for missing data, similar to SPI.

Weaknesses: The Z-score data do not require adjustment by fitting them to gamma or Pearson type II distributions, and it is speculated that because of this, shorter timescales may be less well represented compared with SPI.

Resources: All calculations and explanations of CZI can be found at http://onlinelibrary.wiley.com/doi/10.1002/joc.658/pdf.

References: Edwards and McKee (1997); Wu et al. (2001).

Index name: Crop Moisture Index (CMI)

Ease of use: Yellow

Origins: As part of original work done by Palmer in the early 1960s, CMI is usually calculated weekly along with the Palmer Drought Severity Index (PDSI) output as the short-term drought component in which the impact on agriculture is considered.

Characteristics: As some of the drawbacks associated with PDSI became apparent, Palmer responded to them with the development of CMI. It is intended to be a drought index especially suited to drought impacts on agriculture, in that it responds quickly to rapidly changing conditions. It is calculated by subtracting the difference between potential evapotranspiration and moisture, to determine any deficit.

Input parameters: Weekly precipitation, weekly mean temperature, and the previous week's CMI value.

Applications: Used to monitor droughts in which agricultural impacts are a primary concern.

Strengths: The output is weighted, so it is possible to compare different climate regimes. Responds quickly to rapidly changing conditions.

Weaknesses: As it was developed specifically for grain-producing regions in the United States, CMI may show a false sense of recovery from long-term drought events, as improvements in the short term may be insufficient to offset long-term issues.

Resource: https://www.drought.gov/drought/content/products-current-drought-and-monitoring-drought-indicators/crop-moisture-index.

Reference: Palmer (1968).

Index name: Drought Area Index (DAI)

Ease of use: Yellow

Origins: Developed in the late 1970s by Bhalme and Mooley at the Indian Institute of Tropical Meteorology.

Characteristics: Developed as a method to improve understanding of monsoon rainfall in India, determining both flood and drought episodes using monthly precipitation. By comparing monthly precipitation during the critical monsoon period, the intensities of wet and dry periods are obtained, and the significance of the dryness can be derived based upon the contribution of each month's precipitation to the total monsoon season.

Input parameters: Monthly precipitation during the monsoon season.

Applications: Used to identify when the monsoon season has been adequate or dry, or there is potential for flooding. The drought prediction is a good early warning for the potential of famine development.

Strengths: Very focused on Indian monsoon seasons in the tropics.

Weaknesses: Lack of applicability to other areas or climate regimes.

Resource: The mathematics and associated explanation of this index are in the original paper, http://moeseprints.incois.gov.in/1351/1/large%20scale.pdf.

Reference: Bhalme and Mooley (1980).

Index name: Drought Reconnaissance Index (DRI)

Ease of use: Yellow

Origins: Work was initiated by Tsakiris and Vangelis at the National Technical University of Athens, Greece.

Characteristics: Consists of a drought index that contains a simplified water balance equation considering precipitation and potential evapotranspiration. It has three outputs: the initial value, the normalized value, and the standardized value. The standardized DRI value is similar in nature to SPI and can be compared to it directly. DRI is more representative than SPI, however, as it considers the full water balance instead of precipitation alone.

Input parameters: Monthly temperature and precipitation values.

Applications: Cases where impacts on agriculture or water resources are a primary concern.

Strengths: The use of potential evapotranspiration gives a better representation of the full water balance of the region than SPI provides, which will give a better indication of the drought severity. Can be calculated for many time steps, as with SPI. All the required mathematics are available in the literature.

Weaknesses: Potential evapotranspiration calculations can be subject to errors when using temperature alone to create the estimate. Monthly timescales may not react quickly enough for rapidly developing droughts.

Resource: DRI software is available at http://drinc.ewra.net/.

Reference: Tsakiris and Vangelis (2005).

Index name: Effective Drought Index (EDI)

Ease of use: Yellow

Origins: Developed through work done by Byun and Wilhite, along with staff at NDMC.

Characteristics: Uses daily precipitation data to develop and compute several parameters: effective precipitation (EP), daily mean EP, deviation of EP (DEP) and the standardized value of DEP. These parameters can identify the onset and end of water deficit periods. Using the input parameters, EDI calculations can be performed for any location in the world in which the results are standardized for comparison, giving a clear definition of the onset, end, and duration of drought. At the time of EDI development, most drought indices were being calculated using monthly data, so the switch to daily data was unique and important to the utility of the index.

Input parameters: Daily precipitation.

Applications: A good index for operational monitoring of both meteorological and agricultural drought situations because calculations are updated daily.

Strengths: With a single input required for calculations, it is possible to calculate EDI for any location where precipitation is recorded. Supporting documents explaining the processes are available for the program. EDI is standardized so that outputs from all climate regimes can be compared. It is effective for identifying the beginning, end, and duration of drought events.

Weaknesses: With precipitation alone accounted for, the impact of temperature on drought situations is not directly integrated. Using daily data may make it difficult to use EDI in an operational situation, as daily updates to input data may not be possible.

Resources: The authors state that the code is available by contacting them directly. The calculations are available and described in the original paper referenced below. EDI calculations are part of a suite of indices calculated as part of the Spatial and Time Series Information Modeling (SPATSIM) software package, http://www.preventionweb.net/files/1869_VL102136.pdf.

Reference: Byun and Wilhite (1996).

Index name: Hydro-thermal coefficient of Selyaninov (HTC)

Ease of use: Yellow

Origins: Developed by Selyaninov in the Russia Federation and based on the Russian climate.

Characteristics: Uses temperature and precipitation values and is sensitive to dry conditions specific to the climate regime being monitored. It is flexible enough to be used in both monthly and decadal applications.

Input parameters: Monthly temperature and precipitation values.

Applications: Useful in the monitoring of agricultural drought conditions and has also been used in climate classifications.

Strengths: Simple to calculate, and the values can be applied to agricultural conditions during the growing season.

Weaknesses: The calculations do not take into account soil moisture.

Resources: Information can be found at the website of the Russian National Institute on Agricultural Meteorology, http://cxm.obninsk.ru/index.php?id=154, and at the website of the Interactive Agricultural Ecological Atlas of Russia and Neighboring Countries, http://www.agroatlas.ru/en/content/Climatic_maps/GTK/GTK/index.html.

Reference: Selyaninov (1928).

Index name: NOAA Drought Index (NDI)

Ease of use: Yellow

Origins: Developed in the early 1980s at the Joint Agricultural Weather Facility as part of the United States Department of Agriculture's attempt to use weather and climate data for crop production estimates around the world.

Characteristics: A precipitation-based index in which the actual precipitation measured is compared with normal values during the growing season. Mean precipitation for each week is calculated and a running 8-week average of measured average precipitation is summed and compared. If the actual precipitation is greater than 60 percent of the normal precipitation for the 8-week period, then the current week is assumed to have little or no water stress. If stress is detected, it remains until the actual precipitation is at 60 percent or more of normal.

Input parameters: Monthly precipitation converted to weekly precipitation values.

Applications: Used as an indicator of drought conditions affecting agriculture.

Strengths: The only input is precipitation, in a monthly time step. The calculations and explanation of use are simple.

Weaknesses: At least 30 years' worth of data are required to compute normalized monthly values that are used in the computation of the weekly values.

It has very specific applications related to agriculture, and crop progression and development.

Reference: Strommen and Motha (1987).

Index name: Palmer Drought Severity Index (PDSI)

Ease of use: Yellow

Origins: Developed in the 1960s as one of the first attempts to identify droughts using more than just precipitation data. Palmer was tasked with developing a method to incorporate temperature and precipitation data with water balance information to identify droughts in crop-producing regions of the United States. For many years, PDSI was the only operational drought index, and it is still very popular around the world.

Characteristics: Calculated using monthly temperature and precipitation data along with information on the water-holding capacity of soils. It takes into account moisture received (precipitation) as well as moisture stored in the soil, accounting for the potential loss of moisture due to temperature influences.

Input parameters: Monthly temperature and precipitation data. Information on the water-holding capacity of soils can be used, but defaults are also available. A serially complete record of temperature and precipitation is required.

Applications: Developed mainly as a way to identify droughts affecting agriculture, it has also been used for identifying and monitoring droughts associated with other types of impacts. With the longevity of PDSI, there are numerous examples of its use over the years.

Strengths: Used around the world, and the code and output are widely available. Scientific literature contains numerous papers related to PDSI. The use of soil data and a total water balance methodology makes it quite robust for identifying drought.

Weaknesses: The need for serially complete data may cause problems. PDSI has a timescale of approximately 9 months, which leads to a lag in identifying drought conditions based upon simplification of the soil moisture component within the calculations. This lag may be up to several months, which is a drawback when trying to identify a rapidly emerging drought situation. Seasonal issues also exist, as PDSI does not handle frozen precipitation or frozen soils well.

Resource: http://hydrology.princeton.edu/data.pdsi.php

References: Alley (1984); Palmer (1965).

Index name: Palmer Z Index

Ease of use: Yellow

Origins: The Palmer Z Index responds to short-term conditions better than PDSI and is typically calculated for much shorter timescales, enabling it to identify rapidly developing drought conditions. As part of the original work done by Palmer in the early 1960s, the Palmer Z Index is usually calculated on a monthly basis along with PDSI output as the moisture anomaly.

Characteristics: Sometimes referred to as the moisture anomaly index, and the derived values provide a comparable measure of the relative anomalies of a region for both dryness and wetness when compared to the entire record for that location.

Input parameters: The Palmer Z Index is a derivative of PDSI, and the Z values are part of the PDSI output.

Applications: Useful for comparing current periods to other known drought periods. It can also be used to determine the end of a drought period, when it is used to determine how much moisture is needed to reach the near normal category, as defined by Palmer.

Strengths: Same as for PDSI. The scientific literature contains a number of relevant papers. The use of soil data and a total water balance methodology makes the Palmer Z Index quite robust for identifying drought.

Weaknesses: Same as for PDSI, with the need for serially complete data possibly causing problems. It has a timescale of approximately 9 months, which leads to a lag in identifying drought conditions based upon simplification of the soil moisture component within the calculations. This lag may be up to several months, which is a drawback when trying to identify a rapidly emerging drought situation. Seasonal issues also exist, as the Palmer Z Index does not handle frozen precipitation or frozen soils well.

Resource: Contact NDMC to access the code for the Palmer suite, http://drought.unl.edu/

Reference: Palmer (1965).

Index name: Rainfall Anomaly Index (RAI)

Ease of use: Yellow

Origins: Work began in the early 1960s by van Rooy

Characteristics: Uses normalized precipitation values based upon the station history of a particular location. Comparison to the current period puts the output into a historical perspective.

Input parameters: Precipitation.

Applications: Addresses droughts that affect agriculture, water resources, and other sectors, as RAI is flexible in that it can be analyzed at various timescales.

Strengths: Easy to calculate, with a single input (precipitation) that can be analyzed on monthly, seasonal, and annual timescales.

Weaknesses: Requires a serially complete dataset with estimates of missing values. Variations within the year need to be small compared to temporal variations.

Resources: No resources available.

References: Kraus (1977); van Rooy (1965).

Index name: Self-Calibrated Palmer Drought Severity Index (sc-PDSI)

Ease of use: Yellow

Origins: Initial work was conducted at the University of Nebraska–Lincoln by Wells et al. in the early 2000s.

Characteristics: Accounts for all the constants contained in PDSI, and includes a methodology in which the constants are calculated dynamically based upon the characteristics present at each station location. The self-calibrating nature of sc-PDSI is developed for each station and changes based upon the climate regime of the location. It has wet and dry scales.

Input parameters: Monthly temperature and precipitation. Information on the water-holding capacity of soils can be used, but defaults are also available. A serially complete record of temperature and precipitation data is required.

Applications: Can be applied to meteorological, agricultural, and hydrological drought situations. With the results being tied directly to station location, extreme events are rare, as they are related directly to that station's information and not a constant.

Strengths: With the calculations for sc-PDSI accounting for each individual location, the index reflects what is happening at each site and allows for more accurate comparisons between regions. Different time steps can be calculated.

Weaknesses: As the methodology is not significantly different from PDSI, it has the same issues in terms of time lag, and frozen precipitation and soils.

Resources: The code can be obtained from http://drought.unl.edu/ and https://climatedataguide.ucar.edu/climate-data/cru-sc-pdsi-self-calibrating-pdsi-over-europe-north-america.

Reference: Wells et al. (2004).

Index name: Standardized Anomaly Index (SAI)

Ease of use: Yellow

Origins: Introduced by Kraus in the mid-1970s and was examined closely by Katz and Glantz at the National Center for Atmospheric Research, United States, in the early 1980s. SAI was developed based on RAI, and RAI is a component of SAI. They are similar, but both are unique.

Characteristics: Based upon the results of RAI, and was developed to help identify droughts in susceptible regions, such as the West African Sahel and northeast Brazil. RAI accounts for station-based precipitation in a region and standardizes annual amounts. Deviations are then averaged over all stations in the region to obtain a single SAI value.

Input parameters: Precipitation at monthly, seasonal, or annual time steps.

Applications: Identifying drought events, especially in areas frequented by drought.

Strengths: Single input, which can be calculated for any defined period.

Weaknesses: Only uses precipitation, and calculations are dependent on quality data.

Resources: Equations for the calculations are provided in the literature.

References: Katz and Glantz (1986); Kraus (1977).

Index name: Standardized Precipitation Evapotranspiration Index (SPEI)

Ease of use: Yellow

Origins: Developed by Vicente-Serrano et al. at the Instituto Pirenaico de Ecologia in Zaragoza, Spain.

Characteristics: As a relatively new drought index, SPEI uses the basis of SPI but includes a temperature component, allowing the index to account for the effect of temperature on drought development through a basic water balance calculation. SPEI has an intensity scale in which both positive and negative values are calculated, identifying wet and dry events. It can be calculated for time steps of as little as 1 month up to 48 months or more. Monthly updates allow it to be used operationally, and the longer the time series of data available, the more robust the results will be.

Input parameters: Monthly precipitation and temperature data. A serially complete record of data is required with no missing months.

Applications: With the same versatility as that of SPI, SPEI can be used to identify and monitor conditions associated with a variety of drought impacts.

Strengths: The inclusion of temperature along with precipitation data allows SPEI to account for the impact of temperature on a drought situation. The output is applicable for all climate regimes, with the results being comparable because they are standardized. With the use of temperature data, SPEI is an ideal index when looking at the impact of climate change in model output under various future scenarios.

Weaknesses: The requirement for a serially complete dataset for both temperature and precipitation may limit its use due to insufficient data being available. Being a monthly index, rapidly developing drought situations may not be identified quickly.

Resources: SPEI code is freely available and the calculations are described in the literature: http://sac.csic.es/spei/

Reference: Vicente-Serrano et al. (2010).

Index name: Agricultural Reference Index for Drought (ARID)

Ease of use: Red

Origins: Based upon research done in the southeast United States by Woli at Mississippi State University and Jones et al. at the University of Florida in 2011.

Characteristics: Predicts the status of moisture availability in the soil. It uses a combination of water stress approximations and crop models to identify the impact of water stress on plant growth, development, and yield for specific crops.

Input parameters: Daily temperature and precipitation data. The CERES-Maize model is also used, but other crop simulations models can be used.

Applications: Used for identifying and predicting drought in contexts where agricultural impacts are the primary concern.

Strengths: Crop models and water balance methods prove to be useful in predicting soil moisture and subsequent stress to crops. Can be computed daily so reaction times to drought will be fast.

Weaknesses: Designed and tested in the southeast United States for only a few cropping systems. Not easily transferable.

Resources: The equations and the methodology used are explained in the referenced article below. No source code is publicly available.

Reference: Woli et al. (2012).

Index name: Crop-specific Drought Index (CSDI)

Ease of use: Red

Origins: Developed by Meyer et al. in the early 1990s at the University of Nebraska–Lincoln to examine the impact of drought on actual crop yield.

Characteristics: By calculating a basic soil water balance, it takes into account the impact of drought, but identifies when the drought stress occurred within the development of the crop and what the overall impact to the final yield will be. PDSI and CMI can identify drought conditions affecting a crop, but do not indicate the likely impact on yields.

Input parameters: Daily maximum temperature, daily minimum temperature, precipitation, dew point temperature, wind speed, and global solar radiation are the climatic inputs.

Characteristics of the soil profile are also needed for model development. Yield and phenology data are required for proper correlations to growing days, crop progress, and final yield.

Applications: Developed mainly to help identify the impact of drought on crop yields in the grain-producing regions of the United States, and is very specific to the type of crop being monitored.

Strengths: Very specific to a particular crop and based upon the development of the plant. The model takes into account when the drought stress occurred during plant growth and estimates the overall impact on yield.

Weaknesses: The inputs are quite complex, and many locations will lack the required instruments or period of record needed to properly assess conditions.

Resources: The methodology and calculations are all described thoroughly in the literature, see references below.

References: Meyer et al. (1993a, 1993b).

Index name: Reclamation Drought Index (RDI)

Ease of use: Red

Origins: The United States Bureau of Reclamation developed this drought index in the mid-1990s as a method to trigger drought emergency relief funds associated with public lands.

Characteristics: Developed to define drought severity as well as duration and can also be used to predict the onset and end of drought periods. It has both wet and dry scales and is calculated at the river basin level, in a similar way to the Surface Water Supply Index (SWSI). RDI has water-demand and temperature components, which allow for the inclusion of evaporation into the index.

Input parameters: Monthly precipitation, snowpack, reservoir levels, streamflow, and temperature.

Applications: Used mainly to monitor water supply for river basins.

Strengths: Very specific to each basin. Unlike SWSI, it accounts for temperature effects on climate. Wet and dry scales allow for monitoring of wet and dry conditions.

Weaknesses: Calculations are made for individual basins, so comparisons are hard to make. Having all the inputs in an operational setting may cause delays in the production of data.

Resources: The characteristics and mathematics are provided in the reference below.

Reference: Weghorst (1996).

8.7.2 Soil Moisture

Index name: Soil moisture anomaly (SMA)

Ease of use: Yellow

Origins: Developed by Bergman et al. at the National Weather Service in the United States during the mid-1980s as a way to assess global drought conditions.

Characteristics: Can use weekly or monthly precipitation and potential evapotranspiration values in a simple water balance equation. It is intended to reflect the degree of dryness or saturation of the soil compared with normal conditions, and to show how soil moisture stress influences crop production around the world.

Input parameters: Weekly or monthly temperature and precipitation data along with date and latitude. Values for soil moisture water-holding capacity and site-specific data can be used, although defaults are included.

Applications: Developed and used extensively for monitoring drought impacts on agriculture and crop production around the world.

Strengths: By taking into account the effects of both temperature and precipitation, the water balance aspects that make PDSI so popular are included with the ability to change constants with site-specific data. It considers moisture at different layers of the soil and is more adaptable than PDSI to different locations.

Weaknesses: The data requirements make it challenging to calculate. Potential evapotranspiration estimates can vary quite substantially by region.

Resources: The inputs and calculations are described thoroughly in the literature. No program exists at this time to provide the calculations.

Reference: Bergman et al. (1988).

Index name: Evapotranspiration Deficit Index (ETDI)

Ease of use: Red

Origins: Developed from research at the Texas Agricultural Experiment Station, United States, by Narasimhan and Srinivasan in 2004.

Characteristics: A weekly product that is helpful for identifying water stress for crops. ETDI is calculated along with the Soil Moisture Deficit Index (SMDI), in which a water stress ratio is calculated that compares actual evapotranspiration with reference crop evapotranspiration. The water stress ratio is then compared with the median calculated over a long-term period.

Input parameters: Modeled data from a hydrologic model with the soil and water assessment tool (SWAT) model are used initially to compute soil water in the root zone on a weekly basis.

Applications: Useful for identifying and monitoring short-term drought affecting agriculture.

Strengths: Analyses both actual and potential evapotranspiration and can identify wet and dry periods.

Weaknesses: Calculations are based upon output from the SWAT model, but could be calculated if the appropriate inputs were available. The spatial variability of ETDI increases in the summer months during the period of greatest evapotranspiration and highly variable precipitation.

Resources: Calculations are provided and explained thoroughly in the reference below, along with correlation studies to other drought indices. Information on the SWAT model can be found at http://swat.tamu.edu/software/swat-executables/

Reference: Narasimhan and Srinivasan (2005).

Index name: Soil Moisture Deficit Index (SMDI)

Ease of use: Red

Origins: Developed from research at the Texas Agricultural Experiment Station, United States, by Narasimhan and Srinivasan in 2004.

Characteristics: A weekly soil moisture product calculated at four different soil depths, including the total soil column, at 0.61, 1.23, and 1.83 m, and can be used as an indicator of short-term drought, especially using the results from the 0.61 m layer.

Input parameters: Modeled data from a hydrologic model with the SWAT model are used initially to compute soil water in the root zone on a weekly basis.

Applications: Useful for identifying and monitoring drought affecting agriculture.

Strengths: Takes into account the full profile as well as different depths, which makes it adaptable to different crop types.

Weaknesses: The information needed to calculate SMDI is based upon output from the SWAT model. There are auto-correlation concerns when all the depths are being used.

Resources: The calculations are provided and explained thoroughly in the reference below.

Information on the SWAT model can be found at http://swat.tamu.edu/software/swat-executables/.

Reference: Narasimhan and Srinivasan (2005).

Index name: Soil Water Storage (SWS)

Ease of use: Red

Origins: Unknown—producers have been trying to measure soil moisture accurately since the beginning of agriculture.

Characteristics: Identifies the amount of available moisture within a plant's root zone, which depends upon the type of plant and the type of soil. Precipitation and irrigation both affect the results.

Input parameters: Rooting depth, available water storage capacity of the soil type, and maximum soil water deficit.

Applications: Used mainly for monitoring drought in agricultural contexts, but can also be a component in drought conditions affecting water availability.

Strengths: Calculations are well known and simple to follow, even using defaults. Many soils and crops have been analyzed using this method.

Weaknesses: In areas where soils are not homogeneous, there may be large changes over small distances.

Resources: Calculations and examples are provided in the reference below.

Reference: British Columbia Ministry of Agriculture (2015).

8.7.3 Hydrology

Index name: Palmer Hydrological Drought Index (PHDI)

Ease of use: Yellow

Origins: Part of the suite of indices developed by Palmer in the 1960s with the United States Weather Bureau.

Characteristics: Based on the original PDSI and modified to take into account longer-term dryness that will affect water storage, streamflow, and groundwater. PHDI has the ability to calculate when a drought will end based on precipitation needed by using a ratio of moisture received to moisture required to end a drought. There are four drought categories: near normal, which occurs approximately 28–50 percent of the time; mild to moderate, which occurs approximately 11–27 percent of the time; severe, which occurs approximately 5–10 percent of the time; and extreme, which occurs approximately 4 percent of the time.

Input parameters: Monthly temperature and precipitation. Information on the water-holding capacity of soils can be used, but defaults are also available. A serially complete record of temperature and precipitation data is required.

Applications: Most useful for taking into account drought affecting water resources on longer timescales.

Strengths: Its water balance approach allows the total water system to be considered.

Weaknesses: Frequencies will vary by region and time of year, where extreme drought may not be a rare event during some months of the year. The impact of human influences, such as management decisions and irrigation, are not considered in the calculations.

Resources: The code can be found in the original Palmer paper in the reference below, http://onlinelibrary.wiley.com/doi/10.1002/wrcr.20342/pdf.

Reference: Palmer (1965).

Index name: Standardized Reservoir Supply Index (SRSI)

Ease of use: Yellow

Origins: Developed by Gusyev et al. in Japan as a systematic way to analyze reservoir data in drought conditions.

Characteristics: Similar to SPI in that monthly data are used to compute a probability distribution function of reservoir storage data, to provide information on water supply for a region or basin within a range of −3 (extremely dry) to +3 (extremely wet).

Input parameters: Monthly reservoir inflows and average reservoir storage volumes.

Applications: Takes into account the total inflow and storage associated with any particular reservoir system, and provides information for municipal water supply managers and local irrigation providers.

Strengths: Easy to compute, as it mimics SPI calculations using a standard gamma distribution of the probability distribution function.

Weaknesses: Does not take into account changes due to management of the reservoir and losses due to evaporation.

Resource: The International Centre for Water Hazard and Risk Management has applied the SRSI methodology to several Asian river basins, http://www.icharm.pwri.go.jp/.

Reference: Gusyev et al. (2015).

Index name: Standardized Streamflow Index (SSFI)

Ease of use: Yellow

Origins: Modarres introduced SSFI in 2007, and Telesca et al. investigated it further in 2012. In the original work, Modarres described how SSFI was similar to SPI in that SSFI for a given period was defined as the difference in streamflow from mean to standard deviation.

Characteristics: Developed using monthly streamflow values and the methods of normalization associated with SPI. Can be calculated for both observed and forecasted data, providing a perspective on high and low flow periods associated with drought and flood.

Input parameters: Streamflow data on a daily or monthly timescale

Applications: Monitoring of hydrological conditions at multiple timescales

Strengths: Easy to calculate using the SPI program. A single variable input that allows for missing data makes it easy to use.

Weaknesses: It only accounts for the streamflow in the context of monitoring drought, with no other influences being investigated.

Resources: It is described well in the literature, with mathematics and case studies available. The SPI program is available at http://drought.unl.edu/MonitoringTools/DownloadableSPIProgram.aspx.

References: Modarres (2007); Telesca et al. (2012).

Index name: Standardized Water-level Index (SWI)

Ease of use: Yellow

Origins: Developed by Bhuiyan at the Indian Institute of Technology, India, as a way to assess groundwater recharge deficits.

Characteristics: As a hydrology-based drought indicator, it uses data from wells to investigate the impact of drought on groundwater recharge. Results can be interpolated between points.

Input parameters: Groundwater well levels

Applications: For areas with frequent seasonal low flows on main rivers and streams.

Strengths: The impact of drought on groundwater is a key component in agricultural and municipal water supplies.

Weaknesses: It only takes groundwater into account, and interpolation between points may not be representative of the region or climate regime.

Reference: Bhuiyan (2004).

Index name: Streamflow Drought Index (SDI)

Ease of use: Yellow

Origins: Developed by Nalbantis and Tsakiris using the methodology and calculations of SPI as the basis.

Characteristics: Uses monthly streamflow values and the methods of normalization associated with SPI for developing a drought index based upon streamflow data. With an output similar to that of SPI, both wet and dry periods can be investigated, as well as the severity of these occurrences.

Input parameters: Monthly streamflow values and a historical time series for the streamflow gauge.

Applications: Used to monitor and identify drought events with reference to a particular gauge, which may or may not represent larger basins.

Strengths: The program is widely available and easy to use. Missing data are allowed, and the longer the streamflow record, the more accurate the results. As with SPI, various timescales can be examined.

Weaknesses: A single input (streamflow) does not take into account management decisions, and periods of no flow can skew the results.

Resources: It is described in the literature with mathematical examples provided. The SPI code is available at http://drought.unl.edu/MonitoringTools/DownloadableSPIProgram.aspx. See http://drinc.ewra.net/ for information on SDI.

Reference: Nalbantis and Tsakiris (2008).

Index name: Surface Water Supply Index (SWSI)

Ease of use: Yellow

Origins: Developed by Shafer and Dezman in 1982 to directly address some of the limitations identified in the PDSI.

Characteristics: Takes into account the work done by Palmer with PDSI but adds additional information including water supply data (snow accumulation, snowmelt and runoff, and reservoir data), and is calculated at the basin level. SWSI identifies the approximate frequency of mild drought occurrence at 26–50 percent, moderate drought occurrence at 14–26 percent and severe drought occurrence at 2–14 percent. Extreme drought occurs approximately less than 2 percent of the time.

Input parameters: Reservoir storage, streamflow, snowpack, and precipitation.

Applications: Used to identify drought conditions associated with hydrological fluctuations.

Strengths: Taking into account the full water resources of a basin provides a good indication of the overall hydrological health of a particular basin or region.

Weaknesses: As data sources change or additional data are included, the entire index has to undergo recalculation to account for these changes in the inputs, making it difficult to construct a homogeneous time series. As calculations may vary between basins, it is difficult to compare basins or homogeneous regions.

Resources: Calculations and an explanation of the methodology are provided in the references below.

References: Doesken and Garen (1991); Doesken et al. (1991); Shafer and Dezman (1982).

Index name: Aggregate Dryness Index (ADI)

Ease of use: Red

Origins: The result of work done at California State University, United States, by Keyantash and at the University of California-Berkeley, United States, by Dracup in 2003.

Characteristics: A multivariate regional drought index that looks at all water resources across many timescales and impacts. It was developed to be used across uniform climate regimes.

Input parameters: Precipitation, evapotranspiration, streamflow, reservoir storage, soil moisture content, and snow water content. The inputs are only used if the region for which ADI is being calculated contains the variable.

Applications: Can be used in the context of multiple types of drought impacts. Looking at the total amount of water in a climate regime allows a better understanding of water availability to be made.

Strengths: Takes into account water stored as well as moisture that comes from precipitation.

Weaknesses: Does not take into account temperatures or groundwater, which are accounted for in the description of ADI.

Resources: The methodology and mathematics are explained in the literature, with examples provided. No code was found for this index.

Reference: Keyantash and Dracup (2004).

Index name: Standardized Snowmelt and Rain Index (SMRI)

Ease of use: Red

Origins: Developed to account for frozen precipitation and how it contributes to runoff into streams as snowmelt. The work was conducted by Staudinger et al., and tested over several Swiss basins.

Characteristics: With methods similar to SPI, SMRI takes into account both rain and snow deficits and the associated impact to streamflow, including precipitation stored as snow. It is most widely used as a complement to SPI.

Input parameters: Streamflow data, daily precipitation, and daily temperature data. Gridded data were used in the initial study of SMRI.

Applications: Focuses on the impact of frozen precipitation and the contribution of this stored water to future streamflows. This index is associated with the monitoring of drought situations.

Strengths: Accounting for snow and future contributions to streamflow, it captures all the inputs into a basin. With the ability to use temperature and precipitation to model snow, actual snow amounts are not needed.

Weaknesses: The use of gridded data and the fact that the data used go back only to 1971 is a drawback when investigating performance using point data and longer periods of record. Not using actual snow depths and associated snow water equivalency can lead to errors in runoff projections.

Resources: Background to the methods and calculations is provided in the literature.

Reference: Staudinger et al. (2014).

8.7.4 Remote Sensing

Index name: Enhanced Vegetation Index (EVI)

Ease of use: Green

Origins: Originated from work done by Huete and a team from Brazil and the University of Arizona, United States, who developed a moderate resolution imaging spectroradiometer (MODIS)-based tool for assessing vegetation conditions.

Characteristics: Vegetation monitoring from satellite platforms using the Advanced Very High-Resolution Radiometer (AVHRR) to compute the Normalized Difference Vegetation Index (NDVI) is quite useful. EVI uses some of the same techniques as NDVI, but with the input data from a MODIS-based satellite. Both EVI and NDVI are calculated using the MODIS platform and analyzed on how they perform compared to AVHRR platforms. EVI is more responsive to canopy variations, canopy type and architecture, and plant physiognomy. EVI can be associated with stress and changes related to drought.

Input parameters: MODIS-based satellite information.

Applications: Used to identify stress related to drought over different landscapes. Mainly associated with the development of droughts affecting agriculture.

Strengths: High resolution and good spatial coverage over all terrains.

Weaknesses: Stress to plant canopies could be caused by impacts other than drought, and it is difficult to discern them using only EVI. The period of record for satellite data is short, with climatic studies being difficult.

Resources: Methodology and calculations are provided in the literature, and online resources of products exist: http://www.star.nesdis.noaa.gov/smcd/emb/vci/VH/vh_browse.php.

Reference: Huete et al. (2002).

Index name: Evaporative Stress Index (ESI)

Ease of use: Green

Origins: Developed by a team led by Anderson, in which remotely sensed data were used to compute evapotranspiration over the United States. The team was composed of scientists from the US Department of Agriculture, the University of Alabama–Huntsville, and the University of Nebraska–Lincoln.

Characteristics: Established as a new drought index in which evapotranspiration is compared to potential evapotranspiration using geostationary satellites. Analyses suggest that it performs similarly to short-term precipitation-based indices, but can be produced at a much higher resolution and without the need for precipitation data.

Input parameters: Remotely sensed potential evapotranspiration.

Applications: Especially useful for identifying and monitoring droughts that have multiple impacts.

Strengths: Very high resolution with a spatial coverage of any area.

Weaknesses: Cloud cover can contaminate and affect results. There is not a long period of record for climatological studies.

Resources: Calculations of the index are provided in the literature: http://hrsl.arsusda.gov/drought/.

Reference: Anderson et al. (2011).

Index name: Normalized Difference Vegetation Index (NDVI)

Ease of use: Green

Origins: Developed from work done by Tarpley et al. and Kogan with the National Oceanic and Atmospheric Administration (NOAA) in the United States.

Characteristics: Uses the global vegetation index data, which are produced by mapping 4 km daily radiance. Radiance values measured in both the visible and near-infrared channels are used to calculate NDVI. It measures greenness and vigor of vegetation over a 7-day period as a way of reducing cloud contamination and can identify drought-related stress to vegetation.

Input parameters: NOAA AVHRR satellite data.

Applications: Used for identifying and monitoring droughts affecting agriculture.

Strengths: Innovative in the use of satellite data to monitor the health of vegetation in relation to drought episodes. Very high resolution and great spatial coverage.

Weaknesses: Data processing is vital to NDVI, and a robust system is needed for this step. Satellite data do not have a long history.

Resources: The literature describes the methodology and calculations. NDVI products are available online: http://www.star.nesdis.noaa.gov/smcd/emb/vci/VH/vh_browse.php.

References: Kogan (1995a); Tarpley et al. (1984).

Index name: Temperature Condition Index (TCI)

Ease of use: Green

Origins: Developed from work done by Kogan with NOAA in the United States.

Characteristics: Using AVHRR thermal bands, TCI is used to determine stress on vegetation caused by temperatures and excessive wetness. Conditions are estimated relative to the maximum and minimum temperatures and modified to reflect different vegetation responses to temperature.

Input parameters: AVHRR satellite data.

Applications: Used in conjunction with NDVI and the Vegetation Condition Index (VCI) for drought assessment of vegetation in situations where agricultural impacts are the primary concern.

Strengths: High resolution and good spatial coverage.

Weaknesses: Potential for cloud contamination as well as a short period of record.

Resources: Methodology and calculations are provided in the literature, and online resources of products exist: http://www.star.nesdis.noaa.gov/smcd/emb/vci/VH/vh_browse.php.

Reference: Kogan (1995b).

Index name: Vegetation Condition Index (VCI)

Ease of use: Green

Origins: Developed from work done by Kogan with NOAA in the United States.

Characteristics: Using AVHRR thermal bands, VCI is used to identify drought situations and determine the onset, especially in areas where drought episodes are localized and ill defined. It focuses on the impact of drought on vegetation and can provide information on the onset, duration, and severity of drought by noting vegetation changes and comparing them with historical values.

Input parameters: AVHRR satellite data.

Applications: Used in conjunction with NDVI and TCI for assessment of vegetation in drought situations affecting agriculture.

Strengths: High resolution and good spatial coverage.

Weaknesses: Potential for cloud contamination as well as a short period of record.

Resources: Methodology and calculations are provided in the literature, and online resources of products exist: http://www.star.nesdis.noaa.gov/smcd/emb/vci/VH/vh_browse.php.

References: Kogan (1995b); Liu and Kogan (1996).

Index name: Vegetation Drought Response Index (VegDRI)

Ease of use: Green

Origins: Developed by a team of scientists from NDMC, the United States Geological Survey's Earth Resources Observation and Science Center, and the United States Geological Survey Flagstaff Field Center.

Characteristics: Developed as a drought index that was intended to monitor drought-induced vegetation stress using a combination of remote sensing, climate-based indicators, and other biophysical information and land-use data.

Input parameters: SPI, PDSI, percentage annual seasonal greenness, start of season anomaly, land cover, soil available water capacity, irrigated agriculture, and defined ecological regions. As some of the inputs are derived variables, additional inputs are needed.

Applications: Used mainly as a short-term indicator of drought for agricultural applications.

Strengths: An innovative and integrated technique using both surface and remotely sensed data, and technological advances in data mining.

Weaknesses: Short period of record due to remotely sensed data. Not useful out of season or during periods of little or no vegetation.

Resources: The methods used and a description of the calculations can be found in the reference given below. See also http://vegdri.unl.edu/.

Reference: Brown et al. (2008).

Index name: Vegetation Health Index (VHI)

Ease of use: Green

Origins: The result of work done by Kogan with NOAA in the United States.

Characteristics: One of the first attempts to monitor and identify drought-related agricultural impacts using remotely sensed data. AVHRR data in the visible, infrared, and near-infrared channels are all used to identify and classify stress to vegetation due to drought.

Input parameters: AVHRR satellite data.

Applications: Used to identify and monitor droughts affecting agriculture around the world.

Strengths: Coverage over the entire globe at a high resolution.

Weaknesses: The period of record for satellite data is short.

Resources: The calculations and sample case studies are given in the literature. VHI maps can be found online at http://www.star.nesdis.noaa.gov/smcd/emb/vci/VH/vh_browse.php.

References: Kogan (1990, 1997, 2001).

Index name: Water Requirement Satisfaction Index (WRSI) and Geo-spatial WRSI

Ease of use: Green

Origins: Developed by the Food and Agriculture Organization of the United Nations to monitor and investigate crop production in famine-prone parts of the world. Additional work was done by the Famine Early Warning Systems Network.

Characteristics: Used to monitor crop performance during the growing season and based upon how much water is available for the crop. It is a ratio of actual to potential evapotranspiration. These ratios are crop specific, and are based upon crop development and known relationships between yields and drought stress.

Input parameters: Crop development models, crop coefficients, and satellite data.

Applications: Used to monitor crop development progress and stress related to agriculture.

Strengths: High resolution and good spatial coverage over all terrains.

Weaknesses: Stress related to factors other than available water can affect the results. Satellite-based rainfall estimates have a degree of error that will affect the results of the crop models used and the balance of evapotranspiration.

Resources: http://chg.geog.ucsb.edu/tools/geowrsi/index.html http://iridl.ldeo.columbia.edu/documentation/usgs/adds/wrsi/WRSI_readme.pdf.

Reference: Verdin and Klaver (2002).

Index name: Normalized Difference Water Index (NDWI) and Land Surface Water Index (LSWI)

Ease of use: Green

Origins: Developed from work done by Gao in the mid-1990s at the National Aeronautics and Space Administration (NASA) Goddard Space Center in the United States.

Characteristics: Very similar to the NDVI methodology, but uses the near-infrared channel to monitor the water content of the vegetation canopy. Changes in the vegetation canopy are used to identify periods of drought stress.

Input parameters: Satellite information in the various channels of the near-infrared spectrum.

Applications: Used for monitoring of drought affecting agriculture as a method of stress detection.

Strengths: High resolution and good spatial coverage over all terrains. Different to NDVI, as the two indices look at different signals.

Weaknesses: Stress to plant canopies can be caused by impacts other than drought, and it is difficult to discern them using only NDWI. The period of record for satellite data is short, with climatic studies being difficult.

Resources: The methodology is described in the literature as are the calculations based on the MODIS data being used: http://www.eomf.ou.edu/modis/visualization/.

References: Chandrasekar et al. (2010); Gao (1996).

Note: The NDWI concept and calculations are very similar to those of the Land Surface Water Index (LSWI).

Index name: Soil Adjusted Vegetation Index (SAVI)

Ease of use: Red

Origins: Developed by Huete at the University of Arizona, United States, in the late 1980s. The idea was to have a global model for monitoring soil and vegetation from remotely sensed data.

Characteristics: SAVI is similar to NDVI—spectral indices may be calibrated in such a way that the variations of soils are normalized and do not influence measurements of the vegetation canopy. These enhancements to NDVI are useful because SAVI accounts for variations in soils.

Input parameters: Remotely sensed data, which are then compared to known surface plots of various vegetation.

Applications: Useful for the monitoring of soils and vegetation.

Strengths: High-resolution and high-density data associated with remotely sensed data allow for very good spatial coverage.

Weaknesses: Calculations are complex, as is obtaining data to run operationally. A short period of record associated with the satellite data can hamper climate analyses.

Resources: The methodology and associated calculations are explained well in the literature.

Reference: Huete (1988).

8.7.5 Composite or Modeled

Index name: Combined Drought Indicator (CDI)

Ease of use: Green

Origins: Developed by Sepulcre-Canto et al. at the European Drought Observatory as a drought index for Europe in which SPI, SMA and fraction of absorbed photosynthetically active radiation (fAPAR) are combined as an indicator for droughts affecting agriculture.

Characteristics: Composed of three warning levels (watch, warning, and alert) by integrating three drought indicators: SPI, soil moisture, and remotely sensed vegetation data. A watch is indicated when there is a precipitation shortage, a warning level is reached when the precipitation shortage translates into a soil moisture shortage, and a warning occurs when the precipitation and soil moisture deficits translate into an impact to the vegetation.

Input parameters: SPI computed from station-based precipitation data throughout Europe; in this case, the 3-month SPI is used. Soil moisture data are obtained using the LISFLOOD model, and fAPAR comes from the European Space Agency.

Applications: Used as an indicator of droughts with agricultural impacts.

Strengths: The spatial coverage is good and at a high resolution using a combination of remotely sensed and surface data.

Weaknesses: Using a single SPI value may not be the best option in all situations and does not represent conditions that may carry over from season to season. Hard to replicate and currently not available for areas outside Europe.

Resources: Housed and maintained at the European Drought Observatory within the European Commission's Joint Research Centre: http://edo.jrc.ec.europa.eu/edov2/php/index.php?id=1000.

Reference: Sepulcre-Canto et al. (2012).

Index name: Global Integrated Drought Monitoring and Prediction System (GIDMaPS).

Ease of use: Green

Origins: Developed from work done by Hao et al. at the University of California in Irvine, United States, as a system to monitor and predict drought over the globe.

Characteristics: Provides drought information for SPI, soil moisture and the Multivariate Standardized Drought Index (MSDI). GIDMaPS also uses satellite data combined with data assimilation tools. The output is produced on a gridded basis in near real time, and combines monitoring and prediction as a way to monitor, assess, and anticipate droughts with multiple impacts.

Input parameters: Uses an algorithm in which remotely sensed data are combined with the Global Land Data Assimilation System (GLDAS) index to produce output for three drought indices as well as seasonal predictions.

Applications: Used for monitoring and predicting by producing values for SPI, MSDI, and Standardized Soil Moisture Index. Can be used for agriculture and other sectors.

Strengths: The gridded and global data represent all areas well. With both a wet and a dry scale, GIDMaPS can be used to monitor more than just drought. It is excellent for areas lacking good surface observations with long periods of record. It is relatively easy to use in that it is computed without the need for input from users.

Weaknesses: Grid sizes may not represent all areas and climate regimes equally. A period of record going back to 1980 is very short when considering climatic applications. To modify it, the code and inputs would need to be obtained.

Resources: The literature explains the process well, and online resources and maps are readily available: http://drought.eng.uci.edu/.

Reference: Hao et al. (2014).

Index name: Global Land Data Assimilation System (GLDAS)

Ease of use: Green

Origins: Rodell led the work, which involved scientists from NASA and NOAA in the United States.

Characteristics: Uses a system of surface and remotely sensed data along with land surface models and data assimilation techniques to provide data on terrestrial conditions. Output includes soil moisture characteristics, which are a good drought indicator.

Input parameters: Land surface models, surface-based meteorological observations, vegetation classifications, and satellite data.

Applications: Useful for determining river and streamflow projections as well as runoff components based on current conditions; ideal for monitoring droughts that have multiple impacts.

Strengths: As it is global in nature and available at a high resolution, it can represent most areas. Useful for monitoring developing drought in areas that are data poor.

Weaknesses: The grid size is not sufficiently fine for island nations. Only areas that lack near-real-time surface observations are represented by the data assimilation process.

Resources: The methodology and inputs are described well in the literature. Output is available online.

https://climatedataguide.ucar.edu/climate-data/

nldas-north-american-land-data-assimilation-system-monthly-climatologies http://ldas.gsfc.nasa.gov/nldas/

http://disc.sci.gsfc.nasa.gov/services/grads-gds/gldas

References: Mitchell et al. (2004); Rodell et al. (2004); Xia et al. (2012).

Index name: Multivariate Standardized Drought Index (MSDI)

Ease of use: Green

Origins: Developed by Hao and AghaKouchak at the University of California at Irvine, United States

Characteristics: Uses information on both precipitation and soil moisture to identify and classify drought episodes by investigating precipitation and soil moisture deficits. It is helpful for identifying drought episodes where typical precipitation-based indicators or soil-moisture-based indicators may not indicate the presence of drought.

Input parameters: Monthly precipitation and soil moisture data are needed from the Modern Era Retrospective Analysis (MERRA)-Land systems. MERRA-land data are generated by a 0.66° × 0.50° grid from 1980 onward.

Applications: Useful for the identification and monitoring of drought in cases where precipitation and soil moisture are important contributors to impacts.

Strengths: The gridded and global data represent all areas well. Integrating both a wet and dry scale, it can be used to monitor more than just drought. It is excellent for areas lacking good surface observations with long periods of record. It is relatively easy to use in that it is computed without the need for input from users. Individual indices can be obtained from MSDI output.

Weaknesses: Grid size may not represent all areas and climate regimes equally. A period of record going back to 1980 is very short when considering climatic applications. To modify, the code and inputs would need to be obtained. Not all timescales are produced for SPI and Standardized Soil Moisture Index outputs.

Resources: The literature explains the process well, and online resources and maps are readily available: http://drought.eng.uci.edu/.

Reference: Hao and AghaKouchak (2013).

Index name: US Drought Monitor (USDM)

Ease of use: Green

Origins: Developed by Svoboda et al. in the late 1990s as an analysis of drought conditions using the results of many indicators and inputs, and based on comparing current data with historical conditions. The work was the first operational composite approach applied in the United States.

Characteristics: Uses a method of percentile ranking in which indices and indicators from various periods of record can be compared equivalently. It has a scale of five intensity levels, from abnormally dry conditions that will occur about every three to five years, to exceptional drought conditions that will occur about once every fifty years. It is flexible in that any number of inputs can be used, and it has a level of subjectivity that allows for the inclusion of drought-related impacts in the analysis.

Input parameters: Flexible, as there are no set numbers of indicators. Originally, only a few inputs were used; currently, the construction of USDM involves analysis of 40–50 inputs. Drought indices, soil moisture, hydrological inputs, climatological inputs, modeled inputs, and remotely sensed inputs are all included in the analysis. As new indicators are developed, the USDM is flexible enough to include them.

Applications: Ideal for monitoring droughts that have many impacts especially on agriculture and water resources during all seasons over all climate regimes. It is a weekly product, but can also be adapted for monthly analyses.

Strengths: Uses many indices and indicators, which makes the final results more robust. It is flexible to meet the needs of various users. It was innovative in the way it identified drought and classified intensities, and has the ability to analyze data from various timescales using the percentile ranking methodology.

Weaknesses: Operational data are needed, as most current inputs will provide the best results when the analysis is done. If only a few inputs are available, USDM analysis becomes weaker, but it remains applicable.

Resources: The methodology is explained well in the literature and online, http://droughtmonitor.unl.edu/.

Reference: Svoboda et al. (2002).

References

Alley, W. M. 1984. The Palmer Drought Severity Index: Limitations and assumptions. *Journal of Applied Meteorology* 23:1100–1109.

Anderson, M. C., C. Hain, B. Wardlow, A. Pimstein, J. R. Mecikalski, and W. P. Kustas. 2011. Evaluation of drought indices based on thermal remote sensing of evapotranspiration over the continental United States. *Journal of Climate* 24(8):2025–2044.

Baltas, E. 2007. Spatial distribution of climatic indices in northern Greece. *Meteorological Applications* 14:69–78.

Bergman, K. H., P. Sabol, and D. Miskus. 1988. Experimental indices for monitoring global drought conditions. Proceedings of 13th Annual Climate Diagnostics Workshop, Cambridge, MA, October 31–November 4, pp. 190–197 U.S. Department of Commerce, Washington, DC.

Bhalme, H. N. and D. A. Mooley. 1980. Large-scale droughts/floods and monsoon circulation. *Monthly Weather Review* 108:1197–1211.

Bhuiyan, C. 2004. Various drought indices for monitoring drought condition in Aravalli Terrain of India. Proceedings of the XXth ISPRS Conference. International Society for Photogrammetry and Remote Sensing, Istanbul, Turkey. http://www.isprs.org/proceedings/XXXV/congress/comm7/papers/243.pdf.

British Columbia Ministry of Agriculture. 2015. *Soil water storage capacity and available soil moisture.* Water Conservation Fact Sheet. http://www2.gov.bc.ca/assets/gov/farming-natural-resources-and-industry/agriculture-and-seafood/agricultural-land-and-environment/soil-nutrients/600-series/619000-1_soil_water_storage_capacity.pdf.

Brown, J. F., B. D. Wardlow, T. Tadesse, M. J. Hayes, and B. C. Reed. 2008. The Vegetation Drought Response Index (VegDRI): A new integrated approach for monitoring drought stress in vegetation. *GIScience & Remote Sensing* 45:16–46.

Byun, H. R., and D. A. Wilhite. 1996. Daily quantification of drought severity and duration. *Journal of Climate* 5:1181–1201.

Chandrasekar, K., M. V. R. Sesha Sai, P. S. Roy, and R. S. Dwevedi. 2010. Land Surface Water index (LSWI) response to rainfall and NDVI using the MODIS vegetation index product. *International Journal of Remote Sensing* 31:3987–4005.

De Martonne, E. 1925. *Traité de géographie physique.* Vol. 11. Paris: Armand Colin.

Doesken, N. J., and D. Garen. 1991. Drought Monitoring in the western United States using a surface water supply index. Preprints, Seventh Conference on Applied Climatology, Salt Lake City, Utah, September 10-13, pp. 266–269. American Meteorology Society, Boston, MA.

Doesken, N. J., T. B. McKee, and J. Kleist. 1991. *Development of a surface water supply index for the western United States.* Climatology Report 91-3, Colorado Climate Center. http://climate.colostate edu/pdfs/climo_rpt_91-3.pdf.

Edwards, D. C., and T. B. McKee. 1997. Characteristics of 20th century drought in the United States at multiple time scales. *Atmospheric Science* 634:1–30.

Gao, B. C. 1996. NDWI—A Normalized Difference Water Index for remote sensing of vegetation liquid water from space. *Remote Sensing of Environment* 58(3):257–266.

Gibbs, W. J., and J. V. Maher. 1967. *Rainfall deciles as drought indicators.* Bureau of Meteorology Bulletin No. 48, Melbourne, Australia.

Gusyev, M. A., A. Hasegawa, J. Magome, D. Kuribayashi, H. Sawano, and S. Lee. 2015. Drought Assessment in the Pampanga River Basin, the Philippines. Part 1: A role of dam infrastructure in historical droughts. Proceedings of the 21st International Congress on Modelling and Simulation (MODSIM 2015), Broadbeach, Queensland, Australia , 29 November–4 December 2015, pp. 1586–1592.

Guttman, N. B. 1998. Comparing the Palmer Drought Index and the Standardized Precipitation Index. *Journal of the American Water Resources Association* 34:113–121. doi:10.1111/j.1752-1688.1998. tb05964.

Guttman, N. B. 1999. Accepting the Standardized Precipitation Index: A calculation algorithm. *Journal of the American Water Resources Association* 35:311–322. doi:10.1111/j.1752-1688.1999. tb03592.x.

Hao, Z., and A. AghaKouchak. 2013. Multivariate Standardized Drought Index: A multi-index parametric approach for drought analysis. *Advances in Water Resources* 57:12–18.

Hao, Z., A. AghaKouchak, N. Nakhjiri, and A. Farahmand. 2014. Global integrated drought monitoring and prediction system. *Scientific Data* 1:1–10.

Hayes, M. J. 2006. Drought indices. In *Van Nostrand's Scientific Encyclopedia.* Hoboken, NJ: Wiley. http://onlinelibrary.wiley.com/doi/10.1002/0471743984.vse8593/abst ract;jsessionid=CA39E5A4F67AA81580F505CBB07D2424f01t04.

Hayes, M., M. Svoboda, N. Wall, and M. Widhalm. 2011. The Lincoln Declaration on Drought Indices: Universal meteorological drought index recommended. *Bulletin of the American Meteorological Society* 92:485–488.

Heim, R. R. 2002. A review of twentieth-century drought indices used in the United States. *Bulletin of the American Meteorological Society* 83:1149–1165.

Huete, A., K. Didan, T. Miura, E. P. Rodriguez, X. Gao, and L. G. Ferreira. 2002. Overview of the radiometric and biophysical performance of the MODIS vegetation indices. *Remote Sensing of Environment* 83(1):195–213.

Huete, A. R. 1988. A soil-adjusted vegetation index (SAVI). *Remote Sensing of Environment* 25(3):295–309.

Intergovernmental Panel on Climate Change (IPCC). 2012. *Managing the risks of extreme events and disasters to advance climate change adaptation.* Special Report of Working Groups I and II of the IPCC, eds. C. B. Field, V. Barros, T. F. Stocker, et al. Cambridge: Cambridge University Press.

Katz, R. W., and M. H. Glantz. 1986. Anatomy of a rainfall index. *Monthly Weather Review* 114:764–771.

Keetch, J. J., and G. M. Byram. 1968. *A drought index for forest fire control*. United States Department of Agriculture Forest Service Research Paper SE-38, Southeastern Forest Experiment Station, Asheville, NC.

Keyantash, J., and J. A. Dracup. 2002. The quantification of drought: An evaluation of drought indices. *Bulletin of the American Meteorological Society* 83:1167–1180.

Keyantash, J. A., and J. A. Dracup. 2004. An aggregate drought index: Assessing drought severity based on fluctuations in the hydrologic cycle and surface water storage. *Water Resources Research* 40:W09304. doi:10.1029/2003WR002610.

Kogan, F. N. 1990. Remote sensing of weather impacts on vegetation in non-homogeneous areas. *International Journal of Remote Sensing* 11:1405–1419.

Kogan, F. N. 1995a. Droughts of the late 1980s in the United States as derived from NOAA polar-orbiting satellite data. *Bulletin of the American Meteorology Society* 76(5):655–668.

Kogan, F. N. 1995b. Application of vegetation index and brightness temperature for drought detection. *Advances in Space Research* 15(11):91–100.

Kogan, F. N. 1997. Global drought watch from space. *Bulletin of the American Meteorological Society* 78:621–636.

Kogan, F. N. 2001. Operational space technology for global vegetation assessments. *Bulletin of the American Meteorological Society* 82(9):1949–1964.

Kraus, E. B. 1977. Subtropical droughts and cross-equatorial energy transports. *Monthly Weather Review* 105(8):1009–1018.

Liu, W. T., and F. N. Kogan. 1996. Monitoring regional drought using the Vegetation Condition Index. *International Journal of Remote Sensing* 17(14):2761–2782.

Lyon, B. 2004. The strength of El Niño and the spatial extent of tropical drought. *Geophysical Research Letters* 31:L21204. doi:10.1029/2004GL020901.

McKee, T. B., N. J. Doesken, and J. Kleist. 1993. The relationship of drought frequency and duration to time scales. Proceedings of the 8th Conference on Applied Climatology, 17–23 January 1993, Anaheim, California. American Meteorological Society, Boston, MA.

Meyer, S. J., K. G., Hubbard, and D. A. Wilhite. 1993a. A crop-specific drought index for corn. I. Model development and validation. *Agronomy Journal* 85:388–395.

Meyer, S. J., K. G. Hubbard, and D. A. Wilhite. 1993b. A crop-specific drought index for corn. II. Application in drought monitoring and assessment. *Agronomy Journal* 85:396–399.

Mitchell, K., D. Lohman, P. Houser, et al. 2004. The multi-institution North American Land Data Assimilation System (NLDAS): Utilizing multiple GCIP products and partners in a continental distributed hydrological modelling system. *Journal of Geophysical Research* 109:D07S90. doi:10.1029/2003JD003823.

Modarres, R. 2007. Streamflow drought time series forecasting. *Stochastic Environmental Research and Risk Assessment* 21:223–233.

Nalbantis, I., and G. Tsakiris. 2008. Assessment of hydrological drought revisited. *Water Resources Management* 23(5):881–897.

Narasimhan, B., and R. Srinivasan. 2005. Development and evaluation of Soil Moisture Deficit Index (SMDI) and Evapotranspiration Deficit Index (ETDI) for agricultural drought monitoring. *Agricultural and Forest Meteorology* 133(1):69–88.

Palmer, W. C. 1965. *Meteorological drought*. Research Paper No. 45, US Weather Bureau, Washington, DC.

Palmer, W. C. 1968. Keeping track of crop moisture conditions, nationwide: The Crop Moisture Index. *Weatherwise* 21:156–161.

Rodell, M., P. Houser, U. Jambor, et al. 2004. The Global Land Data Assimilation System. *Bulletin of the American Meteorological Society* 85(3):381–394.

Selyaninov, G. T. 1928. About climate agricultural estimation. *Proceedings on Agricultural Meteorology* 20:165–177.

Sepulcre-Canto, G., S. Horion, A. Singleton, H. Carrao, and J. Vogt. 2012. Development of a combined drought indicator to detect agricultural drought in Europe. *Natural Hazards and Earth Systems Sciences* 12:3519–3531.

Shafer, B. A., and L. E. Dezman. 1982. Development of a surface water supply index (SWSI) to assess the severity of drought conditions in snowpack runoff areas. Proceedings of the Western Snow Conference, April 19–23 (Reno, NV), pp. 164–175. Colorado State University, Fort Collins, CO.

Sivakumar, M. V. K., R. P. Motha, D. A. Wilhite and D. A. Wood, eds. 2011. Agricultural drought indices. Proceedings of a WMO/UNISDR Expert Group Meeting on Agricultural Drought Indices, Murcia, Spain, 2–4 June 2010 (AGM-11, WMO/TD No. 1572; WAOB-2011). Geneva. http://www.droughtmanagement.info/literature/WMO_agricultural_drought_indices_proceedings_2010.pdf.

Staudinger, M., K. Stahl, and J. Seibert. 2014. A drought index accounting for snow. *Water Resources Research* 50:7861–7872. doi:10.1002/2013WR015143.

Strommen, N. D., and R. P. Motha. 1987. An operational early warning agricultural weather system. In *Planning for Drought: Toward a Reduction of Societal Vulnerability*, eds. D. A. Wilhite, W. E. Easterling, and D. A. Wood, pp. 153–162. Boulder, CO: Westview Press.

Svoboda, M., D. LeComte, M. Hayes, et al. 2002. The Drought Monitor. *Bulletin of the American Meteorological Society* 83(8):1181–1190.

Tarpley, J. D., S. R. Schneider, and R. L. Money. 1984. Global vegetation indices from the NOAA-7 meteorological satellite. *Journal of Climate and Applied Meteorology* 23:491–494.

Telesca, L., M. Lovallo, I. Lopez-Moreno, and S. Vicente-Serrano. 2012. Investigation of scaling properties in monthly streamflow and Standardized Streamflow Index time series in the Ebro basin (Spain). *Physica A: Statistical Mechanics and its Applications* 391(4):1662–1678.

Tsakiris, G., and H. Vangelis. 2005. Establishing a drought index incorporating evapotranspiration. *European Water* 9(10):3–11.

van Rooy, M.P. 1965. A rainfall anomaly index independent of time and space. *Notos* 14:43–48.

Verdin, J., and R. Klaver. 2002. Grid-cell-based crop water accounting for the famine early warning system. *Hydrological Processes* 16(8):1617–1630.

Vicente-Serrano, S. M., S. Begueria, and J. I. Lopez-Moreno. 2010. A multi-scalar drought index sensitive to global warming: The Standardized Precipitation Evapotranspiration Index. *Journal of Climate* 23:1696–1718.

Weghorst, K. 1996. *The Reclamation Drought Index: Guidelines and Practical Applications.* Denver, CO: Bureau of Reclamation.

Wells, N., S. Goddard, and M. J. Hayes. 2004. A self-calibrating Palmer Drought Severity Index. *Journal of Climate* 17:2335–2351.

Wilhite, D., and M. Glantz. 1985. Understanding the drought phenomenon: The role of definitions. *Water International* 10:111–120.

WMO (World Meteorological Organization). 2012. *Standardized Precipitation Index User Guide*. WMO-No.1090, Geneva. http://www.droughtmanagement.info/literature/WMO_standardized_precipitation_index_user_guide_en_2012.pdf.

WMO (World Meteorological Organization) and GWP (Global Water Partnership). 2014. National Drought Management Policy Guidelines: A Template for Action (ed. D. A. Wilhite). Integrated Drought Management Programme (IDMP) Tools and Guidelines Series 1. WMO, Geneva. http://www.droughtmanagement.info/literature/IDMP_NDMPG_en.pdf.

WMO (World Meteorological Organization (WMO), UNCCD (United Nations Convention to Combat Desertification) and FAO (Food and Agriculture Organization of the United Nations). 2013. *High Level Meeting on National Drought Policy*, Geneva, March 11–15, 2013. Policy Document: National Drought Management Policy. http://www.wmo.int/pages/prog/wcp/drought/hmndp/documents/PolicyDocumentRev_12-2013_En.pdf.

Woli, P., J. W. Jones, K. T. Ingram, and C. W. Fraisse. 2012. Agricultural Reference Index for Drought (ARID). *Agronomy Journal* 104:287–300.

Wu, H., M. J. Hayes, A. Weiss, and Q. Hu. 2001. An evaluation of the Standardized Precipitation Index, the China-Z Index and the statistical Z-score. *International Journal of Climatology* 21:745–758.

Wu, H., M. J. Hayes, D. A. Wilhite, and M. D. Svoboda. 2005. The effect of the length of record on the Standardized Precipitation Index calculation. *International Journal of Climatology* 25(4):505–520.

Xia, Y., K. Mitchell, M. Ek, et al. 2012. Continental-scale water and energy flux analysis and validation for the North American Land Data Assimilation System project phase 2 (NLDAS-2): 1. Intercomparison and application of model products. *Journal of Geophysical Research* 117:D03109. doi:10.1029/2011JD016048.

Zargar, A., R. Sadiq, B. Naser, and F. I. Khan. 2011. A review of drought indices. *Environmental Reviews* 19:333–349.

Bibliography

Eriyagama, N., V. Smakhtin, and N. Gamage. 2009. *Mapping drought patterns and impacts: A global perspective*. IWMI Research Report No. 133. International Water Management Institute, Colombo, Sri Lanka. http://www.iwmi.cgiar.org/Publications/IWMI_Research_Reports/PDF/PUB133/RR133.pdf.

Hayes, M. J. 2011. *Comparison of Major Drought Indices: Introduction*. National Drought Mitigation Center, University of Nebraska-Lincoln. http://drought.unl.edu/Planning/Monitoring/ComparisonofIndicesIntro.aspx.

Hisdal, H., and L. M. Tallaksen, ed. 2000. *Drought event definition*. Technical Report 6 of the ARIDE Project, Assessment of the Regional Impact of Droughts in Europe, Department of Geophysics, University of Oslo, Norway.

Lawrimore, J., R. R. Heim, M. Svoboda, V. Swail, and P. J. Englehart. 2002. Beginning a new era of drought monitoring across North America. *Bulletin of the American Meteorological Society* 83:1191–1192.

Lloyd-Hughes, B. 2014. The impracticality of a universal drought definition. *Theoretical and Applied Climatology* 117(3):607–611. doi:10.1007/s00704-013-1025-7.

Mishra, A. K., and V. P. Singh. 2010. A review of drought concepts. *Journal of Hydrology* 391:202–216.

Mishra, A. K., and V. P. Singh. 2011. Drought modeling. A review. *Journal of Hydrology* 403:157–175.

Pulwarty, R. S., and M. Sivakumar. 2014. Information systems in a changing climate: Early warnings and drought risk management. *Weather and Climate Extremes* 3:14–21.

Svoboda, M., B. A. Fuchs, C. Poulsen, and J. R. Nothwehr. 2015. The Drought Risk Atlas: Enhancing decision support for drought risk management in the United States. *Journal of Hydrology* 526:274–286. doi:10.1016/j.jhydrol.2015.01.006.

Wardlow, B. D., M. C. Anderson, and J. P. Verdin, eds. 2012. *Remote Sensing of Drought: Innovative Monitoring Approaches*. Boca Raton, FL: CRC.

WMO (World Meteorological Organization). 2006. *Drought monitoring and early warning: Concepts, progress and future challenges*. WMO-No. 1006, Geneva. http://www.droughtmanagement.info/literature/WMO_drought_monitoring_early_warning_2006.pdf.

9

The Application of Triggers in Drought Management: An Example from Colorado

Taryn Finnessey and Nolan Doesken

CONTENTS

9.1 Introduction

This chapter describes a state perspective on drought and explains how one state, Colorado, has learned to use climatic data and monitoring tools to create and use various drought indexes to track water supplies and then trigger actions—all for the purpose of guiding mitigation and response activities.

Colorado's interior continental location means that maritime moisture sources are distant, and atmospheric water supplies are not reliable. This results in a largely semiarid climate where the presence or absence of a few storms each year is the difference between a year with adequate water or a year of shortage. Colorado's mid-latitude location, high overall elevation, and tall mountain ranges mean that snow is a critical part of water supply. The mountain snowpack accumulates over a 6- to 8-month period. For some users, like the state's vast winter recreation industry, the timing of snow accumulation and melt is critical. But for many other users, such as agriculture and municipal water providers, it doesn't matter as much when the snow falls so long as there is enough water in the mountain snowpack by late spring. Then, in several weeks from late April into early July, the mountain snowpack quickly melts, providing the bulk of the entire year's water supplies. Spring rains at lower elevations, summer thunderstorms, and occasional widespread soaking rains in autumn have the potential to make up for occasional and sometimes extreme shortages in winter snowpack. (This is in contrast to California, where there is only one wet season.) But not all of Colorado is watered from melting snow. For many lower-elevation grassland and forest environments, if the spring rains do not materialize and summer thunderstorms are limited, extreme drought can develop quickly.

The result of these factors is that moderate or greater drought is present in some part of Colorado for some portion of the year in more than 9 out of 10 years (McKee 2000). This is the climatic background from which this case study begins.

Perhaps the most complex aspect of drought is that it develops relatively slowly over periods of weeks, months, and even years. Drought commonly extends over multiple seasons and years. Even the professionals whose job it is to track drought often do not know for sure that a drought has started until it is well developed, and they often do not know when it has ended until they look back on it. Other natural disasters (including floods, earthquakes, hurricanes, landslides, and blizzards) have clearly defined beginning and end points and require fairly specific responses. This is not the case with drought, which may mask itself as prolonged nice (i.e., sunny) weather. As a result, the detection of impacts is sometimes the first indication that a drought is occurring. This presents serious challenges as it results in a reactionary response rather than a phased and incremental one. Moreover, a reactionary response is solely focused on crisis management and does not focus on long-term impact reduction through the implementation of risk management actions. This may result in higher costs to individuals, society, and the environment. Evidence shows that dealing with disaster while in the midst of it can be more costly than proactive measures and can lead to less than ideal solutions (Multihazard Mitigation Council 2005). Consequently, there have been significant efforts to promote earlier detection and earlier response to drought.

The State of Colorado adopted its first formal "Drought Response Plan" in 1981 during the second extreme winter snow drought in a 5-year period (1976–1977 and 1980–1981). Both of these droughts occurred during the administration of Governor Richard (Dick) Lamm and during a period of very rapid population growth in Colorado and growth in the ski industry. From the very start, Colorado attempted to use a combination of thorough climate and water supply monitoring in combination with a quantitative approach to triggering actions and responses to drought. From these beginnings, Colorado has gone on to improve approaches and stimulate the development of better measures and indexes, and has developed a much more comprehensive drought mitigation and response plan (Colorado Water Conservation Board 2013) that not only incorporates a multistage response framework but also details monitoring methods, mitigation actions, and vulnerabilities. This multifaceted approach evolved through decades of experience, and was developed with the intent of reducing impacts and costs by enabling earlier response to drought through specific monitoring of indices and predetermined trigger points for action.

9.2 Establishing Triggers

Continual monitoring of selected hydroclimatic variables such as precipitation, snowpack, streamflow, reservoir levels, and evaporation rates, even during nondrought periods, provides baseline data to help detect emerging drought conditions long before impacts are felt. Using indices to determine triggers or thresholds at which actions should be taken provides guidance to decision makers during the onset of an event. These should be viewed as experience-based guidelines rather than strict rules, since droughts differ so much from one event to the next. Some are prolonged and persistent but not initially intense, while others may be short-lived but extremely severe with broad and costly impacts. An appropriate response or triggered action during one event may not be applicable during the next.

9.2.1 Selection of Indices

There are many drought-related indices, each targeting specific types of information. Some provide information on one discrete variable, such as reservoir storage levels or snowpack on a certain date. Other indices blend multiple variables into one composite index to give the user additional information, such as the first index used in Colorado drought response, the Palmer Drought Severity Index (PDSI), which blends information about the temperature and precipitation of a region into one index. Which indices (and how many) are best for monitoring depends entirely on the climate, user, vulnerabilities, and desired outcome (see Chapters 7 and 8).

The selection of the most appropriate indicator(s) is context-specific: it depends partly on the characteristics of a region's climate, but also on the particular societal and ecological vulnerabilities identified in the drought planning process, and the impacts that are desired to be reduced. Ideally, multiple drought indicators will be used (e.g., standardized precipitation index [SPI] and PDSI), since the unique indicators will represent different dimensions of the same drought event. An exception to this rule is observed with water managers, who will rely primarily on reservoir storage levels in key reservoirs to trigger drought response actions, as is the case with Colorado's largest municipal water provider, Denver Water (Denver Water 2016).

9.2.2 Determining Thresholds

Selecting the proper indices is only part of the process. Establishing appropriate and justifiable triggers for each index is critical for guiding and streamlining action during a drought event. Thresholds for action can be both qualitative and quantitative as both provide value, but the use of one without the other can lessen the effectiveness by creating a trigger point that is either hard to define and subjective, or so rigid it triggers a response when one may not yet be warranted.

For instance, a drought plan could state that when the 3-month SPI drops below −1.0, departmental public information officers will work together to establish coordinated media outreach and messaging. In the midst of a dry, hot summer when crops are likely suffering and fires may be burning, this would make a great deal of sense; but what if this occurs over the winter season when there may be fewer impacts and reservoir storage is still high? Implementing a messaging campaign could create alarm when there is not yet a need for it. By using the quantitative data hand-in-hand with qualitative information, response actions can be initiated at an appropriate time and scale.

Because a drought can develop slowly and last for years, or even decades, a multistage response is also important, as the response needs to reflect whether conditions are improving or deteriorating. Having multiple stages, and thus multiple trigger points for each index, enables mitigation and response actions to be phased in and out. This structure also provides a mechanism to guide decision makers during an event, enabling hard decisions to be made objectively by simply following the plan.

9.2.3 Activating Responses to Triggers

Actions associated with each trigger point should also be predetermined and should have multiple phases or stages. True, there will always be unanticipated issues that come up that will need to be addressed, but fleshing out the who, what, and how beforehand helps create clarity when a drought crisis begins to develop and provides a road map for immediate, but gradual, action. It also helps to identify key actors as people move from position to position; as long periods of time can pass between drought events, individuals may not have had to respond to a drought during their tenure, or even know that such a response falls under their purview.

Predetermined triggers and responses tied to those triggers can also help to depoliticize decision-making since there is a clear, phased, data-driven framework for action that addresses a wide array of potential impacts and sectors. Examining possible actions to be taken at each stage of drought and who should be the lead on those actions can eliminate the lag period in which decision makers are trying to determine what to do. It can also reduce hasty, uninformed decisions that may not be the most beneficial use of resources. In short, including drought triggers in the drought preparedness planning efforts will likely reduce impacts and response time to an event.

9.3 Developing a Collaborative Process

Monitoring drought, establishing triggers, and developing a response and mitigation plan requires involvement and collaboration from several different areas of expertise. It is important to include those managing and providing the data, often scientists and researchers, in this partnership as they understand the data and indices, and they are the most capable at continually monitoring drought conditions. In addition to providers of data, the users of the data and indices must also be present. The users of the data are likely the groups who are experiencing and responding to impacts. Finally, those who manage and write policy are essential for bridging the gap between the two other groups, by planning the actions triggered when an index crosses a certain threshold. Having all three parties involved in the selection of indices and development of the trigger points also helps during the onset of an event as all parties have some understanding of the data and have already agreed upon how it will be used.

9.3.1 Understanding Usability

Climate data can be complex and even overwhelming, especially when there are multiple data sources measuring a wide array of elements. This can easily cause confusion among the decision makers and managers whom the data are intended to help. Studies on the use of climate information have identified persistent barriers to effective use; decision makers may misinterpret and misuse data that are unfamiliar to them, or they may find that new data are not suited to their desired application and thus not use them (Rayner et al. 2005; Lemos et al. 2012; Bolson et al. 2013).

To bridge this gap, it is most effective for decision makers and researchers to work collaboratively and iteratively to develop tools that directly inform the planning process (Lemos and Morehouse 2005; Dilling and Lemos 2011). This model, known as coproduction of climate information and services, allows users to express their needs and researchers to build tools that are specifically targeted to meet those needs. Moreover, this allows researchers the opportunity to engage with drought and water professionals to ensure that those researchers understand users, and that those users have greater understanding of the inherent uncertainties that exist in the data which ultimately can help to ensure that the data is being used in an appropriate and reliable manner (Bolson et al. 2013). This coproduction model illustrates that overcoming the barriers mentioned with respect to proper use of data, and broadening the use of climate information, for example, in drought planning, requires repeated engagement (Ferguson et al. 2014). Additionally, those who utilize this method can provide further value by helping to convey the feasibility and benefits of using new information to others as well as highlight potential data sources and analytical approaches (Finnessey 2016).

In addition to working collaboratively with scientists and decision makers to ensure that data are being used properly and effectively, it is also critical that once trigger points for each index are drafted, stakeholders have the opportunity to review and comment on them. While this takes more time during the planning process, it is essential to building trust and credibility, which is necessary to have in order to act swiftly once thresholds are met and action is initiated. Linking appropriate actions to each trigger point is essential to ensuring a systematic, staged response. Without this, the establishment of specific trigger points loses its utility. Providing an opportunity for stakeholder involvement in developing both the thresholds and the associated actions helps to eliminate controversy when an event occurs as everyone is already aware and supportive of the steps that will be taken at each stage to respond.

9.3.2 Protocol for Evaluation and Updates

The purpose of a plan, of any kind, is not to sit on a shelf and collect dust, but rather help chart a clear and coordinated path forward. Plans are composed of a multitude of elements that need to be frequently reevaluated to ensure that they are still applicable and that new knowledge and information is incorporated. The same is true with drought indicators and triggers. New products may become available that provide more utility than past products, or that fill a void that no other existing product could. Thresholds may need to be adjusted to more accurately represent what actually occurred during a drought event, and actions for each of those thresholds may need to be adjusted to reflect improved adaptive capacity or political realities.

There is no right or wrong timeframe for reevaluation, but it should occur on a regular basis and frequently enough that lessons learned can be incorporated while they are still reasonably fresh in users' minds. As an example, the Federal Emergency Management Agency requires that states and local governments update their hazard mitigation plans once every 5 years.

9.4 Examples of Triggers in Action

According to President Dwight Eisenhower, "Plans are useless, but planning is everything." He went on to explain that the definition of an emergency is that it is unexpected, so it will not occur the way you plan it. This is certainly true of drought; but comprehensive, proactive, and integrated drought planning and preparedness can help states and communities to be better equipped to respond when a drought event does occur. As has been shown in Colorado, the combination of active monitoring, proactive mitigation, a qualitative and quantitative vulnerability assessment, and a well-thought-out staged response framework can lessen the overall negative impacts of an event. Furthermore, comprehensive drought planning can inform overall water planning efforts, helping to ensure a more secure water future for regions prone to water scarcity (Finnessey 2016).

9.4.1 Colorado's Drought Mitigation and Response Plan

The State of Colorado has incorporated quantitative trigger points that guide the activation of the staged drought response plan. These trigger points were developed by analyzing observed climate data and overlaying that information with past impacts. This provided quantitative thresholds at which certain impacts are likely to start occurring. The existence of these predetermined decision points has helped to depoliticize the activation process and speed aid to those most impacted by drought. Without the use of long-term observed climate records, it would not have been possible to accurately develop these thresholds.

While the plan has had general triggers for action since its initial development, they were refined after the 2002 drought, which was the driest year on record in Colorado, and then refined again in 2008. In 2012, the state faced another severe statewide drought in what turned out to be the second-driest year on record, but by using the triggers, the state began responding to the drought before the impacts became as severe as in 2002. As a result of this and other changes made after the 2002 drought, the overall drought response in Colorado was more coordinated in 2012 than in 2002 (Ryan and Doesken 2013), with entities such as municipal water providers implementing response measures sooner than previously implemented, and tourism and recreation outfitters diversifying activities to offset revenue losses.

The State of Colorado officially uses a number of indices to guide statewide decision making, including the US Drought Monitor (USDM), the Colorado Modified Palmer Drought Index (CMPDI), the surface water supply index (SWSI), and the SPI. All but one of these indices are described in detail in Chapter 8 in this book, the exception being the CMPDI. The CMPDI is a complex soil moisture calculation, similar to the Palmer drought severity index in that it requires weekly or monthly precipitation and temperature data as inputs. However, the PDSI was initially developed for areas of the country with more precipitation and more homogeneous climates, so Colorado adapted the index by separating the state into 25 climatically similar regions. In recent years, the Colorado Climate Center has added a 26th region—the Sangre de Cristo Mountains, which previously did not have adequate data. The Colorado Modified Palmer Index uses a +4 to −4 scale. It uses a 0 as normal, and drought is shown in terms of negative numbers; for example, −2 is moderate drought, −3 is severe drought, and −4 is extreme drought (Colorado Water Conservation Board 2013).

9.4.2 Implementing Task Forces

Continual monitoring is essential when using triggers for drought response, as you need to be aware of current conditions at all times. In Colorado, the Governor's Water Availability Task Force remains active regardless of drought conditions, as this is the group responsible for continuous monitoring. All other task forces are activated only when conditions meet the predetermined thresholds outlined in the drought mitigation and response plan. Activation can be tailored to specific regions of the state and for specific sectors based on where the impacts are most prevalent. For example, in 2011 the comprehensively revised plan was activated for the agricultural sector in response to extreme and exceptional drought conditions, as classified by the USDM, in southeastern and south central Colorado. It was later expanded to the entire state in 2012 as conditions continued to decline, and the Municipal Water Impact Task Force was also activated.

Once triggers for action are met, a staged drought response is initiated. This includes the activation of the Drought Task Force (DTF), which is composed of cabinet members from the Colorado Department of Agriculture, Department of Local Affairs, Department of Natural Resources, and Department of Public Safety. These members meet monthly during activation to coordinate response efforts and address cross-agency issues that arise. They also have a direct line to the governor, which can help to streamline communication and improve response times. This group is crucial for providing some of the qualitative context necessary for decision-making, while also responding on the quantitative data to guide that process.

9.4.3 Case Study—Colorado 2011–2013 Drought

In May 2011, it became apparent that drought conditions had begun to develop in south central and southeast Colorado and that the agricultural communities in those areas were likely to experience severe impacts as a result. Consequently, the Governor activated the state's drought mitigation and response plan at a phase 2 level for the agricultural sector in that region of the state, which resulted in a number of actions, as highlighted in Table 9.1.

TABLE 9.1

Drought Response Plan Summary

Severity Indicators and Impacts (Colorado Modified Palmer Drought Index [CMPDI] or SWSI, SPI, and US Drought Monitor)	Drought Phase and Response Summary	Actions to be Considered
−1 to positive indices in all river basins or modified Palmer climate division −0.5 to positive SPI (6 months) D0 Abnormally Dry D0 ranges: CMPDI or SWSI: −1.0 to −1.9 SPI: −0.5 to −0.7 Indicator blend percentile: 21–30 Impacts: Short-term dryness slowing planting, growth of crops or pastures	Normal conditions Regular monitoring	• CWCB/WATF monitors situation on monthly basis, discusses trends with National Weather Service (NWS), State Climatologist, State Engineer, Natural Resource Conservation Service (NRCS), and others as appropriate. • Data reviewed for drought emergence and summarized in Governor's Drought Situation Report. • Implement long-term mitigation actions identified in drought mitigation plan. • ITF chairs meet twice yearly to monitor progress on long-term drought mitigation and review any lessons from previous drought periods, and review the response plan.
−1.0 to −2.0 in any river basin or modified Palmer climate division −0.6 to −1.0 SPI (6 months) D1 Moderate drought D1 ranges: CMPDI or SWSI: −2.0 to −2.9 SPI: −0.8 to −1.2 Indicator blend percentile: 11–20 Impacts: Some damage to crops, pastures; streams, reservoirs, or wells low, some water shortages developing or imminent; voluntary water-use restrictions requested	**Phase 1** More close monitoring of conditions for persisting or rapidly worsening drought; official drought not yet declared	• ITF chairs alerted of potential for activation, monitoring of potential impacts. • Assess need for formal ITF and DTF activation depending on timing, location, or extent of drought conditions, existing water supply, and recommendation of WATF; DTF is comprised of WATF, ITF chairs, and lead agencies. • DTF lead agencies (CDA/DOLA/DNR) notified of need for potential activation.

(Continued)

TABLE 9.1 *(Continued)*

Drought Response Plan Summary

Severity Indicators and Impacts (Colorado Modified Palmer Drought Index [CMPDI] or SWSI, SPI, and US Drought Monitor)	Drought Phase and Response Summary	Actions to be Considered
Less than −2.0 in any river basin or modified Palmer climate division Less than −1.0 SPI (6 months) D2 Severe Drought D2 ranges: CMPDI or SWSI: −3.0 to −3.9 SPI: −1.3 to −1.5 Indicator blend percentile: 6–10 Impacts: Crop or pasture losses likely; water shortages common; water restrictions likely to be imposed	**Phase 2** Drought task force and impact task force are activated; potential drought emergency declared	• DTF chairs prepare Governor's Memorandum of potential drought emergency based on recommendations from WATF. • Governor's Memorandum activates the drought task force and necessary impact task forces. • Department of Agriculture initiates Secretarial Disaster Designation process if appropriate. • The DTF Chairs and CWCB meet with activated impact task force chairs to outline phase 2 activity. • Activated ITFs make an initial damage or impact assessment (physical and economic). • ITFs recommend opportunities for incident mitigation to minimize or limit potential impacts. • Periodic reports are made by the ITF chairs to the DTF chairs. • ITF chairs designate their respective department public information officer (PIO) to interface with media for their relative area of concern and develop media messages. • Relevant state agencies undertake response and incident mitigation actions with their normal programs with available resources. • The DTF conducts a gap analysis identifying any unmet needs that cannot be handled through normal channels.

(Continued)

TABLE 9.1 *(Continued)*

Drought Response Plan Summary

Severity Indicators and Impacts (Colorado Modified Palmer Drought Index [CMPDI] or SWSI, SPI, and US Drought Monitor)	Drought Phase and Response Summary	Actions to be Considered
Lowest reading at −2.0 to −3.9 in any river basin or modified Palmer climate division Less than −1.0 to −1.99 SPI (6 months) D3 Extreme Drought to D4 Exceptional Drought D3 Ranges CMPDSI or SWSI: −3.0 to −4.9 SPI: −1.3 to −1.9 Indicator blend Percentile: 3–5 Impacts: Major crop/pasture losses; widespread water shortages or restrictions very likely to be imposed D4 Ranges: CMPDI or SWSI: −5.0 or less SPI: −2.0 or less Indicator blend Percentile: 0–2 Impacts: Exceptional and widespread crop/pasture losses; shortages of water in reservoirs, streams, and wells creating water emergencies	**Phase 3** Drought emergency is declared by Proclamation of the Governor	• Governor's Memorandum updated to activate additional Impact Task Forces as necessary. • DTF Chairs prepares a Governor's Proclamation of drought emergency. • Governor's Proclamation activates the GDEC. • DTF briefs GDEC. • Activated ITFs continue to assess, report, and recommend response measures and incident mitigation. • Unmet needs are reported to the DTF Chairs. • DTF Chairs determine the unmet needs that can be met by reallocation of existing resources. Those which cannot are forwarded to the GDEC with recommendations. • The GDEC assembles the data provided to advise the Governor with recommendations to support a request for a Presidential Drought Declaration. • Governor requests a Presidential Declaration. • If approved, Federal-State Agreement establishes the Colorado Division of Emergency Management Director as the state coordinating officer (SCO). • Long-term recovery operations commence.

(Continued)

TABLE 9.1 *(Continued)*

Drought Response Plan Summary

Severity Indicators and Impacts (Colorado Modified Palmer Drought Index [CMPDI] or SWSI, SPI, and US Drought Monitor)	Drought Phase and Response Summary	Actions to be Considered
Lowest reading at −1.6 in any river basin or modified Palmer climate division −0.8 SPI (6 months)	Return to **phase 2**	• DTF Chairs and the GDEC determine if all requirements for assistance are being met within the DTF and State agency channels.
D1 Moderate drought Coming out of drought: some lingering water deficits; pastures or crops not fully recovered	Return to **phase 1**	• GDEC briefs the Governor and prepares Proclamation to end drought emergency. • Long-term recovery operations continue. • ITFs continue assessments. • ITFs issue final report and conclude formal regular meetings. • The DTF issues a final report and is deactivated.
Lowest reading at −1.0 in any river basin −0.5 SPI (6 months)	Return to normal conditions	CWCB/WATF resume normal monitoring.

Source: Adapted from the Colorado Water Conservation Board, 2013, *Colorado Drought Mitigation and Response Plan*, Denver, CO, http://cwcbweblink.state.co.us/WebLink/0/doc/173111/Electronic.aspx?searchid=45a1d11c-9ccf-474b-bed4-2bccb2988870.

These actions enabled faster execution of secretarial drought designations, better coordination among local, state, and federal agencies charged with providing assistance, and more timely transfer of data and information. It also made limited emergency agricultural drought grant funds available for the affected area.

Figure 9.1 depicts the evolution of this drought in one hard-hit county of southeastern Colorado (Otero) where precipitation for 2011–2013 was actually less than in any consecutive 3-year period any time during recorded history back to 1890. The suite of SPI values for timescales of 1–24 months is shown (top of figure) alongside a similar depiction of drought for Otero County based on the USDM. Short-duration SPI values began indicating early onset of drought already in the fall of 2010. The drought worsened and spread during the 2011 growing season and then eased a little during the winter before crescendoing to extreme to exceptional drought at all time scales during summer 2012. The USDM lagged the SPI by a few weeks but eventually showed similar severity.

US drought moniter depiction Otero County, Colorado

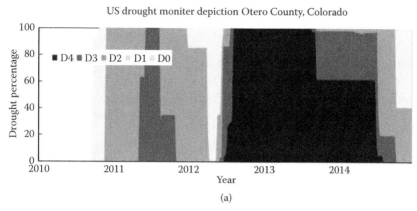

(a)

Standardized precipitation index at the Rocky Ford Weather Station in Otero County, Colorado

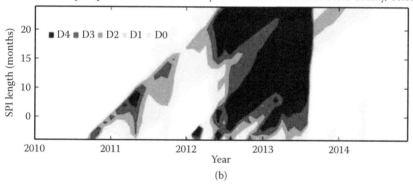

(b)

FIGURE 9.1
Standardized Precipitation Index (SPI) for all time scales from 1 month through 24 months, color-scaled to mimic the USDM for the Rocky Ford National Weather Service Cooperative weather station in Otero County, Colorado (a), and coinciding USDM drought classification categories for Otero County, Colorado (b), for the period January 2010 through December 2014. Colors range from white = no drought and yellow = abnormally dry to deep dark red = D4 (Exceptional Drought).

By May 2012, conditions had not only deteriorated but also expanded state-wide, and continued to worsen throughout the irrigation season. The conditions at that time are summarized below:

• The whole (100 percent) of Colorado was classified as experiencing severe, extreme, or exceptional drought as determined by the USDM. Of that, 69 percent of the state was experiencing extreme drought while 2 percent was classified as in exceptional drought conditions.

- Modified Palmer Drought Severity indices ranged from –1.01 to –5.3, with most in the –3 to –4 range. This represents severe to extreme drought.
- The 6-month SPI across much of the state was –2 (extremely dry), with pockets of –1 (moderately dry).
- Much of the west slope of Colorado saw 0 percent of normal precipitation in June; other areas of Colorado saw between 0 and 70 percent of average June precipitation.
- The previous 3 months' temperatures had been well above average for most of Colorado, with the month of June running 4 to 8 degrees above average. Statewide, June 2012 was the warmest June on record (1895–2012).
- Nearly all major reservoirs in Colorado had seen declines in storage levels in the previous few months.
- Irrigation abandonments were being reported.
- Sixty-two out of 64 counties had been granted a secretarial disaster declaration for crop disaster, with the remaining two eligible as contiguous counties.

Based on this data and information, the DTF recommended an expansion of activation of the state's drought mitigation and response plan for the agricultural sector to cover the entire state. The DTF also recommended increasing the activation to phase 3 for the agricultural sector only, the highest level of activation possible within the plan. These recommendations were necessary to comply with the state drought mitigation and response plan, support the Division of Emergency Management in their continued justification for secretarial declarations, and respond to serious concerns about drought conditions throughout Colorado. The governor agreed and expanded activation. In June 2013, 2 years after the initial activation of the state's drought mitigation and response plan, the municipal sector was added. This is a great example of the creeping effect that drought can have, slowly expanding into larger regions and multiple sectors. Thankfully in the case of Colorado, unprecedented rains began to fall later that year and into the next, allowing for complete deactivation of the plan. While drought impacts could not be avoided, the overall response of the state was praised as proactive, timely, and responsible.

9.5 Final Remarks

Establishing triggers creates clear guidelines that increase preparedness for drought, and helps to mitigate impacts at the onset of, during, and coming out of drought. These triggers also provide a basis for greater collaboration among data providers, users, and decision makers by creating a response framework that is data-driven, incremental, and logical.

References

Bolson, J. B., C. J. Martinez, P. Srivastava, N. Breuer, and P. Knox. 2013. Climate information use among Southeast U.S. water managers: Beyond barriers and toward opportunities. *Regional Environmental Change* 13(1):141–151. doi:10.1007/s10113-013-0463-1.

Colorado Water Conservation Board. 2013. *Colorado Drought Mitigation and Response Plan.* Denver, CO. http://cwcbweblink.state.co.us/WebLink/0/doc/173111/Electronic.aspx?searchid=45a1d11c-9ccf-474b-bed4-2bccb2988870 Accessed December 12, 2016.

Denver Water. 2016. *Drought Response Plan.* Denver, CO. http://www.denverwater.org/docs/assets/CECFBC95-E611-03E5-FD2B05E1B8A6B497/DroughtResponsePlan.pdf Accessed December 12, 2016.

Dilling, L., and M. C. Lemos. 2011. Creating usable science: Opportunities and constraints for climate knowledge use and their implications for science policy. *Global Environmental Change* 21:680–689.

Ferguson, D. B., J. Rice, and C. Woodhouse. 2014. *Linking Environmental Research and Practice: Lessons from the Integration of Climate Science and Water Management in the Western United States.* Climate Assessment for the Southwest, University of Arizona, Tucson, AZ.

Finnessey, T., M. Hayes, J. Lukas, and M. Svoboda. 2016. Using climate information for drought planning. *Climate Research* 70:251–263. doi:10.3354/cr01406.

Lemos, M. C., C. J. Kirchhoff, and V. Ramprasad. 2012. Narrowing the climate information usability gap. *Nature Climate Change* 2:789–794.

Lemos, M. C., and B. J. Morehouse. 2005. The co-production of science and policy in integrated climate assessments. *Global Environmental Change* 15:57–68.

McKee, T., N. Doesken, J. Kleist, and C. Shrier. 2000. *A History of Drought in Colorado: Lessons Learned and What Lies Ahead.* http://www.cwi.colostate.edu/publications/wb/9.pdf Accessed December 15, 2016.

Multihazard Mitigation Council. 2005. *Natural Hazard Mitigation Saves: An Independent Study to Assess the Future Savings from Mitigation Activities. Vol. 1, Findings, Conclusions and Recommendations.* National Institute of Building Sciences, Washington, DC. http://c.ymcdn.com/sites/www.nibs.org/resource/resmgr/MMC/hms_vol1.Pdf. Accessed December 12, 2016.

Rayner, S., D. Lach, and H. Ingram. 2005. Weather forecasts are for wimps: Why water resource managers do not use climate forecasts. *Climatic Change* 69:197–227.

Ryan, W., and N. Doesken. 2013. *Drought of 2012 in Colorado.* Climatology Report 13–01. Department of Atmospheric Science, Colorado State University. http://ccc.atmos.colostate.edu/pdfs/. Accessed December 15, 2016.

10

Advancements in Satellite Remote Sensing for Drought Monitoring

Brian D. Wardlow, Martha C. Anderson, Christopher Hain,
Wade T. Crow, Jason Otkin, Tsegaye Tadesse, and Amir AghaKouchak

CONTENTS

10.1 Introduction

Drought is a complex climatic phenomenon of global importance with major, wide-ranging impacts to many sectors of society, including agriculture, the economy, energy, health and water, and other natural resources. In many regions of the world, drought is a common, recurring natural event that can have significant, detrimental economic, social, and environmental impacts. For example, the annual impact of drought in the United States is estimated at $6–8 billion (NCDC 2014), with even farther-reaching effects in developing regions that can result in famine, malnutrition, loss of life, and social and political conflict. Changes in climate and the projected increase in climatic extremes such as drought (Dai 2012) coupled with increasing demands on finite water supplies and food production capabilities have raised the significance of developing effective drought early warning and mitigation strategies.

Drought monitoring is a key component for effective drought preparedness strategies, providing critical information on current conditions that can be used to trigger mitigation actions to lessen the impact of this natural hazard. However, drought can be both complex and challenging to monitor because it lacks a single universal definition, which makes the detection and assessment of key drought characteristics such as severity, geographic extent, and duration difficult (Mishra and Singh 2010). Three operational, physically based definitions were developed by Wilhite and Glantz (1985) to differentiate and describe different types of drought: meteorological, agricultural, and hydrologic. The temporal length of dryness needed to initiate and recover from a drought event and the specific environmental factors affected (e.g., rainfall deficits vs. plant health vs. reservoir water levels) are primary factors distinguishing among these different types of drought. In general, the time period associated with the manifestation or cessation of drought increases as we progress from meteorological through to hydrologic drought. As a result, a period of dryness may result in the emergence of one type of drought (e.g., meteorological) but not the others, while in the case of more prolonged or more severe dry events, several types of drought may be occurring at the same time. As a result, a number of drought indicators related to precipitation, soil moisture, vegetation health, and surface and groundwater have been developed to characterize specific types of drought and have been analyzed collectively in efforts such as the US Drought Monitor (USDM) (Svoboda et al. 2002) to more fully describe drought conditions (see also Chapters 7 and 9).

10.2 Traditional Drought Monitoring Tools

In situ-based observations of meteorological (e.g., precipitation and temperature) and hydrologic (e.g., soil moisture, streamflow, groundwater, and reservoir levels) parameters have provided the basis for most traditional indicators used for drought monitoring. Prime examples are the climate-based Palmer Drought Severity Index (PDSI, Palmer 1968) and Standardized Precipitation Index (SPI) (McKee et al. 1995), as well as anomalies of streamflows and reservoir levels from gauging stations and soil moisture from probes in the soil. For most of these indicators, an extended record of historical observations is used to calculate an "anomaly" measure to identify how the current conditions compare with the historical average conditions for drought detection and assessment of severity. *In situ* observations are point based and represent a measurement of conditions at a discrete geographic location.

To describe conditions between measurement locations, traditional spatial interpolation techniques have been applied to *in situ*-based data and derived drought indicators, or alternatively all *in situ* data within a specified geographic unit (e.g., county) may be areally averaged into a single value to represent the entire spatial unit. In either case, the ability to resolve detailed variations in drought conditions is limited by the number and spatial distribution of observing locations, which can vary considerably among regions and countries, as well as different environmental observations (e.g., rainfall vs. soil moisture measurements). In the United States, for example, the number and spatial density of National Weather Service (NWS) automated weather station locations measuring precipitation and temperature varies considerably, with a higher density of stations in the eastern United States compared to parts of the western United States. For hydrologic variables such as soil moisture, ground-based measurements in the United States are even more limited than are meteorological observations, and many countries around the world have few to no soil moisture observations.

The temporal length of *in situ* data records can also present a challenge, given that drought monitoring requires an extended record of observations to calculate historically meaningful anomalies that can be used to detect and measure drought events. The periods of record can vary among stations measuring a specific environmental parameter (e.g., temperature), and these varying lengths of data records must be considered and reconciled before the data can be transformed into a drought indicator. The length of record can also vary considerably among environmental parameters, as many weather stations measuring precipitation and temperature have data records spanning many decades while soil moisture probe locations often have data records no longer than 10–15 years.

Data quality and consistency are additional factors that can affect the applicability of *in situ* data for drought monitoring. Datasets with long histories of observations such as precipitation commonly have periods of missing observations that can range from few random days to longer blocks of weeks or months within the data record. In such cases, temporal interpolation methods have to be employed to fill data gaps, and may result in estimates that may or may not be representative of conditions during that period. Data consistency can also be an issue across sites that make *in situ* measurements. Data can be collected using different types of instrumentation and/or methods, as well as different data protocols, which can lead to data inconsistencies when observations are combined for different networks. For example, soil moisture observations available across the United States are collected by a series of networks at the national and state/regional scales, such as the US Department of Agriculture (USDA) Soil Climate Analysis Network (SCAN) and various state mesonets. Although these various soil moisture networks provide valuable measurements for drought monitoring, their data collection methods (e.g., type of soil moisture probe and different soil depths) and formats are often not consistent, leading to disparity in measurement quality and vertical support (Diamond et al. 2013).

10.3 Traditional Remote Sensing Methods for Drought Monitoring

Satellite-based remote sensing provides a unique perspective on drought, providing spatially distributed information that can be used in tandem with traditional, *in situ*-based measurements to gain a more complete view of drought conditions across the landscape. The space-borne earth observation era began in 1960 with the launch of the Television Infrared Observation Satellite (TIROS) by the National Aeronautics and Space Administration (NASA), which was designed to determine the utility of satellite-based imagery for the study of the earth. Although TIROS was designed for meteorological and climatological observations, the value of the satellite-based observations of the earth's environment were realized through this effort and provided the foundation for subsequent development of satellite-based, land observation remote sensing instruments in the following decades.

Remotely sensed satellite imagery provides a "big picture" view of the spatial patterns and conditions of the earth's land and water surfaces and atmosphere. The digital image data acquired by these space-borne remote sensing instruments overcome several of the issues related to *in situ*-based observations highlighted in the previous section. Satellite imagery provides a spatially continuous series of spectral measurements across large geographic areas that are acquired in the form of pixel-based grids. The ground area measured in these image pixels varies by satellite-based sensor, ranging from several meters (e.g., 1–30 m) to several kilometers (e.g., 1–25 km). The complete spatial coverage of satellite imagery can fill in the spatial gaps within and between *in situ*-based observational networks and provide invaluable information in many parts of the world where such networks may be sparse or nonexistent. Another benefit is that satellite imagery is collected in an objective and quantitative manner, yielding spatially and temporally consistent datasets that are required for environmental monitoring activities such as drought detection and severity assessment. Most satellite-based sensors record the reflected or emitted signal of electromagnetic radiation (EMR) from multiple regions of the EM spectrum spanning the visible, infrared, and microwave wavelengths. EMR in different spectral regions is responsive to different environmental parameters and can collectively be used to estimate and assess different drought-related environmental conditions such as plant stress and soil moisture. As a result, satellite-based estimates of these types of environmental conditions provide a valuable source of internally consistent historical data for accurate anomaly detection required for drought monitoring.

Historically, the application of satellite remote sensing for operational drought monitoring has primarily involved the use of Normalized Difference Vegetation Index (NDVI) data from the National Oceanic and Atmospheric Administration (NOAA) advanced very high resolution radiometer (AVHRR). The NDVI, which was developed in the early 1970s by Rouse et al. (1974), is a simple mathematical transformation of data from two spectral bands commonly available on most satellite-based sensors, the visible red and near infrared (NIR). The visible region is sensitive to changes in plant chlorophyll content, while the NIR region responds to changes in the intercellular spaces of the spongy mesophyll layers of the plants' leaves. Based on these interactions, the NDVI was developed as a general indicator of the state and condition of vegetation, with index values increasing with the amount of healthy green photosynthetically active vegetation. A large body of research has shown that NDVI has a strong relationship with several biophysical vegetation characteristics (e.g., green leaf area and biomass) (Asrar et al. 1989; Baret and Guyot 1991) and temporal changes in index values are highly correlated with interannual climate variations (Peters et al. 1991; Yang et al. 1998; McVicar and Bierwith 2001; Ji and Peters 2003). As a result, negative deviations in NDVI values for a given time period during the growing season compared to the long-term historical average NDVI value for that same period is indicative of vegetation stressed from an event such as drought. This concept has formed the basis for many NDVI-based drought monitoring efforts throughout the world, which have primarily relied upon the analysis of historical NDVI time series data collected by the satellite-based AVHRR sensor. AVHRR has provided a near-daily global coverage of 8 km gridded NDVI products dating back to the 1980s (*note:* 1-km AVHRR NDVI data are available since 1989 over the continental United States). The long multidecade time series of AVHRR NDVI data has proved valuable for drought monitoring because NDVI anomaly measures calculated from this dataset reflect the degree of departure of "current" vegetation conditions from longer-term historical average NDVI values for that same date over a period of more than 25 years. The use of AVHRR NDVI-based measures for drought monitoring can be traced back more than 20 years to the early work of Hutchinson (1991); Tucker et al. (1986); Kogan (1990); Burgan et al. (1996); and Unganai and Kogan (1998). Key examples of current operational drought monitoring using AVHRR NDVI anomaly products include the US Agency for International Development (USAID) Famine Early Warning System (FEWS) and the United Nations Food and Agricultural Organization (FAO) Global Information and Early Warning System (GIEWS) on Food and Agriculture.

The Vegetation Health Index (VHI) (Kogan 1995), which builds upon the NDVI concept and incorporates a temperature component through the use remotely sensed thermal infrared (TIR) data, is another traditional remote sensing indicator used for drought monitoring. The VHI integrates two indices into its calculation: the NDVI-based Vegetation Condition Index (VCI; Kogan and Sullivan 1993) and the TIR-based Temperature Conditions Index (TCI; Kogan 1995). The VCI is based on the assumption that the historical maximum and minimum NDVI values represent the upper and lower bounds of possible vegetation at a specific location (i.e., areas within an image pixel), with anonymously low NDVI values indicative of vegetation stress. The complimentary TCI is based on a similar concept where the historical maximum and minimum TIR values represent the upper and lower bounds of thermal response of vegetation for a specific location. Higher TIR anomalies expressed in the TCI should correspond to drought-stressed vegetation because more energy is being partitioned into the sensible heat flux rather than the latent heat flux because there is less moisture available to be transpired and evaporated from the vegetation and soil background. Kogan (1995) unified the VCI and TCI into the VHI to provide a remotely sensed indicator representative of both the NDVI and thermal response of vegetation. The VHI has been derived globally since 2005 at 8- and 16-km spatial resolutions using AVHRR NDVI and TIR data inputs.

10.4 Recent Remote Sensing Advancements

Although NDVI and VHI have both proved useful for drought monitoring, they provide only a partial view of drought conditions, focusing on vegetation health and agricultural drought. Saturation effects at high levels of NDVI, background contamination at low levels, and the empirical nature of the NDVI-TIR combination in the VHI further limit their capabilities over a broad range of surface and drought conditions (Karnieli et al. 2010). Given that drought is a complex natural hazard and several components of hydrologic cycle influence drought conditions, additional information regarding other hydrologic parameters such as evapotranspiration, soil moisture, groundwater, and precipitation is needed to provide a more comprehensive picture of drought conditions. Historically, the capability to estimate these types of hydrologic variables operationally from satellite remote sensing for drought monitoring has been limited because either the available satellite-based sensors did not acquire the necessary observations to retrieve such information or the historical record of appropriate satellite observations lacked sufficient length to calculate meaningful drought anomalies. However, since the early 2000s, a number of new satellite-based sensors have been launched, providing new types of earth observations acquired at a high temporal frequency (in some cases, with 1- to 2-day revisit time) and over a broader spectral extent, expanding the suite of remote sensing tools that can functionally monitor these various components of the hydrologic cycle.

The Moderate Resolution Imaging Spectroradiometer (MODIS), for example, is on board NASA's Terra and Aqua platforms and collects 1-km spectral observations globally on a near daily basis in the visible and NIR regions, extending the global time series of the NDVI data record that was established with the AVHRR. MODIS spectral observations also extend into the middle infrared (MIR) region, which can be used to assess plant water content, as well as the TIR region, which can be used to develop thermal-based tools for evapotranspiration estimation. Microwave sensors such as the Advanced Microwave Scanning Radiometer for Earth Observing System (EOS) (AMSR-E) and Quick Scatterometer (QuikSCAT) collect key observations that can be used to estimate soil moisture (Bolten et al. 2010). Gravity field observations from NASA's Gravity Recovery and Climate Experiment (GRACE) also provide new insights into water cycle variables including soil moisture and groundwater (Rodell and Famiglietti 2002). Collectively, the availability of this suite of new remote sensing observations, with time-series datasets extending for more than a decade, coupled with advancements in environmental models and algorithms and computing capabilities, has resulted in the rapid emergence of many new remote sensing tools that monitor different aspects of the hydrologic cycle that influence drought conditions.

This chapter will discuss several satellite-based remote sensing tools that have been developed for drought monitoring and early warning. Key examples will be presented that characterize different components of the hydrologic cycle related to drought that include vegetation status and health, ET, soil moisture, groundwater, and precipitation. The tools highlighted in this chapter are either currently operational or hold the potential to be operational in the near future. The chapter will include a brief introduction to remote sensing-related prediction tools that can provide drought early warning information. A short discussion of upcoming satellite missions that hold considerable potential to further advance drought monitoring, as well as directions of future research in this field, will also be presented.

10.5 Vegetation Condition

Historically, the development of satellite-based drought monitoring tools has focused on assessing general vegetation health conditions through the analysis of vegetation indices (VIs) such as the NDVI and VHI, which were discussed previously. While these VIs have proved valuable for this application, as demonstrated by their widespread use in monitoring systems such as FEWS, they still have several challenges for adequately characterizing drought-related vegetation stress. These VI-based methods rely on comparing the departure of the VI values to the historical average VI value for a given time period during the year, with drought stress represented by below-average VI values. As a result, a VI value threshold must be established that signifies a drought-stress vegetation signal within the range of negative VI anomaly values. In addition, other thresholds must be established to classify different drought severity levels. For example, does a 25 percent negative VI anomaly distinguish between drought and nondrought conditions and, if so, what negative percent VI values represent different drought severity levels (e.g., 25–35 percent = moderate, 35–50 percent = severe, and >50 percent = extreme)? Selection of such thresholds is often arbitrary and can be challenging given that they can vary by vegetation type, geographic region, and season. In addition, other environmental factors such as flooding, fire, frost, pest infestation, plant viruses, and land use/land cover change can result in negative VI anomalies that mimic a drought signal. As a result, traditional VI-based anomaly products can be misinterpreted as drought if analyzed in isolation. More recently, the remote sensing community has placed an emphasis on developing composite drought indicators (CDIs) that integrate various types of information, including VIs, into a single indicator that is representative of drought-specific vegetation conditions.

A prime example of a remote sensing-based CDI approach is the vegetation Drought Response Index (VegDRI), which integrates satellite-based vegetation condition observations, climate-based drought index data, and other environmental information to characterize drought stress on vegetation (Brown et al. 2008; Wardlow et al. 2012). VegDRI builds upon the traditional VI-based approach using satellite-based NDVI anomaly information as a general indicator of vegetation health, which is analyzed in concert with climate-based indicators that reflect the dryness conditions (i.e., SPI) for the same time period. Collectively, drought stress would be manifested as both below-average NDVI conditions and abnormally dry conditions in data inputs for VegDRI, with the NDVI anomaly decreasing and the climatic dryness conditions increasing as drought conditions intensified. VegDRI also incorporates several environmental characteristics of the landscape (land use/land cover, soils, topography, and ecological setting) that can influence climate–vegetation interactions at a given location. An empirical-based regression tree analysis method is used to analyze a historical record of satellite, climate, and environmental data to build the VegDRI models. VegDRI characterizes the severity of drought stress on vegetation using a modified version of the PDSI classification scheme (Palmer 1968). More technical details about the VegDRI methodology are presented by Brown et al. (2008) and Wardlow et al. (2012).

A VegDRI map for June 11, 2012, is presented in Figure 10.1, showing the widespread severe to extreme drought conditions that covered parts of the Rocky Mountains region and western United States at that time. VegDRI has three drought severity classes (moderate, severe, and extreme), as well as a predrought stress class reflecting areas of potential drought emergence. VegDRI maps have been operationally produced over the CONUS

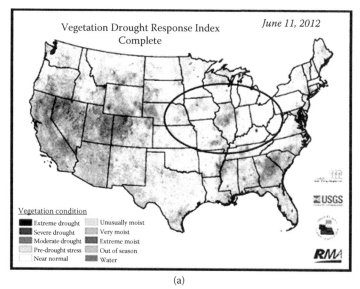

(a)

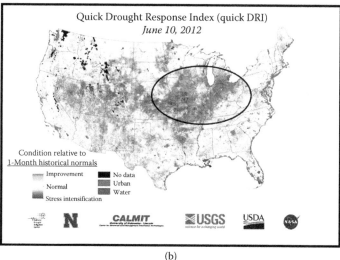

(b)

FIGURE 10.1

The VegDRI map (a) shows the general seasonal drought conditions on June 11, 2012, over the continental United States, with large areas of moderate to severe drought depicted across the western and parts of the southeastern United States. The QuickDRI map (b) highlights shorter-term drought stress intensification over many parts of the drought-stricken western United States, but also reveals the rapid, shorter-term intensification of drought conditions that were beginning to occur across the US Corn Belt region (highlighted in a black oval), which were not detected by the longer-term VegDRI or in other key tools such as the USDM until later June or early July. This example illustrates that QuickDRI represents an early warning alarm tool of rapidly changing drought conditions that can complement existing drought monitoring tools such as VegDRI and the USDM.

since 2008, with a continuous time series of historical 1-km spatial resolution maps dating back to 1989. VegDRI information is routinely used by the USDM authors, US Bureau of Land Management (BLM) rangeland assessment programs, NWS drought bulletins, and several state drought task forces in the western United States. The VegDRI concept has gained interest internationally, with a similar VegDRI tool being developed in Canada for the North American Drought Monitor (NADM) and modified versions of the tool in China, Czech Republic, and India. VegDRI has proved to be a useful agricultural drought indicator that reflects longer-term seasonal conditions on the scale of several months (Brown et al. 2008; Tadesse et al. 2015).

The Quick Drought Response Index (QuickDRI) (B. D. Wardlow [The Quick Drought Response Index (QuickDRI)] pers. comm., January 20, 2017) is another CDI that has recently been developed to characterize shorter-term drought intensification on the order of a few weeks to a month. QuickDRI uses a similar modeling approach to VegDRI to integrate several new extended time-series remote sensing datasets and climate indictors that are sensitive to shorter-term changes in environmental conditions that influence drought stress. These variables include the thermal-based Evaporative Stress Index (ESI) (Anderson et al. 2007, 2011) representing the ET component of the hydrologic cycle (see Section 10.6), modeled root-zone soil moisture data from the North American Land Data Assimilation System-2 (NLDAS-2) (Xia et al. 2012) representing subsurface moisture conditions, and the climate-based Standardized Precipitation Evapotranspiration Index (SPEI) (Vicente-Serrano et al. 2010) and SPI (McKee et al. 1995) representing precipitation and air temperature conditions. Additional input variables include the Standardized Vegetation Index (SVI) (Peters et al. 2002) derived from time-series NDVI data to represent general vegetation health and the same set of environmental variables used in VegDRI. The same regression-tree-based analysis technique used for VegDRI was adopted for QuickDRI model development with models based on anomaly data for the ESI, soil moisture, and climate inputs representative of conditions on a 1-month time step. As a result, QuickDRI is designed to monitor the level of drought intensification over a monthly time period to serve as an "alarm" indicator of rapidly emerging drought conditions that are not detected by the longer seasonal VegDRI. Figure 10.1 shows the rapidly intensifying drought signal over the US Corn Belt region that is captured by QuickDRI for the flash drought conditions that occurred across that area during the early summer. By comparison, the longer-term seasonal QuickDRI showed most of the same area in normal or predrought conditions and did not represent the emerging severe to extreme drought conditions until mid- to late July. The completion of an operational QuickDRI tool for the CONUS is planned for summer 2017.

The Combined Drought Indicator (Sepulcre et al. 2012) was developed for operational monitoring across Europe that combines anomaly information from a climate indicator (i.e., SPI), modeled soil moisture (i.e., output from

LISFLOOD model [De Roo et al. 2000]), and remotely sensed vegetation conditions (i.e., fraction of Absorbed Photosynthetically Active Radiation [fAPAR]) observed from the Medium Resolution Imaging Spectrometer (MERIS). The index is based on the historical analysis of the relationship between precipitation deficits expressed by the SPI and the response of soil moisture and fAPAR and inherent time-lag relationships among these variables. The drought categories classified by the Combined Drought Indicator include a watch and warning category and two alert categories that are classified based on whether there is a precipitation and/or soil moisture deficit and the vegetation stress that is detected in the fAPAR data. The European Drought Observatory (EDO) operational program produces a 0.25-degree spatial resolution Combined Drought Indicator map over Europe on a 10-day time step.

10.6 Evapotranspiration (ET)

In addition to green biomass amount, as sampled by standard VIs, another indicator of vegetation health is the rate at which plants consume and transpire water. As available soil moisture in the root zone depletes toward the permanent wilting level, plants reduce their transpiration rates to conserve remaining water. This reduces evaporative cooling of the leaf surfaces, resulting in a detectable thermal signal of elevated canopy temperature that can be measured from space using thermal infrared sensors (Moran 2003). In such cases, the exposed soil surface is typically dry and also elevated in temperature—further enhancing the composite thermal signature of drought (Anderson et al., 2008, 2012; Kustas and Anderson, 2009). Land surface temperature (LST) is, therefore, a valuable remote sensing diagnostic of drought conditions and their impacts on vegetation health. This fact was exploited in the empirical VHI described in Section 10.5, which is a linear combination of scaled VI and LST anomalies contributing with opposite sign, with negative VI and positive LST departures interpreted as a signal of drought stress (Kogan 1995). Karnieli et al. (2010) demonstrated, however, that under energy-limited circumstances (e.g., at higher latitudes and elevations), positive LST departures can be a sign of beneficial plant growth conditions and therefore a more physically based interpretation of LST anomalies may be useful for more definitively attributing causal factors to stress.

One approach has been to use remotely sensed LST to diagnose evaporative fluxes using a physical model of the land surface energy balance (see, e.g., review by Kalma et al. 2008). Such models estimate the evaporative cooling required to keep the land surface at the observed temperature given the radiative load (solar plus atmospheric radiation) prescribed over the modeling domain. The derived evapotranspiration (ET) includes water extracted from the soil profile and transpired by the vegetation as well as water evaporated directly from the soil and other surfaces, and is therefore a valuable metric of both soil moisture status and vegetation health.

One example of an ET-based drought indicator is the ESI, describing standardized anomalies in the ratio of actual-to-potential ET (fPET = ET/PET) computed with the Atmosphere-Land Exchange Inverse (ALEXI) energy balance model (Anderson et al. 2007, 2012). ALEXI uses time changes in LST, obtained from geostationary satellites or day-night polar orbiter overpasses, to estimate time-integrated fluxes of daytime sensible and latent heating. Anderson et al. (2012) showed that ESI in general agrees well with spatiotemporal patterns in standard precipitation-based drought indicators and with the USDM, but can be generated at significantly higher spatial resolution (meter to kilometer scale) by using the LST proxy for rainfall information. Recent work reveals that Landsat-scale ESI can be effectively used to separate moisture response from different land cover types (e.g., crops, forest patches, and surface water bodies), thereby better capturing agricultural drought impacts over heterogeneous surfaces. Otkin et al. (2016) showed the LST inputs to ESI convey effective early warning of rapid stress development during flash drought events, with the crops showing an elevated thermal signal several weeks before significant changes in VI can be detected from space.

Figure 10.2 shows the time evolution of ESI, developed at 4-km resolution using GOES-E and W TIR imagery, during the flash drought that affected the US Corn Belt in 2012. In addition, maps of temporal changes in the ESI (ΔESI) convey valuable information about the rate at which vegetation and soil moisture conditions are deteriorating and recovering during periods of drought intensification and abatement (Anderson et al. 2012; Otkin et al. 2013). Early signals of significant drought intensification were visible in ESI and ΔESI in late May/early June—well before the USDM recorded extreme drought in the region starting in mid-July. Real-time ESI products are generated daily on an 8-km grid over most of North America as part of NOAA's GOES ET and Drought (GET-D) Information system (http://www.ospo.noaa.gov/Products/land/getd/). Efforts are underway to transition a prototype 5-km global ESI product, generated using MODIS or VIIRS day-night LST differences, to operational status.

The Famine Early Warning Systems Network (FEWSNET) also produces an ET anomaly product, generated with the Simplified Surface Energy Balance Operational (SSEB-op) modeling system, which is a simplified LST-based ET retrieval methodology designed specifically for operational applications (Senay et al. 2013). Primary remote sensing inputs include time-composited MODIS LST (8-day) and NDVI (16-day) products, as well as topographic information from the Shuttle Radar Topographic Mission (SRTM). SSEB-op ET anomaly products are generated routinely in several climate-sensitive global regions to support early detection of regional crop failure and threats to food security. Domestically, SSEB-op ET products are used within the USGS Water for Sustaining and Managing America's Resources for Tomorrow (WaterSMART) program for accounting of regional water use and availability.

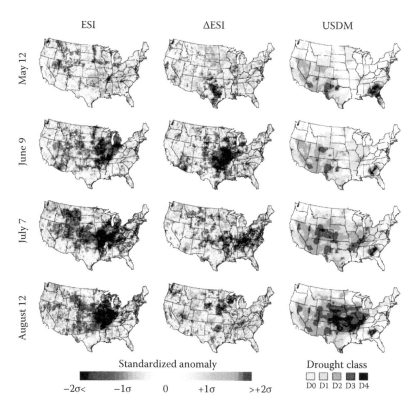

FIGURE 10.2
Monthly evolution of ESI (1-month composite), DESI (1-month difference), and USDM drought classifications during the 2012 US flash drought event. ESI detected signals of intensifying stress in the central United States in late May. DESI highlights areas where drought intensification (red) or recovery (green) is most significant.

ET diagnosed using remotely sensed LST via energy balance provides complementary yet independent information in comparison with estimates derived using prognostic land surface or hydrological models using water balance constraints, such as the NLDAS suite of soil moisture and ET indices (Xia et al. 2012). Hain et al. (2015) demonstrated that LST-based ET can inherently reflect ancillary sources of moisture besides local rainfall, including water applied through irrigation or extracted phreatically by vegetation growing over shallow water tables. These moisture sources may help mitigate drought impacts locally, but are difficult to capture in water balance models without extensive a priori knowledge of the complete hydrologic system, including human manipulations. Senay et al. (2012) report similar advantages in a comparison of the SSEB framework with a related water balance approach to ET mapping (forced by precipitation data); however, they note the latter may be more useful in some hydrologic applications that require detailed information on the temporal variability in soil moisture and runoff.

Remotely sensed actual ET is also a complement to drought indices based on potential ET such as the evaporative demand drought index (EDDI) (Hobbins 2016; McEvoy 2016), which describes the desiccating power of the local atmospheric conditions. High evaporative demand can be an effective early indicator of rapid drought onset, although it does not always result in actual drought impacts materializing on the ground, for example, because of amelioration by ancillary moisture sources. Taken together, there is much to be learned about ecosystem resilience and susceptibility to drought by comparing actual ET anomalies derived through the energy and water balance with anomalies in potential evapotranspiration (PET) deduced from meteorological conditions. A multi-index early warning system following signal progression from EDDI to ESI could be used to track if/how atmospheric precursors of drought evolve into vegetative stress that can negatively impact crop yields or rangeland condition. See further discussion of this topic in Chapter 11.

10.7 Precipitation

While there are different types of droughts (Wilhite 2005), they all are caused or intensified by varying periods of precipitation deficit. Ground-based precipitation measurements have been widely used for monitoring and understanding droughts (Hayes et al. 1999; Sheffield et al. 2012; AghaKouchak et al. 2014). A number of global gauge-based precipitation products have been developed for drought monitoring; however, they have several major limitations, including temporal inconsistencies, spatial inhomogeneities, and limitations in observational support in remote and undeveloped areas (Diamond et al. 2013). Furthermore, ground-based observations are often collected using different types of instruments, which can present challenges in creating temporally and spatially consistent drought information (Sorooshian et al. 2011) when long-term historical records are needed to calculated precipitation-based drought indicators such as the SPI.

Remote sensing of precipitation has offered a unique avenue for global drought monitoring, providing spatially distributed estimates of global precipitation over extended periods of time (Wardlow et al. 2012; AghaKouchak et al. 2015). Several algorithms have been developed to retrieve precipitation from one or more satellite sensors in geostationary earth orbit (GEO) and low earth orbit (LEO) (Kiladze and Sochilina 2003). For example, rainfall can be estimated using cloud-top temperature information available from satellite thermal infrared and visible data (Arkin et al. 1994; Joyce and Arkin 1997; Turk et al. 1999). By measuring the emitted microwave energy from the earth and the atmosphere, passive microwave sensors can be used to estimate instantaneous rainfall rates (Kummerow et al. 1996, 2001).

GEO infrared-based precipitation estimates offer more frequent (every 15–30 minutes) observations and larger spatial coverage relative to micro-wave-based precipitation estimates (Sorooshian et al. 2011). However, microwave-based estimates are known to be more accurate relative to infrared-based precipitation estimates, mainly because microwave radiation penetrates into clouds and provides more physically based estimates of the cloud's water content (Kummerow et al. 2001). Several studies have used remotely sensed image data from both infrared and microwave sensors to improve precipitation estimation (Joyce et al. 2004). Currently, there are a number of multisensor precipitation products, including:

- Tropical Rainfall Measuring Mission (TRMM) multi-satellite pre-cipitation analysis (TMPA) (Huffman et al. 2007)
- Climate Prediction Center (CPC) morphing technique (CMORPH) (Joyce et al. 2004)
- Precipitation estimation from remotely sensed information using artificial neural networks (PERSIANN) (Hsu et al. 1997; Sorooshian et al. 2000; Hong et al. 2004)
- Global precipitation climatology project (GPCP) (Adler et al. 2003)
- Integrated multisatellite retrievals for global precipitation mission (GPM) (IMERG) (Hou et al. 2014])
- Precipitation estimation from remotely sensed information using artificial neural networks—climate data record (PERSIANN-CDR) (Ashouri et al. 2015)
- Climate hazards group infrared precipitation with stations (CHIRPS) (Funk et al. 2015)

Some of these data records are microwave based (e.g., TMPA, CMORPH), while others are mainly infrared based (e.g., PERSIANN). However, each incorporates data from multiple sensors, either directly or for calibration/adjustment.

Satellite-based precipitation estimates have been evaluated extensively against ground observations (Hossain and Anagnostou 2004; Ebert et al. 2007; Tian et al. 2009; Anagnostou et al. 2010; Gebremichael 2010; AghaKouchak and Mehran 2013; Chappell et al. 2013; Katiraie-Boroujerdy et al. 2013; Nasrollahi et al. 2013; Maggioni et al. 2014). While there are uncertainties in satellite-observed precipitation products, they offer unique opportunities for exploring droughts from space (AghaKouchak et al. 2015), including model-based and data-driven drought monitoring (Anderson and Kustas 2008; Hao et al. 2014). The main limitation of most available satellite-based precipitation data records is their relatively short length (around 15–20 years). To address this issue, PERSIANN-CDR and CHIRPS combine long-term infrared-based estimates and ground-based observations to create a long-term climatology (30+ years) for drought monitoring and assessment. GPCP also offers a long-term precipitation dataset by combining satellite observations and ground-based measurements. The spatial resolutions of these precipitation datasets with long-term records suitable for drought monitoring range from 0.05 degrees (CHIRPS) to 2.5 degrees (GPCP).

Combined satellite observations and ground-based measurements provide more reliable precipitation estimates; however, they are not available in near real-time, which limits their usefulness to operational drought early warning systems. To address this limitation, a number of algorithms have been proposed to combine near real-time satellite observations (e.g., PERSIANN, TMPA, and IMERG) with long-term gauge-corrected records (e.g., GPCP; AghaKouchak and Nakhjiri 2012).

Satellite-based precipitation information has been integrated into different drought monitoring and prediction systems such as the African drought monitor (Sheffield et al. 2006) and the global integrated drought monitoring and prediction system (GIDMaPS) (Hao et al. 2014). Precipitation data are used to generate drought indicators such as the SPI (McKee et al. 1993) and the percent of normal precipitation (PNP) (Werick et al. 1994). Alternatively, precipitation information can be used as forcing to simulate hydrologic conditions (e.g., Sheffield et al. 2006) relevant to drought. Figure 10.3 displays the drought condition in July 2011 based on a 6-month SPI generated using GPCP data produced from GIDMaPS (Hao et al. 2014). The gridded SPI data are based on the remotely sensed precipitation climatology of the CPCP dataset and clearly delineate the regional-scale droughts that impacted several areas around the world in 2011, including the US southern Great Plains and northern Mexico, the Greater Horn of Africa, and southeast China.

The recently developed GPM IMERG (Hou et al. 2014) combines TMPA, CMORPH, and PERSIANN algorithms (Huffman et al. 2014; Yong et al. 2015). Preliminary assessments indicate that IMERG improves multisensor precipitation estimation though integrating multiple algorithms (Prakash et al. 2016a, 2016b; Tang et al. 2016a, 2016b). Combining multiple algorithms allows taking advantage of the strengths of both IR-based (more frequent sampling) information and microwave-based (more accurate and physically based) observations. Currently, the length of record for the IMERG data is too short for drought monitoring, but the developers plan to generate IMERG for the TRMM era, extending the historical record to support drought applications.

10.8 Terrestrial Water Storage

10.8.1 Soil Moisture

Soil moisture status is a key parameter for drought monitoring because it is a primary driver of drought-related vegetation stress when soil moisture levels approach the wilting point. Given that plants respond to soil moisture

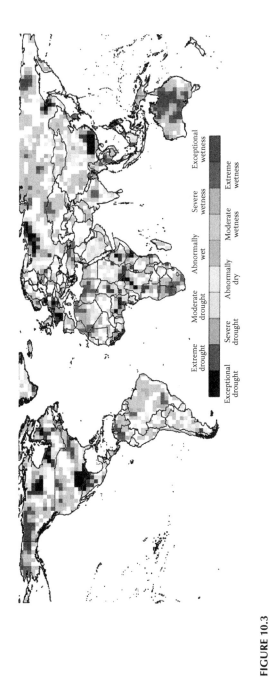

FIGURE 10.3

Drought condition in July 2011 based on 6-month Standardized Precipitation Index (SPI) generated using the global precipitation climatology project (GPCP) data. Downloaded from the global drought monitoring and prediction system (GIDMaPS; Hao et al. 2014).

conditions and regulate their water consumption by balancing moisture availability versus evaporative demand, soil moisture represents an indicator of early-stage vegetation drought stress. As discussed earlier, *in situ* measurements of soil moisture are somewhat limited in the United States and lacking or nonexistent in many parts of the world. Microwave remote sensing has proved useful for estimating soil moisture conditions because the microwave emissivity of soil is strongly impacted by the amount of soil water present. This sensitivity has been leveraged for the development of techniques for inferring surface soil moisture content via satellite-based observations in the microwave spectrum. Two decades of field campaign and aircraft studies support the conclusion that the microwave L-band (near 1.4 GHz) represents the preferred frequency band for such retrievals. As a result, the remote sensing of soil moisture entered a new era with the launch of the first two L-band satellite missions designed specifically for soil moisture retrieval: the European Space Agency's soil moisture ocean salinity (SMOS) mission in 2009 and the NASA soil moisture active/passive (SMAP) mission in 2015.

Since 2010, the SMOS mission (Kerr et al. 2010) has produced a global 45-km soil moisture product with approximately 2–3 day revisit. In particular, SMOS level 3 soil moisture retrievals are publicly available (following registration) at http://www.catds.fr/Products/Products-access, and extensive validation of SMOS soil moisture products is described in Kerr et al. (2016). The SMOS mission is based solely on passive microwave radiometry. In contrast, the NASA SMAP mission was designed to merge soil moisture information acquired simultaneously from both passive radiometry and active radar observations (Entekhabi et al. 2010). The SMAP radar failed in July 2015, but the radiometer has continued to function well and produce high-quality soil moisture products. SMAP has benefited from the application of a sophisticated radio frequency interference (RFI) mitigation strategy developed in response to L-band radio RFI discovered during early portions of the SMOS mission. SMAP 36-km resolution soil moisture products are currently available (with ~24 hours latency and 2–3 day update cycle) at https://nsidc.org/data/smap/smap-data.html. Ground-based validation of these products is described in Chan et al. (2017).

Regardless of their source, microwave-based surface soil moisture retrievals suffer from three primary shortcomings: (1) lower spatial resolutions compared to standard VI datasets (typically greater than 30 km), (2) limited vertical sampling depth, and (3) reduced accuracy over heavily vegetated surfaces. A broad range of spatial downscaling strategies is currently being applied for SMOS and SMAP soil moisture products, and, in mid-2017, the SMAP mission will begin operational production of a 9-km soil moisture product based on downscaling observations from the SMAP radiometer using backscattering observations acquired from the ESA Sentinel-1 satellite. The relatively shallow vertical soil penetration depth (~2–5 cm) of these products also represents an obvious limitation for agricultural drought monitoring because the soil moisture information is not indicative of the entire root zone condition, which influences plant stress. However, recent progress has been made in the development of land data assimilation systems that vertically extrapolate surface soil moisture across the entire vegetation root zone. In addition, the simple exponential filtering of remotely sensed surface soil moisture time series has shown promise for effectively recovering agricultural drought information contained in deeper root-zone soil moisture observations (Qiu et al. 2014). Finally, the attenuation of soil signals by the vegetation canopy is minimized by the use of longer microwave wavelengths, which have relatively less scattering and absorption interaction with vegetation than shorter wavelengths. As a result, the relative transparency of the vegetation canopy is greater for L-band SMOS and SMAP soil moisture products than for higher-frequency X- and C-band sensors used in earlier soil moisture remote sensing products.

10.8.2 Groundwater

In contrast to microwave-based retrievals, which are limited to only the top couple of centimeters of the vertical soil column, gravity-based remote sensing provides an integrated measure of variations in total terrestrial water storage (TWS)—including variations in shallow surface and root-zone soil moisture and groundwater storage. Gravity measurements detect changes in total terrestrial mass, including those due to long-term hydrologic variability of the various TWS components. GRACE gravity observations have been incorporated into land data assimilation models such as the catchment land surface model (CLSM; Koster et al. 2000) to estimate soil moisture and groundwater conditions (Zaitchik et al. 2008). Houborg et al. (2012) developed soil moisture and groundwater anomaly indicators from the GRACE data assimilation results tailored for operational drought monitoring over the CONUS. These indicators included percentile products of shallow soil moisture, root-zone soil moisture, and groundwater that are operationally produced for the CONUS on a weekly time step (GRACE drought products are available at http://drought.unl.edu/monitoringtools/nasagracedataassimilation.aspx). GRACE percentile products are generated from a long-term, 60+ year climatology of soil moisture and groundwater climatology developed using an open loop simulation of the CLSM. Figure 10.4 presents the GRACE root zone and shallow groundwater anomaly maps for June 12, 2012, as severe drought conditions were beginning to impact the US Corn Belt. The swath of lower percentile value in the root-zone soil moisture map spanning northern Kansas through northern Indiana reflects the emerging drought conditions during early summer 2012. The shallow groundwater percentile map follows a similar pattern across this area (Figure 10.4), but has less widespread low percentiles as the impact to the deeper water resources is lagged compared to the soil moisture above it. Work is ongoing to globally extend the GRACE drought products currently produced over the CONUS. Recent research has also focused on isolating the groundwater component of the monthly TWS signal and led to significant advances in

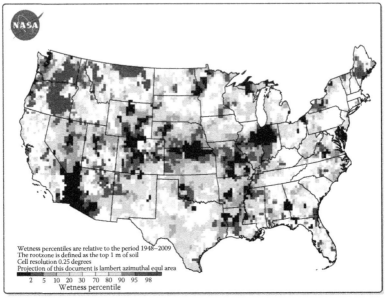

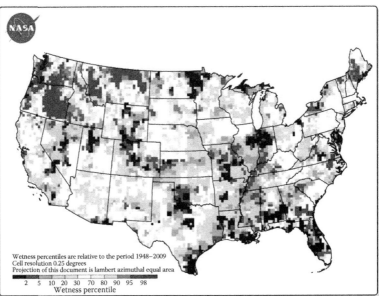

FIGURE 10.4
GRACE-Based root-zone soil moisture and shallow groundwater percentile maps are presented for June 11, 2012, when the early stages of severe drought over the US Corn Belt region began to emerge. Low percentile values (dark reds) are shown for the low soil moisture conditions occurring from northern Kansas through northern Indiana. The groundwater percentiles follow a similar pattern across this year, but have less widespread low percentiles because of the long response time to deeper groundwater conditions. Available at: http://drought.unI. edu/MonitoringTools/NASAGRACEDataassimilation.aspx

the ability to track the large-scale impact of long-term hydrologic drought (and/or the development of large-scale irrigation) on groundwater storage (Thomas et al. 2014).

10.8.3 Data Assimilation

In addition to the GRACE TWS products, data assimilation techniques have been applied for other emerging drought monitoring tools, particularly in the assimilation of surface soil moisture retrievals derived from microwave remote sensing into the water balance component of a land surface model. *Data assimilation* refers to mathematical techniques designed to integrate sparse observations (in both space and time) into more continuous dynamic models. Relative to the use of observations alone, a data assimilation analysis has three main advantages: (1) it produces estimates that are more temporally and spatially continuous, (2) it provides for the optimal merger of observations and models such that the impact of independent errors in both are minimized, and (3) it allows for the efficient extrapolation of information between observed and unobserved land surface states. The third advantage is particularly relevant for the assimilation of microwave soil moisture because the vertical support of microwave soil moisture retrievals is limited to only the first few centimeters of the soil column. As a result, land data assimilation systems are commonly employed to produce (deeper) root-zone soil moisture predictions that are constrained by a time series of surface soil moisture observations (Kumar et al. 2009).

Recently, a number of these data assimilation systems have been implemented operationally to produce soil moisture information, which is relevant for drought monitoring. Collectively, these systems currently provide the best available representation of continuous global variations in root-zone soil moisture availability. A primary example is the 9-km SMAP surface and root-zone soil moisture products (http://nsidc.org/data/docs/daac/smap/sp_l4_sm/index.html) based on the assimilation of the SMAP radiometer brightness temperature into the NASA global modeling and assimilation office catchment model (Reichle et al. 2016). In addition, near-real-time SMOS retrievals are currently being assimilated into the USDA's 2-layer Palmer model to produce a global 0.25-degree root-zone soil moisture analysis for large-scale crop conditions assessment by the USDA Foreign Agricultural Service (Bolten and Crow 2012). Resulting root-zone imagery from this analysis is regularly posted at http://www.pecad.fas.usda.gov/cropexplorer.

10.8.4 Forecasting and Prediction

The remote sensing tools and products that have been presented in the previous sections of this chapter characterize current drought conditions that are appropriate for drought monitoring purposes, but are somewhat limited for early warning applications that typically require information about future conditions to give decision makers lead time to implement drought mitigation actions (Tadesse et al. 2016). Satellite remote sensing has proved to be an effective means for real-time drought monitoring, as demonstrated by the numerous tools summarized in this chapter that have emerged over the past decade. However, the development of complementary remote sensing-based tools for drought forecasting and prediction has been very limited until recently, with efforts built upon the previous work of several monitoring tools (presented earlier) to provide projections of future drought conditions.

The vegetation outlook (VegOut) is a vegetation condition prediction tool that was developed experimentally for the CONUS by Tadesse et al. (2010) and is currently being implemented for Ethiopia and the Great Horn of Africa region in work underway by Tadesse et al. (2014). VegOut builds upon the VegDRI modeling methodology presented earlier by applying the same regression tree analysis technique to satellite-based VI observations, climate-based drought index data, biophysical information (e.g., land use/land cover, soils, and elevation), and several oceanic indicators. The empirical VegOut models are based on the historical analysis of the relationship between the VI-based vegetation conditions and the preceding climate conditions represented in the climate-based SPI input and the teleconnection signal from several oceanic indicators (e.g., Pacific decadal oscillation [PDO] and Atlantic multidecadal oscillation [AMO] indices) while considering the biophysical characteristics of a location such as land cover and elevation. The rationale underlying the VegOut is that time-lagged relationships exist between vegetation response and prior climatic and oceanic conditions. The historical analysis of these variables for a 20+ year period using a regression-tree-based data mining method is used to reveal these historical interactions and develop models that predict future vegetation conditions at multiple time steps (e.g., upcoming 1–3 months). The specific vegetation condition measure being estimated by VegOut is the standardized NDVI (SDNDVI), which is a standardized NDVI value calculated from the historical time-series NDVI data using a z-score approach that represents seasonal greenness of a given time period during the growing season compared to the historical average greenness conditions for the same period. Work by Tadesse et al. (2014) for the central United States and Tadesse et al. (2010) for East Africa showed that VegOut had a reasonable predictive accuracy of forecasting future vegetation conditions for outlook periods spanning 1–3 months. This work found that correlations between predicted and observed conditions were generally greater than 0.70 for the shorter outlook periods and 0.60 for the longer 3 month period over the central United States and Ethiopia. An operational VegOut-Africa tool producing dekadal (10-day) updates of 1-, 2-, and 3-month outlook maps is currently being developed for the Greater Horn of Africa region.

Another drought forecasting approach using remotely sensed data inputs has been developed by Otkin et al. (2015) using the rapid change index (RCI) data derived from the ET-based ESI presented earlier in this chapter. The RCI is designed to detect unusually rapid decreases in the ESI that are indicative of either rapid onset drought events or changes in drought severity. Otkin et al. (2015) expanded the RCI concept to include a forecasting component by using linear regression between the RCI and USDM drought severity classes to compute drought intensification probabilities based on the current RCI value. The initial RCI forecasting work showed that the probability of drought development and/or intensification over subseasonal time scales is higher than normal when the RCI is negative. Figure 10.5 demonstrates the utility of these forecasts as a drought early warning tool for a drought event that occurred across the central United States during 2002. On June 3, high drought intensification probabilities were present across most of South Dakota where the drought severity subsequently intensified by up to three USDM categories during the next 4 weeks. By July 1, a long band of high probabilities had developed from southwestern South Dakota into eastern Kansas. As was the case earlier in the summer, these high probabilities occurred several weeks before a period of rapid drought intensification, with some locations experiencing a two-category increase in USDM drought severity by the end of July. This example also illustrates the strong correspondence between regions experiencing rapid decreases in the ESI (as depicted by negative RCI values) and rapidly deteriorating crop conditions. This example demonstrates that statistical regression methods that combine drought early warning signals in remote sensing datasets such as the ESI with information from other variables can produce useful probabilistic forecasts of drought development.

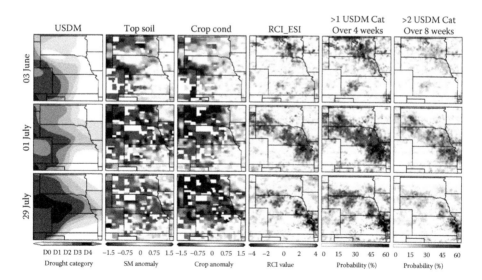

FIGURE 10.5
Evolution of the USDM drought depiction, USDA topsoil moisture and crop condition anomalies, rapid change index, and the probability of at least a one-category increase in USDM drought severity over a 4-week period or at least a two-category increase over an 8-week period from June 3 to July 29, 2002.

As remote sensing of the land surface has led to an improved characterization of the initial land surface states through advanced data assimilation techniques that were discussed earlier, a need to improve short-term forecasts of drought intensification or recovery has emerged. While the drought community has largely focused on seasonal forecasts of drought, which are mainly tied to the prediction of the large-scale circulation patterns (e.g., El Niño southern oscillation [ENSO]), forecasts in the 1- to 3-week range can provide actionable information, especially in the agricultural sector, where decisions are still made within such a timeframe. Through the combination of advanced land data assimilation methods and ensemble-based numerical weather prediction models (e.g., from operational centers such as National Climate Prediction Center [NCEP], European Centre for Medium-Range Weather Forecasts [ECMWF], Canadian Meteorological Centre [CMC], and United Kingdom Meteorological Office [UKMET]), a probabilistic forecast of drought improvement or recovery is possible. Operational numerical weather prediction (NWP) systems currently produce hundreds of 10- to 15-day forecasts each day, which can be used to force a land surface model ensemble toward improving drought forecasts. Such a system would ingest all available land surface remote sensing inputs through data assimilation in a land surface modeling framework that would then be forced with all available bias-corrected NWP ensemble-based forecasts. This land surface model ensemble could then be used to produce probabilistic forecasts of variables such as changes in soil moisture, which are closely tied to drought intensification or recovery.

10.9 Conclusion and Future Directions

Over the past decade, the application of satellite remote sensing for operational drought monitoring and early warning has rapidly advanced with the development of a suite of new tools, such as those discussed in this chapter. This advancement can be attributed to several catalysts, including the availability of new types of earth observations acquired by various space-borne sensors that have been launched since 2000; the development of extended time series of these observations, with many now spanning 15+ years; and the development of advanced computing capabilities and analytical methods to analyze and integrate these remotely sensed data into new drought indicators. Collectively, these advancements have allowed different components of the hydrologic cycle and biophysical characteristics of the landscape relevant to drought to be estimated and monitored, enabling a more complete view of drought to be obtained via satellite remote sensing.

Currently, the remote sensing community has several ongoing or planned efforts to continue expanding the types of tools and information that can be obtained from satellite observations for drought applications.

One emerging area is solar-induced fluorescence (SIF) of vegetation, which represents emitted radiation from chlorophyll pigments in a plant that occurs as part of photosynthesis. SIF can be an early-stage indicator of stress. The premise is that as a plant becomes stressed by drought, it will reduce its photosynthetic capacity (i.e., productivity), and the proportion of absorbed solar radiation emitted through fluorescence decreases as demonstrated by Sun et al. (2015). SIF has been shown to be a direct proxy of changes in photosynthetic capacity (Flexas et al. 2002; Damm et al. 2015) in response to the earliest stages of plant stress from events such as drought, which was demonstrated in the work of Sun et al. (2015) for the 2011 and 2012 droughts over the south-central United States and the US Corn Belt, respectively. Work has focused on the retrieval of SIF information from remotely sensed data acquired by the European Global Ozone Monitoring Instrument 2 (GOME-2) and the Japanese Greenhouse gases Observing Satellite (GOSAT; Frankenberg et al. 2011). Although both sensors were designed to measure atmospheric conditions, they have spectral bands placed in the visible red regions where SIF affects the absorption features over very narrow wavelength ranges called Fraunhofer lines. The satellite SIF data record for GOME-2 data ranges from 2007 to present, GOSAT data from 2009 to 2015, and the recent NASA orbiting carbon observatory-2 data (OCO-2; Frankenberg et al. 2014) from 2014 to present. In 2018, a follow-on OCO-3 sensor will be deployed on the international space station (ISS), providing data over a range of diurnal sampling times. Planned codeployment of the ecosystem space-borne thermal radiometer on the space station (ECOSTRESS) during a similar timeframe as OCO-3 will allow exciting opportunities for investigating synergies in SIF-TIR signals of plant stress.

Another emerging area of satellite remote sensing for drought monitoring is developing an indicator reflective of the vapor pressure deficit (VPD), which represents the difference between the amount of moisture in the air and the amount of moisture the air can hold when saturated. As the atmosphere dries, the VPD increases (i.e., decreasing humidity levels) and can be a precursor to drought onset or the intensification of existing drought conditions. Recent work by Behrangi et al. (2015) demonstrated that VPD can be calculated over large areas using remotely sensed near-surface temperature and relative humidity observations from the NASA atmospheric infrared sounder (AIRS) in combination with *in situ* dew point temperature and relative humidity data. Behrangi et al. (2016) found that the co-occurrence of high temperatures and low atmospheric humidity, which was expressed by high VPD, was an important factor in the development and evolution of the 2011 and 2012 droughts in the south-central and Corn Belt regions of the United States. For both events, the VPD showed marked increases during the formation and rapid intensification in drought conditions, demonstrating that remotely sensed VPD holds considerable potential to offer new atmospheric insights for drought early warning and assessment that complements the terrestrial information provided by the other tools presented in this chapter.

The application of satellite-based remote sensing for operational drought monitoring and early warning has significantly expanded and matured since the early 2000s, resulting in a suite of tools that characterize several hydrologic dimensions of drought. Given that several remote-sensed drought indicators now have relatively long-term (20–30 years) historical records, additional research on evaluating their spatial and temporal accuracy in characterizing drought patterns and conditions is a critical next step in effectively integrating this new information into drought decision-making activities. Evaluation will require comparing drought indicators' response to relevant observed impacts (e.g., crop yield reductions, reservoir levels, soil moisture depletion, economic losses, and reduction of ecosystem services) from historical drought events that have occurred within the satellite observational record. Drought impact/ remote sensing indicator comparisons will be challenging given that drought impact documentation is limited for many sectors; the impacts may be either direct or indirect, and often the reported impacts are collected at suboptimal spatial (e.g., a real county or district report vs. an individual pixel of satellite product) and temporal (e.g., annual impact report vs. weekly update of satellite product) scales. Some preliminary efforts have been undertaken to compare remotely sensed drought indicators with observed impacts (Otkin et al. 2015; Tadesse et al. 2015; Otkin et al. 2016), as well as more formal efforts to systematically collect drought impacts, as demonstrated by the drought impact reporter (droughtreporter.unl.edu/) for the United States. Work is needed in this area to establish triggers based on these indicators (e.g., three consecutive weeks of extreme drought detected by an indicator) that can be used by decision makers to implement a specific drought mitigation action (e.g., eligibility for assistance, demand reduction measures). Another key area of work is the long-term maintenance of these remotely sensed indicators into the future. As remote sensing tools and products such as those highlighted in this chapter become formally integrated into operational drought monitoring and early warning systems, sustaining the required satellite observations to maintain them will be key, as decision-making activities will be reliant on this information. This poses a challenge to the remote sensing community because a series of satellite sensors will be needed over time to replace aging instruments that degrade. This will require dedicated efforts to intercalibrate remote sensing observations between sensors to ensure comparable data inputs are used in the calculation of these indicators, resulting in consistent long-term data records.

Given the multifaceted nature of drought, it is clear that a single index is unlikely to tell the complete story of drought evolution, and so the question of interindex synergy is also raised. Ideally we would deploy a suite of diagnostic remote sensing tools that allow us to watch agricultural drought as it moves through its various phases—from atmospheric demand to enhanced evaporative loss to soil moisture depletion to canopy stress and degradation, and finally to yield loss and associated impacts. Such a multi-index screening, much like those used in the medical fields, may help us to catch signs of developing drought early and trace the progression to more and more serious consequences, allowing more effective and proactive adaptation to the evolving conditions.

References

Adler, R., G. J. Huffman, A. Chang, et al. 2003. The version-2 Global Precipitation Climatology Project (GPCP) monthly precipitation analysis (1979–present). *Journal of Hydrometeorology* 4(6):1147–1167. doi: http://dx.doi.org/10.1175/1525-7541.

Aghakouchak, A., L. Cheng, O. Mazdiyasni, and A. Farahmand. 2014. Global warming and changes in risk of concurrent climate extremes: Insights from the 2014 California drought. *Geophysical Research Letters* 41:8847–8852. doi: http://dx.doi.org/10.1002/2014GL062308.

AghaKouchak, A., A. Farahmand, J. Teixeira, et al. 2015. Remote sensing of drought: Progress, challenges and opportunities. *Reviews of Geophysics* 53(2):452–480. doi: http://dx.doi.org/10.1002/2014RG000456.

Anagnostou, E. N., V. Maggioni, E. Nikolopoulos, T. Meskele, F. Hossain, and A. Papadopoulos. 2010. Benchmarking high-resolution global satellite rainfall products to radar and rain-gauge rainfall estimates. *IEEE Transactions on Geoscience and Remote Sensing* 48(4):1667–1683.

AghaKouchak, A., and A. Mehran. 2013. Extended contingency table: Performance metrics for satellite observations and climate model simulation. *Water Resources Research* 49:7144–7149. doi: http://dx.doi.org/10.1002/wrcr.20498.

AghaKouchak, A., and N. Nakhjiri. 2012. A near real-time satellite-based global drought climate data record. *Environmental Research Letters* 7(4):044037. doi: http://dx.doi.org/10.1088/1748-9326/7/4/044037.

Anderson, M. C., C. R. Hain, B. Wardlow, J. R. Mecikalski, and W. P. Kustas. 2011. Evaluation of drought indices based on thermal remote sensing of evapotranspiration over the continental U.S. *Journal of Climate* 24:2025–2044.

Anderson, M. C., C. R. Hain, B. Wardlow, A. Pimstein, J. R. Mecikalski, and W.P. Kustas. 2012. A thermal-based Evaporative Stress Index for monitoring surface moisture depletion. In *Remote Sensing for Drought: Innovative Monitoring Approaches*, eds. B. Wardlow, M. C. Anderson, and J. Verdin, 145–167. Boca Raton, FL: CRC Press.

Anderson, M. C., and W. P. Kustas. 2008. Thermal remote sensing of drought and evapotranspiration. *Eos Transactions American Geophysical Union* 89(26): 233–234.

Anderson, M. C., J. M. Norman, J. R. Mecikalski, J. A. Otkin, and W. P. Kustas. 2007. A climatological study of evapotranspiration and moisture stress across the continental U.S. based on thermal remote sensing: II. *Surface moisture climatology. Journal of Geophysical Research* 112: D11112. doi: http://dx.doi.org/11110.11029/12006JD007507.

Arkin, P. A., R. Joyce, and J. E. Janowiak. 1994. The estimation of global monthly mean rainfall using infrared satellite data: The GOES Precipitation Index (GPI). *Remote Sensing Reviews* 11(1–4):107–124. doi: http://dx.doi.org/10.1080/02757259409532261.

Ashouri, H., K.-L. Hsu, S. Sorooshian, et al. 2015. PERSIANN-CDR: Daily precipitation climate data record from multisatellite observations for hydrological and climate studies. *Bulletin of the American Meteorological Society* 96(1):69–83.

Asrar, G., R. B. Myneni, and E. T. Kanemasu. 1989. Estimation of plant canopy attributes from spectral reflectance measurements. In *Theory and Applications of Optical Remote Sensing*, eds. G. Asrar, 252–296. New York: Wiley.

Baret, F., and G. Guyot. 1991. Potentials and limits to vegetation indices for LAI and APAR assessments. *Remote Sensing of Environment* 35:161–173.

Behrangi, A., E. J. Fetzer, and S. L. Granger. 2016. Early detection of drought onset using near surface temperature and humidity observed from space. *International Journal of Remote Sensing* 37:3911–3923.

Behrangi, A., P. Loikith, E. Fetzer, H. Nguyen, and S. Granger. 2015. Utilizing humidity and temperature data to advance monitoring and prediction of meteorological drought. *Climate* 3:999.

Bolten, J. D., and W. T. Crow. 2012. Improved prediction of quasi-global vegetation conditions using remotely-sensed surface soil moisture. *Geophysical Research Letters* 39:L19406. doi: http://dx.doi.org/10.1029/2012GL053470.

Bolten, J. D., W. T. Crow, X. Zhan, T. J. Jackson, and C. A. Reynolds. 2010. Evaluating the utility of remotely sensed soil moisture retrievals for operational agricultural drought monitoring. *IEEE Journal of Selected Topics in Applied Earth Observations and Remote Sensing* 3(1):57–66.

Brown, J. F., B. D. Wardlow, T. Tadesse, M. J. Hayes, and B. C. Reed. 2008. The Vegetation Drought Response Index (VegDRI): A new integrated approach for monitoring drought stress in vegetation. *GIScience and Remote Sensing* 45(1):16–46.

Burgan, R. E., R. A. Hartford, and J. C. Eidenshink. 1996. *Using NDVI to Assess Departure from Average Greenness and Its Relation to Fire Business.* Gen. Tech. Rep. INT-GTR-333, U.S. Department of Agriculture, Forest Service, Intermountain Research Station, Ogden, Utah.

Chan, S., R. Bindlish, P. E. O'Neill, et al. 2017. Assessment of the SMAP level 2 passive soil moisture product. *IEEE Transaction on Geoscience and Remote Sensing* 54(8):4994–5007.

Chappell, A., L. J. Renzullo, T. H. Raupach, and M. Haylock. 2013. Evaluating geostatistical methods of blending satellite and gauge data to estimate near real-time daily rainfall for Australia. *Journal of Hydrology* 493(17):105–114.

Dai, A. 2012. Increasing drought under global warming in observations and models. *Nature Climate Change* 3(1):52–58.

Damm, A., L. Guanter, E. Paul-Limoges, et al. 2015. Far-red sun-induced chlorophyll fluorescence shows ecosystem-specific relationships to gross primary production: An assessment based on observational and modeling approaches. *Remote Sensing Environment* 166:91–105.

De Roo, A. P. J., C. G. Wesseling, and W. P. A. Van Deursen. 2000. Physically based river basin modelling within a GIS: The LISFLOOD model. *Hydrological Processes* 14(11–12):15–30.

Diamond, H. J., T. R. Karl, M. A. Palecki, et al. 2013. U.S. climate reference network after one decade of operations: Status and assessment. *Bulletin of the American Meteorological Society* 94(4):485–489.

Ebert, E., J. Janowiak, and C. Kidd. 2007. Comparison of near real time precipitation estimates from satellite observations and numerical models. *Bulletin of the American Meteorological Society* 88(1):47–64.

Entekhabi, D., E. Njoku, P. O'Neill, et al. 2010. The Soil Moisture Active and Passive (SMAP) mission. *Proceedings of the IEEE* 98(5):704–716.

Flexas, J., J. M. Escalona, S. Evain, et al. 2002. Steady-state chlorophyll fluorescence (Fs) measurements as a tool to follow variations of net CO_2 assimilation and stomatal conductance during water-stress in C3 plants. *Physioligia Plantarum* 114(2):231–240.

Frankenberg, C., J. B. Fisher, J. Worden, et al. 2011. New global observations of the terrestrial carbon cycle from GOSAT: Patterns of plant fluorescence with gross primary productivity. *Geophysical Research Letters* 38:L17706. doi: http://dx.doi.org/10.1029/2011GL048738.

Frankenberg, C., C. O'Dell, J. Berry, et al. 2014. Prospects for chlorophyll fluorescence remote sensing from the Orbiting Carbon Observatory-2. *Remote Sensing of Environment* 147:1–12.

Funk, C., P. Peterson, M. Landsfeld, et al. 2015. The climate hazards infrared precipitation with stations—A new environmental record for monitoring extremes. *Scientific Data* 2:150066.

Gebremichael, M. 2010. Framework for satellite rainfall product evaluation. *Geophysical Monograph Series* 191:265–275.

Hain, C. R., W. T. Crow, M. C. Anderson, and M. T. Yilmaz. 2015. Diagnosing neglected moisture source/sink processes with a thermal infrared-based two-source energy balance model. *Journal of Hydrometeorology* 16:1070–1086.

Hao, Z., A. AghaKouchak, N. Nakhjiri, and A. Farahmand. 2014. Global integrated drought monitoring and prediction system. *Scientific Data* 1:140001.

Hayes, M. J., M. D. Svoboda, D. A. Wilhite, and O. V. Vanyarkho. 1999. Monitoring the 1996 drought using the standardized precipitation index. *Bulletin of the American Meteorological Society* 80(3):429–438.

Hobbins, M. T., A. Wood, D. J. McEvoy, et al. 2016. The evaporative demand drought index. Part 1: Linking drought evolution to variations in evaporative demand. *Journal of Hydrometeorology* 17(6):1745–1761.

Hong, Y., K. Hsu, X. Gao, and S. Sorooshian. 2004. Precipitation estimation from remotely sensed imagery using an artificial neural network cloud classification system. *Journal of Applied Meteorology and Climatology* 43(12):1834–1853.

Hossain, F., and E. N. Anagnostou. 2004. Assessment of current passive-microwave- and infrared-based satellite rainfall remote sensing for flood prediction. *Journal of Geophysical Research: Atmospheres* 109(D7): 1–14.

Hou, A. Y., R. K. Kakar, S. Neeck, et al. 2014. The global precipitation measurement mission. *Bulletin of the American Meteorological Society* 95:701–722.

Houborg, R., M. Rodell, B. Li, R. Reichle, and B. F. Zaitchik. 2012. Drought indicators based on model-assimilated Gravity Recovery and Climate Experiment (GRACE) terrestrial water storage observations. *Water Resources Research* 48(7):W07525. doi: http://dx.doi.org/10.1029/2011WR011291.

Hsu, K., X. Gao, S. Sorooshian, and H. Gupta. 1997. Precipitation estimation from remotely sensed information using artificial neural networks. *Journal of Applied Meteorology* 36(9):1176–1190.

Huffman, G., R. Adler, D. Bolvin, et al. 2007. The TRMM multi-satellite precipitation analysis: Quasi-global, multiyear, combined-sensor precipitation estimates at fine scale. *Journal of Hydrometeorology* 8(1):38–55.

Huffman, G. J., D. T. Bolvin, D. Braithwaite, K. Hsu, R. Joyce, and P. Xie. 2014. NASA global precipitation measurement (GPM) integrated multi-satellite retrievals for GPM (IMERG). *NASA Algorithm Theoretical Basis Document (ATBD). Version 4.4.* http://pmm.nasa.gov/sites/default/files/document_files/IMERG_ATBD_V4.4.pdf Accessed June 23, 2017.

Hutchinson, C. F. 1991. Use of satellite data for famine early warning in sub-Saharan Africa. *International Journal of Remote Sensing* 12:1405–1421.

Ji, L., and A. J. Peters. 2003. Assessing vegetation response to drought in the northern Great Plains using vegetation and drought indices. *Remote Sensing of Environment* 87:85–98.

Joyce, R., and P. A. Arkin. 1997. Improved estimates of tropical and subtropical precipitation using the GOES Precipitation Index. *Journal of Atmospheric and Oceanic Technology* 14(5):997–1011.

Joyce, R., J. Janowiak, P. Arkin, and P. Xie. 2004. CMORPH: A method that produces global precipitation estimates from passive microwave and infrared data at high spatial and temporal resolution. *Journal of Hydrometeorology* 5(3):487–503.

Kalma, J. D., T. R. McVicar, and M. F. McCabe. 2008. Estimating land surface evaporation: A review of methods using remotely sensed surface temperature data. *Surveys in Geophysics* 29(4):421–469.

Karnieli, A., N. Agam, R. T. Pinker, et al. 2010. Use of NDVI and land surface temperature for drought assessment: Merits and limitations. *Journal of Climate* 23:618–633.

Katiraie-Boroujerdy, P. S., N. Nasrollahi, K. Hsu, and S. Sorooshian. 2013. Evaluation of satellite-based precipitation estimation over Iran. *Journal of Arid Environments* 97:205–219.

Kerr, Y., A. Al-Yarri, N. Rodriguez-Fernandez, et al. 2016. Overview of SMOS performance in terms of global soil moisture monitoring after six years in operation. *Remote Sensing of Environment* 180:40–63.

Kerr, Y. H., P. Waldteufel, J. P. Wigneron, et al. 2010. The SMOS mission: New tool for monitoring key elements of the global water cycle. *Proceedings of the IEEE* 98(5):666–687.

Kiladze, R. I., and A. S. Sochilina. 2003. On the new theory of geostationary satellite motion. *Astronomical and Astrophysical Transactions* 22(4–5):525–528.

Kogan, F. 1990. Remote sensing of weather impacts on vegetation. *International Journal of Remote Sensing* 11:1405–1419.

Kogan, F. 1995. Application of vegetation index and brightness temperature for drought detection. *Advances in Space Research* 15:91–100.

Kogan, F., and J. Sullivan. 1993. Development of global drought-watch system using NOAA/AVHRR data. *Advances in Space Research* 13:219–222.

Koster, R. D., M. J. Suarez, A. Ducharne, M. Stieglitz, and P. Kumar. 2000. A catchment-based approach to modeling land surface processes in a general circulation model 1. Model structure. *Journal of Geophysical Research* 105:24809–24822.

Kumar, S. V., R. H. Reichle, R. D. Koster, W. T. Crow, and C. D. Peters-Lidard. 2009. Role of subsurface physics in the assimilation of surface soil moisture observations. *Journal of Hydrometeorology* 10:1534–1547.

Kummerow, C., Y. Hong, W. Olson, et al. 2001. The evolution of the Goddard profiling algorithm (GPROF) for rainfall estimation from passive microwave sensors. *Journal of Applied Meteorology* 40(11):1801–1820.

Kummerow, C., W. S. Olson, and L. Giglio. 1996. A simplified scheme for obtaining precipitation and vertical hydrometeor profiles from passive microwave sensors. *IEEE Transactions on Geoscience Remote Sensing* 34(5):1213–1232.

Kustas, W., and M. Anderson. 2009. Advances in thermal infrared remote sensing for land surface modeling. *Agricultural and Forest Meteorology* 149(12):2071–2081.

Maggioni, V., M. R. Sapiano, R. F. Adler, Y. Tian, and G. J. Huffman. 2014. An error model for uncertainty quantification in high-time-resolution precipitation products. *Journal of Hydrometeorology* 15(3):1274–1292.

McEvoy, D. J., J. L. Huntington, M. T. Hobbins, et al. 2016. The evaporative demand drought index. Part II: CONUS-wide assessment against common drought indicators. *Journal of Hydrometeorology* 17(6):1763–1779.

McKee, T. B., N. J. Doesken, and J. Kleist. 1995. Drought monitoring with multiple time scales. *Ninth Conference on Applied Climatology*, Dallas, Texas, January 15–20, pp. 233–236.

McVicar, T. R., and P. B. Bierwirth. 2001. Rapidly assessing the 1997 drought in Papua New Guinea using composite AVHRR imagery. *International Journal of Remote Sensing* 22:2109–2128.

Mishra, A. K., and V. P. Singh. 2010. A review of drought concepts. *Journal of Hydrology* 391(1):202–216.

Moran, M. S. 2003. Thermal infrared measurement as an indicator of plant ecosystem health. In *Thermal Remote Sensing in Land Surface Processes*, eds. D. A. Quattrochi, and J. Luvall, 257–282. Philadelphia, PA: Taylor and Francis.

Nasrollahi, N., K. Hsu, and S. Sorooshian. 2013. An artificial neural network model to reduce false alarms in satellite precipitation products using MODIS and CloudSat observations. *Journal of Hydrometeorology* 14(6):1872–1883.

NCDC (National Climatic Data Center). 2014. *Billion Dollar U.S. Weather Disasters*. http://www.ncdc.noaa.gov/oa/reports/billionz.html (accessed January 20, 2017).

Otkin, J. A., M. C. Anderson, C. Hain, I. E. Mladenova, J. B. Basara, and M. Svoboda. 2013. Examining rapid onset drought development using the thermal infra-red-based evaporative stress index. *Journal of Hydrometeorology* 14(4):1057–1074.

Otkin, J. A., M. C. Anderson, C. Hain, and M. Svoboda. 2015. Using temporal changes in drought indices to generate probabilistic drought intensification forecasts. *Journal of Hydrometeorology* 16:88–105.

Otkin, J. A., M. C. Anderson, C. Hain, et al. 2016. Assessing the evolution of soil moisture and vegetation conditions during the 2012 United States flash drought. *Agricultural and Forest Meteorology* 218–219:230–242.

Palmer, W. C. 1968. Keeping track of crop moisture conditions, nationwide: The new crop moisture index. *Weatherwise* 21(4):156–161.

Peters, A. J., D. C. Rundquist, and D. A. Wilhite. 1991. Satellite detection of the geographic core of the 1988 Nebraska drought. *Agricultural and Forest Meteorology* 57:35–47.

Peters, A. J., E. A. Walter-Shea, L. Ji, A. Vina, M. Hayes, and M. D. Svoboda. 2002. Drought monitoring with NDVI-based Standardized Vegetation Index. *Photogrammetric Engineering and Remote Sensing* 68(1):71–75.

Prakash, S., A. K. Mitra, A. AghaKouchak, Z. Liu, H. Norouzi, and D. S. Pai. 2016b. A preliminary assessment of GPM-based multi-satellite precipitation estimates over a monsoon dominated region. *Journal of Hydrology* 1–12 doi: http://dx.doi.org/10.1016/j.jhydrol.2016.01.029.

Prakash, S., A. K. Mitra, D. S. Pai, and A. AghaKouchak. 2016a. From TRMM to GPM: How well can heavy rainfall be detected from space? *Advances in Water Resources* 88:1–7.

Qiu, J., W. T. Crow, G. S. Nearing, X. Mo, and S. Liu. 2014. The impact of vertical measurement depth on the information content of soil moisture times series data. *Geophysical Research Letters* 41(14):4997–5004.

Reichle, R., G. De Lannoy, R. Koster, W. Crow, and J. Kimball. 2016. *SMAP L4 9 km EASE-Grid Surface and Root Zone Soil Moisture Geophysical Data.* Version 2. Boulder, Colorado USA: NASA National Snow and Ice Data Center Distributed Active Archive Center. doi:http://dx.doi.org/10.5067/YK70EPDHNF0L.

Rodell, M., and J. S. Famiglietti. 2002. The potential of satellite-based monitoring of groundwater storage changes using GRACE: The High Plains aquifer, Central US. *Journal of Hydrology* 263(1–4):245–256.

Rouse, J. W. Jr., R. H. Haas, J. A. Schell, D. W. Deering, and J. C. Harlan. 1974. *Monitoring the Vernal Advancement and Retrogradation (Green Wave Effect) of Natural Vegetation.* NASA/GSFC Type III Final Report, Greenbelt, MD.

Senay, G. B., S. Bohms, R. K. Singh, et al. 2013. Operational evapotranspiration mapping using remote sensing and weather datasets: A new parameterization for the SSEB approach. *Journal of the American Water Resources Association* 49(3):577–591.

Senay, G., S. Bohms, and J. P. Verdin. 2012. Remote sensing of evapotranspiration for operational drought monitoring using principles of water and energy balance. In *Remote Sensing of Drought: Innovative Monitoring Approaches,* eds. B. D. Wardlow, M. Anderson, and J. P. Verdin, 123–144. Boca Raton, FL: CRC.

Sepulcre, G., S. Horion, A. Singleton, H. Carrao, and J. Vogt. 2012. Development of a combined drought indicator to detect agricultural drought in Europe. *Natural Hazards and Earth Systems* 12(11):3519–3531.

Sheffield, J., G. Goteti, and E. Wood. 2006. Development of a 50-year high resolution global dataset of meteorological forcings for land surface modeling. *Journal of Climate* 13:3088–3111.

Sheffield, J., E. Wood, and M. Roderick. 2012. Little change in global drought over the past 60 years. *Nature* 491(7424):435–438.

Sorooshian, S., P. Arkin, J. Eylander, et al. 2011. Advanced concepts on remote sensing of precipitation at multiple scales. *Bulletin of the American Meteorological Society* 92(10):1353–1357.

Sorooshian, S., K. Hsu, X. Gao, H. Gupta, B. Imam, and D. Braithwaite. 2000. Evolution of the PERSIANN system satellite-based estimates of tropical rainfall. *Bulletin of the American Meteorological Society* 81(9):2035–2046.

Sun, Y., R. Fu, R. Dickinson, J. Joiner, and C. Frankenberg. 2015. Drought onset mechanisms revealed by satellite solar-induced fluorescence: Insights from two contrasting extreme events. *Journal of Geophysical Research: Biogeosciences* 102:2427–2440.

Svoboda, M., D. LeComte, M. Hayes, et al. 2002. The Drought Monitor. *Bulletin of the American Meteorological Society* 83(8):1181–1190.

Tadesse, T., G. B. Demisse, B. Zaitchik, and T. Dinku. 2014. Satellite-based hybrid drought monitoring tool for prediction of vegetation condition in Eastern Africa: A case study for Ethiopia. *Water Resources Research* 50:2176–2190. doi:10.1002/2013WR014281.

Tadesse, T., T. Haigh, N. Wall, et al. 2016. Linking seasonal predictions into decision-making and disaster management in the Greater Horn of Africa. *Bulletin of the American Meteorological Society* 96(8): ES89–ES92. doi: http://dx.doi.org/10.1175/BAMS-D-15-00269.1.

Tadesse, T., B. D. Wardlow, J. D. Brown, and K. Callahan. 2015. Assessing the vegetation condition impacts of the 2011 drought across the U.S. Southern Great Plains using the Vegetation Drought Response Index (VegDRI). *Journal of Applied Meteorology and Climatology* 54(1):153–169.

Tadesse, T., B. D. Wardlow, M. J. Hayes, M. D. Svoboda, and J. F. Brown 2010. The Vegetation Condition Outlook (VegOut): A new method for predicting vegetation seasonal greenness. *GIScience and Remote Sensing* 47(1):25–52.

Tang, G., Y. Ma, D. Long, L. Zhong, and Y. Hong. 2016a. Evaluation of GPM Day-1 IMERG and TMPA Version-7 legacy products over Mainland China at multiple spatiotemporal scales. *Journal of Hydrology* 533:152–167.

Tang, G., Z. Zeng, D. Long, et al. 2016b. Statistical and hydrological comparisons between TRMM and GPM level-3 products over a midlatitude basin: Is day-1 IMERG a good successor for TMPA 3B42V7? *Journal of Hydrometeorology* 17(1):121–137.

Thomas, A. C., J. T. Reager, J. S. Famiglietti, and M. Rodell. 2014. A GRACE-based water storage deficit approach for hydrological drought characterization. *Geophysical Research Letters* 41(5):1537–1545.

Tian, Y., C. Peters-Lidard, J. Eylander, et al. 2009. Component analysis of errors in satellite-based precipitation estimates. *Journal of Geophysical Research* 114:D24101. doi: http://dx.doi.org/10.1029/2009JD011949.

Tucker, C. J., C. O. Justice, and S. D. Prince. 1986. Monitoring the grasslands of the Sahel 1984–1985. *International Journal of Remote Sensing* 7:1571–1581.

Turk, F. J., G. D. Rohaly, J. Hawkins, et al.1999. Meteorological applications of precipitation estimation from combined SSM/I, TRMM and infrared geostationary satellite data. In *Microwave Radiometry and Remote Sensing of the Earth's Surface and Atmosphere,* eds. P. Pampaloni, and S. Paloscia, 353–363. Utrecht, The Netherlands: VSP Int. Sci. Publisher.

Unganai, L. S., and F. N. Kogan. 1998. Drought monitoring and corn yield estimation in southern Africa from AVHRR data. *Remote Sensing of Environment* 63:219–232.

Vicente-Serrano, S. M., S. Begueria, and J. I. Lopez-Moreno. 2010. A multiscalar drought index sensitive to global warming: The Standardized Precipitation Evapotranspiration Index. *Journal of Climate* 23:1696–1718.

Wardlow, B. D., T. Tadesse, J. F. Brown, K. Callahan, S. Swain, and E. Hunt. 2012. The Vegetation Drought Response Index (VegDRI): An integration of satellite, climate, and biophysical data. In *Remote Sensing of Drought: Innovative Monitoring Approaches,* eds. B. D. Wardlow, M. A. Anderson, and J. Verdin, 51–74. Boca Raton, FL: CRC.

Werick, W., G. Willeke, N. Guttman, J. Hosking, and J. Wallis. 1994. National drought atlas developed. *EOS Transactions American Geophysical Union* 75(8):89.

Wilhite, D. A., ed. 2005. *Drought and Water Crises: Science, Technology, and Management Issues*. Boca Raton, FL: CRC.

Wilhite, D. A., and M. H. Glantz. 1985. Understanding the drought phenomenon: The role of definitions. *Water International* 10:111–120.

Xia, Y., K. Mitchell, M. Ek, et al. 2012. Continental-scale water and energy flux analysis and validation for the North American Land Data Assimilation System project phase 2 (NLDAS-2): 1. Intercomparison and application of model products. *Journal of Geophysical Research* 117:D03109. doi: http://dx.doi.org/10.1029/2011JD016048.

Yong, B., D. Liu, J. J. Gourley, et al. 2015. Global view of realtime TRMM multisatellite precipitation analysis: Implications for its successor global precipitation measurement mission. *Bulletin of the American Meteorological Society* 96:283–96.

Yang, L., B. Wylie, L. L. Tieszen, and B. C. Reed. 1998. An analysis of relationships among climatic forcing and time-integrated NDVI of grasslands over the U.S. northern and central Great Plains. *Remote Sensing of Environment* 65:25–37.

Zaitchik, B. F., M. Rodell, and R. H. Reichle. 2008. Assimilation of GRACE terrestrial water storage data into a land surface model: Results for the Mississippi River Basin. *Journal of Hydrometeorology* 9(3):535–548. doi: http://dx.doi.org/10.1175/2007JHM951.1.

11

Evapotranspiration, Evaporative Demand, and Drought

Mike Hobbins, Daniel McEvoy, and Christopher Hain

CONTENTS

11.1 Introduction and Motivation

The monitoring and forecasting of drought is undergoing a paradigm shift with regard to the treatment of evapotranspiration (ET), evaporative demand (E_0), and the consideration of temperature impacts on drought in a changing climate. Previously, these have been poorly estimated by physically flawed parameterizations with detrimental effects on the estimation of hydrological variables central to drought analysis. Recently, however, our understanding of ET and E_0 and their interrelations under drought have advanced considerably; further, data availability to drive their improved treatment is no longer a constraint as reanalyses, remotely sensed data, and forecast products have increased in accuracy, resolution, and temporal extent. These advances are permitting development of ET- and E_0-related drought products that are physically based and that accurately represent drought dynamics. However, this paradigm shift is more evident in the science of drought than it is in the practice of its monitoring, and particularly forecasting.

Fundamentally, the enhanced treatment of ET and E_0 we describe in this chapter permits a more holistic approach to understanding drought: one that is based on the water balance. As a hydrological phenomenon, drought may broadly be defined as a sustained, large-scale imbalance of supply to and demand for moisture at the land–atmosphere interface. Precipitation ($Prcp$) supplies the former flux, while ET comprises the latter. This may be expressed in a general water balance over a given period as follows:

$$\Delta S = Prcp - ET - Runoff \qquad (11.1)$$

where ΔS represents the change in storage within, and $Runoff$ the net transfer of liquid water out of, the control volume above and below ground. ET is defined as the sum of transpiration from vegetation and evaporation from bare soil and open water (including sublimation from ice and snow). ET is limited by either moisture availability (θ) at the land–atmosphere interface or the demand for moisture in the atmosphere defined by E_0, an idealized flux that measures the "thirst of the atmosphere." Thus, examination of ET also entails explication of the concept of E_0.

E_0 represents the maximal rate of ET—that is, the rate that would occur with ample, or nonlimiting, moisture availability ($\theta = 1$). E_0 can be estimated or observed using various techniques—using equations for potential evaporation (E_p; sometimes called "PET") or reference evapotranspiration (ET_0), or observations of pan evaporation (E_{pan})—and in various conceptualizations varying from fully physical to simple equations based on air temperature alone. At its best, E_0 is estimated as a fully physical function of radiative and meteorological forcings:

$$E_0 = f\{T, q, R_n, U_z, P_a\} \qquad (11.2)$$

where T is the 2-m air temperature, q is a measure of humidity (such as relative humidity, dew point, or specific humidity), R_n is net radiation (the sum of net shortwave and net longwave radiation at the surface), U_z is the wind speed at z meters above the surface (z is commonly 2 m, so U_2 is used), and P_a is surface atmospheric pressure. The most commonly used fully physical estimators are based on the Penman (1948) equation, such as the Penman–Monteith ET_0 equation (Monteith 1965).

11.1.1 Current Treatment of *ET* and *E₀* in Drought

There are fundamental conceptual problems with traditional and current practices with respect to ET and E_0. First, we recognize that the treatment of ET can be more burdensome than that of E_0. ET is a difficult quantity to retrieve from remote sensing (RS): generally, various RS-derived datasets (e.g., land surface temperature [LST] or vegetation information) and meteorological datasets (e.g., U_z and T) must be combined in a physically based framework to estimate ET, usually indirectly from methods based on the relationship between sensible heat flux (H) and LST. Additionally, ET is difficult to validate as observations are not readily available, and when they are (e.g., from eddy-covariance [EC] tower platforms), they are only representative of small areas. The comparison of ET observations with RS retrievals is further complicated by scaling considerations of the EC tower footprint and incomplete energy closure of the EC observations. It should be noted that for drought applications, some of these issues can be minimized by transformation of ET retrievals into anomaly space; however, the need for long-term time series of RS-derived ET to ascertain the background mean state can be a further complication.

These complications explain why many traditional and current drought-monitoring operations estimate ET indirectly, using land surface models (LSMs) that constrain an estimate of E_0 by some measure of θ (such as soil moisture [SM]), or:

$$ET = f\{E_0, \theta\}, \tag{11.3}$$

where θ may commonly be parameterized in an LSM (as in the Sacramento soil moisture accounting model). Many current or legacy hydrological and drought products use simplified parameterizations of E_0 that are often buried deep in LSMs and that neglect much of the variety of forcings shown in Equation 11.2, instead relying solely on T or on some combination of T, q, and/or incident solar radiation (R_d). At most, such treatments examine drought as a function of $Prcp$ and T only. These cannot adequately represent the physics of ET and E_0 or their interrelations in drought, and this has severe consequences for the estimation of short-term variability vital to analysis of drought and of secular and climate-scale trends. The most charitable reading of such treatments of E_0 is that they still examine drought as an imbalance of supply (*Prcp*) and demand, where T is used as a proxy for E_0 and hence for demand. This short-cutting treatment of the drivers of drought is the central issue with current state-of-the-art monitoring and outlooks (as exemplified in the monthly and seasonal drought outlooks from the NOAA Climate Prediction Center [CPC]).

11.1.2 Goals of This Chapter

In this chapter, we aim to support the movement toward an improved treatment of the evaporative aspects of drought. Our main thrusts are to describe the various problems with the current definition, estimation, and use of both ET and E_0 in operational drought monitoring and/or forecasting, to outline the current state of the science, and to suggest ways forward to address the issues, highlighting some of the advantages of using either ET or E_0, or both. We will demonstrate that ET and E_0 are important in drought dynamics and that their accurate and fully physical representations offer new opportunities to improve monitoring and forecasting. This will necessitate our showing that E_0 sets the stage for drought—for example, the potential for vegetation stress in agricultural drought—while ET shows the onset of actual stress and that the specific interrelations of ET and E_0 show promise in drought early warning, spotting flash drought onset, and monitoring of ongoing drought. Further, we will show that forecasting E_0 leads to the opportunity to forecast stress from E_0 and hence, in part, to achieve the long-sought goal of forecasting drought itself. With respect to drought and climate change, we caution that E_0 must be estimated properly, and that we must understand what trends in ET and E_0 mean for drought vulnerability under a changing climate.

11.2 E_0, ET, and Their Physical Relationships to Drought

The relationship between E_0 and ET is particularly important in terms of monitoring and forecasting drought. Separately, each estimate provides a unique contribution to the understanding of drought development; when used together, they provide a particular characterization of the role of the coupling between the atmosphere and land surface. From a theoretical standpoint, increased E_0 precedes an observable land surface response (e.g., a decrease in *SM*), with increased E_0 leading to stronger coupling between E_0 and the land surface. While anomalies in E_0 can be considered a leading indicator in drought development (potential for stress), an observation of the onset of vegetation stress, through an estimation of ET (actual stress), is still needed to properly characterize the role of the coupling between atmospheric demand and evolution of the land-surface response.

11.2.1 Evaporative Demand, E_0

As noted earlier, E_0 is defined as the maximal rate of ET that would occur if moisture availability at the evaporating surface were not limiting. This rate is constrained only by the drying power of the air, the energy available for evaporation, and the ability to bear evaporated moisture away. E_0 is an umbrella term for a measure that takes on various conceptions that differ in their surface assumptions and/or measurement. E_0 may be estimated either as a physical observation of E_{pan} (generally daily from US Class-A evaporation pans) or as ET_0 (ET from a well-watered, highly specified reference crop) at weighing lysimeters. More commonly, it is synthesized as either E_p or ET_0 from observed radiative and meteorological drivers, with the prime difference between these two rates being their surface assumptions: E_p is more loosely defined as the ET rate that would pertain given ample moisture supply at the surface; ET_0 is the ET rate given a well-watered and otherwise specifically defined, ideal surface, including a specific crop cover.

Estimators of these fluxes range in data requirements, adherence to physics and therefore in quality. Physically based approaches synthesize the effects of all physical drivers shown in Equation 11.2: these are Penman (1948)-based approaches, such as Penman–Monteith (Montieth 1965) or PenPan (Rotstayn et al. 2006). However, the development of the E_p concept (Penman 1948) arose when the stringent data and computational requirements of fully physical parameterizations (Equation 11.2) precluded their estimation on useful scales. Instead, simplified approaches to E_0 developed at around the same time became popular, such as those based on T and R_n (e.g., the Priestley–Taylor equation for partial equilibrium evaporation, or evaporation from extensive wet surfaces; Priestley and Taylor 1972). Finally, many approaches are based solely on T, the most well-known being the Thornthwaite (1948) equation, which was developed as a tool to derive climate classifications and so involved mean conditions, not the deviations from the mean necessary for drought analysis. Other popular T-based methods include the Hamon (1961), Blaney-Criddle (Blaney and Criddle 1962), and Hargreaves-Samani (Hargreaves and Samani 1985) equations.

E_0 parameterizations with physical underpinnings are crucial in drought analyses. For example, a rigorous decomposition of daily E_0 variability (using Penman–Monteith ET_0, widely considered the best estimator of E_0 when all drivers are available and for nonsecular timescales) into its contributions from each of the drivers demonstrated that the drivers of daily variability—the most appropriate timescale for drought analyses—vary across the continental United States (CONUS) and with seasons (Hobbins 2016). Contrary to the basic assumptions of T-based approaches to E_0, T is not the dominant driver for many regions across CONUS and many seasons: in summer, U_z dominates variability in the Desert Southwest and R_d in the Southeast; in winter, q dominates along the Eastern Seaboard. Similar conclusions were drawn by Roderick et al. (2007) in their decomposition of E_0 trends across Australia: U_z

was found to dominate the mean trend at stations, with T playing little role. These studies underline the dangers of using T-based E_0 in drought analyses, both in the short-term variability central to drought dynamics, and at secular and climate timescales (see Section 11.4).

The eradication of poor parameterizations of E_0 must overcome significant paradigmatic inertia within the operational drought-monitoring community. This is demonstrated by the fact that new parameterizations of T-based E_0 were being developed and/or installed as measure of E_0 in new drought metrics (e.g., the PDSI as originally conceived by Palmer [1965], or the original SPEI [Vicente-Serrano et al. 2010]) even as our understanding of the dynamics of both E_0 and drought deepened, and long after data to support estimation of fully physical E_0 became available. It is this inertia we seek to overcome here.

11.2.2 Evapotranspiration, *ET*

Accurate knowledge of ET is an essential component in efforts to monitor the global water cycle, climate variability, agricultural productivity, floods, and droughts. Model-based estimates of ET from global landmasses range from 58 to 85 10^3 km^3 yr^{-1}, although the exact magnitude and spatial distribution is still in question (Dirmeyer et al. 2006). Thermal infrared (TIR) RS has proved to be an invaluable asset in modeling spatially distributed evaporative fluxes (Kalma et al. 2008). Most prognostic LSMs determine ET through a water-balance approach, relying on spatially distributed estimates of *Prcp* interpolated from coarse-resolution gauge networks or mapped using satellite techniques, neither of which currently provides adequate accuracy at scales useful for drought monitoring. Nonetheless, a number of diagnostic, RS-based methods to estimate ET have been developed in the past few decades, mainly estimating ET as a residual of the surface energy balance (Kalma et al. 2008):

$$ET = R_n - H - G, \tag{11.4}$$

where R_n and H are already defined, and G is soil heat flux.

Diagnostic ET methods based on TIR RS require no information regarding antecedent *Prcp* or *SM* storage capacity—the current surface moisture status is deduced directly from the RS-derived temperature signal. In general, dry soil or stressed vegetation heats up faster than wet soil or well-watered vegetation. TIR RS data sources provide multiscale information that can be used to bridge between the observation scale (~100 m) and global model pixel scale (10–100 km), facilitating direct model accuracy assessment. Examples of diagnostic ET methods include the Surface Energy Balance Algorithm for Land (SEBAL; Bastiaanssen et al. 1998) model, the Mapping EvapoTranspiration at high Resolution with Internalized Calibration (METRIC; Allen et al. 2007) model, the Simplified Surface Energy Balance (SSEB; Su 2002) model, the Atmosphere Land Exchange Inverse (ALEXI;

Anderson et al. 1997, 2011b) model, and the Operational Simplified Surface Energy Balance (SSEBop; Senay et al. 2011, 2013) model. Some of these methods (SEBAL, METRIC, SEBS, and SSEBop) focus generally on the use of a single RS observation of TIR *LST* and provide a scaling between a "hot" pixel (where $ET = 0$) and a "cold" pixel (where $ET = E_0$), providing *ET* estimates when accurate representations of the "hot/cold" pixels can be made. Few of these methods have been employed in drought monitoring, as much of their focus has been placed on high-resolution, field-scale estimation of consumptive water use (i.e., actual *ET*). However, methods such as ALEXI use a time-integrated measure of *LST* during the mid-morning hours, a time when *LST* and *SM* have been shown to be strongly correlated (Anderson et al. 1997). The ability of TIR *LST* to assess the current *SM* state and the effects of vegetation stress on *ET* provides a unique opportunity to augment current drought-monitoring methods, providing an estimate of actual water availability that can be used with estimates of E_0 to estimate anomalies in water use and/or vegetation stress.

11.2.3 *ET-E_0* Relations under Drought

The actual flux of *ET* that supplies moisture from the land surface to the atmosphere and the idealized flux of E_0 that demands it are linked across the land surface–atmosphere interface, particularly in drought. Here we discuss the use of these linkages to monitor and predict drought.

The relationships between long-term mean water balance components and their water and energy limits can be illustrated across the hydroclimatic spectrum using the Budyko (1974) framework shown in Figure 11.1. This framework is also informative for understanding their relationships under transient dry anomalies or droughts, which push regions toward the water limit (i.e., toward the right in Figure 11.1). How *ET* and E_0 respond to drying anomalies or drought depends on the regional mean behavior or on the regional hydroclimatic state at drought onset. For the case of regions that are energy-limited either climatologically or as a transient anomaly, recall that E_0 defines the upper energy limit on *ET* so variations in E_0 are matched by variations in *ET* (until *ET* becomes constrained by water availability, at which point the dynamic changes). Until this point, *ET* and E_0 vary together (see Figure 11.1b) with forcing from E_0 to *ET*, as both increase under increasing energy availability or advection. This is a *parallel relationship*, increasing in the case of developing drought and decreasing in drought mediation. This dynamic dominates in flash (or rapid onset) drought under energy-limited initial conditions.

On the other hand, in water-limited conditions (or when energy limitations on *ET* give way to tighter water limitations) *ET* is constrained by the availability of water: further decreasing the availability of water results in decreasing *ET* and less of the energy available at the land surface being expended as latent heat, so leaving more energy available for *H*. This increased *H* flux

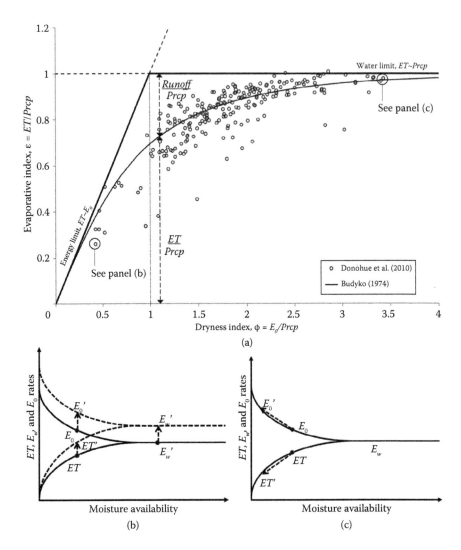

FIGURE 11.1
Interrelations of *ET* and E_0 in drought expressed in the context of the Budyko (1974) concept.
In (a), solid lines denote the water limit on *ET* (horizontal line) and energy limit on *ET* (sloped
line), to which the behavior of *ET* is asymptotic in arid and humid hydroclimatic extremes,
respectively. The dashed curve shows Budyko's (1974) idealized relation between dryness
index (Φ) and the evaporative index (ε). Vertical arrows represent the *ET* and *Runoff* portions
of *Prcp*, respectively. Circles represent typical observed climatological annual behavior, from
229 Australian basins (from Donohue, R.J., et al., *J. Hydrol.*, 390, 23–34, 2010). (Adapted from
Hobbins, M. T., and J. Huntington, *Handbook of Applied Hydrology*, 42–1–42–18, McGraw-Hill
Education, New York, 2016.) The interrelations of drying anomalies from either extreme of
the hydroclimatic spectrum are shown in (b) for an energy-limited basin and (c) for a water-
limited basin.

raises the temperature of the overpassing air, increasing its vapor pressure deficit, and thereby increases E_0. Thus, as *ET* decreases, E_0 increases in what is known as the *complementary relationship* (see Figure 11.1c); it was first proposed by Bouchet (1963) and observed across CONUS by Hobbins et al. (2004). This complementarity is also evident at larger space and time scales: lower *ET* leads to less cloudy conditions with attendant increases in surface energy and *H*, which again results in higher E_0. In these conditions, forcing is from *ET* to E_0, with the fluxes varying oppositely in both drying (drought intensification) and wetting (drought mediation). This dynamic dominates in sustained drought.

In summary, we observe a complementary relationship between *ET* and E_0 developing in sustained droughts and parallel relationships developing in flash drought onset (Hobbins et al. 2016; McEvoy et al. 2016a). The key here is that E_0 increases in both types of drought, whereas *ET* increases in flash drought onset and decreases thereafter, and is suppressed in sustained drought. This suggests the opportunity to treat E_0 as a robust precursor of both types of drought (as in the EDDI tool described in Section 11.3.2). It should be noted that high-E_0 events will always precede actual onset of stress, but not all high-E_0 events develop into drought, nor should we expect each event to, since meteorological and radiative factors also control the persistence of a high-E_0/low-*Prcp* regime. Nonetheless, it may be useful to conceptualize the persistence of high E_0 as a stage-setter for moisture deficits or vegetative stress, and depressed *ET* as reflecting the onset of such stress. The interactions of *ET* and E_0 with energy and water availability are summarized in Figure 11.2.

Figure 11.3 shows a hypothetical evolution of agricultural and/or meteorological drought. In this case, sufficient *SM* conditions prevail in a meteorological regime highlighted by below-normal *Prcp* and above-average atmospheric demand for *ET* (e.g., from high temperatures, solar loading, or wind speeds). In this case, one would expect a metric that estimates anomalies in atmospheric demand to start to show the potential for *SM* stress developing if large-scale meteorological conditions were to remain the same. As the meteorological

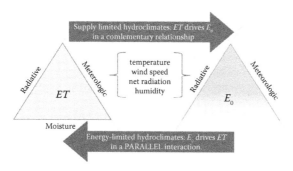

FIGURE 11.2
Interrelations of *ET* and E_0 summarized with respect to drought. The sides of the triangles represent the constraints, or drivers, of each flux.

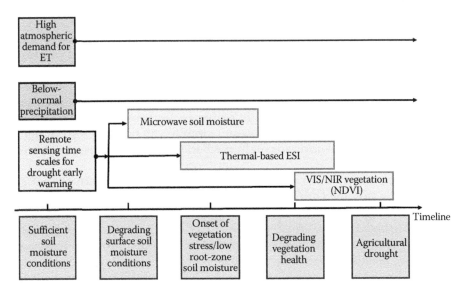

FIGURE 11.3
Schematic showing the temporal evolution of agricultural and/or meteorological drought and typical remote sensing observational timescales for drought early warning capabilities.

forcing continues, surface *SM* conditions begin to degrade, a condition that can be remotely sensed by active and passive microwave sensors. As surface *SM* conditions continue to degrade, root-zone *SM* conditions will also begin to show below-normal conditions and, at this point, the onset of vegetation stress due to inadequate moisture supplies will occur. The onset of actual vegetation stress can be determined through the use of methods that determine actual *ET* from satellite observations of TIR *LST*. Finally, as conditions continue, surface and root-zone *SM* conditions are well below normal, and damage to vegetation health commences. Anomalies in vegetation health, usually determined by below-normal "greenness," are usually observed through vegetation indices based on visible and near-infrared wavelengths (e.g., the Normalized Difference Vegetation Index [NDVI]). In summary, the introduction of E_0- and *ET*-based tools for drought monitoring provide a necessary augmentation to current drought-monitoring systems, especially in terms of providing an understanding of early drought evolution that can aid in the further development of drought early warning systems.

11.3 E_0- and *ET*-based Drought-Monitoring Tools

The availability of RS-derived data in both thermal and visible bands is maturing, as is the necessary climatology required to place analyses in statistical contexts, particularly from reanalyses. Further, the interactions between ET and E_0 in drought are becoming better understood while awareness is increasing of the dangers of simplistic parameterizations of E_0 as either an end in itself, or as a driver of ET from LSMs. This has led the drought and ET communities to begin to coalesce behind new approaches to drought monitoring that examine the flux of moisture from the earth's surface to the atmosphere—the demand side of the surface water balance (or imbalance, in the case of drought). Further, new tools are being developed that use RS observations and/or a better understanding of ET and E_0 relations in drought. Here we examine some of these emerging tools and their promise.

11.3.1 Existing Tools

As an example of an existing tool that has been shown to be fraying with age, the Palmer Drought Severity Index (PDSI; Palmer 1965) has long been commonly used in the United States in monitoring, generally on weekly or monthly timescales. It gained popularity due to its minimal data requirements: only *Prcp* and T are needed. In the United States, it is central to the suite of tools informing the US Drought Monitor's (USDM) assessment of developing drought conditions. It is also used in NOAA's operational PDSI. The PDSI derives the water balance using a simple hydrology bucket model with a two-layer soil column, where the difference between *Prcp* and the sum of *Runoff* and the ET results in a moisture anomaly then used to derive a non-dimensional drought index. In the bucket model, the maximum possible ET is the minimum of the available water in the soil column and E_0.

Despite its popularity and longevity, the PDSI only poorly resolves the evaporative aspects of drought, due to its use of T-based E_0, which Palmer (1965) originally estimated from the Thornthwaite (1948) equation and then by another T-based model (of unknown origin) further divorced from observed E_0-T relations. While T-based E_0 correlates well with humidity and net radiation on subannual timescales, drought-scale anomalies in observed E_0 are often forced by drivers other than T (Equation 11.2), particularly across CONUS (Hobbins 2016). The PDSI is not suited for widespread application, particularly in more arid regions or at high latitudes or in cooler seasons in which neglected cold processes dominate (Sheffield et al. 2012). Further, the empirical parameters characterizing local climate and drought timing were only calibrated for the midwestern United States. These issues pertaining to its widespread use in time and space may be obscured to casual users seeking a simple off-the-shelf index with minimal data requirements. Indeed, the PDSI is often used in long-term drought analyses worldwide, to questionable effect (see Section 11.4).

11.3.2 Emerging Tools

McKee et al. (1993) were the first to recognize the value of examining drought at numerous timescales, implementing this concept in the now-popular

Standardized Precipitation Index (SPI). Development of the SPI was a major breakthrough in the drought-monitoring community and allowed users to see that a drought could be occurring in the short term (e.g., 1–3 month *Prcp* deficits) even while long-term conditions were wet (e.g., a 24–48 month *Prcp* surplus). The primary limitation of SPI is that it only considers *Prcp* and ignores other atmospheric drivers of drought. To improve upon SPI, the Standardized Precipitation Evapotranspiration Index (SPEI) (Vicente-Serrano et al. 2010) was developed with the original goal of having a multiscalar drought index that could account for a warming climate. To accomplish this, SPEI uses a simple water balance ($Prcp - E_0$) as the accumulating variable. The T-based Thornthwaite (1948) E_0 approach was initially used but, as with PDSI, caution must be taken when using any T-based E_0. Beguería et al. (2014) tested several different E_0 approaches in computing global SPEI, and recommended the fully physical Penman–Monteith model if data are available.

A signal feature of E_0 is that it increases in droughts initiated across the hydroclimatic spectrum (i.e., in energy- and water-limited hydroclimates) and across timescales (i.e., in sustained and flash droughts), as noted in Section 11.2.3. This robust signal lies at the heart of the **Evaporative Demand Drought Index** (EDDI; Hobbins et al. 2016; McEvoy et al. 2016a), an emerging drought-monitoring and early warning tool. EDDI works by ranking E_0 depths (using Penman–Monteith ET_0 for E_0) accumulated over a given timescale relative to same-period depths drawn from a climatology. Periods ranking higher (lower) than the median indicate drier (wetter) than normal conditions. The rank is converted to percentiles of the standard normal distribution, which are then categorized and mapped. It is multiscalar in time and can operate at the spatial resolution of the drivers of E_0. Users report that multiple timescales are needed for a convergence-of-evidence approach (e.g., Nolan Doesken, Colorado State Climatologist, pers. comm.,), as dynamics specific to drying impacts on sectors within a region operate at various timescales. EDDI shows promise as an early warning indicator (Figure 11.4) of hydrological drought, and for ongoing monitoring of agricultural drought, both in dryland farming and rangeland environments. Ongoing research will reveal whether the strong relations between forest physiology and E_0 permit the index to improve fire weather risk prediction. As EDDI relies solely on the radiative and meteorological forcings of atmospheric E_0 and their feedbacks with the state of the land surface, it requires no *SM*, *Prcp*, or land surface data, enabling EDDI to operate in ungauged areas. While EDDI is simple to estimate, one must use a fully physical E_0 to properly reflect the ET-E_0 interrelations and land surface drying anomalies. Fortunately, requisite drivers (Equation 11.2) are available across CONUS and globally.

While the use of E_0 focuses on relating atmospheric demand to developing drought conditions (e.g., EDDI) and highlighting the potential for vegetation stress, other indicators are necessary to estimate a direct response of the land surface to drought and to estimate the onset of actual

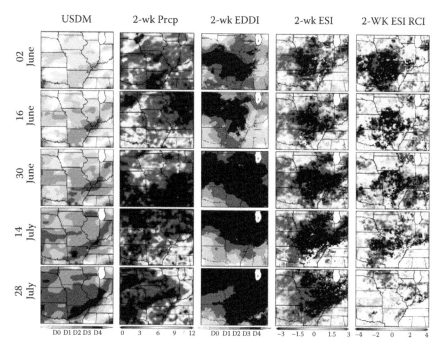

FIGURE 11.4

Temporal evolution of USDM drought categories, 2-week *Prcp* totals, and 2-week EDDI, ESI, and ESI RCI during a "flash" drought in the midwestern United States, from June 2 to July 28, 2012. (Adapted from Otkin et al. 2014.)

vegetation stress. This has led to the development of the **Evaporative Stress Index** (ESI) (Anderson et al. 2011a), which is based on RS-derived estimates of *ET* retrieved via energy balance principles using observations of *LST*. The ESI represents standardized anomalies in the ratio of *ET* to E_0, and normalization by E_0 serves to minimize variability in *ET* due to seasonal variations in available energy and vegetation cover, further refining focus on the relationships between *SM* and *ET*. As an indicator of actual *ET*, the ESI requires no information regarding *Prcp* or *SM* storage—the current available moisture to vegetation is deduced directly from the *LST*. This signal also inherently accounts for both *Prcp*- and non-*Prcp*-related sources and sinks of plant-available moisture (e.g., irrigation, tile drainage, vegetation tied to groundwater reserves; Hain et al. 2015), which can modify the vegetation response to *Prcp* anomalies. Rapid onset of vegetation and/or water stress can occur when extreme atmospheric anomalies persist for an extended period of time (e.g., several weeks) over a given location (Otkin et al. 2013, 2014, 2016; see example from 2012 in Figure 11.4). Therefore, the development of ESI Rapid Change Indices (ESI RCI) (Otkin et al. 2014) based on weekly changes in the ESI have been developed and evaluated over CONUS. The ESI RCI is designed to capture the accumulated rate of moisture stress change occurring over the full duration of a rapidly changing event. The use of the ESI has been increasing in the drought-monitoring community following the development of an operational system, the Geostationary Operational Environmental Satellite (GOES) EvapoTranspiration and Drought product system (GET-D), which provides daily, operational ESI datasets over all of North America at an 8-km spatial resolution. Additional background information on ESI and ESI RCI is provided in Chapter 10.

11.3.3 Early Warning

A significant advantage of examining drought from the evaporative perspective is the early warning that doing so affords. Because of the timing of the physical processes that link atmospheric forcing to, and signals from, the land surface (see Section 11.2) and our increased capabilities in monitoring these processes and the state of moisture stress in vegetation, we find that many of these tools provide information about drying anomalies or drought potential well before products that form the existing monitoring (and outlook) suite. As an example, Figure 11.4 shows the development of the flash drought of 2012 across the midwestern United States at regular intervals as monitored by the USDM, 2-week EDDI, 2-week ESI, and 2-week ESI RCI. The 2-week EDDI was clearly observing atmospheric drying for many weeks before the onset of the drought as measured by the USDM. In early June, EDDI, ESI, and ESI RCI were all showing rapidly deteriorating conditions across much of the central United States (e.g., Missouri, eastern Kansas, eastern Oklahoma, northern Arkansas, and Iowa), while the USDM only showed D0 and D1 drought classification (abnormally dry and moderate drought, respectively) over much of the same region. It was not until mid to late July, that the USDM introduced D3 and D4 categories (extreme drought and exceptional drought, respectively) over the region. This case study highlights the potential of metrics such as EDDI, ESI, and ESI RCI in providing early warning information about rapidly developing drought events (Hobbins et al. 2016; McEvoy et al. 2016a; Otkin et al. 2014, 2015, 2016).

11.3.4 Attribution

One of the advantages of using a physically based E_0 that incorporates the effects of all radiative and meteorological drivers (such as Penman-Montieth ET_0) is that the changes in E_0 that result in or reflect a drying anomaly may be attributed explicitly, that is, they may be diagnosed as to the relative strength of the drivers (e.g., whether T, q, R_d, or U_2, in the case of the ET_0 that underpins EDDI). Briefly, at a timescale of interest, the effects of changes in the drivers combine to generate changes in the E_0 (ΔE_0); according to Equation 11.5:

$$\Delta E_0 = \frac{\partial E_0}{\partial T}\Delta T + \frac{\partial E_0}{\partial R_d}\Delta R_d + \frac{\partial E_0}{\partial q}\Delta q + \frac{\partial E_0}{\partial U_2}\Delta U_2 \qquad (11.5)$$

where each of the summed terms on the right represents the effect on E_0 of changes in a driving variable (T, R_d, q, and U_2, respectively), with the driver anomalies (e.g., ΔU_2) coming from observations at the timescale of interest, and the sensitivities (e.g., $\partial E_0 / \partial U_2$) having been derived explicitly for an estimate of E_0 from Penman–Monteith ET_0 (Hobbins 2016).

The power of this attribution technique is demonstrated in Figure 11.5, where the development of an E_0 anomaly is tracked during a period of drought intensification and attributed into the relative contributions from each driver. This plot demonstrates the 12-week anomalies (i.e., deviations from the climatological mean accumulated across a 12-week period) of each of the four drivers of the Penman–Monteith E_0, and the top panel shows

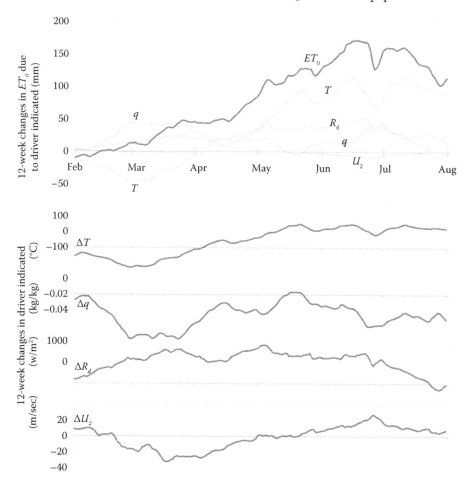

FIGURE 11.5

E_0 signal of drought intensification in the Sacramento river basin, CA, from February through July, 2014. The top panel shows the changes in 12-week ET_0 and its components due to each driver; lower panels show the 12-week anomalies in the drivers, as indicated.

the effects of the anomalies in each of these drivers on the overall change in 12-week E_0. Interestingly, the contributions from some of the drivers are negative (contributing to wetting), while some are positive (contributing to drying), while some change sign throughout the period. In this example, the drought intensification (increasing E_0) is forced from February through March by a combination of below-normal q and above-normal R_d, mediated to some degree by below-normal T. In April, T becomes anomalously high as q returns to near-normal conditions and, in combination with consistently high R_d, drives E_0 to peaks in June and July. U_2 plays little role in the event. This sort of analysis demonstrates which of these drivers is the most important in a particular event, and—enough events considered—a climatology thereof, and will therefore assist in knowing which drivers are regionally important to drought. Of note is that each of these drivers is forecastable, permitting a forecast of the effects of coming anomalies on drought, and so forecasting of drought itself, as discussed in the next Section (11.3.5).

11.3.5 Forecasting E_0

Forecasts of E_0 at timescales ranging from daily to seasonal are increasingly desired by stakeholders and managers in a number of sectors, including agriculture, water-resource management, and wildland-fire management, largely driven by recent developments highlighting the value of E_0 for drought monitoring. However, few such E_0 forecast tools currently exist.

Weather-scale forecasts of E_0 (i.e., for lead times of up to 2 weeks) can be valuable to agricultural producers to assist in irrigation scheduling. Dynamical weather forecast models, such as those used by the US National Weather Service (NWS), output all the necessary variables to compute a physically based E_0 as a post-processing step; ET estimates can then be derived using crop coefficients. Users need to be aware that raw dynamical model output will often contain biases, and the spatial resolution of the models (particularly global models) can be quite low. Some recent studies have examined bias-correction and downscaling methods to improve raw E_0 forecasts (e.g., Ishak et al. 2010; Silva et al. 2010). Other potential improvements include the use of retrospective forecast analogs (Tian and Martinez 2012a, 2012b) and ensemble forecasting to improve skill over single deterministic forecast runs (Tian and Martinez 2014). The only operational E_0 weather forecast product is the Forecast Reference ET (FRET) developed by the NWS. FRET produces E_0 forecasts of values and anomalies for Days 1–7 over CONUS. While more research is needed, the 7-day accumulated E_0 anomalies could be useful to provide early warning of developing flash drought conditions in the growing season.

E_0 forecasts at subseasonal (3 weeks to 3 months) and seasonal (3 to 9 months) scales can be used for long-term planning purposes as opposed to day-to-day operations. As yet, only a few studies have examined the skill and potential application of seasonal E_0 forecasts. Tian et al. (2014) used the Climate

Forecast System Version 2 (CFSv2; Saha et al. 2014), a global dynamical seasonal forecast model, to evaluate bias-corrected and downscaled E_0 quantities over the southeast US. Moderate skill was found during the cold season, but little forecast skill was found during the growing season due to the inability of the coarse global model to resolve convective processes. A broader analysis over CONUS examined the skill of using E_0 anomalies derived from CFSv2 to forecast droughts (McEvoy et al. 2016b) and found E_0 forecasts to be nearly universally more skillful than *Prcp* forecasts in predicting drought, with the greatest skill during the growing season in major agricultural regions of the central and northeast United States. Figure 11.6 demonstrates that, averaged over CONUS, E_0 forecast skill is greater than that for *Prcp* during all seasons. The east north central region in Figure 11.6 shows large differences between E_0 and *Prcp* forecast skill with E_0 having much greater skill during the growing season, while the southeast region in Figure 11.6 shows one region where E_0 forecast skill is quite weak and often similar to that of *Prcp*.

The question then arises: Why are E_0 forecasts typically much more skillful than *Prcp* forecasts? In general, T is more predictable than *Prcp*, and several studies have linked a multidecadal warming trend to improved T predictions in seasonal forecast models (Jia et al. 2015; Peng et al. 2013). The ability of CFSv2 to change atmospheric CO_2 over time leads to more consistent above-normal T forecasts (Peng et al. 2013), which has been realized over the last

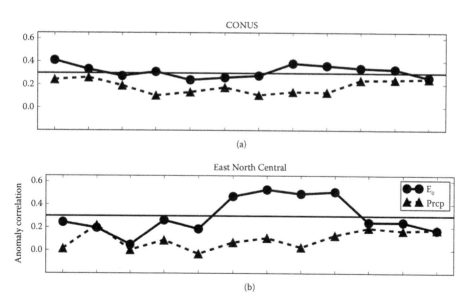

FIGURE 11.6
One-month lead time forecast skill (based on the correlation between forecasted and observed anomalies) for each seasonal period area-averaged over CONUS (a) and East North Central (b; Iowa, Minnesota, Wisconsin, and Michigan). *(Continued)*

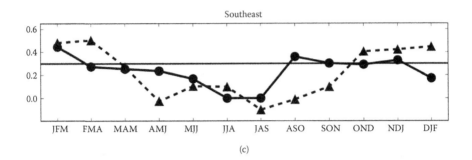

(c)

FIGURE 11.6 (Continued)
One-month lead time forecast skill (based on the correlation between forecasted and observed anomalies) for each seasonal period area-averaged over and southeast (c; Florida, Alabama, Georgia, South Carolina, North Carolina, and Virginia) regions. One-month lead forecasts were initialized during the month prior to the target date (e.g., JFM forecasts were initialized in December). The straight reference line indicates an anomaly correlation of 0.3, which represents the start of moderate skill according to NOAA's Climate Prediction Center. Correlations were computed between CFSv2 reforecast and the University of Idaho's gridded meteorological data (Abatzoglou 2013) for the period 1982–2009. (Adapted from McEvoy et al. 2016b.)

several decades. However, T is not the only E_0 driver (nor the most important in some regions), and other factors could be contributing to increased forecast skill over $Prcp$. SM memory is the primary land surface variable (with El Niño Southern Oscillation [ENSO] being the primary oceanic variable) contributing to seasonal predictability. Dirmeyer and Halder (2016) used CFSv2 to show that T, humidity, surface heat fluxes, and daytime boundary layer development all respond to variations in SM, while $Prcp$ is unresponsive. The strong link to the land surface, as opposed to the free atmosphere, thus leads to improved seasonal E_0-forecast skill.

A major hurdle in producing operational seasonal E_0 forecasts is the availability of all E_0 drivers from operational forecast producers. At the time of writing, the CFSv2 is the only model in the North American Multimodel Ensemble (NMME; a state-of-the-art seasonal forecast tool that uses an ensemble of nine models from the United States and Canadian institutions) to provide public access to forecasts of all E_0 drivers in real time.

11.4 E_0, ET, and Drought at Climate Scales

In this section, we do not attempt to predict how drought will evolve under climate change (see Chapter 4). Rather we summarize the state of the science with regard to the roles of ET and E_0 in climate-scale drought analyses and forcing.

11.4.1 General Expectations

Under anthropogenic climate change heat is being added to the climate engine, and our general expectations are for increased E_0 due to increased T and a resulting increase in ET in regions where moisture is available. In general, climate model projections indicate that wet areas will get wetter and dry areas drier, in line with latitudinal increases in subtropical dry zones (the descending arm of the Hadley cell); also projected are poleward shifts in the main mid-latitudinal storm tracks and changing *Prcp* seasonality (Trenberth et al. 2014).

11.4.2 The "Warming Is Drying" Message

The scientific community is not yet in agreement on the effects of these expected dynamics on drought. The main current issues with the estimation and use of ET and E_0 in drought analyses at climate scales are exemplified in a recent arc of papers. Here, we start with one of the most influential papers on the topic: that of Dai et al. (2004), who ran the PDSI model across the globe at a 2.5-degree resolution from 1870 to 2002, forced by *Prcp* and T. Briefly, they reported a doubling of very dry areas (PDSI < −3.0) since the 1970s, and a decrease in very wet areas (PDSI > 3.0) during the 1980s, with regions at both extremes combined almost doubling in area since 1972. They concluded that the risk of drought had increased with global warming, and that this was likely the result of a higher water-holding capacity of the air due to warming, which raised E_0 and hence ET. This "warming is drying" message resonated deeply within the community, particularly within the Intergovernmental Panel on Climate Change's (IPCC) *Fourth Assessment Report (AR4)* (Meehl et al. 2007), though it has been revisited by the IPCC *Special Report on Extremes* (Seneviratne et al. 2012).

However, the simple E_0-parameterization at the basis of the evaporative driver of the PDSI bucket hydrology model used in Dai et al. (2004)—and operationally to this date—suffers from two issues: it does not incorporate changes in the other, non-T drivers of E_0 (R_n, U_z, and humidity), and it cannot reflect changes in how plants will uptake carbon—determining the rate of the transpiration component of ET—in a CO_2-enriched atmosphere.

To highlight the former issue, Hobbins et al. (2008) compared the effects of a T-based E_0 to observed E_0 (from E_{pan}) on trends of water balance components of the PDSI at 35 stations across Australia and New Zealand from 1975 to 2004, finding that T-based E_0 increased over the period almost everywhere (in line with T increases). Indeed, particularly in energy-limited regions, these trends were opposite to declines in observed E_0, with the trends in E_0 found to result, in the mean, from trends in U_z rather than in T (Roderick et al. 2007). When applied in the PDSI, the contrasting T-based and observed E_0 trends resulted in SM trends that bore no relation to each other (Hobbins et al. 2008). Similarly, comparing global PDSI changes forced by E_0 from the T-based

Thornthwaite (1948) equation traditionally used in the PDSI to those forced by the physically based Penman–Monteith ET_0, Sheffield et al. (2012) found that, contrary to previous warnings (e.g., the IPCC AR4; Meehl et al. 2007), there had been little change in long-term (1950–2008) global average drought trends: the physically based E_0-forced PDSI showed drying over only 58 percent of the globe, whereas the T-based E_0 PDSI showed drying almost uniformly across the globe. Again, in energy-limited regions, differences in E_0 trend directions translated into differences in wetting versus drying between the two E_0 types (in water-limited regions, PDSI trends are driven by trends in $Prcp$, not in E_0). These studies clearly demonstrate the importance of proper selection of E_0 parameterization for climate-scale drought analyses.

The reliance on T as a proxy for E_0 also has deleterious effects on the reconstruction of paleoclimate drought records (Sheffield et al. 2012). Generally, tree-ring data are scaled to T-based PDSI in their overlapping period. However, not only does this relationship assume that tree growth relates to T alone, and not, crucially, to atmospheric CO_2, but also it breaks down at high elevations and latitudes, particularly during the recent decades of rapid warming. These assumptions lead to an overestimation of past changes and so to an underestimation of recent changes. Clearly, then, T-based parameterizations of E_0 in the PDSI—which effectively take a state variable (T) as a proxy for a flux (ET)—may be of use in the short term, but not in long-term analyses of drought or drought-relevant fluxes (e.g., ET, SM) of the future (Trenberth et al. 2014) or the past (Sheffield et al. 2012).

This is not to say that the sole source of the "warming is drying" message is misconceptions due to the use of T-based E_0 parameterizations. Dai (2013) attempted to reconcile patterns of observed and modeled (from global climate models, GCMs) drying trends in the last part of the twentieth century using the self-calibrated PDSI forced by Penman–Monteith E_p (a fully physical measure). He concluded that while GCMs were able to capture ENSO's influence on drought and its recent trends, the differences between modeled and observed aridity changes result from natural SST variations not captured by the GCMs, and that more severe and widespread droughts should be expected over many land areas during the coming decades, due to either decreased $Prcp$ or increased ET. This study implies that regional patterns important to infrastructure planning decisions to mitigate drought vulnerability are not being captured by GCMs. Further, Cook et al. (2015) used PDSI forced with Penman–Monteith E_p and $Prcp$ from 17 GCMs running Representative Concentration Pathways (RCPs) 4.5 and 8.5 in June through August 2050–2099 and gave dire warnings of drought worse than those in the medieval climate anomaly (1100–1300) over the US Central Plains and Southwest—concluding that "megadroughts," or multidecadal droughts, will become more likely in both regions.

The question as to whether and where the future will become drier or wetter is becoming more nuanced—the IPCC *Special Report on Extremes* (Seneviratne et al. 2012) notes probable overestimation of drought and an overreliance on PDSI-based results—and its resolution will require more regional analyses.

11.4.3 Addressing Carbon-Plant Relations in Climate-Scale Analyses

There is more to capturing all the dynamics necessary to fully reflect drought or drying/wetting trends at climate timescales than simply turning to a fully physical E_0 measure: at these timescales, plant-carbon (P-C) relations will change, with significant hydrological consequences. The heart of the issue is that using current fully physical E_0 formulations, including either of the widely used Penman–Monteith equations for E_0 (E_p or ET_0), implicitly fixes P-C relations: such formulations lock in the response of vegetation to the current climate for which the wind function (in the case of E_p) and the stomatal conductance (in the case of ET_0) are calibrated. Doing so ignores the effects of climate-scale physiological changes of vegetation to increased CO_2: that increased water-use efficiency (WUE) leads to a similar carbon uptake and photosynthesis for lower transpiration losses from the plant. A stark warning on using E_0 formulations with fixed P-C relations in climate-scale analyses is sounded by Roderick et al. (2015), who survey the contradictory claims and assumptions of past analyses.

Exemplifying the danger of assuming fixed P-C relations, Feng and Fu (2013) found that the Penman–Monteith ET_0 resulted in dramatic (230 mm yr^{-1}) increases in ET_0 under coupled model intercomparison project (CMIP) phase 3/RCP8.5 drivers (from mean drivers from 27 models) in the period 2070–2099 relative to 1970–1999. Combined with a low *Prcp* increase (41 mm yr^{-1}), this led to decreases in the aridity index (*Prcp*/E_0) over much of the terrestrial land surface and an increase in aridity at a global scale. Rising T was shown to have led to an ET_0 increase via an increase in vapor pressure deficit of 7–9 percent/K over land with the other drivers remaining inconsequential. However, Roderick et al. (2015) consider that this type of analysis lies at the root of the "warming is drying" message previously discussed. In contrast, they examined the aridity of the land surface at climate scales using the aridity index, noting that, despite regional increases and decreases, there has been little overall change in global aridity since 1948. They contest the commonly held notion that a warmer climate leads to a more arid land surface, and claim to have resolved the mismatch between observations and modeling in such a way as to align with the geological record. In doing so, their study highlights an example of a hidden modeling problem in at least the CMIP3 GCMs discussed: that of a fixed stomatal conductance.

Instead, Roderick et al. (2015) suggest assessing agricultural and ecological drought or aridity trends at climate scales using a new approach drawn from scientific communities in agricultural, ecology, and forestry. They propose using vegetation data from GCMs that do not rely on a fixed stomatal conductance parameter or wind function for estimating E_0. Instead, the aridity index is based on the ratio of gross primary productivity (GPP) to WUE. In doing so, they note that CMIP3 and CMIP5 models show, on global average, a warmer climate being less arid for both meteorological (low *Prcp*) and hydrological (low *Runoff*) drought, and they suggest, that as GPP increases with increasing atmospheric CO_2, global agro/ecological aridity will decrease. Their results resolve GCM modeling with remotely sensed observations and the geological record (i.e., they resolve the global aridity paradox). But questions remain (see Section 11.5).

11.4.4 Summary of Climate-Scale Analysis Issues

In a broad assessment of climate-scale drought analyses, Trenberth et al. (2014) highlight the following issues that may lead to conflicting conclusions: uncertainties in input forcing—particularly *Prcp* and U_z—and in long-term *ET* estimates, where the differences between global land and regional estimates may be significant; and the difficulties in capturing the role of natural variability. Further, decadal trend assessments are unreliable without base periods long enough to capture natural variability—particularly when accounting for the effects of ENSO, the Pacific Decadal Oscillation, and the Interdecadal Pacific Oscillation. While Trenberth et al (2014) state that simple, *T*-based E_0 parameterizations may have merit (so long as their shortcomings are recognized), we feel that this may be an acceptable compromise position only for informed hydrological and agricultural science communities. Indeed, we have seen other communities—such as ecological science communities and water-resource managers and their decision-making stakeholders—drawing conclusions from simple off-the-shelf models without appreciating their data and modeling nuances. We recommend against the use of PDSI for long-term climate analyses and, further, that studies that use *T*-based E_0 in long-term analyses should be ignored. Indeed, with the near-ubiquity of long-term data for all drivers of E_0 there is little cause to use such metrics at almost any timescale.

11.5 Research Directions

In this section, we highlight some of the questions and directions to guide future research into *ET*, E_0, and their roles in monitoring and forecasting drought.

11.5.1 Operational Products

As always, scientists and their products' users clamor for improvements in the spatial and temporal resolutions of data for driving models and of the resulting derived information. This aligns with a call from the agricultural and drought-monitoring communities for an increase in satellite overpass frequency, perhaps by future additional LANDSAT missions. A current drive to improve the estimation of *ET* from lakes and reservoirs should make significant contributions both to water-resources planning and response to hydrological drought, particularly in the western United States. With all the recent developments in *ET*- and E_0-based drought monitoring, the next steps toward advancing the science of predictions would be to develop real-time forecast products. These would build on the FRET product (Section 11.3.5) and expand to subseasonal and seasonal forecast timescales.

11.5.2 ESI as a Predictor of Agricultural Yield Anomalies

The utility of the ESI for monitoring crop stress and predicting agricultural drought impacts on yields has been evaluated in several case studies in the United States, Brazil, Tunisia, and Europe using ALEXI driven by GOES- and moderate resolution imaging spectroradiometer (MODIS)-derived *LST* time differences (e.g., Anderson et al. 2015, 2016; Mladenova et al. 2017; Otkin et al. 2016). An example of the potential capability of ESI is highlighted during the 2012 flash drought in the central United States, when ESI and ESI RCI identified the area of largest corn-yield impacts early in the season (May), before significant drought appeared in the USDM or VegDRI (related to NDVI anomalies). The ESI also agreed with county observer data collected by the USDA National Agricultural Statistics Service (NASS), who recorded visual topsoil moisture and crop conditions as they degraded through the growing season. Following Anderson et al. (2015, 2016), the annual yield data serve as a proxy indicator of agricultural drought impacts and enable assessment of relative timing and strength of correlation with multiple drought indicators. ESI and ESI RCI performance will be assessed in comparison with standard global *Prcp*- and vegetation index-based drought indices to develop a better understanding of conditions and locations where ESI adds unique value as an early indicator of developing crop stress.

11.5.3 Forecasting E_0

Historically, seasonal drought forecasting has been achieved through *T* and *Prcp*, and understanding the relationships these variables have to large-scale coupled oceanic-atmospheric processes such as ENSO. To advance E_0 forecasting (at any timescale; weather or seasonal) beyond post-processing of dynamical model output, we should strive to achieve better physical understanding of the relationships between E_0 and sources of predictability (Tian et al. [2014] and McEvoy et al. [2016b] touch on this for E_0). This includes ENSO for seasonal forecasts but other indices including the Madden-Julian oscillation for subseasonal forecasts. Using modern, high-resolution datasets to build upon past work establishing *T* and *Prcp* relationships to large-scale climate patterns (e.g., Cayan et al. 1999; Redmond and Koch 1991) for E_0 and the individual drivers is a logical step toward improving E_0 forecasting. Such real-time E_0 seasonal forecast products are currently in development.

11.5.4 Climate-Scale Analyses

For climate-scale drought vulnerability and ecological assessments requiring evaporative estimates, we must develop a robust climate-scale E_0 or similar metric of aridity, and it must incorporate all physically relevant drivers—including the vegetative effects of increased atmospheric CO_2—with uncertainties that are well expressed and transmissible through the analyses to users in a useful manner. Further, decisions at strategic planning timescales require support from more regional analyses.

Such advances will help us address more-specific questions. To what degree have ENSO, the Pacific Decadal Oscillation, and the Interdecadal Pacific Oscillation been affected by climate change? What are their effects on long-term ET and E_0? Will nutrient cycling constrain increases in GPP with atmospheric CO_2? Are the results of Roderick et al. (2015) robust across GCMs? How will their global conclusions regionalize or seasonalize?

11.5.5 Research-to-Operations/Applications (R2O/R2A) and Operations-to-Research (O2R) Arcs

As shown in Figure 11.3, observations related to the evolution of agricultural or meteorological drought can be represented by a number of different indicators. As many of these indicators represent different temporal signatures of developing droughts or different physical responses to drought, an urgent need exists for the education of users, and such stakeholder engagement with the drought, agriculture, and water resources communities would have as a benefit the verification of these indicators. Such efforts are being undertaken by the National Integrated Drought Information System (NIDIS). One NIDIS study shows that greater benefits can be realized for indicators based on E_0 and ET (e.g., EDDI and ESI) where assessments of the users' awareness of such products generally scores low, even for users in agricultural regions (McNie 2014). For these products to become widely accepted, the integration of ET-based monitoring and forecast products into operational decision support systems should be undertaken with robust, well-developed training programs based on educating stakeholders on how to use new ET-based drought indicators.

11.6 Concluding Remarks

Moving from reliance on monitoring near-past conditions toward incorporating complementary streams of information provided by closer to real-time monitoring, warning on drought-incipient conditions derived from observations, and drought forecasting will provide a more complete evaporative perspective of drought. This improved perspective not only closes the physical water balance but also migrates the monitoring of drought closer to the mission of drought early warning information systems. Shifting the focus from drought indicators that are solely based on a reactive assessment of the current drought state (e.g., NDVI anomalies) toward indicators based on E_0 and ET (e.g., EDDI and ESI) can provide an opportunity for proactive drought monitoring and generate valuable tools for land and water managers, and decision makers. However, special care should be taken to adequately work with stakeholders on how to integrate these datasets and their attendant information and uncertainties into their decision-making processes so that their full potential may be realized. Our goal here has been to contribute to that effort.

References

Abatzoglou, J. T. 2013. Development of gridded surface meteorological data for ecological applications and modelling. *International Journal of Climatology* 33:121–131. doi:10.1002/joc.3413.

Allen, R. G., M. Tasumi, and R. Trezza. 2007. Satellite-based energy balance for mapping evapotranspiration with internalized calibration (METRIC)—Model. *Journal of Irrigation Drainage Engineering* 133(4):380–394. doi:10.1061/(ASCE) 0733-9437(2007)133:4(380).

Anderson, M. C., C. R. Hain, J. Otkin, et al. 2013. An intercomparison of drought indicators based on thermal remote sensing and NLDAS-2 simulations with U.S. drought monitor classification. *Journal of Hydrometeorology* 14(4):1035–1036. doi:10.1175/JHM-D-12-1040.1.

Anderson, M. C., C. Hain, B. Wardlow, A. Pimstein, J. R. Mecikalski, and W. P. Kustas. 2011a. Evaluation of drought indices based on thermal remote sensing of evapotranspiration over the continental United States. *Journal of Climate* 24:2025–2044. doi:10.1175/2010JCLI3812.1.

Anderson, M. C., W. Kustas, J. M. Norman, et al. 2011b. Mapping daily evapotranspiration at field to continental scales using geostationary and polar orbiting satellite imagery. *Hydrology and Earth System Sciences* 15:223–239. doi:10.5194/hess-15-223-2011.

Anderson, M. C., J. M. Norman, G. R. Diak, W. P. Kustas, and J. R. Mecikalski. 1997. A two-source time-integrated model for estimating surface fluxes using thermal infrared remote sensing. *Remote Sensing of Environment* 60(2):195–216. doi:10.1016/S0034-4257(96)00215-5.

Anderson, M. C., C. Zolin, C. R. Hain, K. Semmens, M. T. Yilmaz, and F. Gao. 2015. Comparison of satellite-derived LAI and precipitation anomalies over Brazil with a thermal infrared-based evaporative stress index for 2003–2013. *Journal of Hydrology* 526:287–302. doi:/10.1016/j.jhydrol.2015.01.005.

Anderson, M. C., C. A. Zolin, P. C. Sentrelhas, et al. 2016. The evaporative stress index as an indicator of agricultural drought in Brazil: An assessment based on crop yield impacts. *Remote Sensing of Environment* 174:82–99. doi:10.1016/j.rse.2015.11.034.

Bastiaanssen, W., M. Merenti, R. A. Feddes, and A. Holtslag. 1998. A remote sensing surface energy balance algorithm for land (SEBAL). 1. Formulation. *Journal of Hydrology* 212:198–212. doi:10.1016/S0022-1694(98)00253-4.

Beguería, S., S. M. Vicente-Serrano, F. Reig, and B. Latorre. 2014. Standardized precipitation evapotranspiration index (SPEI) revisited: Parameter fitting, evapotranspiration models, tools, datasets and drought monitoring. *International Journal of Climatology* 34:3001–3023. doi:10.1002/joc.3887.

Blaney, H. F., and W. D. Criddle. 1962. *Determining Consumptive Use and Irrigation Water Requirements.* USDA Technical Bulletin 1275. Vol. 59. U.S. Department of Agriculture, Beltsville, MD.

Bouchet, R. J. 1963. *Évapotranspiration réelle et potentielle, signification climatique.* Vol. 62. IAHS Publications, Berkeley, CA, 134–142.

Budyko, M. I. 1974. *Climate and life.* Orlando, FL: Academic Press.

Cayan, D. R., K. T. Redmond, and L. G. Riddle. 1999. ENSO and hydrologic extremes in the western United States. *Journal of Climate* 12(9):2881–2893. doi:10.1175/1520-0442(1999)012<2881:EAHEIT>2.0.CO;2.

Cook, B. I., T. R. Ault, and J. E. Smerdon. 2015. Unprecedented 21st century drought risk in the American Southwest and Central Plains. *Science Advances* 1:e1400082. doi:10.1126/sciadv.1400082.

Dai, A. 2013. Increasing drought under global warming in observations and models. *Nature Climate Change* 3:52–58. doi:10.1038/nclimate1633.

Dai, A., K. E. Trenberth, and T. Qian. 2004. A global dataset of Palmer drought severity index for 1870–2002: Relationship with soil moisture and effects of surface warming. *Journal of Hydrometeorology* 5:1117–1130. doi:10.1175/JHM-386.1.

Dirmeyer, P., and S. Halder. 2016. Application of the land-atmosphere coupling paradigm to the operational Coupled Forecast System (CFSv2). *Journal of Hydrometeorology* 18:85–108. doi:10.1175/JHM-D-16-0064.s1.

Dirmeyer, P. A., X. Gao, M. Zhao, Z. Guo, T. Oki, and N. Hanasaki. 2006. GSWP-2: Multimodel analysis and implications for our perception of the land surface. *Bulletin of the American Meteorological Society* 87:1381–1397. doi:10.1175/BAMS-87-10-1381.

Donohue, R. J., M. L. Roderick, and T. R. McVicar. 2010. Can dynamic vegetation information improve the accuracy of Budyko's hydrological model? *Journal of Hydrology* 390:23–34. doi:10.1016/j.jhydrol.2010.06.025.

Feng, S., and Q. Fu. 2013. Expansion of global drylands under a warming climate. *Atmospheric Chemistry and Physics* 13(19):10081–10094. doi:10.5194/acp-13-10081-2013.

Hain, C. R., W. T. Crow, M. C. Anderson, and M. T. Yilmaz. 2015. Diagnosing neglected soil moisture source/sink processes via a thermal infrared-based two-source energy balance model. *Journal of Hydrometeorology* 16:1070–1086. doi:10.1175/JHM-D-14-0017.1.

Hamon, W. R. 1961. Estimating potential evapotranspiration. *Journal of the Hydraulics Division* 87(3):107–120.

Hargreaves, G. H., and Z. A. Samani. 1985. Reference crop evapotranspiration from temperature. *Applied Engineering in Agriculture* 1(2):96–99. doi:10.13031/2013.26773.

Hobbins, M. T. 2016. The variability of ASCE standardized reference evapotranspiration: A rigorous, CONUS-wide decomposition and attribution. *Transactions of the ASABE* 59(2):561–576. doi:10.13031/trans.59.10975.

Hobbins, M. T., A. Dai, M. L. Roderick, and G. D. Farquhar. 2008. Revisiting the parameterization of potential evaporation as a driver of long-term water balance trends. *Geophysical Research Letters* 35(12):L12403. doi:10.1029/2008GL033840.

Hobbins, M. T., and J. Huntington. 2016. Evapotranspiration and Evaporative Demand. In *Handbook of Applied Hydrology*, ed. V. P. Singh, 42–1–42–18. New York: McGraw-Hill Education.

Hobbins, M. T., J. A. Ramírez, and T. C. Brown. 2004. Trends in pan evaporation and actual evapotranspiration across the conterminous US: Paradoxical or complementary? *Geophysical Research Letters* 31(13):L13503. doi:10.1029/2004GL0198426.

Hobbins, M. T., A. Wood, D. J. McEvoy, et al. 2016. The evaporative demand drought index. Part I: Linking drought evolution to variations in evaporative demand. *Journal of Hydrometeorology* 17:1745–1761. doi:10.1175/JHM-D-15-0121.1.

Ishak, A. M., M. Bray, R. Remesan, and D. Han. 2010. Estimating reference evapotranspiration using numerical weather modelling. *Hydrological Processes* 24:3490–3509. doi:10.1002/hyp.7770.

Jia, L., X. Yang, G. A. Vecchi, et al. 2015. Improved seasonal prediction of temperature and precipitation over land in a high-resolution GFDL climate model. *Journal of Climate* 28:2044–2062. doi:10.1175/JCLI-D-14-00112.1.

Kalma, J. D., T. R. McVicar, and M. F. McCabe. 2008. Estimating land surface evaporation: A review of methods using remotely sensed surface temperature data. *Surveys in Geophysics* 29(4):421–469. doi:10.1007/s10712-008-9037-z.

McEvoy, D. J., J. L. Huntington, M. T. Hobbins, et al. 2016a. The evaporative demand drought index. Part II: CONUS-wide assessment against common drought indicators. *Journal of Hydrometeorology* 17:1763–1779. doi:10.1175/JHM-D-15-0122.1.

McEvoy, D. J., J. L. Huntington, J. F. Mejia, and M. T. Hobbins. 2016b. Improved seasonal drought forecasts using reference evapotranspiration anomalies. *Geophysical Research Letters* 43:377–385. doi:10.1002/2015GL067009.

McKee, T. B., N. J. Doesken, and J. Kleist. 1993. The relationship of drought frequency and duration to time scales. In *Proceedings of Eighth Conference on Applied Climatology*, 17–22 January. American Meteorological Society, Anaheim, California, 179–184.

McNie, E. 2014. *Evaluation of the NIDIS Upper Colorado River Basin Drought Early Warning System*. Western Water Assessment, Cooperative Institute for Research in Environmental Sciences, University of Colorado, Boulder, p. 40.

Meehl, G. A., T. F. Stocker, W. D. Collins, et al. 2007. Global Climate Projections. In *Climate Change 2007: The Physical Science Basis. Contribution of Working Group I to the Fourth Assessment Report of the Intergovernmental Panel on Climate Change*, eds. S. Solomon, D. Qin, M. Manning, et al., 747–845. Cambridge: Cambridge University Press.

Mladenova, I. E., J. D. Bolten, W. T. Crow, et al. 2017. Intercomparison of soil moisture, evaporative stress and vegetation indices for estimating corn and soybean yields over the U.S. *IEEE Journal of Selected Topics in Applied Earth Observations and Remote Sensing (JSTARS)* 99:1–16. doi:10.1109/JSTARS.2016.2639338.

Monteith, J. L. 1965. Evaporation and environment. *Symposia of the Society for Experimental Biology* 19:205–234.

Otkin, J. A., M. C. Anderson, C. R. Hain, I. E. Mladenova, J. B. Basara, and M. Svoboda. 2013. Examining rapid onset drought development using the thermal infrared based Evaporative Stress Index. *Journal of Hydrometeorology* 14(4):1057–1074. doi:10.1175/JHM-D-12-0144.1.

Otkin, J. A., M. C. Anderson, C. R. Hain, and M. Svoboda. 2014. Using temporal changes in drought indices to generate probabilistic drought intensification forecasts. *Journal of Hydrometeorology* 16:88–105. doi:10.1175/JHM-D-14-0064.1.

Otkin, J. A., M. C. Anderson, C. R. Hain, et al. 2016. Assessing the evolution of soil moisture and vegetation conditions during the 2012 United States flash drought. *Agricultural and Forest Meteorology* 218:230–242. doi:10.1016/j.agrformet.2015.12.065.

Otkin, J. A., M. Shafer, M. Svoboda, B. Wardlow, M. C. Anderson, and C. R. Hain. 2015. Facilitating the use of drought early warning information through interactions with agricultural stakeholders. *Bulletin of the American Meteorological Society* 96(7):1073–1078. doi:10.1175/BAMS-D-14-00219.1.

Palmer, W. C. 1965. *Meteorological Drought*. Research Paper 45, U.S. Department of Commerce, Washington, DC., 58.

Peng, P., A. G. Barnston, and A. Kumar. 2013. A comparison of skill between two versions of the NCEP Climate Forecast System (CFS) and CPC's operational short-lead seasonal outlooks. *Weather Forecast* 28:445–462. doi:10.1175/WAF-D-12-00057.1.

Penman, H. L. 1948. Natural evaporation from open water, bare soil, and grass. *Proceedings of the Royal Society of London, Series A* 193(1032):120–145.

Priestley, C. H. B., and R. J. Taylor. 1972. On the assessment of surface heat flux and evaporation using large-scale parameters. *Monthly Weather Review* 100:81–92. doi:10.1175/1520-0493(1972)100,0081:OTAOSH.2.3.CO;2.

Redmond, K. T., and R. W. Koch. 1991. Surface climate and streamflow variability in the western United States and their relationship to large-scale circulation indices. *Water Resources Research* 27:2381–2399. doi:10.1029/91WR00690.

Roderick, M. L., P. Greve, and G. D. Farquhar. 2015. On the assessment of aridity with changes in atmospheric CO_2. *Water Resources Research* 51:5450–5463. doi:10.1002/2015WR017031.

Roderick, M. L., L. D. Rotstayn, G. D. Farquhar, and M. T. Hobbins. 2007. On the attribution of changing pan evaporation. *Geophysical Research Letters* 34:L17403. doi:10.1029/2007GL031166.

Rotstayn, L. D., M. L. Roderick, and G. D. Farquhar. 2006. A simple pan-evaporation model for analysis of climate simulations: Evaluation over Australia. *Geophysical Research Letters* 33:L17715. doi:10.1029/2006GL027114.

Saha, S., S. Moorthi, X. Wu, et al. 2014. The NCEP climate forecast system version 2. *Journal of Climate* 27:2185–2208. doi:10.1175/JCLI-D-12-00823.1.

Senay, G. B., S. Bohms, R. Singh, et al. 2013. Operational evapotranspiration mapping using remote sensing and weather datasets: A new parameterization for the SSEB approach. *Journal of the American Water Resources Association* 49(3):577–591. doi:10.111/jawr.12057.

Senay, G. B., M. E. Budde, and J. P. Verdin. 2011. Enhancing the Simplified Surface Energy Balance (SSEB) approach for estimating landscape ET: Validation with the METRIC model. *Agricultural Water Management* 98(4):606–618. doi:10.1016/j.agwat.2010.10.014.

Seneviratne, S. I., N. Nicholls, D. Easterling, et al. 2012. Changes in climate extremes and their impacts on the natural physical environment. In *Managing the Risks of Extreme Events and Disasters to Advance Climate Change Adaptation*, eds. C. B. Field, V. Barros, T. F. Stocker, et al., 109–230. A Special Report of Working Groups I and II of the Intergovernmental Panel on Climate Change (IPCC). Cambridge: Cambridge University Press.

Sheffield, J., E. F. Wood, and M. L. Roderick. 2012. Little change in global drought over the past 60 years. *Nature* 491(7424):435–438. doi:10.1038/nature11575.

Silva, D., F. J. Meza, and E. Varas. 2010. Estimating reference evapotranspiration (ET_0) using numerical weather forecast data in central Chile. *Journal of Hydrology* 382:64–71. doi:10.1016/j.jhydrol.2009.12.018.

Su, Z. 2002. The Surface Energy Balance System (SEBS) for estimation of turbulent heat fluxes. *Hydrology and Earth System Sciences* 6(1):85–100. doi:10.5194/hess-6-85-2002.

Thornthwaite, C. W. 1948. An approach toward a rational classification of climate. *Geographical Review* 38(1):55–94. doi:10.2307/210739.

Tian, D., and C. J. Martinez. 2012a. Forecasting reference evapotranspiration using retrospective forecast analogs in the southeastern United States. *Journal of Hydrometeorology* 13:1874–1892. doi:10.1175/JHM-D-12-037.1.

Tian, D., and C. J. Martinez. 2012b. Comparison of two analog-based downscaling methods for regional reference evapotranspiration forecasts. *Journal of Hydrology* 475:350–364. doi:10.1016/j.jhydrol.2012.10.009.

Tian, D., and C. J. Martinez. 2014. The GEFS-based daily reference evapotranspiration (ET_0) forecast and its implication for water management in the southeastern United States. *Journal of Hydrometeorology* 15(3):1152–1165. doi:10.1175/JHM-D-13-0119.1.

Tian, D., C. J. Martinez, and W. D. Graham. 2014. Seasonal prediction of regional reference evapotranspiration based on climate forecast system version 2. *Journal of Hydrometeorology* 15:1166–1188. doi:10.1175/JHM-D-13-087.1.

Trenberth, K. E., A. Dai, G. van der Schrier, et al. 2014. Global warming and changes in drought. *Nature Climate Change* 4:17–22. doi:10.1038/nclimate2067.

Vicente-Serrano, S. M., S. Beguería, and J. I. López-Moreno. 2010. A multiscalar drought index sensitive to global warming: The standardized precipitation evapotranspiration index. *Journal of Climate* 23:1696–1718. doi:10.1175/2009JCLI2909.1.

12

A Role for Streamflow Forecasting in Managing Risk Associated with Drought and Other Water Crises

Susan M. Cuddy, Rebecca (Letcher) Kelly, Francis H. S. Chiew, Blair E. Nancarrow, and Anthony J. Jakeman

CONTENTS

12.1 Introduction

Climatic variability is a significant factor influencing agricultural production decisions. Historically in Australia, farmers and governments have invested heavily in reducing the influence of this variability on agricultural production. This investment has included construction of large dams on major river systems throughout the country, primarily for irrigation purposes, and allocation and development of groundwater resources. This development policy has placed undue pressures on ecosystems and has significantly modified river systems. In 1994, the Council of Australian Governments began a period of water reform, entering a new management phase for water resources. A major turning point in this reform was promulgation of the National Water Initiative of 2004, where protection of river systems has since been underpinned by the three pillars of regulation, water planning, and water trading. These reforms have included assessment of the sustainable yield from aquifer systems, often found to be below current allocation and even extraction levels, as well as allocation of a proportion of flows to the environment. In many catchments, these water reforms have not only reduced irrigators' access to some types of water but also implicitly increased the effect of climate variability on their decision-making by increasing their reliance on pumping variable river flows.

These management and allocation pressures are compounded by streamflow (and, to a lesser extent, climate) being much more variable in Australia than elsewhere. The interannual variability of river flows in temperate Australia (and southern Africa) is about twice that of river flows elsewhere in the world (Figure 12.1; Peel et al. 2001). This means that temperate Australia is more vulnerable than other countries to river flow-related droughts and floods. In such a challenging environment, forecasting tools that support improved decision-making resulting in efficiencies in water use and reduced risk-taking are highly desirable. The development and use of such tools is the focus of considerable research and extension activity in government and industry.

12.1.1 Hydroclimate Variability and Seasonal Streamflow Forecast

Relationships between sea surface temperatures and climate are well documented. The relationship between Australia's hydroclimate and the El Niño southern oscillation (ENSO) is among the strongest in the world (Chiew and McMahon 2002). El Niño describes the warm phase of a naturally occurring sea surface temperature oscillation in the tropical Pacific Ocean. Southern oscillation refers to a seesaw shift in surface air pressure at Darwin, Australia, and the South Pacific island of Tahiti. Several indices have been

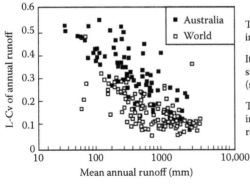

FIGURE 12.1
Interannual variability of Australian streamflow relative to the rest of the world.

derived from this relationship, in particular the southern oscillation index (SOI), which describes the Tahiti minus Darwin sea-level pressure and is commonly used as an indicator of ENSO. The strong relationships that exist between climate, streamflow, and ENSO form the scientific basis for forecast tools developed throughout Australia and other parts of the world. In the Australian context, the Bureau of Meteorology routinely provides seasonal climate outlooks (e.g., probability that the total rainfall over the next three months will exceed the median), and climate and streamflow forecasting tools such as *Rainman* (Clewett et al. 2003) are promoted. More recently, significant research over the past decade and translation of research to operations through the Water Information Research And Development Alliance (WIRADA) between the Bureau of Meteorology and the Commonwealth Scientific Industrial and Research Organisation (CSIRO) has led to state-of-the-art operational seasonal streamflow forecasts provided routinely online by the Bureau of Meteorology (http://www.bom.gov.au/water/ssf/).

12.1.2 Adoption Constraints

A major issue for the designers of decision support tools is the degree of likely uptake by the potential users, and this is no different for seasonal forecasting. The farming community, which is traditionally conservative when it comes to changing well-entrenched behaviors, is particularly reticent to adopt such tools. Many factors play a part in users' decisions to adopt these tools and the information they provide.

Knowledge, awareness, and understanding of the potential outcomes available through the use of the tools vary. Confidence in the outcomes is often lacking, especially when the tools may be replacing well-tried and comfortable practices. These practices may be seen to be adequate for the decisions they are assisting, and hence users do not perceive a need for new technologies.

Previous experiences associated with the technologies being used by the tools will also be a factor. These may be first-hand experiences or purely word of mouth in the community. Local opinion will frequently be more powerful than information from outsiders. Naturally, if past experiences have resulted in negative consequences, the uptake of the new technology will be even less likely. Confidence in the new technology and trust in the provider of the technology are, therefore, likely to be highly influential. In fact, the human factor frequently can be less certain than the technologies themselves.

12.2 Estimating the Potential of Seasonal Streamflow Forecast

Most investigations of the potential of forecast tools compare their predictions against "no knowledge." This section describes the coupling of forecast models to models that simulate a range of water management behaviors within a constrained problem definition. Quantification of the net financial return to irrigators of adopting climate forecasts as part of their decision-making process would provide a strong measure of the benefit of these forecasts. This is tempered by an analysis of the potential market, which reveals that a significant improvement in reliability and relevance is required before widespread adoption can be considered.

12.2.1 Case Study Context

To consider the potential benefits to agricultural production of seasonal forecasts, we investigated their potential impact on farm-level decisions and returns in an irrigated cropping system. We premised that the potential benefit of seasonal forecasts was probably greatest in a farming system subject to significant uncertainty. For this reason, the farming system represented in the decision-making models is that of an irrigated cotton producer operating on an unregulated river system, relying on pumping variable river flows for irrigation purposes during the season. This type of farm is typical in unregulated areas of the Namoi basin in the northern Murray–Darling basin, particularly the Cox's Creek area (Figure 12.2). However, for this analysis, the modeling should be considered to represent a theoretical or model farm rather than a farm from a particular system; the value of forecasts was tested on this farm using forecasts and flows from many different river systems in eastern New South Wales. We did this to test the sensitivity of the results and recommendations to the hydrology and climate of the river system.

Given that the model farm is assumed to be pumping from the river for irrigation supply, production and water availability are limited by the number of days on which the farm can pump flows from the river. To mimic the types of flow rules on these unregulated systems and to test the sensitivity of results to these rules, two pumping thresholds were considered, the 20th and 50th percentile of flow (i.e., flow that is exceeded 20 percent or 50 percent of the time).

The forecast provided for each year is the number of days that are above these pumping thresholds (i.e., the number of days on which pumping is allowed). The model farmer factors this forecast and the total volume of water allowed to be pumped on each such day (the daily extraction limit, defined by policy as a fixed volume of water) into the planting decision.

Streamflow forecasts were constructed over an 86-year period for seven catchments and the two pumping threshold regimes using three forecast methods. Farmer decisions were then simulated using these three forecast methods as the basis of the decision, as well as using three decision alternatives for comparison. This section describes the catchments considered in the analysis and the streamflow forecasting results for each. The decision models used in the analysis of these forecasts are then described before results are presented. These results should be considered indicative of the

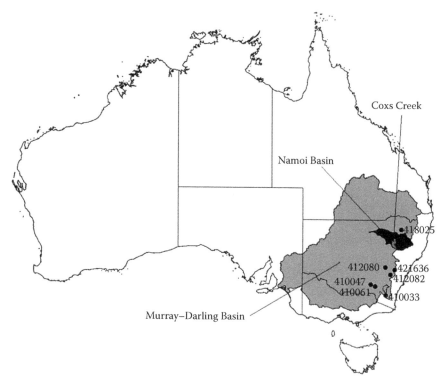

FIGURE 12.2
Map of the case study area highlighting the Cox's Creek region of the Namoi basin within the Murray–Darling basin system of eastern Australia.

potential benefits of seasonal forecasting in eastern Australia. The complexity of different production systems and many of the influences on real-life decisions have not been considered for this preliminary analysis. However, this analysis does provide an interesting insight into the potential for forecasting methods to help farmers adjust away from the impacts of climate variability.

12.2.2 Seasonal Forecast Models

The relationship between streamflow and ENSO and the serial correlation in streamflow can be exploited to forecast streamflow several months ahead. These relationships are well described in Chiew and McMahon (2003), Wang et al. (2009), and Kirono et al. (2010). They demonstrate statistical significance in the lag correlation between hydroclimate and the SOI (and other ENSO indicators), particularly in late spring and through summer. Using this relationship, we can forecast summer streamflow throughout most of eastern Australia from spring indicators of ENSO. Serial correlation in streamflow must also be considered when forecasting streamflow because it is generally stronger than the streamflow–ENSO relationship and is persistent throughout the year.

To make risk-based management decisions, we must express forecasts as exceedance probabilities (e.g., probability of getting at least 10 pumping days). In this study, exceedance probability forecasts are derived at tributary scale for seven unimpaired catchments in the Murray–Darling basin. The derivation of the forecasts is based on categorization and consequent nonparametric modeling of streamflow distributions and their antecedent conditions (e.g., discrete SOI categories) (see, e.g., Sharma 2000; Piechota et al. 2001). Catchments were selected because of their relative proximity to the Namoi basin (all within the Murray–Darling basin in New South Wales) and to reflect a range of rainfall–runoff conditions and forecast skills. Proximity to the Namoi basin is to support coupling with the decision-making models that have been developed by Letcher (2002) within the water management regulatory framework in the Namoi basin, although they simulate representative farmer behavior.

Daily streamflow data from the period 1912 to 1997 are used. The data include extended streamflow estimates using a conceptual daily rainfall–runoff model (Chiew et al. 2002). The catchment locations and long-term average rainfall–runoff characteristics are summarized in Table 12.1.

Forecasts are made for the number of days in October–February that the daily flow exceeds the two pumping thresholds under consideration. The thresholds are calculated based on flow days only, defined as days when the daily flow exceeds 0.1 mm. The forecast is derived by relating the number of days in October–February that the daily flow exceeds a threshold to explanatory variables available at the end of September. The explanatory variables used are the SOI value averaged over August and September and the total flow volume in August and September. We derive the forecast using the nonparametric seasonal forecast model described in Piechota et al. (2001) and express it as exceedance probabilities. Such forecasts closely approximate low-risk decision-making behavior and can be used as a direct input into the decision-making models.

Three forecast models are used:

1. FLOW: Forecast derived from flow volume in August–September
2. SOI: Forecast derived from SOI value in August–September
3. FLOW+SOI: Forecast derived from flow volume and SOI value in August–September

12.2.3 Forecast Model Results

All models exhibit significant skill in the forecast, summarized in Table 12.2. Two measures of forecast skill are used—Nash-Sutcliffe E and LEPS scores—for illustration and broadly to reflect the performance objectives of the modeling (Bennett et al. 2013).

TABLE 12.1

Summary of Characteristics for Catchments Used in the Analysis

Catchment	Lat.	Long.	Area (km²)	Rainfall (mm)	Runoff (mm)	Runoff Coef. (%)	% Days Flow >0.1 mm	Percentile Flows (mm)	
								20%	50%
410033 Murrumbidgee River at Mittagang Crossing	36.17	149.09	1891	882	134	10–15	71	0.55	0.28
410047 Tarcutta Creek at Old Borambola	35.15	147.66	1660	818	110	10–15	50	0.68	0.31
410061 Adelong Creek at Batlow Road	35.33	148.07	155	1138	256	>20	89	0.97	0.44
412080 Flyers Creek at Beneree	33.50	149.04	98	915	106	10–15	50	0.65	0.29
412082 Phils Creek at Fullerton	34.23	149.55	106	821	124	10–15	62	0.58	0.27
418025 Halls Creek at Bingara	29.91	150.58	156	755	44	6	24	0.22	0.14
421036 Duckmaloi River at below dam site	33.77	149.94	112	967	244	>20	80	0.95	0.40

Catchment and Rainfall—Runoff Characteristics

TABLE 12.2

Summary of Forecast Skills for Catchments Used in the Analysis

		Forecast Skill					
		FLOW		SOI		FLOW+SOI	
Catchment	Case	E	LEPS	E	LEPS	E	LEPS
410033 Murrumbidgee	Days >20%	0.35	27.1	0.23	11.6	0.58	41.7
River at Mittagang Crossing	Days >50%	0.36	23.1	0.19	12.2	0.60	39.7
410047 Tarcutta Creek at	Days >20%	0.41	32.8	0.23	17.6	0.57	46.4
Old Borambola	Days >50	0.39	26.2	0.18	11.2	0.50	36.0
410061 Adelong Creek	Days >10%	0.54	41.4	0.16	12.0	0.64	49.5
at Batlow Road	Days >20%	0.63	42.0	0.17	11.1	0.71	50.4
412080 Flyers Creek at	Days >20%	0.34	25.8	0.22	10.2	0.54	37.6
Beneree	Days >50%	0.42	28.8	0.22	10.9	0.56	40.0
412082 Phils Creek at	Days >20%	0.40	19.2	0.22	12.3	0.59	32.1
Fullerton	Days >50%	0.54	30.0	0.22	12.2	0.64	39.7
418025 Halls Creek at	Days >20%	0.13	12.4	0.16	11.7	0.29	26.3
Bingara	Days >50%	0.26	15.3	0.16	13.0	0.44	31.5
421036 Duckmaloi River	Days >20%	0.16	12.3	0.24	13.5	0.45	28.1
at below dam site	Days >50%	0.24	16.7	0.27	17.7	0.51	34.0

Nash and Sutcliffe (1970) provide a measure of the agreement between the "mean" forecast (close to the 50 percent exceedance probability forecast) and the actual number of days in October–February that the daily flow exceeds a threshold. A higher E value indicates a better agreement between the forecast and actual values, with an E value of 1.0 indicating that all the "mean" forecasts for all years are exactly the same as actual values.

The LEPS score (Piechota et al. 2001) attempts to compare the distribution of forecast (forecast for various exceedance probabilities) with the number of days in October–February that the daily flow exceeds a threshold. A LEPS score of 10 percent generally indicates that the forecast skill is statistically significant. A forecast based solely on climatology (same forecast for every year based on the historical data) has a LEPS score of 0.

The LEPS scores in all the forecast models are greater than 10 percent, indicating significant skill in the forecast. The SOI model has similar skill in the seven catchments, with E values of about 0.2 and LEPS scores of 10–15 percent. The FLOW model is considerably better than the SOI model in five catchments (410033, 410047, 410061, 412080, 412082; E generally greater than 0.35 and LEPS generally greater than 25 percent), whereas at the gauge sites of the other two catchments (418025, 421036), the FLOW and SOI models have similar skill. In all seven catchments, the FLOW+SOI model has greater skill than the FLOW or SOI model alone. In the five catchments where the FLOW model has greater skill than the SOI model, the E and LEPS for the FLOW+SOI model are generally greater than 0.5 and 40 percent, respectively (compared to 0.35 and 25 percent in the FLOW model). In the two catchments where the FLOW model and SOI model have similar skill, the E and LEPS for the FLOW+SOI model are generally greater than 0.3 and 25 percent (compared to less than 0.25 and 20 percent in the FLOW or SOI model alone).

12.2.4 Decision-Making Models

All decisions were modeled using a simple farm model that assumed farmers act to maximize gross margin each year, given constraints on land and water available to them in the year. This model is a modified version of a decision model for the Cox's Creek catchment developed by Letcher (2002) and also reported in Letcher et al. (2004). Total farm gross margin was analyzed for all catchments, pumping thresholds, and forecast methods using four possible decision methods:

1. *Seasonal forecast decision.* The decision is made assuming that the 20th and 50th percentile exceedance probability forecasts (using SOI, FLOW, and SOI+FLOW) for the number of pumping days are correct.

2. *Naïve decision.* The decision is made assuming that the number of pumping days this year is equal to the number of pumping days observed last year.

3. *Average climate decision.* The decision is made assuming that the number of days for which pumping is possible in each year is the same and equal to the average number of days pumping is permitted over the entire 86-year period.

4. *Perfect knowledge decision.* The decision is made with full knowledge of the actual number of days on which pumping is possible in each year. This is essentially used to standardize the results, because it is a measure of the greatest gross margin possible in each year, given resource constraints.

The same simple farm model is used in all cases. This model allows the farm to choose among three cropping regimes—irrigated cotton with winter wheat rotation, dryland sorghum and winter wheat rotation, and dryland cotton and winter wheat rotation. Production costs are incurred on crop planting, so areas planted for which insufficient water is available over the year generate a loss. For such crops, it is assumed that the area irrigated is cut back and a dryland yield is achieved on the remaining area planted.

12.2.5 Modeling Results

We ran models for each catchment over the 86-year period for every combination of pumping threshold, forecast, and decision-making method.

The total gross margin achieved by the farm over the entire simulation period under each of the decision models and forecasting methods is charted for the 20th percentile (Figure 12.3) and the 50th percentile (Figure 12.4) pumping threshold, respectively. In each figure, the x-axis labels the seven catchment identifiers and the y-axis is the total gross margin in Australian dollars. These figures lead to a consistent set of observations:

- Use of any of the three forecast methods leads to a greater gross margin than either the average or the naïve decision methods.
- In general, the SOI+FLOW method gives the greatest gross margin of the three forecast methods, with SOI generally providing the lowest gross margin.
- The forecast methods provide a substantial return in gross margin relative to the total achievable gross margin (via the perfect decision model) in each case (on average, 55 percent of the possible maximum).

To investigate the consistency of the forecast skill, we derived the percent of time during the simulation period during which different income levels were exceeded for each decision model and forecast method. Results for a single catchment (410033) and the 20th percentile pumping threshold are presented in Figure 12.5.

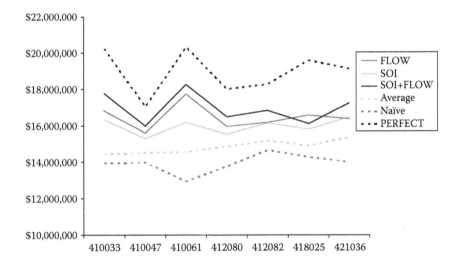

FIGURE 12.3
Total profit (annual gross margin) over 86-year simulation period for each catchment using different decision methods for pump threshold at the 20th percentile of flow.

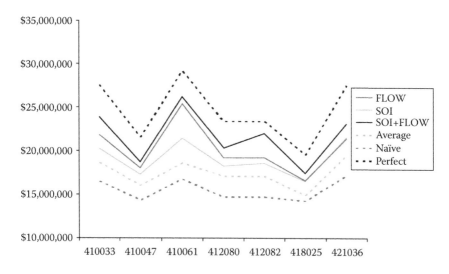

FIGURE 12.4
Total profit (annual gross margin) over 86-year simulation period for each catchment using different decision methods for pump threshold at the 50th percentile of flow.

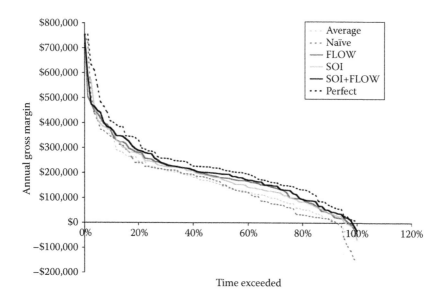

FIGURE 12.5
Exceedance probability for annual gross margin for one catchment (410033) with a pumping threshold at the 20th percentile of flow.

Several observations can be made about the consistency of the forecasts:

- Negative gross margins (losses) are experienced in a greater number of years for both the average and naïve decision methods (>7 percent of time) than for any of the seasonal forecast methods (<3.5 percent).
- The naïve and average decision methods give a lower income at almost all exceedance probabilities, and for those areas where they are greater, the difference is very small.
- The naïve decision method gives a greater gross margin for very high gross margin years (2.4 percent of the time).

12.3 Reality Bites

The integrated modeling approach developed for this study demonstrates the gains that can be made by the routine incorporation of seasonal forecasting into water management decision.

To test the likeliness of farmers to use the seasonal forecasting tools, we conducted a series of 10 semistructured scoping interviews. The interviewees were irrigators on a nonregulated tributary of the Namoi River (in the Upper Murray–Darling basin) and were therefore highly dependent on the river flows. With the uncertainty of annual water supplies to irrigate their crops, it was thought that this group might be more positive about seasonal forecasting than those on regulated rivers.

Given the small sample number, we compared the data collected with that from a similar study conducted by others in the southern Murray–Darling basin (URS Australia 2001). That study consisted of 29 interviews followed by a workshop with six participants. The findings of both studies were similar, thus providing confidence in the outcomes of these limited interviews.

Knowledge and understanding of the term and the forms of seasonal forecasting were highly variable, ranging from a good understanding to little or misguided awareness. Participants seemed to misunderstand the difference between types of forecasting and sources of forecasting. However, considerable support exists for natural signals rather than the use of technology—for example:

> Some of the best indicators in times of drought have been the ants' nests around the house—if there is a lot of movement by the ants, it's generally going to rain soon. There are many natural signs that are more useful than the scientific information we are given.

The degree to which people understood probabilities, or thought they could be useful to them, was also variable. Many were skeptical of the probabilities, given their derivation by the extrapolation of past data to the present. Usefulness was also questioned in view of past experiences:

> At the last meeting we were told, "There will be a 50/50 chance that we will get above-average rainfall and 50/50 chance we will get below-average rainfall." This told us nothing.

The degree to which these farmers incorporated seasonal forecasting information into their decision-making was also variable. Although no one used it as a regular aid, some said they sometimes used it, others considered it but rarely used it, and some said they did not use it at all.

Those who said they did use it indicated that it could affect changes in planning for the timing of seeding, spraying times, planting rates, the type of crops planted, and the number of stock purchased. However, they stressed that the seasonal forecasting was only one piece of information they used, combining it with natural indicators and sources of information used in the past. Decisions were still very conservative.

> If they say it is going to be a dry year I won't buy more cattle. If it is going to be a wet year, I may decide to buy more cattle.

Those who did not use seasonal forecasting in their decision-making were reluctant to do so because they had bad experiences in the past, or knew of someone who had. Other reasons included a lack of understanding or restricted access to the information. They seemed to pay more attention to short-term forecasting than to seasonal. The consequences of poor short-term decisions were not seen to be as dire as the consequences of seasonal mistakes.

> Really if your gut feeling tells you it is going to be dry then it probably will be. If it tells you it will be wet, it probably will be.
> I'm an old-time farmer and I feel that you take what you get.
> I don't have much confidence in the information. It is usually only 50 percent accurate which is the same as tossing a coin.

Rainfall probabilities were considered to be the most useful information seasonal forecasting could provide. However, decision-making would still be highly conservative.

> I would pay attention if they told me there was a 75 percent chance that we will go into a drought. However, if they told me that there was a 25 percent chance of below-average rainfall, with a 75 percent chance of above-average rainfall, I would pay more attention to the prediction of below-average rainfall.

The farmers were asked if they would be more willing to use a tool that predicted only extreme events with better than usual reliability, rather than more frequent rainfall predictions with lesser certainty. Generally, it was agreed that this would be preferable, but there was considerable cynicism that sufficient reliability could be obtained for their purposes.

They did acknowledge, however, the difficulty associated with forecasting, especially given the limited recorded weather history in Australia. Then again, it seemed there was little likelihood that any latitude would be given to the scientists if the forecasts were mistaken. Reliability was very important, and until this could be achieved to help farmers in decision-making, uptake of the technology would be limited. And memories can be very long.

> Indigo Jones was a long-range forecaster a while back, and he was considered to be very good. In 1974 he predicted it would be wet and we had some of the biggest floods in history. However, in 1975 he predicted it would be wetter still, and we had one of the worst droughts on record. After that I lost faith in long-range forecasters.

It is, therefore, apparent that the potential market for seasonal forecasting tools in the farming community will be limited in the short term. One must understand the likely users of the technology and exactly what decisions they believe it can assist them with.

12.4 Summary

Regions with high interannual variability of streamflow present challenges for managing the risks associated with the use of their water resource systems. Reliable forecasts of streamflow months in advance offer opportunities to manage this risk. Knowledge of more or less streamflow than average has the potential to influence farmer decision-making and/or the allocation of water to the environment or to replace groundwater stocks under threat. In the long term, streamflow forecasting has the potential to improve the viability of agricultural production activities and increase water use efficiency while maintaining desirable environmental flows.

The ability of the integrated modeling approach described above to provide a comparison with alternate heuristic forecasting techniques has demonstrated the practical advantage of forecasting methods over these alternate techniques. In the overwhelming number of cases, where water management and consequent planting decisions were based on seasonal forecasts, the enterprise would have returned a better result in terms of water use efficiency and net gains in profit. Yet adoption is slow, perhaps reflecting the general conservative nature of farmers, at least in Australia, and the need for them to see real and sustained benefit before they will consider incorporating such tools into their decision-making.

The social analysis confirmed that knowledge and understanding of the term and the forms of seasonal forecasting were highly variable. There seemed to be a misunderstanding of the difference between types of forecasting and sources of forecasting. However, there was considerable support for natural signals rather than the use of technology. The degree to which people understood probabilities, or thought they could be useful, was also variable. Many were skeptical of the probabilities given their derivation by the extrapolation of past data to the present. Their usefulness was also questioned in view of past experiences.

An aggressive water reform agenda, underpinned by an acknowledgment of the finite size of the water resource and recognition of the legitimacy of the environment as a water user, is driving research in and development of tools that can fine-tune critical decisions about water allocation.

12.5 Future Directions

In spite of recent improvements, adoption of climate variability management tools such as seasonal forecasting is low among the farming community. Farmers have expressed negativity about the reliability of the tools and their benefits.

It could be argued that many farmers, particularly those in large irrigated enterprises, have already reduced the risk associated with timely access to water by building large on-farm water storages and installing more efficient water reticulation systems—that is, they have invested (at a significant cost) out of the uncertainty for which seasonal forecasting tools are trying to compensate. So, it would seem that the need to consider the use of technology such as seasonal forecasting is directly related to degree of exposure and risk management behavior.

A consequence of the current water reforms—as timely access to instream water is no longer guaranteed and on-farm water storage is increasingly regulated—is an increase in risk exposure. This may force users to invest in tools that provide marginal gains. To identify where these marginal gains are, and the different levels of benefit that are possible, forecasting tools need to be tailored to a range of niche markets whose needs, decision-making behaviors, and current resistance must be clearly articulated. These niche markets need to be identified. This case study gives some strong leads—for example, irrigation versus dryland and high-equity versus mortgage participants. Significant benefit from research and development in climate risk management can only be realized if it produces tools that match users' needs and expectations and that can be incorporated into their decision-making and risk assessment processes.

In recognition of this, and the wider challenge in water resources management generally, the Australian Government provided additional resources to the Bureau of Meteorology under the Water Act 2007 to deliver water information for Australia. Through this, and the research in the WIRADA (http://www.csiro.au/en/Research/LWF/Areas/Water-resources/Assessing-water-resources/WIRADA) between the Bureau of Meteorology and the CSIRO, Australia is now one of only a few countries in the world that routinely provide online state-of-the-art seasonal streamflow forecasts (http://www.bom.gov.au/water/ssf/). With continued improvement in forecast science and forecast skill, and readily accessible and specifically tailored products, adoption of seasonal streamflow forecast is likely to increase to help inform and better manage Australia's highly variable water resources systems.

Acknowledgments

The opportunity to undertake the research described in this chapter was made possible via an 8-month research grant from the managing climate variability program within Land and Water Australia, the government agency that was responsible at that time for land and water resources research and development in Australia. Since that time, authors have progressed the science through their involvement in the WIRADA and other research initiatives.

References

Bennett, N.D., B. F. W. Croke, G. Guariso, et al. 2013. Characterising performance of environmental models. *Environmental Modelling and Software* 40:1–20.

Chiew, F. H. S., and T. A. McMahon. 2002. Global ENSO-streamflow teleconnection, streamflow forecasting and interannual variability. *Hydrological Sciences Journal* 47:505–522.

Chiew, F. H. S., and T. A. McMahon. 2003. Australian rainfall and streamflow and El Niño/Southern Oscillation. *Australian Journal of Water Resources* 6:115–129.

Chiew, F. H. S., M. C. Peel, and A. W. Western. 2002. Application and testing of the simple rainfall–runoff model SIMHYD. In *Mathematical Models of Small Watershed Hydrology and Applications*, eds. V. P. Singh and D. K. Frevert, 335–367. Littleton, CO: Water Resources Publications.

Clewett, J. F., N. M Clarkson, D. A. Gorge, et al. 2003. Rainman StreamFlow (version 4.3): A Comprehensive Climate and Streamflow Analysis Package on CD to Assess Seasonal Forecasts and Manage Climatic Risk. QI03040. Department of Primary Industries, Queensland.

Kirono, D.G.C., F. H. S. Chiew, and D. M. Kent. 2010. Identification of best predictors for forecasting seasonal rainfall and runoff in Australia. *Hydrological Processes* 24:1237–1247, doi:10.1002/hyp.7585.

Letcher, R. A. 2002. Issues in integrated assessment and modeling for catchment management. PhD diss., The Australian National University, Canberra.

Letcher, R. A, A. J. Jakeman, and B. F. W. Croke. 2004. Model development for integrated assessment of water allocation options. *Water Resources Research* 40:W0552.

Nash, J. E., and J. V. Sutcliffe. 1970. River forecasting using conceptual models, 1. A discussion of principles. *Journal of Hydrology* 10:282–290.

Peel, M. C., T. A. McMahon, B. L. Finlayson, and F. G. R. Watson. 2001. Identification and explanation of continental differences in the variability of annual runoff. *Journal of Hydrology* 250:224–240.

Piechota, T. C., F. H. S. Chiew, J. A. Dracup, and T. A. McMahon. 2001. Development of an exceedance probability streamflow forecast using the El Niño–Southern Oscillation and sea surface temperatures. *ASCE Journal of Hydrologic Engineering* 6:20–28.

Sharma, A. 2000. Seasonal to interannual rainfall probabilistic forecasts for improved water supply management: Part 3—A nonparametric probabilistic forecast model. *Journal of Hydrology* 239:249–259.

URS Australia. 2001. *Defining Researching Opportunities for Improved Applications of Seasonal Forecasting in South-Eastern Australia with Particular Reference to the Southern NSW and Victorian Grain Regions.* Report to the Climate Variability in Agriculture Program, Land and Water Australia, Canberra.

Wang, Q. J., D. E. Robertson, and F. H. S. Chiew. 2009. A Bayesian joint probability modeling approach for seasonal forecasting of streamflows at multiple sites. *Water Resources Research* 45:W05407. doi:10.1029/2008WR007355.

13

Drought Mitigation: Water Conservation Tools for Short-Term and Permanent Water Savings

Amy L. Vickers

CONTENTS

13.1 Introduction: A New *Era* of Water Scarcity or an Old *Error* of Water Waste?

The 1882 gravestone of a child that had lain for 129 years beneath the Lake Buchanan reservoir in Bluffton, Texas, resurfaced during a severe drought that began in 2011. For 5 years that memorial (along with remnants of an abandoned bank, a cotton gin, and house foundations in Old Bluffton), which had been abandoned to build nearby Buchanan Dam, baked in the sun during a withering multiyear drought (Hlavaty 2016). Some say the drought brought back the old ghost town, but others shake their heads and point to the water managers and politicians whose actions to reduce water demands were too little and too late. Had officials acted earlier and more aggressively to impose mandatory lawn watering restrictions and other proven water-saving strategies, the reservoir's cool, dark waters would have been preserved over the buried stone structures during the drought.

There have been profound advances in water efficiency, technologies, and conservation practices over the past 20 years that are capable of reducing many urban and agricultural water demands by at least one-third. Despite this, the potential for large-scale water-saving strategies to mitigate if not overcome the impacts of drought and long-term supply shortages has yet to be fully tapped by more than just a few water systems. Those demand-side water supply options are too often ignored—and at our peril. For how long can humanity afford to err in pursuing this most accessible, cost-effective, and environmentally friendly option to help meet our current water needs, let alone those to ensure a water secure future?

"If an alarm bell was needed to focus global attention on water security, it has rung," warns Sandra Postel, international water policy expert, author, and director of the Global Water Policy Project based in Los Lunas, New Mexico (Postel 2016). Exactly how the increasing water demands of the twenty-first century's growing population will be met amidst declining freshwater availability even during nondrought times is, indeed, a formidable challenge. The signs of water stress are daunting: half of the world's more than 7 billion people live in urban environments—and by 2050, global population is projected to grow 25 percent, reaching over 9 billion people (United Nations World Water Assessment Programme 2016, 3). Yet, nearly half of the world's largest cities and 71 percent of global irrigated area are already in regions that experience at least periodic water shortages (Brauman et al. 2016). In the United States, 80 percent of states report that by the early 2020s, they expect water shortages even under nondrought "average" conditions (US General Accountability Office 2014). Nature has granted us a fixed freshwater budget; either we live within its limits, or we pay a high price for its lesser alternatives.

While many now cast their eyes to the vast oceans and the allure of desalination to solve the world's water supply problems, the brakes often slam hard on that dream when tallying up its formidable costs. The process of rendering ocean and brackish water into drinkable and usable water can cost 10 to 20 times more than that for freshwater development. And the financial costs of desalination may dim in comparison to its environmental burdens. Despite its abundance, desalting ocean water is far from free: vacuuming seawater into desalination plants destroys marine biota, generates large quantities of membrane filter solid waste, demands copious amounts of chemicals that become hazardous waste, and requires more energy than conventional water treatment plants. The controversial Carlsbad desalination plant in San Diego, California, cost over $1 billion to build (higher than its original budget) and comes with an annual $50 million electricity bill just to run the plant—all at a cost that will more than double the water bills of San Diego residents and businesses compared to what their nearby Southern California neighbors pay. Within its first year of operation, the Carlsbad plant racked up a dozen environmental violations for "chronic toxicity" for a chemical waste it was piping into the ocean. And for all that cost, the Carlsbad plant will only meet about 10 percent of the San Diego area's water demands—hardly enough to satisfy the many irrigated green lawns planted in that semiarid region—in a service area that has yet to maximize its water savings potential from conservation (Gorn 2016).

After more than a century of water supply development and accompanying exploitation of the natural ecosystems, the goal of quenching humanity's thirst for more water seems as elusive as ever. The severity and cost of the world's droughts and chronic water supply problems continue to worsen, in tandem with declining groundwater and surface water supplies. Yet, on every continent and in nearly every water system facing drought or long-term water shortage, an obvious but chronically neglected antidote has existed for more than a century: the minimization of water waste.

> Year 1900: [I]t is evident that there must be a great amount of water wasted in many cities. Millions of dollars are being spent by many of our larger cities to so increase their supply that two-thirds of it may be wasted. This waste is either intentional, careless, or through ignorance. (Folwell 1900, 41)
> Today: Estimates indicate that about 30% of global water abstraction is lost through leakage. (United Nations World Water Assessment Programme 2016, 12)

"Fix leaks," the most basic and oft-repeated admonition by water utilities to the public, is not always advice that they follow themselves. A study conducted by the American Water Works Association's Water Loss Control Committee of the real (leakage and other physical) losses in 11 water utilities found that they averaged 83 gallons/connection/day in 2015—an increase from the average 70 gallons/connection/day reported for those same systems in 2011 (Sayers et al. 2016). Given that US residential use averages about 88 gallons per capita per day (gpcd) (Maupin et al. 2014), the water lost through leakage in those 11 systems is nearly equivalent to the water demands of an additional person at every connection in their service area.

Avoidable and costly water waste—from leaking, neglected underground pipes to green lawns in deserts, and archaic flooding and inefficient sprinkler methods to grow food crops—remains so prevalent that it is typically considered normal if not inevitable. But is that a mindset we can continue to afford to guide drought response and water management today? To be sure, all water systems will have some leaks, humans need water for its functional value as well as its aesthetic and inspirational qualities, and beneficial reuse is a component of some irrigation losses. But to what extent have we defined our *true water needs* in contrast to our *water wants, demands, and follies*?

The contrast between water-tight systems and leaky ones is glaring, particularly in the face of reservoir-draining droughts and other water supply constraints. Cities such as Singapore and Lisbon report water losses of less than 10 percent, yet recently London has reported losses of 25 percent and Norway 32 percent (United Nations World Water Assessment Programme 2016, 25). Ongoing maintenance and repair of aging and leaking distribution water pipes and mains, many of which have a useful life of about 100 years before they need total replacement, is often a major source of avoidable system losses for water supply utilities. But too often suppliers neglect basic maintenance of their water infrastructure, and sometimes to an extreme. For example, Suez Water (formerly known as United Water), a private water company that serves over 300,000 residents in Rockland County, New York, for many years has reported its infrastructure leakage and other losses to exceed 20 percent. In one recent year, it was revealed that the "snail's pace" of Suez's main replacement program put it on an astonishing 704-year schedule, a clue as to why that system's high water losses have made it a source of public ridicule for so long. Despite its failure to implement an aggressive water loss recovery program (in tandem with New York state regulators who for years have flouted enforcement of their own water loss standards) to increase available supplies from its existing sources, for several years Suez, a player in the global desalination industry, has claimed that it "needs" to build a costly desalination plant on the Hudson River. Much to the consternation of water ratepayers and local officials who have challenged Suez's proposal, as well as the company's poor efforts at promoting conservation, the water demands of Rockland residents and businesses have been largely flat and under that system's safe yield since the early 2000s—hardly conditions that justify incurring public debt for a costly new water supply (*Our Town News* 2015, 6).

Despite declining domestic per capita water use in the United States (averaging about 88 gpcd, due in large part to national water efficiency standards for plumbing fixtures and appliances established first by the US Energy Policy Act of 1992 and in recent years by updated standards developed by the US Environmental Protection Agency's WaterSense and Energy Star programs [Vickers and Bracciano 2014]), not all Americans are using the national average amount of water. Excessive outdoor water use for lawn irrigation, much of it inefficient and too often leading to hardscape runoff and turf diseases, remains a vexing problem in countless US communities that doubles, triples, and sometimes quadruples average indoor winter demands. Does the average resident in Scottsdale, Arizona, really *need* to use three times more water than someone in Santa Fe, New Mexico, both desert communities that receive less than 15 inches of average annual rainfall? And how can water-scarce western US cities such as Henderson (Nevada), Denver and Fort Collins (Colorado), and Santa Barbara (California) seriously complain about water shortages and consider raising public debt for desalination and wastewater reclamation facilities when more than 50 percent of their single family home water use is seasonal, much of it typically devoted to lawn irrigation (Figure 13.1)?

Many point to the west and southwest regions of the United States for examples of excessive water use, such as the large volumes of water (over 50 percent of summertime demand in many places) devoted to residential lawn watering in large swathes of California, Nevada, Arizona, Colorado, and Texas cities and suburbs (Figure 13.1). Unfortunately, such practices are becoming more prevalent, including in regions such as precipitation-rich New England, and they are taking a toll. Such demands can tax the ecological balance of reservoirs, rivers, and aquifers even during times of normal precipitation, but they incur even more severe impacts during drought.

Although it is argued that raising water rates and sending a strong pricing signal about the value of water will curb abusive water use, some people,

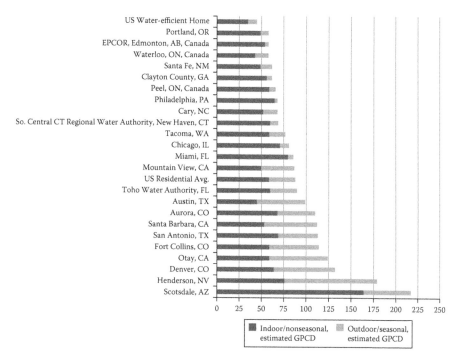

FIGURE 13.1

Indicators of single family residential water use in the United States and Canada, average gallons per capita per day (gpcd). (From Maupin, M.A., et al., Estimated use of water in the United States in 2010, *U.S. Geological Survey Circular* 1405, 2014; Water Research Foundation, *Residential end uses of water*, Version 2, Water Research Foundation, Denver, CO, 2016; Vickers, A., *Water efficiency and drought management practices: Comparison of national perspectives and local experiences*, Annual Water Conference on Agricultural & Urban Water Use: Drought Management Practices, Water Efficiency, and Energy: sponsored by Southern California Edison, Tulare, CA, 2014.)

particularly the affluent, are price insensitive when it comes to wanting a perfect-looking green lawn. As Postel and Richter (2003, 176) pointed out more than a decade ago in *Rivers for Life: Managing Water for People and Nature,* "hefty water bills may not be enough: outright bans on lawn watering when river flows drop below ecological thresholds may be necessary" to preserve healthy streamflows and fish stocks.

Public officials and water managers typically remain stubbornly resistant to calls for mandatory lawn watering bans during droughts, opting for less effective voluntary approaches that may only shift but not reduce demand, even during the most severe conditions when water supplies can fall perilously low. And they are just as loath to admit the reason why they are averse to watering restrictions: excessive lawn irrigation may drain the town reservoir, but it fills the town coffers with revenues, especially during hot and dry weather.

Failure to take appropriate drought response actions early, in particular implementing mandatory cutbacks and bans on lawn watering, too often causes needless wildlife suffering and death. The official and media narrative on the damage and deaths associated with drought commonly distorts such losses as being caused by the drought, overlooking how officials were too little or too late in imposing restrictions on nonessential water demands. For example, in the major drought of 2016 that hit New England and the Ipswich River region of eastern Massachusetts most severely, the river reached historic lows and stopped flowing that summer. In its last gasp before drying up completely, the river's meager flows reversed course toward the town wells of North Reading and Wilmington, swallowed up in large part to sprinkle the last drops of Ipswich River water on nearby suburban lawns. As thousands of fish began dying while the Ipswich collapsed, the local Ipswich River Watershed Association declared it a "river in crisis" and implored officials to take aggressive conservation actions to save the river's flows (Ipswich River Watershed Association 2016). Yet, most of the water managers in the towns that draw water from the Ipswich waited until the river was completely dried up and the damage done, in late summer and after fish, turtles, and other wildlife were dead, to start imposing restrictions on outdoor water use. One local media outlet headlined the story as "Drought killing Ipswich River wildlife, forcing water restrictions across region" (MacNeill 2016). Did drought kill the wildlife, or was it the water managers and public officials who failed to impose watering restrictions early to keep the river flowing so that the wildlife could survive?

On the spectrum of water use, how wide is the stretch of inefficiency and waste? When we compute the simple equation that subtracts our *true water needs* from our *total water demands*, the sum—water waste and inefficiency—reveals an expansive "new" source of freshwater capacity that can not only relieve the effects of drought but also help offset the adverse impacts of long-term shortages.

13.2 Water Conservation: The Great Untapped Water Supply

Water conservation is a powerful yet underutilized drought mitigation tool that can stave off the severe water shortages, financial losses, and public safety risks that historically have been assumed to be an inevitable consequence of drought. Hundreds of hardware technologies and behavior-driven measures are available to boost the efficiency of water use; when implemented and put into action, they can drive down short-term as well as long-term water demands (Vickers 2001, 2014).

For nearly every example of water waste and inefficiency that can be found in water systems, homes, landscapes, industries, businesses, and farms, there is a water conservation device, technology, or practice that will save water (Table 13.1) (AIQUEOUS and Amy Vickers & Associates, Inc. 2016;

TABLE 13.1

Overview of Water Conservation Incentives, Measures, and Potential Savings

End User Category	Examples of Conservation Incentives and Measures	Potential Water Savings Range (%)[a]
System (water utility)	System water losses/nonrevenue water are a maximum 10 percent of total production	Varies
	AWWA Water Audit (software) report of water utility system water losses, with independent audit data validation, prepared annually and reported to the public	
	Ongoing leak detection and repair, including emerging technologies such as continuous acoustic monitoring (CAM), satellite earth mapping using remote sensing technology for leak detection, and internal pipe repair technologies that avoid excavation and pipe replacement	
	Infrastructure rehabilitation and replacement (mains, services, valves, and hydrants)—ongoing and on schedule	
	Metering and meter maintenance (e.g., correct sizing, calibration, timely replacement)	
	District metered areas (DMAs)	
	Automatic metering infrastructure (AMI)	
	Pressure reduction and management	
Residential and domestic (indoor)	Financial incentives for conservation such as coupons/rebates, bill credits, and inclining tiered water rates to promote installation of high-efficiency plumbing fixtures and appliances	15–50
	Toilets (maximum 0.8 gpf to 1.28 gpf, composting, and retrofit devices for older high-volume fixtures)	
	Urinals (maximum 0.125 gpf to 0.5 gpf, nonwater, composting, and retrofit devices for older high-volume fixtures)	
	Showerheads (maximum 2.0 gpm)	
	Faucets (kitchen—maximum 2.0 gpm; bathroom sinks—maximum 0.5 gpm to 1.0 gpm; aerators for older high-volume faucets)	
	Clothes washers (maximum 10 gpl to 20 gpl, depending on washer capacity; wash full loads only)	
	Dishwashers (maximum 3 gpl to 5 gpl; wash full loads only)	
	Point-of-use hot water heaters	
	Leak repair and maintenance (e.g., leaking toilets and dripping faucets)	
Lawn and landscape irrigation	Nondrought: Permanent mandatory maximum once or twice weekly watering rule (temporary variance for new turf and plant establishment, e.g., 30 days), irrigation during evening and early morning hours only (minimize evaporative losses due to higher daytime temperatures and wind)	15–100

(Continued)

TABLE 13.1 *(Continued)*

Overview of Water Conservation Incentives, Measures, and Potential Savings

End User Category	Examples of Conservation Incentives and Measures	Potential Water Savings Range (%)[a]
	Drought: No or limited watering (e.g., once or twice weekly or once every 10 days)	
	Inclining tiered water rates and surcharges for excessive lawn and landscape watering and other nonessential outdoor uses, outdoor water use evaluations, "cash for grass" rebates and bill credits for turf removal, and irrigation system efficiency upgrades	
	Water-efficient landscape designs (e.g., functional turf areas only) that thrive on natural rainfall or very limited irrigation	
	Native and/or drought-tolerant adaptive turf and plants ("Right plant, right place")	
	Organic soil, mulches, turf, and plants, and natural pest control to boost plant vitality, retain soil and plant moisture, and avoid introducing harmful chemicals into local drinking water sources and the environment	
	Avoidance of synthetic fertilizers and harmful chemicals that require "watering in" and which can lead to excessive lawn growth (and increased mowing and irrigation demands) and pollution into waterways	
	Efficient irrigation systems and devices (e.g., limited irrigation areas, automatic rain shut-off, high-efficiency rotating spray nozzles, drip and micro-spray irrigation for plant beds and gardens)	
	Soil moisture sensors and "smart" irrigation controllers—only if they will reduce irrigation use (Note: "smart" controllers cannot reduce water waste that results from poorly maintained not-smart irrigation systems that are badly designed, leaking, and have broken sprinkler heads; controllers must also be properly installed and maintained or else they may *increase* water use)	
	Rainwater harvesting (e.g., essential uses and efficient irrigation only)	
	Leak repair and maintenance (e.g., broken sprinkler heads and hoses)	
Pools and other outdoor	Swimming pool covers at night and when not in use	
	Avoid installation of water decorations and faux water "features"—fake ponds, rivulets, and waterfalls—which waste water through evaporative losses and leaks, and create artificial environments that disrupt natural ecosystems and wildlife habitat	
Commercial, industrial, and institutional	Conservation-oriented inclining water rates; rebates and bill credits that promote installation of high-efficiency devices, equipment, and fixtures	15–50

(Continued)

TABLE 13.1 *(Continued)*

Overview of Water Conservation Incentives, Measures, and Potential Savings

End User Category	Examples of Conservation Incentives and Measures	Potential Water Savings Range (%)[a]
	Submetering of cooling towers, heating systems, process, and other large end uses to monitor efficiency and detect leaks	
	Efficient cooling and heating systems (e.g., recirculating, point-of-use, green roofs, and elimination of once-through cooling systems)	
	Process and wastewater reuse, improved flow controls	
	High-efficiency, composting, and water-free plumbing fixtures, appliances, and equipment	
	Point-of-use hot water heaters Leak repair and maintenance (e.g., hose repair, broom, air blower, and other dry cleaning methods)	
Agricultural	Metering of on-farm water uses (e.g., irrigation, livestock)	10–50
	Efficient irrigation systems and practices (e.g., surge valves, drip/trickle, low energy precision or spray application [LEPA or LESA], laser leveling, furrow diking, tailwater reuse, canal and conveyance system lining and management)	
	Drones for targeted monitoring of plant water needs—soil, nutrient, and pest conditions—and crop growth and yield evaluation	
	Advanced irrigation scheduling (e.g., customized to real-time weather, soil moisture and crop data, nighttime irrigation)	
	Native and drought-adaptive crop conversions (including younger, higher yielding plants)	
	Keyline design ("amplified contour ripping")	
	Dry farming, deficit irrigation	
	Land conservation methods (e.g., conservation tillage, organic farming, integrated pest management)	

Sources: AIQUEOUS and Amy Vickers & Associates, Inc., Energy efficiency potential for real water loss reduction in the Pacific Northwest, Prepared for the Northwest Energy Efficiency Alliance, Portland, Oregon, 2016; American Water Works Association, *M36 Water Audits and Loss Control Programs*, 4th ed., American Water Works Association, Denver, CO, 2016; U.S. Environmental Protection Agency, Best practices to consider when evaluating water conservation and efficiency as an alternative for water supply expansion, EPA-810-B-16-005, 2016; Vickers, A., *Handbook of Water Use and Conservation: Homes, Landscapes, Businesses, Industries, Farms*, WaterPlow Press, Amherst, MA, 2001; Vickers, A., NOFA's New 5th Edition Organic Land Care Standards: Water Perspectives, *Proceedings of the WaterSmart Innovations Conference*, Las Vegas, NV, 2011; Vickers, A., Water efficiency and drought management practices: Comparison of national perspectives and local experiences, *Annual Water Conference—Agricultural & Urban Water Use: Drought Management Practices, Water Efficiency, and Energy*; sponsored by Southern California Edison, Tulare, CA, 2014.

Note: gpf, gallons per flush; gpm, gallons per minute; gpl, gallons per load.

[a] Actual water savings by individual users will vary depending on existing efficiencies of use, number and type of measures implemented, and related factors.

American Water Works Association 2016; US Environmental Protection
Agency 2016; Vickers 2001, 2011, 2014). Hardware measures, such as leak
repairs, high-efficiency toilets and other fixtures, and more efficient cooling
and heating systems, will result in long-term demand reductions and typically require one action only (installation or repair) to realize ongoing water
savings. Behavior-oriented measures, such as turning off the faucet while
brushing teeth, and other actions involving human decision making typically
realize savings on a short-term basis but not over the long term. That is why
hardware, technology-based efficiency measures are favored by conservation
managers, whose goal is permanent, long-term water reductions (Vickers
2001). Examples of efficiency measures implemented by individual end users
among each major customer sector document not only water reductions but
also financial savings and other benefits (Table 13.2) (American Society of
Landscape Architects 2016; Donnelly 2015; Austin Water 2016;Dupré 2016;
Florendo and Wuelfing 2016; Postel 2014; Purington 2016; Southern Nevada
Water Authority 2016; *The Economist* 2016; Vickers 2016).

How much water can be saved by instituting restrictions during drought?
A lot, and probably much more than we know. The 35–50 percent water
demand reductions achieved in 2016 by several towns in Fairfield County,
Connecticut, during the severe drought exemplify how implementation of
mandatory watering restrictions can be both necessary and highly effective,
particularly in comparison to voluntary requests for conservation. While
many residents and businesses in several drought-afflicted Connecticut
towns cooperated with an initial request for voluntary watering restrictions
by local officials beginning in July of that year, at least one town, Greenwich
(an affluent community with high outdoor water demands), largely ignored
the appeal. As the drought persisted and reservoir levels plummeted, by
late summer an emergency declaration and ban on outdoor watering was
imposed (Borsuk and Oliveira 2016). The lawn watering ban quickly resulted
in demand reductions, with Greenwich's water use dropping over 50 percent, from about 18 mgd just before the ban to less than 8 mgd 1 month later
(Aquarion Water Company 2017). While lawn watering bans are not viewed
as a permanent conservation measure, their results shed light on at least
some of the potential savings that may be achieved by adopting permanent
mandatory restrictions (i.e., maximum twice- or once-weekly watering).

The implementation of large-scale water conservation programs in
response to drought and in particular long-term water shortages demonstrates the profound role that water-saving measures can play in abating
supply shortfalls. Beyond temporary drought responses, in some cases
the water demand reductions from multiyear conservation programs have
served to minimize or cancel major water and wastewater infrastructure
expansion plans and related long-term capital debt. For example, the 2016
annual average 209 million gallons a day (mgd) demands of the approximately 2.5 million people in the Boston metropolitan area served by the
Massachusetts Water Resources Authority (MWRA) was 37 percent lower

TABLE 13.2

Examples of Water Savings from Conservation

End User Category	Measures Implemented	Reported Savings
System (water utility)	Water loss and leak reduction were addressed by Portugal's largest water utility, Portuguesa das Aguas Livres (EPAL) which serves about 3 million people in 34 communities around Lisbon, Empresa. Reductions in nonrevenue water (NRW) were achieved through district metering areas (DMAs), continuous pressure and flow monitoring and management, advanced data analytics of night flows and pressure variations, and active "find and fix" leak detection using ground microphones and acoustic correlation technology with leak location registered on GIS. Cost savings from leak recovery in two DMAs paid for the entire DMA project.	System NRW was reduced from 24 percent in 2005 to 8 percent by 2014, from an average 154 gallons/connection/day down to 52 gallons/connection/day, with cumulative savings of 31 billion gallons system wide.
Residential (indoor and outdoor)	A representative sample of single family residential (SFR) customers in the 430,000 population service area of Solano County Water Agency in California was selected to participate in a pilot study of potential water savings from conservation measures: home water use efficiency surveys (audits) and rebate programs for high-efficiency (HE) toilets, clothes washers, and turf replacement programs, typically the highest home indoor and outdoor end uses of water.	Average annual savings by SFR customers who adopted conservation measures in many cases exceeded the EPA WaterSense program's minimum 20 percent savings goal: residential water use surveys (26,000 gallons), HE toilets (15,000 gallons), HE washers (8,600 gallons), and turf replacement (18,600 gallons).
Lawn and landscape irrigation	Two demonstration gardens (1,900 square feet each), a "native garden" using only native California plants and a "traditional garden" using lawn and exotic plant species, were established to compare their water demands, costs, and maintenance requirements (Santa Monica, California).	Native garden usages showed 77 percent less water, 68 percent less labor, and 66 percent less waste than the traditional garden. Installation costs were $16,700 for the native garden and $12,400 for the traditional garden—but the native garden saved $2,200 annually in reduced maintenance.

(Continued)

TABLE 13.2 (Continued)

Examples of Water Savings from Conservation

End User Category	Measures Implemented	Reported Savings
	Maximum twice/weekly lawn watering restriction adopted as a permanent rule by Dallas Water Utilities (DWU), a major supplier to over 2 million residents in North Texas, has achieved broad community and political support. Despite its often hot and drought-prone climate, Dallas's lawns and landscapes continue to thrive with limited irrigation. "If twice weekly works in North Texas, it can work almost anywhere," observed one Texas water manager.	Overall demand is 13 percent lower since rule adoption; 16 percent lower demands on nonwatering days; 10 percent lower demand on watering days (average).
	Replacing high-water-demand turf with native plant ground cover and other water-efficient landscaping at homes, businesses, and golf courses is a high priority for the Southern Nevada Water Authority (SNWA) and its popular water smart landscape rebate program. The water supplier pays $1 to $2 per square foot of grass removed, depending on the site's total landscaped area, and up to $300,000 to customers who complete the conversion process. SNWA's conservation programs save over one billion gallons of water annually, owing in large part to the more than 175 million square feet of water-thirsty turf removed by customers.	Large landscapes with an average 15,000-square-foot conversion to water-thrifty landscaping are saving thousands of gallons daily, as much as 825,000 gallons per year, on top of reduced water, mowing, and irrigation maintenance costs.
Commercial, industrial, and institutional	School (Deerfield Academy, Deerfield, Massachusetts): A private boarding school with 1,200 students, faculty, and staff on a 280-acre campus with 90-plus buildings adopted water-saving opportunities that quickly produced water, energy, and cost savings: Air-cooled ice machines replaced old water-cooled machines; high-efficiency pre-rinse spray valves in the kitchen, using only 0.6 gallons per minute (gpm) in the salad room and 1.1 gpm in the bakery; HE clothes washers in dorms (11–24 gallons per load); ultra high-efficiency 0.8 gallon per flush toilets in most new renovations (exceptions: long sanitary drain lines in isolated building areas and old cast iron sewer lines); pint-flush urinals; 2.0 gpm showerheads in dorms, locker rooms, and faculty housing; and faucet aerators in many dorm (1.5 gpm) and public (0.5 gpm) restrooms.	Total annual school demand was reduced by 2 million gallons (over 15 percent) following installation of "easy" recommendations from campus water audits; simple payback of less than 1 year for many measures.

(Continued)

TABLE 13.2 (Continued)

Examples of Water Savings from Conservation

End User Category	Measures Implemented	Reported Savings
	Cooling towers: Advanced chemical-free resin water electrodeionization (RW-EDI) technology developed by the US Department of Energy's Argonne National Laboratory uses electricity, ion exchange membranes, and resin for water deionization and impurities separation from water that causes scaling, fouling, and corrosion in cooling tower make-up water. The treatment system allows for use of lower-cost reclaimed water and other lower-quality water sources for tower make-up supply. The system's modular design allows for easy sizing and retrofitting of existing cooling towers. There is annual replacement of membrane packs instead of ongoing maintenance.	Water use was reduced by nearly 30 percent, with cycles of concentration increased from 3 to 8. Chemical use decreased by nearly 40 percent. Operation and maintenance costs reduced about 30 percent and wastewater discharges reduced by over 50 percent.
Agricultural	Almonds (Central Valley of California): Moisture sensors in nut groves monitor soil and other data to a cloud-based program that schedules "smart" irrigation runs to a grid of drip tapes that irrigate the almond trees. A calibrated pulse of water customized for each tree is delivered every half-hour, alternating from one side of the tree trunk to the other to optimize water uptake.	Irrigation requirements are reduced up to 30 percent.
	Grain (Tucson, Arizona): Upgrades to irrigation ditches and precision leveling of fields.	Improved precision leveling of field reduced irrigation demands 20–30 percent.

Sources: American Society of Landscape Architects, 2016, Designing our future: Sustainable landscapes: Garden/Garden—A comparison in Santa Monica, https://www.asla.org/sustainablelandscapes/gardengarden.html; Austin Water, Commercial Conservation Program (Austin Water, TX), Austin commercial water conservation news, *Argonne Labs Develops New Cooling Water Treatment*, 2016; Donnelly, A., Effective tools for reducing NRW within a major water utility: EPAL Portugal case study, *Proceedings of the North American Water Loss Conference*, Atlanta, Georgia, 2015; Drinkwine, M., et al., SNWA's Targeted Site Visit Research Study, *Proceedings of the WaterSmart Innovations Conference*, Las Vegas, NV, 2015; Dupré, Y., *Twice Weekly Watering: The Dallas Experience*, City of Dallas Water Utilities, Dallas, TX, 2016; Florendo, A., and K. Wuelfing, Yes, conservation programs do save water! Here's how much..., *Proceedings of the WaterSmart Innovations Conference*, Las Vegas, NV, 2016; Nagappan, P., California farmers innovate to fight drought, *Water Deeply*, 2016; Postel, S., Young farmers in the western U.S. adapt to a water-scarce future, *Water Currents* (blog), National Geographic Society, 2014; Purington, D., Project profile: Water conservation—A school looks again and the savings may surprise you, *Water Efficiency Magazine*, March–April, 2016; Southern Nevada Water Authority, *Water Smart Landscape Rebate for Business*, 2016, https://www.snwa.com/biz/rebates_wsl.html; *The Economist*, The future of agriculture: Factory Fresh, 2016, http://www.economist.com/technology-quarterly/2016-06-09/factory-fresh; Vickers, A., Water audits and savings at Deerfield Academy, *American Water Works Association Sustainable Water Management Conference*, Providence, RI, 2016.

than that system's high of 334 mgd in 1987, when it served only 2.1 million (Massachusetts Water Resources Authority 2017). The MWRA's decreasing water demands in spite of its 400,000 service area population increase during that time is largely the result of a comprehensive, multiyear conservation program that implemented permanent "hardware" water-saving measures. Instrumental to this achievement were aggressive leak repairs (the city of Boston could not account for as much as 50 percent of its water during the 1980s), innovations in industrial water use efficiency, and the installation of water-saving toilets and plumbing fixture retrofit devices. Significant water savings for the MWRA and other water systems in Massachusetts have also been realized by that state as a result of it being the first in the nation to require low-volume, maximum 1.6 gallon per flush (gpf) toilets, the most water-efficient standard in the United States at that time (Vickers 1989). The MWRA's conservation savings not only transformed that system's supply status from shortfall to abundance but also averted construction of a controversial dam project on the Connecticut River that was projected to incur a debt of more than $500 million (1987 dollars) (Amy Vickers & Associates, Inc. 1996). Should the MWRA need to reduce demands even further (i.e., respond to a drought, supply new users, or meet emergency water demands), a plethora of additional water efficiency measures can be implemented to increase water savings beyond the 37 percent already achieved.

13.3 Conclusions

Reducing our water demands and waste should always be an obvious response to drought: using less during times of shortfall, enjoying more in periods of natural abundance. Water conservation should not be just an emergency response to drought, but a long-term approach to managing and alleviating stresses on the world's finite water supplies so that water systems are more resilient in the face of droughts when they do occur.

Water conservation is a powerfully effective short-term drought mitigation tool that is also a proven approach to better managing long-term water demands. Conservation-minded water systems have demonstrated that the efficient management of public, industrial, and agricultural water use during drought is critical to controlling and minimizing the adverse effects of reduced precipitation on water supplies. If we understand where and how much water is used and apply appropriate efficiency practices and measures to reduce water waste, we can more easily endure—economically, environmentally, and politically—drought and projected water shortages. The lessons of effective drought management strategies (i.e., early implementation of comprehensive conservation measures, often requiring use restrictions) show that conservation can also be tapped to help overcome current and projected supply shortfalls that occur during nondrought times as well. The implementation of water waste reduction and efficiency measures can lessen the adverse impacts of excessive water demands on the natural water systems (rivers, aquifers, and lakes) and the ecological resources on which they depend. The notable demand reductions achieved by water-efficiency-minded cities and water systems prove the significant role conservation can play in not only coping with drought but also overcoming supply limitations and bolstering drought resistance through the preservation of water supply capacity. Like any savvy investor, efficiency-minded public officials and water managers who minimize their system water losses and invest in customer water conservation programs will yield a treasure trove of "new" water supplies in reservoirs and aquifers that protect them from future shortages and blunt the effects of drought. Human activities play a key role in our experience of drought. A water-rich or water-poor future will be determined largely by our water conservation actions today.

References

AIQUEOUS and Amy Vickers & Associates, Inc. 2016. *Energy efficiency potential for real water loss reduction in the Pacific Northwest.* Prepared for the Northwest Energy Efficiency Alliance, Portland, Oregon.

American Society of Landscape Architects. 2016. Designing our future: Sustainable landscapes: Garden/Garden—A comparison in Santa Monica. https://www.asla.org/sustainablelandscapes/gardengarden.html. Accessed January 17, 2017.

American Water Works Association. 2016. *M36 Water Audits and Loss Control Programs*, 4th ed. Denver, CO: American Water Works Association.

Amy Vickers & Associates, Inc. 1996. *Final report: Water Conservation Planning USA Case Studies Project.* Prepared for the United Kingdom Environment Agency, Demand Management Centre. U.S. Environmental Agency, Amherst, MA.

Aquarion Water Company (Bridgeport, Connecticut). Figure 1 Daily Water Demands-Regional Water Demands, 8/29/2016 to 1/16/2017, January 20, 2017. http://www.aquarion.com/CT/emergency-update

Austin Water. 2016. Commercial Conservation Program (Austin Water, TX). Austin commercial water conservation news, *Argonne Labs Develops New Cooling Water Treatment*, Fall 2016, p.3. http://us3.campaign-archive2.com/?u=f218bd2d60b9 22e71eb4b372&id=6a88f38048. Accessed January 27, 2017.

Borsuk, K., and N. Oliveira. 2016. Stamford, Greenwich officials questions timing of water emergency. *Stamford Advocate*, October 12. http://www.stamfordadvocate.com/local/article/Stamford-Greenwich-officials-question-timing-of-9965248.php. Accessed January 27, 2017.

Brauman, K. A., B. D. Richter, S. Postel, M. Malsy, and M. Flörke. 2016. Water depletion: An improved metric for incorporating seasonal and dry-year water scarcity into water risk assessments. *Elementa: Science of the Anthropocene* 4: 83.

Donnelly, A. 2015. Effective tools for reducing NRW within a major water utility: EPAL Portugal case study. *Proceedings of the North American Water Loss Conference*, Atlanta, Georgia, December 8–9.

Drinkwine, M., K. Sovocool, and M. Morgan. 2015. SNWA's Targeted Site Visit Research Study. *Proceedings of the WaterSmart Innovations Conference*, Las Vegas, Nevada, October 8.

Dupré, Y. 2016. *Twice weekly watering: The Dallas experience*. Dallas, TX: City of Dallas Water Utilities.

Florendo, A., and K. Wuelfing. 2016. Yes, conservation programs do save water! Here's how much... *Proceedings of the WaterSmart Innovations Conference*, Las Vegas, Nevada, October 5.

Folwell, A. P. 1900. *Water-supply engineering*, 1st ed. New York: Wiley.

Gorn, D. 2016. Desalination's future in California is clouded by cost and controversy. *KQED Science*, October 31. https://ww2.kqed.org/science/2016/10/31/desalination-why-tapping-sea-water-has-slowed-to-a-trickle-in-california/. Accessed January 27, 2017.

Hlavaty, C. 2016. Five years after the height of Texas drought, Lake Buchanan ghost town back underwater. *Houston Chronicle*, July 14. http://www.chron.com/news/houston-texas/texas/article/Five-years-after-height-of-Texas-drought-8377911.php. Accessed January 27, 2017.

Ipswich River Watershed Association. 2016. *River in crisis*. http://www.ipswichriver.org/featured/river-in-crisis/. Accessed August 10, 2016.

MacNeill, A. 2016. *Drought killing Ipswich River wildlife, forcing water restrictions across region*. *Salem News* (Salem, MA), August 9. http://www.salemnews.com/news/local_news/drought-killing-ipswich-river-wildlife-forcing-water-restrictions-across-region/article_78f99e19-7840-52bf-a7ec-421eeb91e413.html. Accessed January 27, 2017.

Massachusetts Water Resources Authority. 2017. Archive - Water Use and System Demand. http://www.mwra.state.ma.us/monthly/wsupdat/archivedemand.htm (accessed January 24, 2017).

Maupin, M. A., J. F. Kenny, S. S. Hutson, J. K. Lovelace, N. L. Barber, and K. S. Linsey. 2014. Estimated use of water in the United States in 2010. *U.S. Geological Survey Circular* 1405: 21. doi: 10.3133/cir1405.

Nagappan, P. 2016. California farmers innovate to fight drought. *Water Deeply*, May 13. https://www.newsdeeply.com/water/articles/2016/05/13/california-farmers-innovate-to-fight-drought. Accessed January 27, 2017.

Our Town News. 2015. Leaks are us: United Water's fuzzy data. *Our Town News* (Pearl River, NY), July 1, p. 6.

Postel, S. 2014. Young farmers in the western U.S. adapt to a water-scarce future. *Water Currents (blog)*, National Geographic Society, August 21. http://newswatch.nationalgeographic.com/2014/08/21/young-farmers-in-the-western-u-s-adapt-to-a-water-scarce-future/. Accessed January 27, 2017.

Postel, S. 2016. Water risks are growing; here's a tool to help us prepare. *Water Currents (blog)*, National Geographic Society, January 28. http://voices.nationalgeographic.com/2016/01/28/water-risks-are-growing-heres-a-tool-to-help-us-prepare/.

Postel, S., and B. Richter. 2003. *Rivers for life: Managing water for people and nature*. Washington, DC: Island Press.

Purington, D. 2016. Project profile: Water conservation—A school looks again and the savings may surprise you. *Water Efficiency Magazine,* March–April. http://foresternetwork.com/water-efficiency-magazine/project-profile-water-efficiency-magazine/water-conservation-a-school-looks-again-and-the-savings-may-surprise-you/. Accessed January 27, 2017.

Sayers, D., W. Jernigan, G. Kunkel, and A. Chastain-Howley. 2016. The water audit data initiative: Five years and counting. *Journal of the American Water Works Association* 108(11): E598–E605.

Southern Nevada Water Authority. 2016. *Water smart landscape rebate for business.* https://www.snwa.com/biz/rebates_wsl.html. Accessed September 2016.

The Economist. 2016. The future of agriculture: Factory Fresh. http://www.economist.com/technology-quarterly/2016-06-09/factory-fresh. Accessed January 27, 2017.

United Nations World Water Assessment Programme. 2016. *The United Nations World Water Development Report 2016: Water and jobs.* Paris: UNESCO.

U.S. Environmental Protection Agency. 2016. *Best practices to consider when evaluating water conservation and efficiency as an alternative for water supply expansion.* EPA-810-B-16-005. Washington, DC: US Environmental Protection Agency. Office of Water.

U.S. Government Accountability Office. 2014. *FRESHWATER: Supply concerns continue, and uncertainties complicate planning. Report to Congressional Requesters, GAO-14-430.* Washington, DC: US Government Accountability Office.

Vickers, A. 1989. New Massachusetts toilet standard sets water conservation precedent. *Journal of the American Water Works Association* 81(3): 48–51.

Vickers, A. 2001. *Handbook of water use and conservation: Homes, landscapes, businesses, industries, farms.* Amherst, MA: WaterPlow Press.

Vickers, A. 2011. NOFA's New 5th Edition Organic Land Care Standards: Water Perspectives. *Proceedings of the WaterSmart Innovations Conference,* Las Vegas, NV, October.

Vickers, A. 2014. Water efficiency and drought management practices: Comparison of national perspectives and local experiences. *Annual Water Conference—Agricultural & Urban Water Use: Drought Management Practices, Water Efficiency, and Energy; sponsored by Southern California Edison,* Tulare, CA, November 5.

Vickers, A. 2016. Water audits and savings at Deerfield Academy. *American Water Works Association Sustainable Water Management Conference,* Providence, RI, March 9.

Vickers, A., and D. Bracciano. 2014. Low-volume plumbing fixtures achieve water savings. *Opflow* 40(7): 8–9.

Water Research Foundation. 2016. *Residential end uses of water,* Version 2. PDF Report #43096 b. Denver, CO: Water Research Foundation.

14

The Role of Water Harvesting and Supplemental Irrigation in Coping with Water Scarcity and Drought in the Dry Areas

Theib Y. Oweis

CONTENTS

14.1 Introduction

Water scarcity and drought are among the most serious obstacles to agricultural development and a major threat to the environment in the dry areas. Agriculture in the dry areas accounts for more than 75 percent of the total consumption of water. With rapid increases in demand, water will be increasingly reallocated away from agriculture and the environment.

Despite scarcity, water continues to be misused. Mining groundwater is now a common practice, risking both water reserves and quality.

Land degradation is another challenge in the dry areas, closely associated with drought-related water shortage. Climatic variation and change, mainly as a result of human activities, are leading to depletion of the vegetation cover and loss of biophysical and economic productivity. This happens through exposure of the soil surface to wind and water erosion, and shifting sands, salinization of land, and water logging. Although these are global problems, they are especially severe in the dry areas.

Two major environments occupy the dry areas. The first is the wetter rainfed areas, where rainfall is sufficient to support economical dry farming. However, because rainfall amounts and distribution are suboptimal, drought periods often occur during one or more stages of crop growth, causing very low crop yields. Variation in rainfall amounts and distribution from 1 year to the next causes substantial fluctuations in production. This situation creates instability and negative socioeconomic impacts. The second environment is the drier environment (steppe or *badia*), characterized by an annual rainfall too low to support economical dry farming. Most of the dry areas lie in this zone. Small and scattered rainstorms in these regions fall on lands that are generally degraded with poor vegetative cover. Rainfall, although low, may accumulate through runoff from vast areas in a large volume of ephemeral water and largely be lost through direct evaporation or in salt sinks.

With scarcity, it is essential that available water be used at highest efficiency. Many technologies are available to improve water productivity and management of scarce water resources. Among the most promising technologies are (1) supplemental irrigation (SI) for rainfed areas and (2) rainwater harvesting for the drier environments (Oweis and Hachum 2003). Improving scarce water productivity, however, requires exploiting not only water management but also other inputs and cultural practices. This chapter addresses the concepts and potential roles of supplemental irrigation and water harvesting in improving water productivity and coping with increased scarcity, drought, and climate change in the dry areas.

14.2 Supplemental Irrigation

Precipitation in the rainfed areas is low in amount and suboptimal in distribution, with great year-to-year fluctuation. In a Mediterranean climate, rainfall occurs mainly during the winter months. Crops must rely on stored soil moisture when they grow rapidly in the spring. In the wet months, stored water is ample, plants sown at the beginning of the season are in early growth stages, and the water extraction rate from the root zone is limited. Usually little or no moisture stress occurs during this period (Figure 14.1). However, during spring, plants grow faster, with a high evapotranspiration

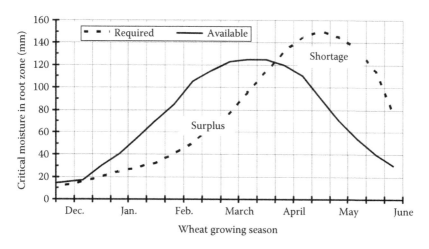

FIGURE 14.1
Typical soil moisture pattern over the growing season of a Mediterranean-type wheat. (From Oweis, T. and Hachum, A. 2012. *Supplemental irrigation, a highly efficient water-use practice.* 2nd edition. ICARDA, Aleppo, Syria. iv + 28 pp. With permission.)

rate and rapid soil moisture depletion due to higher evaporative demand. Thus, a stage of increasing moisture stress starts in the spring and continues until the end of the season. As a result, rainfed crop growth is poor and yield is low. The mean grain yield of rainfed wheat in the dry areas is about 1 t/ha, far below the yield potential of wheat (more than 5–6 t/ha).

Supplemental irrigation aims to overcome the effects of drought periods as soil moisture drops and halts crop growth and development. Limited amounts of water, if applied during critical times, can result in substantial increases in yield and water productivity.

Research results from the International Center of Agricultural Research in the Dry Areas (ICARDA) and other organizations, as well as harvests from farmers' fields, have demonstrated substantial increases in crop yield in response to the application of relatively small amounts of irrigation water. Table 14.1 shows increases in wheat grain yields under low, average, and high rainfall in northern Syria, with application of limited amounts of SI. By definition, rainfall is the major source of water for crop growth and production; thus, the amount of water added by SI cannot by itself support economical crop production. In addition to yield increases, SI also stabilized wheat production over years (i.e., reduced the interannual variability of yields).

The impact of SI goes beyond yield increase to substantially improving water productivity. The productivity of irrigation water and rainwater is improved when they are used conjunctively. Average rainwater productivity of wheat ranges from 0.35 to 1.0 kg/m^3. It was found that 1 m^3 of water applied as SI at the proper time could produce more than 2.0 kg of wheat (Oweis et al. 1998, 2000).

TABLE 14.1

Yield and Water Productivity (WP) for Wheat under Rainfed and Supplemental Irrigation (SI) in Dry, Average, and Wet Seasons in Tel Hadya, North Syria

Season/Annual Rainfall (mm)	Rainfed Yield (t/ha)	Rainfall WP (kg/m³)	Irrigation Amount (mm)	Total Yield (t/ha)	Yield Increase Due to SI (t/ha)	Irrigation WP (kg/m³)
Dry (234 mm)	0.74	0.32	212	3.38	3.10	1.46
Average (316 mm)	2.30	0.73	150	5.60	3.30	2.20
Wet (504 mm)	5.00	0.99	75	6.44	1.44	1.92

*Source:*Adapted from Oweis, T. and Hachum, A. 2012. *Supplemental irrigation, a highly efficient water-use practice.* 2nd. edition. ICARDA, Aleppo, Syria. iv + 28 pp.

Using irrigation water conjunctively with rain was found to produce more wheat per unit of water than if used alone in fully irrigated areas where rainfall is negligible. In fully irrigated areas, water productivity for wheat ranges from 0.5 to about 0.75 kg/m³, one-third of that achieved with SI. This difference suggests that allocation of limited water resources should be shifted to more efficient practices (Oweis and Hachum 2012.). Food legumes, which are important for providing low-cost protein for people of low income and for improving soil fertility, have shown similar responses to SI in terms of yield and water productivity.

In the highlands of the temperate dry areas in the northern hemisphere, frost occurs between December and March. Field crops go into dormancy during this period. In most years, the first rainfall sufficient to germinate seeds comes late, resulting in a poor crop stand when the crop goes into dormancy. Rainfed yields can be significantly increased if the crop achieves good early growth before dormancy. This can be achieved by early sowing with the application of a small amount of SI. A 4-year trial, conducted at the central Anatolia plateau of Turkey, showed that applying 50 mm of SI to wheat sown early increased grain yield by more than 60 percent, adding more than 2 t/ha to the average rainfed yield of 3.2 t/ha (Ilbeyi et al. 2006). Water productivity reached 5.25 kg grain/m³ of consumed water, with an average of 4.4 kg/m³. These are extraordinary values for water productivity with regard to the irrigation of wheat.

14.2.1 Optimization of Supplemental Irrigation

Optimal SI in rainfed areas is based on the following three criteria: (1) water is applied to a rainfed crop that would normally produce some yield without irrigation; (2) because rainfall is the principal source of water for rainfed crops, SI is applied only when rainfall fails to provide essential moisture for improved and stable production; and (3) the amount and timing of SI are scheduled not to provide moisture stress-free conditions throughout the growing season, but to ensure a minimum amount of water available during the critical stages of crop growth that would permit optimal instead of maximum yield (Oweis and Hachum 2012).

14.2.1.1 Deficit Supplemental Irrigation

Deficit irrigation is a strategy for optimizing production. Crops are deliberately allowed to sustain some degree of water deficit and yield reduction (English and Raja 1996). The adoption of deficit irrigation implies appropriate knowledge of crop water use and responses to water deficits, including the identification of critical crop growth periods, and of the economic impacts of yield reduction strategies. In a Mediterranean climate, rainwater productivity increased from 0.84 to 1.53 kg grain/m³ of irrigation water when only one-third of the full crop water requirement was applied (Figure 14.2). It further increased to 2.14 kg/m³ when two-thirds of the requirement was applied, compared to 1.06 kg/m³ at full irrigation. The results show greater water productivity at deficit than at full irrigation. Water productivity is a suitable indicator of the performance of irrigation management under deficit irrigation of cereals (Zhang and Oweis 1999), in analyzing the water saving in irrigation systems and management practices, and in comparing different irrigation systems.

There are several ways to manage deficit irrigation. The irrigator can reduce the irrigation depth, refilling only part of the root zone soil water capacity, or reduce the irrigation frequency by increasing the interval between successive irrigations. In surface irrigation, wetting furrows alternately or placing them farther apart is one way to implement deficit irrigation. However, not all crops

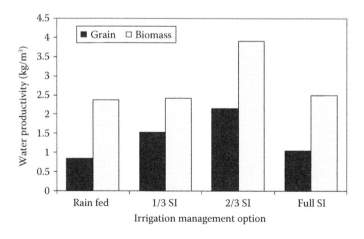

FIGURE 14.2
Water productivity of wheat under rainfed, deficit, and full SI conditions. (Adapted from Oweis, T. and Hachum, A. 2012. *Supplemental irrigation, a highly efficient water-use practice.* 2nd edition. ICARDA, Aleppo, Syria. iv + 28 pp.)

respond positively to deficit irrigation. This should be examined for local conditions and under different levels of water application and quality.

14.2.1.2 Maximizing Net Profits

An increase in crop production per unit of land or per unit of water does not necessarily increase farm profit because of the nonlinearity of crop yield with production inputs. Determining rainfed and SI production functions is the basis for optimal economic analysis. SI production functions for wheat (Figure 14.3) may be developed for each rainfall zone by subtracting rainwater production function from total water production function. Because the rainfall amount cannot be controlled, the objective is to determine the optimal amount of SI that results in maximum net benefit to the farmers. Knowing the cost of irrigation water and the expected price per unit of the product, we can see that maximum profit occurs when the marginal product for water equals the price ratio of the water to the product. Figure 14.4 shows the amount of SI to be applied under different rainfall zones and various price ratios to maximize net profit of wheat production under SI in a Mediterranean climate.

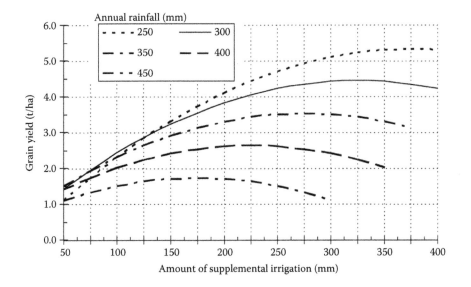

FIGURE 14.3
SI production functions for wheat in different rainfall zones in Syria. (Adapted from Oweis, T. and Hachum, A. 2012. Supplemental irrigation, a highly efficient water-use practice. 2nd. edition. ICARDA, Aleppo, Syria. iv + 28 pp.)

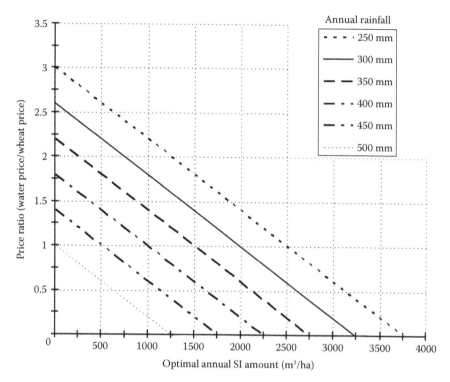

FIGURE 14.4
Optimal economical annual SI amount (m³/ha) in different rainfall zones in Syria. (Adapted from Oweis, T. and Hachum, A. 2012. *Supplemental irrigation, a highly efficient water-use practice.* 2nd. edition. ICARDA, Aleppo, Syria. iv + 28 pp.)

14.2.1.3 Cropping Patterns and Cultural Practices

Among the management factors for more productive farming systems are the use of suitable crop varieties, improved crop rotation, sowing dates, crop density, soil fertility management, weed control, pest and disease control, and water conservation measures. SI requires crop varieties adapted to or suitable for varying amounts of water application. An appropriate variety manifests a strong response to limited water application and maintains some degree of drought tolerance. In addition, the varieties should respond to higher fertilization rates than are generally required under SI.

Given the inherent low fertility of many dry-area soils, judicious use of fertilizer is particularly important. In northern Syria, 50 kg N per hectare is sufficient under rainfed conditions. However, with water applied by SI, the crop responds to nitrogen up to 100 kg/ha, after which no further benefit is obtained. This rate of nitrogen uptake greatly improves water productivity. There must also be adequate available phosphorus in the soil so that response to nitrogen and applied irrigation is not constrained.

To obtain the optimum output of crop production per unit input of water, the monocrop water productivity should be extended to a multicrop water productivity. Water productivity of a multicrop system is usually expressed in economic terms such as farm profit or revenue per unit of water used. Although economic considerations are important, they are not adequate as indicators of sustainability, environmental degradation, and natural resource conservation.

14.2.2 Water versus Land Productivity

Land productivity (yield) and water productivity are indicators for assessing the performance of supplemental irrigation. Higher water productivity is linked with higher yields. This parallel increase in yields and water productivity, however, does not continue linearly. At some high level of yield, greater amounts of irrigation water are required to achieve additional incremental yield increase. Water productivity of wheat (Figure 14.5) starts to decline as yield per unit of land increases above certain levels.

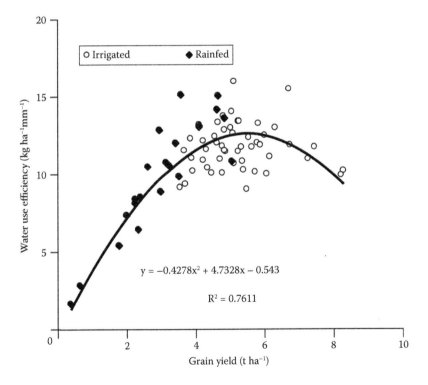

FIGURE 14.5
Relationship between crop water productivity and crop grain yield for durum wheat under SI in Syria. (Adapted from Zhang, H., and T. Oweis, *Agric. Water Manage.* 38, 195, 1999.)

It is clear that the amount of water required to achieve yield increases above 5 t/ha is much higher than that needed at lower yield levels. It would be more efficient to produce only 5 t/ha with lower water application than to achieve maximum yield with application of excessive amounts of water. The saved water would be used more efficiently if applied to new lands. This, of course, applies only when water, not land, is the limiting resource and is insufficient to fully irrigate all available land.

The association of high water productivity values with high yields has important implications for crop management in achieving efficient use of water resources in water-scarce areas (Oweis et al. 1998). Attaining higher yields with increased water productivity is economical only when the increased gains in crop yield are not offset by increased costs of other inputs. The curvilinear water productivity–yield relationship reflects the importance of attaining relatively high yields for efficient use of water. Policies for maximizing yield should be considered carefully before they are applied under water-scarce conditions. Guidelines for recommending irrigation schedules under normal water availability may need to be revised when applied in water-scarce areas.

14.3 Water Harvesting

14.3.1 The Concept and Components of the System

The drier environments, "the steppe" (or, as they are called in the Arab world, *Al Badia*), occupy the vast majority of the dry areas. The disadvantaged people, who depend mainly on livestock grazing, generally live there. The natural resources of these areas are fragile and subject to degradation. Because of harsh natural conditions and the occurrence of drought, people increasingly migrate from these areas to the urban areas, with the associated high social and environmental costs.

Precipitation in the drier environments is generally low relative to crop requirements. It is unfavorably distributed over the crop-growing season and often comes with high intensity. It usually falls in sporadic, unpredictable storms and is mostly lost to evaporation and runoff, leaving frequent dry periods. Part of the rain returns to the atmosphere directly from the soil surface by evaporation after it falls, and part flows as surface runoff, usually joining streams and flowing to "salt sinks," where it loses quality and evaporates. A small portion of the rain joins groundwater. The overall result is that most of the rainwater in the drier environments is lost, with no benefits or productivity. As a result, rainfall in this environment cannot support economical dry farming like that in rainfed areas (Oweis et al. 2001).

Water harvesting can improve the situation and substantially increase the portion of beneficial rainfall. In agriculture, water harvesting is based on depriving part of the land of its share of rainwater to add to the share of another part. This brings the amount of water available to the target area closer to the crop water requirements so that economical agricultural production can be achieved. Water harvesting may be defined as the process of concentrating precipitation through runoff and storing it for beneficial use.

Water harvesting is an ancient practice supported by a wealth of indigenous knowledge. Indigenous systems such as *jessour* and *meskat* in Tunisia; *tabia* in Libya; *cisterns* in north Egypt; *hafaer* in Jordan, Syria, and Sudan; and many other techniques are still in use (Oweis et al. 2004). Water harvesting may be developed to provide water for human and animal consumption, domestic and environmental purposes, and plant production. Water harvesting systems have three components:

1. *The catchment area* is the part of the land that contributes some or all of its share of rainwater to another area outside its boundaries. The catchment area can be as small as a few square meters or as large as several square kilometers. It can be agricultural, rocky, or marginal land, or even a rooftop or a paved road.

2. *The storage facility* is a place where runoff water is held from the time it is collected until it is used. Storage can be in surface reservoirs, in subsurface reservoirs such as cisterns, in the soil profile as soil moisture, or in groundwater aquifers.

3. *The target area* is where the harvested water is used. In agricultural production, the target is the plant or animal, whereas in domestic use, it is the human being or the enterprise and its needs.

14.3.2 Water Harvesting Techniques

Water harvesting techniques may be classified into two major types, based on the size of the catchment (Figure 14.6): microcatchment systems and macrocatchment systems (Oweis et al. 2001).

14.3.2.1 Microcatchment Systems

Surface runoff in microcatchment systems is collected from small catchments (usually less than 1000 m^2) and applied to an adjacent agricultural area, where it is stored in the root zone and used directly by plants. The target area may be planted with trees, bushes, or annual crops. The farmer has control, within the farm, over both the catchments and the target areas. All the components of the system are constructed inside the farm boundaries, which provides a maintenance and management advantage. But because of the loss of productive land, it is practiced only in the drier environments,

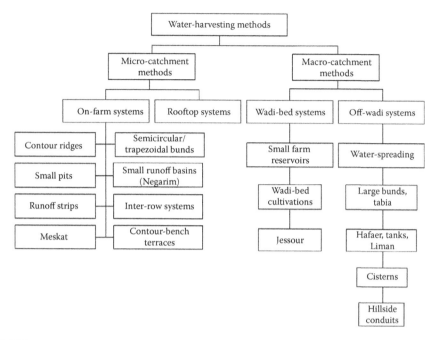

FIGURE 14.6
Classification of major rainwater harvesting systems in the dry areas. (From Oweis, T., et al., *Water Harvesting: Indigenous Knowledge for the Future of the Drier Environments*, ICARDA, Aleppo, Syria, 2001. With permission.)

where cropping is so risky that farmers are willing to allocate part of their farm to be used as a catchment. They are simple in design and may be constructed at low cost. Therefore, they are easy to replicate and adapt. They have higher runoff efficiency than the macrocatchment systems and usually do not need a water conveyance system. Soil erosion may be controlled and sediment directed to settle in the cultivated area. These systems generally require continuous maintenance, with relatively high labor input. The most important microcatchment water harvesting systems in the dry areas are described below.

14.3.2.1.1 Contour Ridges

Contour ridges consist of bunds, or ridges, constructed along the contour line at an interval of, usually, between 5 and 20 m. A 1- to 2-m strip upstream of the ridge is for cultivation, and the rest constitutes the catchment. The height of the ridges varies according to the slope and the expected depth of the runoff water retained behind it. The bunds may be reinforced by stones when necessary. This is a simple technique, which can be implemented by the farmers themselves. Bunds can be formed manually, with animal-driven equipment, or by tractors fitted with suitable implements. Ridges may be constructed on a wide range of slopes, from 1 to 50 percent.

Contour ridges are important for supporting the regeneration and new plantations of forage, grasses, and hardy trees on mild to steep slopes in the steppe (*badia*). In the semiarid tropics, they are used for the arable cropping of sorghum, millet, cowpeas, and beans. This system is sometimes combined with other techniques (such as the *zay* system) or with *in situ* water conservation techniques (such as the tied-ridge system) in the semiarid tropics.

14.3.2.1.2 Semicircular and Trapezoidal Bunds

Semicircular and trapezoidal bunds are earthen bunds created with spacing sufficient to provide the required runoff water for the plants. Usually, they are built in staggered rows. The technique can be used not only on an even, flat slope, but also on slopes up to 15 percent. The technique is used mainly for rangeland rehabilitation or fodder production, but can also be used for growing trees, shrubs, and, in some cases, field crops and vegetables.

14.3.2.1.3 Small Pits

The most famous pitting system is the *zay* system used in Burkina Faso. This form of pitting consists of digging holes 5–15 cm deep. Manure and grasses are mixed with some of the soil and put into the *zay*. The rest of the soil is used to form a small dike, downslope of the pit. Pits are used in combination with bunds to conserve runoff, which is slowed by the bunds. Pits are excellent for rehabilitating degraded agricultural lands. However, labor requirements for digging the *zay* are high and may constitute a large financial investment, year after year. This is because the pits have to be restored after each tillage operation. A special disk plow may be adjusted to create small pits for range rehabilitation.

14.3.2.1.4 Small Runoff Basins

Sometimes called *negarim*, these runoff basins are small and of a rectangular or elongated diamond shape; they are surrounded by low earth bunds. *Negarim* work best on smooth ground, and their optimal dimensions are 5–10 m wide by 10–25 m long. They can be constructed on almost any slope, including very gentle ones (1–2 percent slopes), but on slopes above 5 percent, soil erosion may occur, and the bund height should be increased. They are most suitable for growing tree crops like pistachios, apricots, olives, almonds, and pomegranates, but they may be used for other crops. When used to grow trees, the soil should be deep enough to hold sufficient water for the entire dry season.

14.3.2.1.5 Runoff Strips

This technique is applied on gentle slopes and is used to support field crops in drier environments (such as barley in the *badia*), where production is usually risky or has a low yield. In this technique, the farm is divided into strips following contour lines. One strip is used as a catchment and the strip downstream is cropped. The cropped strip should not be too wide (1–3 m), and the catchment width should be determined with a view to providing the required runoff water to the cropped area. The same cropped strips are cultivated every year. Clearing and compaction may be implemented to improve runoff.

14.3.2.1.6 Contour Bench Terraces

Contour bench terraces are constructed on very steep sloping lands and combine soil and water conservation with water harvesting techniques. Cropping terraces are usually built to be level. Supported by stone walls, they slow water and control erosion. Steeper, noncropped areas between the terraces supply additional runoff water. The terraces contain drains to safely release excess water. They are used to grow trees and bushes but are rarely used for field crops. Some examples of this technique can be seen in the historic bench terraces in Yemen. Because they are constructed in steep mountain areas, most of the work is done by hand.

14.3.2.1.7 Rooftop Systems

Rooftop and courtyard systems collect and store rainwater from the surfaces of houses, large buildings, greenhouses, courtyards, and similar impermeable surfaces. Farmers usually avoid storing the runoff provided by the first rains to ensure cleaner water for drinking. If water is collected from soil surfaces, the runoff has to pass through a settling basin before it is stored.

The water collected is used mainly for drinking and other domestic purposes, especially in rural areas where there is no tap water. Extra water may be used to support domestic gardens. It provides a low-cost water supply for humans and animals in remote areas.

14.3.2.2 Macrocatchment Systems

Macrocatchment systems collect runoff water from relatively large catchments, such as natural rangeland or a mountainous area, mostly outside farm boundaries, where individual farmers have little or no control. Water flows in temporary (ephemeral) streams called *wadi* and is stored in surface or subsurface reservoirs, but it can also be stored in the soil profile for direct use by crops. Sometimes water is stored in aquifers as a recharge system. Generally, runoff capture, per unit area of catchment, is much lower than for microcatchments, ranging from a few percent to 50 percent of annual rainfall.

One of the most important problems associated with these systems involves water rights and the distribution of water, both between the catchment and cultivated areas and between various users in the upstream and downstream areas of the watershed. An integrated watershed development approach may overcome this problem. The most common macrocatchment systems are discussed below.

14.3.2.2.1 Small Farm Reservoirs

Farmers who have a *wadi* passing through their lands can build a small dam to store runoff water. The water can subsequently be used to irrigate crops or for domestic and animal consumption. These reservoirs are usually small, but may range in capacity from 1,000 to 500,000 m³. The most important aspect of this system is the provision of a spillway with sufficient capacity to allow for the excessive peak flows. Most of the small farm reservoirs built by farmers in the rangelands (*badia*) have been washed away because they lacked spillway facilities or because their spillway capacity was insufficient. Small farm reservoirs are very effective in the *badia* environment. They can supply water to all crops, thus improving and stabilizing production. Moreover, the benefits to the environment are substantial.

14.3.2.2.2 Wadi-Bed Cultivation

Cultivation is very common in *wadi* beds with slight slopes. Because of slow water velocity, eroded sediment usually settles in the *wadi* bed and creates good agricultural lands. This may occur naturally or result from the construction of a small dam or dike across the *wadi*. This technique is commonly used with fruit trees and other high-value crops. It can also be helpful for improving rangelands on marginal soils. The main problems associated with this type of water harvesting system are the costs and the maintenance of the walls.

14.3.2.2.3 Jessour

Jessour is an Arabic term given to a widespread indigenous system in southern Tunisia. Cross-wadi walls are made of either earth or stones, or both, and always have a spillway—usually made of stone. Over a period of years, while water is stopped behind these walls, sediment settles and accumulates, creating new land that is planted with figs and olives, but which may also be used for other crops. Usually, a series of *jessour* are placed along the *wadi*, which originates from a mountainous catchment. These systems require maintenance to keep them in good repair. Because the importance of these systems for food production has declined recently, maintenance has also been reduced and many systems are losing their ability to function.

14.3.2.2.4 Water-Spreading Systems

The water-spreading technique is also called floodwater diversion. It entails forcing part of the *wadi* flow to leave its natural course and go to nearby areas, where it is applied to support crops. This water is stored solely in the root zone of the crops to supplement rainfall. The water is usually diverted by building a structure across a stream to raise the water level above the areas to be irrigated. Water can then be directed by a levee to spread to farms at one or both sides of the *wadi*.

14.3.2.2.5 Large Bunds

Also called *tabia*, the large bund system consists of large, semicircular, trapezoidal, or open V-shaped earthen bunds with a length of 10 to 100 m and a height of 1–2 m. These structures are often aligned in long staggered rows facing up the slope. The distance between adjacent bunds on the contour is usually half the length of each bund. Large bunds are usually constructed using machinery. They not only support trees, shrubs, and annual crops but also support sorghum and millet in sub-Saharan Africa.

14.3.2.2.6 Tanks and Hafaer

Tanks and hafaer usually consist of earthen reservoirs, dug into the ground in gently sloping areas that receive runoff water either as a result of diversion from wadi or from a large catchment area. The so-called Roman ponds are indigenous tanks usually built with stone walls. The capacity of these ponds ranges from a few thousand cubic meters in the case of the hafaer to tens of thousands of cubic meters in the case of tanks. Tanks are very common in India, where they support more than 3 million hectare of cultivated lands. Hafaer are mostly used to store water for human and animal consumption. They are common in West Asia and North Africa.

14.3.2.2.7 Cisterns

Cisterns are indigenous subsurface reservoirs with a capacity ranging from 10 to 500 m^3. They are basically used for human and animal water consumption. In many areas they are dug into the rock and have a small capacity. In northwest Egypt, farmers dig large cisterns (200–300 m^3) in earth deposits, underneath a layer of solid rock. The rock layer forms the ceiling of the cistern, whereas the walls are covered by impermeable plaster materials. Modern concrete cisterns are being constructed in areas where a rocky layer does not exist. In this system, runoff water is collected from an adjacent catchment or is channeled in from a more remote one. The first rainwater runoff of the season is usually diverted from the cistern to reduce pollution. Settling basins are sometimes constructed to reduce the amount of sediment. A bucket and rope are used to draw water from the cistern.

Cisterns remain the only source of drinking water for humans and animals in many dry areas, and the role they play in maintaining rural populations in these areas is vital. In addition to their more usual domestic purposes, cisterns are now also used to support domestic gardens. The problems associated with this system include the cost of construction, the cistern's limited capacity, and influx of sediment and pollutants from the catchment.

14.3.2.2.8 Hillside-Runoff Systems

In Pakistan, this technique is also called *sylaba* or *sailaba*. Runoff water flowing downhill is directed, before joining *wadi* by small conduits, to flat fields at the foot of the hill. Fields are leveled and surrounded by levees. A spillway is used to drain excess water from one field to another farther downstream. When all the fields in a series are filled, water is allowed to flow into the *wadi*. When several feeder canals are to be constructed, distribution basins are useful. This is an ideal system with which to utilize runoff from bare or sparsely vegetated hilly or mountainous areas.

14.3.3 Water Harvesting for Supplemental Irrigation

Where groundwater or surface water is not available for supplemental irrigation, water harvesting can be used to provide the required amounts during
the rainy season. The system includes surface or subsurface storage facilities
ranging from an on-farm pond or tank to a small dam constructed across the
flow of a *wadi* with an ephemeral stream. It is highly recommended when
interseasonal rainfall distribution and/or variability are so high that crop
water requirements cannot be reasonably met. In this case, the collected
runoff is stored for later use as supplemental irrigation (Oweis et al. 1999).
Important factors include storage capacity, location, and safety of storage
structures. Two major problems associated with storing water for agriculture are evaporation and seepage losses. Following are management options
proven to be feasible in this regard (Oweis and Taimeh 2001):

1. Harvested water should be transferred from the reservoir to be
   stored in the soil as soon as possible after collection. Storing water
   in the soil profile for direct use by crops in the cooler season saves
   substantial evaporation losses that normally occur during the high
   evaporative demand period. Extending the use of the collected water
   to the hot season reduces its productivity because of higher evaporation and seepage losses.
2. Emptying the reservoir early in the winter provides more capacity
   for following runoff events. Large areas can be cultivated with reasonable risk.
3. Spillways with sufficient capacity are vital for small earth dams
   constructed across the stream.

14.4 Adaptation to Climate Change

The potential impacts of climate change on drylands ecosystems are very
complex and vary from place to place. The main changes expected in addition to rising temperature and CO_2 levels include decreasing annual precipitation, increasing rainfall variability, and higher intensity and frequency
of extreme events, such as droughts, rainstorms/floods, and hurricanes.
Generally, ecosystems will be directly affected by climate change in three
ways (HLPE 2015; IPPC 2014):

1. Increased temperature and CO_2 levels will increase evapotranspiration and reduce soil water, which will further stress crops, shorten
   the crop-growing periods, and reduce yields.
2. Rainfall characteristics are likely to change. The Intergovernmental
   Panel on Climate Change (IPCC) indicates that it is likely that total

rainfall in the Mediterranean and subtropics will decrease by up to 20 percent by the end of the century.

3. The intensity and distribution of the rainfall will be negatively changed. Methods to estimate trends in precipitation extremes at the local level are still challenging, but drought may be intensified because a smaller total amount of precipitation is expected in a more intensive pattern.

Increased intensity will encourage more runoff with higher soil erosion and lower opportunity for infiltration, especially on slopes in degraded drylands. This will cause more moisture stress and reduced recharge of aquifers. While this may increase the availability of surface water, it may also result in increased floods with associated soil water erosion. Changes in rainfall distribution are likely to intensify drought spells and the duration of droughts, exposing vegetation to increased moisture stress (HLPE 2015; IPCC 2014). As a result, there will be less support for vegetation and further land degradation. Water harvesting can help in adapting to climate change in two ways (Oweis 2016):

1. More intensive rainstorms will lead to higher runoff. More intense storms imply shorter storm durations, and hence less opportunity time for infiltration and soil water storage. This provides a golden opportunity for adaptation with water harvesting. Small contour ridges and pits, constructed with short spacing on slopes, will slow down runoff and allow water and soil to be directed and deposited behind the bunds. Furthermore, the reduction of opportunity time for infiltration will be offset by allowing water to stay longer in the pits to infiltrate. In other words, water harvesting increases the opportunity time for infiltration and storage in the soil profile. This system also has the potential to enhance groundwater recharge, providing a chance for the conjunctive use of surface and groundwater resources to alleviate drought and provide more resilience to the communities in this environment.

2. It can provide more water storage. Drought is already a challenge in arid environments. Especially in low-rainfall areas, there could be several years of effectively no runoff. People in dry environments adapt to prolonged drought by increasing storage of water and food or in extreme cases by migration, relocation, modification of dwellings, and so forth. Water harvesting allows adaptation by enhancing soil water conservation to support plants over more than one season.

In rainfed systems, less and more erratic precipitation is expected as a result of global warming. Lower precipitation will cause a further moisture stress on already stressed rainfed crops, and in some areas on the peripheries of the rainfed zones, dryland agriculture may diminish as a result. It is also expected that rainfall will be more erratic and intensive and the season will have prolonged drought spells. Crop yield losses are mainly associated with soil moisture stress during such drought spells. Prolonged drought spells during the rainy seasons resulting from global warming will make the crop situation even worse, and a further drop in yields is expected (HLPE 2015).

Supplemental irrigation can support adaptation to climate change in several ways:

1. It adds irrigation water to compensate for lower rainfall and less moisture storage, which alleviates soil water stress during dry spells. It is, however, important to consider the changes in rainfall characteristics and the durations of drought when designing interventions.

2. Increased temperature will increase evapotranspiration at times when rainfall in decreasing. Supplemental irrigation can alleviate the stress by applying small amounts of water at critical stages of growth.

3. Drought spells combined with CO_2 increases will put further stress on crops. Supplemental irrigation can reverse this impact by providing better soil water, where CO_2 can act as a fertilizer to substantially increase yields.

4. The crop calendar, depending on the onset of rain, can be modified by supplemental irrigation to avoid drought, heat, and/or frost associated with climate change during crop-sensitive stages of growth. Early sowing of crops in a Mediterranean highlands winter season helped boost wheat water productivity and yields (Oweis and Hachum 2012).

14.5 Conclusions

In the dry areas, where water is most scarce, land is fragile and drought can inflict severe hardship on already poor populations. Using water most efficiently can help alleviate the problems of water scarcity and drought. Among the numerous techniques for improving water use efficiency, the most effective are supplemental irrigation and water harvesting.

Supplemental irrigation has great potential for increasing water productivity in rainfed areas. Furthermore, it can be a basis for water management strategies to alleviate the effects of drought. Reallocating water resources to rainfed crops during drought can save crops and reduce negative economic consequences in rural areas. However, to maximize the benefits of SI, other inputs and cultural practices must also be optimized. Limitations to implementing supplemental irrigation include availability of irrigation water, cost of conveyance and application, and lack of simple means of water scheduling. In many places, high profits have encouraged farmers to deplete groundwater aquifers. Appropriate policies and institutions are needed for optimal use of this practice.

Water harvesting is one of the few options available for economic agricultural development and environmental protection in the drier environments. Furthermore, it effectively combats desertification and enhances the resilience of the communities and ecosystems under drought. Success stories are numerous and technical solutions are available for most situations. The fact that farmers have not widely adopted water harvesting has been attributed to socioeconomic and policy factors, but the main reason has been lack of community participation in developing and implementing improved technologies. Property and water rights are not favorable to development of water harvesting in most of the dry areas. New policies and institutions are required to overcome this problem. It is vital that concerned communities be involved in development from the planning to the implementation phases. Applying the integrated natural resource management approach helps integrate various aspects and avoid the conflicts of water harvesting and supplemental irrigation.

Both supplemental irrigation and water harvesting are effective in the adaptation to climate change. Alleviating the impacts of dry spells on crops and increasing soil moisture storage will boost yields and water productivity and combat land degradation.

References

English, M., and S. N. Raja. 1996. Perspectives on deficit irrigation. *Agricultural Water Management* 32:1–4.

High Level Panel of Experts (HLPE). 2015. *Water for food security and nutrition.* A report by the High Level Panel of Experts on Food Security and Nutrition of the Committee on World Food Security. FAO, Rome, Italy.

Ilbeyi, A. Ustun, H., Oweis T., Pala, M, and Benli, B. 2006. Wheat water productivity in a cool highland environment: Effect of early sowing with supplemental irrigation. *Agricultural Water Management* 82:399–410.

Intergovernmental Panel on Climate Change (IPCC). 2014. *Fifth assessment report: Climate change 2014.* IPCC Technical Paper VI. Climate Change and Water. IPCC, Geneva, Switzerland.

Oweis, Theib. 2016. Rainwater harvesting for restoring degraded dry agro-pastoral ecosystems; a conceptual review of opportunities and constraints in a changing climate. Accepted and published online in Environmental Reviews, http://www.nrcresearchpress.com/doi/abs/10.1139/er-2016-0069#.WDp1e-Z95PZ. Accessed January 15, 2017.

Oweis, T., and A. Hachum. 2003. Improving water productivity in the dry areas of West Asia and North Africa. In *Water productivity in agriculture: Limits and opportunities for improvement*, eds. W. J. Kijne, R. Barker, and D. Molden, 179. Wallingford, UK: CABI Publishing.

Oweis, T., and A. Hachum. 2012. *Supplemental irrigation: A highly efficient water-use practice*. 2nd edition. Aleppo, Syria: ICARDA.

Oweis, T., A. Hachum, and A. Bruggeman, eds. 2004. *Indigenous water harvesting systems in West Asia and North Africa*. Aleppo, Syria: ICARDA.

Oweis, T., A. Hachum, and J. Kijne. 1999. *Water harvesting and supplemental irrigation for improved water use efficiency in the dry areas*. SWIM Paper 7, International Water Management Institute, Colombo, Sri Lanka.

Oweis, T., M. Pala, and J. Ryan. 1998. Stabilizing rain-fed wheat yields with supplemental irrigation and nitrogen in a Mediterranean-type climate. *Agronomy Journal* 90:672.

Oweis, T., D. Prinz, and A. Hachum. 2001. *Water harvesting: Indigenous knowledge for the future of the drier environments*. Aleppo, Syria: ICARDA.

Oweis, T., and A. Taimeh. 2001. Farm water harvesting reservoirs: Issues of planning and management in the dry areas. In Proceedings of a Joint UNU-CAS International Workshop, Beijing, China, September 8–13, 165–183.

Oweis, T., H. Zhang, and M. Pala. 2000. Water use efficiency of rainfed and irrigated bread wheat in a Mediterranean environment. *Agronomy Journal* 92:231.

Zhang, H., and T. Oweis. 1999. Water-yield relations and optimal irrigation scheduling of wheat in the Mediterranean region. *Agricultural Water Management* 38:195.

Section IV

Case Studies in Integrated Drought and Water Management: The Role of Science, Technology, Management, and Policy

15

Floods Punctuated by Drought: Developing an Early Warning System for the Missouri River Basin in the Midst of Alternating Extremes

Chad McNutt, Doug Kluck, Dennis Todey, Brian A. Fuchs, Mark D. Svoboda, and Courtney Black

CONTENTS

The Missouri River is an enigma. Despite over a century of human attempts to understand and control it, the longest waterway in the United States remains perplexing and mysterious. The river eludes the efforts of modern technology to tame it fully and continues to provide water of uncertain quantity and quality. It either furnishes too much of it, as it did during the disastrous floods of 1844, 1881, 1943, 1951, and 1993, or too little of it, as was the case during the severe droughts of the 1860's, 1890's, 1920's, 1930's, late 1980's through early 1990's, and early 2000's.

Lawson 2009

15.1 Introduction

The Missouri River is the longest river in North America. Its drainage basin incorporates the states of Montana, North and South Dakota, Wyoming, Nebraska, Iowa, Colorado, Kansas, Minnesota, and Missouri, an area of more than 520,000 square miles. The Missouri River starts its course in western Montana and terminates over 2,300 miles downstream near St. Louis, Missouri. Between 1944 and 1964, the US Army Corps of Engineers (Corps) and the Bureau of Reclamation (BoR) built a series of dams and reservoirs under the Pick–Sloan program. The geographic area of the Missouri River Basin along with the major Corps dams is shown in Figure 15.1. The projects were developed to account for the Missouri Basin's tendency to deliver too much water in some years and not enough in others. The climate along the Missouri Basin can also vary greatly between the upper and lower parts of the basin. For example, the climate in the Upper Basin (North Dakota, South Dakota, Wyoming, and Montana) is semiarid with annual precipitation of 254 mm–508 mm (10 in–20 in) and widely varying temperatures due to exposure to Arctic and Pacific air masses. The Lower Basin (Nebraska, Colorado, Kansas, Iowa, and Missouri), however, is more consistent with a humid-continental climate (with the exception of Colorado) and can receive upwards of 1,000 mm (~40 in) of precipitation annually. The historical feature that has so often defined the Missouri Basin is the extreme year-to-year variation in runoff and sensitivity to precipitation.

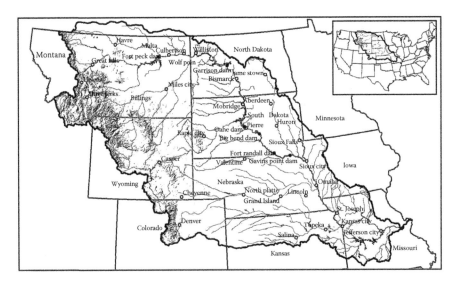

FIGURE 15.1
Geographic area of the Missouri Basin, with states, cities, and the six major Corps dams on the Missouri mainstem labeled.

A good example of this variation is the record flood in the basin in 2011 and the subsequent severe drought in 2012. While the two extremes coming in quick succession were not necessarily unique for the Missouri Basin, the fact that the record drought emerged so quickly from record flooding the previous year and with such magnitude when many were forecasting another flood makes this particularly salient for understanding and improving how we provide early warning for drought in a basin as variable as the Missouri. Figure 15.2 shows the historical annual runoff for the Upper Basin of the Missouri and helps put into context just how anomalous 2011 and 2012 were. Following these events in 2014, the National integrated drought information system (NIDIS), an interagency program lead by the National Oceanic and Atmospheric Administration (NOAA), began developing a drought early warning information system (DEWS) for the Missouri Basin. The Missouri DEWS is just one of several DEWS that NIDIS has been developing since late 2006, when the US Congress created NIDIS and charged it with developing a series of regional drought early warning systems across the United States.

The objective of this case study is to highlight the challenges of developing a drought early warning system in a basin that exhibits such extreme annual variability in runoff and sensitivity to precipitation variability. This includes how improvements in monitoring and prediction can be leveraged to address both needs to understand slow-onset disasters like drought and rapidly developing extremes like large runoff events. This topic is particularly timely given new evidence that the Missouri Basin is becoming even more variable in terms of runoff. Livneh et al. (2016) assessed meteorological trends

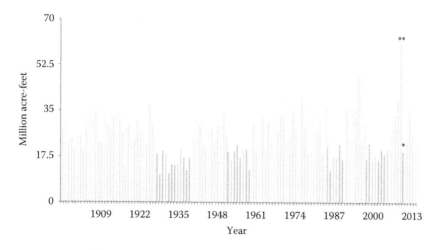

FIGURE 15.2

Missouri Basin (mainstem) annual runoff in million acre-feet above Sioux City, Iowa. Light shaded bars indicate normal or above-normal years while dark shaded years show below-normal runoff years. Double asterisk (**) indicates runoff for 2011 and asterisk (*) shows runoff for 2012.

favorable for high runoff events in the Missouri Basin, finding that annual runoff variability has nearly doubled in the last 20 years. The variability has mostly come from high runoff events and in particular from above-normal precipitation in the Upper Missouri Basin (above Gavins Point Dam) during the October–March timeframe. For example, the authors note that nine of the ten largest runoff events have occurred since 1975 (Livneh et al. 2016). While the basin may be tilting toward more frequent large runoff years, drought will still punctuate these events. The 2012 drought, while not prolonged, was an extremely intense event, and a good example of a climate surprise (Hoerling et al. 2013) that had large impacts across the basin. A chilling prospect is that a more prolonged drought similar to what occurred in the 1930s and 1950s and more recently from 2000 to 2007 in the basin is still a possibility, and it could come while we are preparing for a flood.

15.2 Alternating Extremes: The 2011 Flood and 2012 Drought

While the flood of 2011 was not predicted, there was a moment as the snowmelt accelerated and anomalously large rainfall fell in the upper part of the basin that it became clear the Corps could not evacuate water in the Upper Basin dams fast enough. This led *Nebraska Life Magazine* (Bartels and Spencer 2012) to characterize the flood as a hungry lion waiting to devour large areas of the basin. The conditions that led to the 2011 flood were a confluence of events that were difficult to predict. The basin had experienced wet conditions in the previous 4 years, and 2010–2011 continued this trend. The cold and wet winter resulted in above-average snowpack above the Fort Peck and Garrison dams and across the Great Plains. In particular, the cold conditions decreased loss of the snowpack in the plains from evaporation and delayed the snowmelt. The final component that contributed to the flood was record late spring rainfall in Montana, Wyoming, and the Dakotas (Hoerling et al. 2013; NWS 2012). This resulted in runoff of more than 48 million acre-feet above Sioux City, Iowa, from March through July, which is 20 percent more than the design runoff for the system (USACE 2012a). January–May 2011 was the wettest period on record (since 1895) for the Missouri Basin (Hoerling et al. 2013).

In late 2011, as the floodwaters were still receding, NOAA issued a La Niña advisory for the coming winter (2011–2012). The 2010–2011 winter had also been a La Niña winter, and many were concerned that a second, or "double dip," La Niña winter could cause another high runoff year. What occurred instead was a rapidly developing drought in late spring 2012, which intensified in the summer. Figure 15.3 illustrates how rapidly the drought emerged, peaking on October 2, 2012, with approximately 92 percent of the basin in some level of drought. The 2012 water year (beginning in October 2011)

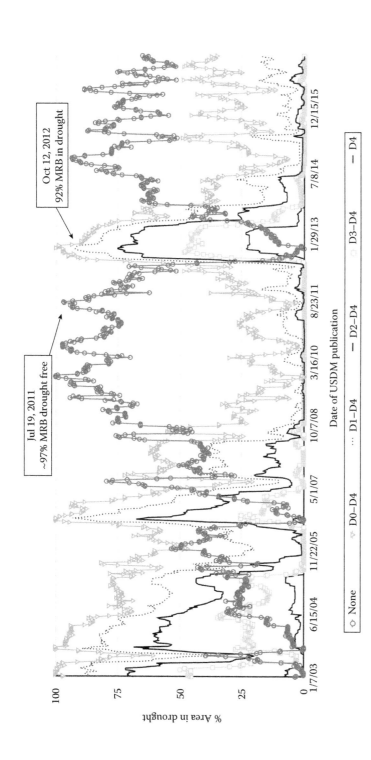

FIGURE 15.3

Percentage of the Missouri River Basin in drought based on the US Drought Monitor (USDM) classification scheme from 2003 to present. The dates on the X-axis correspond to the weekly publication of the USDM. D0 is equivalent to abnormally dry conditions and is not considered a drought category. D1 is considered "moderate drought" (20th percentile); D2 is "severe drought" (10th percentile); D3 is "extreme drought" (5th percentile); and D4 is considered "exceptional drought" (2nd percentile).

started poorly. Fall through spring (October–May) temperatures were the second warmest on record, and the March–May 2012 period was the warmest on record (since 1895) for the Northern Rockies/Great Plains region (Montana, Wyoming, North Dakota, South Dakota, and Nebraska). In May, the spring rains for the region largely failed, causing the May–August period to be the third driest on record (since 1895). Hoerling et al. (2014) assessed the causes of the drought and found it was due largely to a reduction in atmospheric moisture transport into the Great Plains from the Gulf of Mexico and was difficult to predict. Based on climate simulations, the authors also concluded the Great Plains may have shifted toward warmer and drier summertime conditions compared to the 1980s and 1990s, and that this shift may have increased the risk of drought in the of summer 2012. Impacts from the drought were extreme. For example, Nebraska hay production decreased by 28 percent, corn and soybean production decreased by 16 and 21 percent, respectively, and indemnity payments to producers totaled $1.49 billion (NDMC 2014). In South Dakota, the number of pheasant roosters harvested was the lowest since 2002, and by the end of the 2012 growing season, 80 percent of the pasture and range conditions were rated as poor or very poor statewide, which caused feed shortages and increases in feed costs. Like the flood of 2011, the drought of 2012 caught many by surprise. Because of its rapidly evolving nature and extreme impacts to the agricultural sector, many were referring to this as a "flash drought." Flash droughts are characterized by rapidly deteriorating crop and range conditions and are caused by anomalously warm surface temperatures coupled with short-term dryness (Svoboda et al. 2002). Flash droughts were a relatively uncommon phenomenon, but given the magnitude and intensity of the 2012 drought event, many are using it as a case study to develop drought indicators to improve how we monitor these rapidly emerging drought events (Hobbins et al. 2016; Otkin et al. 2015, 2016).

15.3 The Interaction of Floods and Droughts in the Missouri River Basin

Following the 2011 flood, the Corps conducted two post-flood analyses. The first analysis (USACE 2012a) focused on a review by an independent panel of experts that examined the causes and response to the flooding and provided recommendations to mitigate or prevent impacts from similar or larger floods in the future. The second report (USACE 2012b) examined whether keeping more flood storage in the Missouri River mainstem reservoir system would reduce flood risk for subsequent floods. Flood control in the Missouri River Basin is based on runoff entering the reservoirs during the spring/summer when runoff is at its peak. Under ideal conditions, the mainstem reservoir system begins the runoff season with its conservation storage full, which is

78 percent of the total storage, and the flood control storage empty, which is the remaining 22 percent of total storage. The large volume of conservation storage was provided in the system's design so that the reservoir system could serve authorized purposes, such as navigation and water supply, through long extended droughts, such as the 1930–1941 drought. The water stored in the flood control zone in the spring/summer is then released at slower rates over the remainder of the year to meet authorized purposes. The water in the reservoir system's flood control zone is not saved from one year to the next, and efforts are made to evacuate all stored floodwaters prior to the next runoff season. Evacuating stored floodwaters during the fall and winter, however, is complicated by the fact that during the winter months, large releases are not possible because of ice, which reduces the capacity of the river channel. This limits the number of days the Corps has to release large amounts of stored flood waters prior to the runoff season and results in decisions about how much water to evacuate being made in the fall, before ice begins to form, and prior to knowing the amount of precipitation in the upcoming water year.

This presents a problem for both floods and droughts in that the Corps must balance releases from the mainstem reservoir system to ensure enough flood control storage space is available to capture runoff during flood events but not so much that it cannot meet the other authorized purposes of the reservoirs: navigation, hydropower, water quality control, water supply, irrigation, fish and wildlife, and recreation. The significance of the Corp's post-flood analysis (USACE 2012b) on creating more flood control storage space was that in addition to determining whether additional storage would reduce flood risk, the analysis also assessed the potential economic impacts of increasing flood storage space at the expense of some of the other authorized purposes. The analysis concluded that the economic impacts to the other authorized purposes would be significant if more flood control storage space was made, but when single large flood events were considered, like 2011, the benefits of more flood control storage space would be an important consideration. The ability of the Corps to reflect on singular large runoff events to make this trade-off between flood control storage space and meeting the requirements of the other authorized purposes, though, is predicated on the ability to forecast their occurrence at lead times of 6 months or more. To understand the feasibility of using forecasts at such long lead times, the Corps commissioned the NOAA Earth System Research Laboratory and its Physical Sciences Division (Hoerling et al. 2013) to assess the meteorological conditions in which the flood occurred and whether it could have been predicted. The report concluded that the record spring rainfall was the most critical variable in the series of events that led to the flooding. The report also noted that the below-normal temperatures and above-normal precipitation in the winter of 2010 were consistent with what would be expected in a La Niña year, but it could not explain the wet spring in the upper part of the basin. The report also concluded that there was no way to predict, in the fall of 2010, the record spring rainfall, which would have been necessary for the Corps to have had enough time to evacuate additional water prior to the runoff in 2011. The lack of a skillful seasonal forecast was confirmed in a subsequent study by Pegion and Webb (2014) that focused specifically on the seasonal forecast skill for the 2011 flood as well as the predictability of the 2012 drought. The authors concluded forecast skill was only good at short lead timescales for the lower Missouri Basin and only during El Niño events, and that neither the 2011 flood nor the 2012 drought could have been predicted at seasonal lead times using existing forecast methodologies and models. Based on this study and the Hoerling et al. (2013 and 2014) assessments, forecasts alone would not be sufficient to inform the Corps' water management decisions at the necessary timescale and improve their ability to anticipate these large annual extremes in runoff.

15.4 Outcomes of the 2011 Flood and 2012 Drought: Implications for Improving Drought Early Warning

15.4.1 Improved Monitoring and Indicator Development

The expert review panel commissioned by the Corps in the first post-flood analysis had six recommendations to mitigate or prevent impacts from similar high runoff events. One of these recommendations identified the need for better monitoring across the lower elevations of the Great Plains Upper Basin (above Sioux City, Iowa) and, in particular, improved understanding of the plains snowpack and soil moisture. The independent panel found that the Corps underestimated the volume of water in the plains in their forecasts and the amount of runoff that would result specifically from the plains snow. Monitoring both of these variables is difficult, however. For example, estimating runoff from plains snow can be confounded by issues with blowing snow, sublimation, and fine-scale differences in topography while soil moisture can vary based on soil type and several other variables that can differ over relatively small spatial scales. Estimating soil moisture is further complicated by the fact that there is currently a very limited number of soil moisture observing sites. In the winter of 2010 and early spring of 2011, the Corps knew the plains snowpack was above average and soils were wetter than normal, but it was difficult to know how this would ultimately affect runoff and runoff efficiency.

Both of these variables are critical (USACE 2012a) to understanding drought, as overprediction of runoff can result in potentially large impacts to streamflow and water supply, among other issues (such as impacts to rangelands). Improving understanding of both variables then was identified as a critical gap in knowledge, and this led to the formation of a team of climatologists and federal scientists from across the region to develop a series of recommendations (USACE 2013) for improving the snow and soil moisture monitoring in the Upper Missouri Basin.

The goal of the assessment, and the ultimate recommendations, was to improve the monitoring infrastructure and methodology so that real-time estimates of snow water equivalent and soil moisture, and the run-off from the melted snow, could be generated specifically for the northern plains (i.e., North Dakota, South Dakota, and parts of eastern Wyoming and Montana). The data would be used to inform forecasts of both floods and droughts. Recommendations from this report were ultimately included in the Water Resources Reform and Development Act of 2014 (WRRDA, P.L. 113–121) and subsequently in the NIDIS strategic plan for developing a drought early warning system in the Missouri Basin. As of publication, funding is still being sought to build the network in the Upper Basin.

15.4.2 Improved Communication and Coordination

Following the 2011 flood, the Corps' Northwestern Division, which covers a large part of the United States and includes the Columbia River Basin in addition to the Missouri, committed to better communication with congressional delegations, states, tribes, and stakeholders. Part of this effort was to expand the webinars the Corps was holding on the status of conditions and management of the Missouri Basin system. The webinars continue to the present day and are held on a monthly basis from January through July. They include representatives from NOAA's Regional Climate Services Director and the National Weather Service (NWS). The NOAA representative provides updates on recent weather and short-term weather forecasts as well as long-term climate outlooks. The NWS partners provide information regarding the Missouri Basin snowpack and streamflow conditions, and flooding conditions and/or the potential for flooding. The Corps provides an update on basin conditions and runoff forecasts as well as reservoir conditions and forecasted reservoir operations. The webinar also provides a question-and-answer opportunity for all participants to query the Corps and their NOAA and NWS partners. The inclusion of NOAA in the webinars is significant in that it allows the stakeholders on the webinar to hear directly from the experts as opposed to the Corps being the only messenger. This new structure for the webinars added more credibility to the process and improved the perception the Corps was coordinating well with its partners (Kevin Grode, personal communication, January 23, 2017). To complement the Corps webinars, the NOAA Regional Climate Services Director and the South Dakota State Climatologist started a webinar series in late 2011 that elaborated on the weather and climate conditions in the region. The webinars were initially created to deliver information that would support the Corps and to track conditions following to the 2011 flood. As the drought emerged in 2012, however, the webinars were modified to cover current drought conditions, impacts, and both short-term and seasonal forecasts. While the 2012 drought was unfolding, and the Corps, NOAA, and many other groups were working to track it, the Corps was still working to better understand the 2011 flood, which would ultimately have consequences for how it anticipated drought as well.

In 2012, the Corps' Northwestern Division commissioned NOAA's Earth System Research Laboratory in Boulder, Colorado, to address key questions identified in the post-flood assessment process. The first report (Hoerling et al. 2013) assessed the conditions that led to the flood; the second report (Pegion and Webb 2014) addressed whether the flood and drought could have been predicted with a 6-month lead time; and the last report considered why 9 of the 10 highest historic annual runoff years in the Missouri Basin have occurred since 1975 (Livneh et al. 2016). The last example of how coordination and communication increased following the flood and drought events is evidenced by the fact that the recommendations from the *Snow Sampling and Instrumentation Recommendations* report (USACE 2013) were included in the Water Resources Reform and Development Act (WRRDA) of 2014. The WRRDA is an authorization bill that sanctions a number of water resource projects and capabilities for the Corps. While an authorization bill does not necessarily result in funding, it was nevertheless impressive that the joint effort was incorporated in the 2014 WRRDA so soon after it was completed. Most likely this was the result of having a strong technical team and good participation from the states in the region.

15.5 Conclusion

As is common with extreme climate events, neither the flood of 2011 nor the drought of 2012 was predicted. Despite their obvious differences, they are nevertheless linked by our inability to fully anticipate and respond to such extremes. In other words, the fact that they are linked is a reflection of the limits of our science and our understanding of their causes and also the infrastructure and institutional capabilities to cope with each. Our experience with each event, though, has been an opportunity to improve the science, our observation and monitoring infrastructure, our management policies, and the way we collaborate and communicate among government agencies, academic institutions, and the public. Following the 2011 and 2012 extreme events, an exceptional effort was made by the Corps, NOAA and NIDIS, state climate offices and agencies, and academics to come together and conduct assessments (Hoerling et al. 2013; Livneh et al. 2016; Piegon and

Webb 2014) on the state of our knowledge and how we could better anticipate these opposite extremes. As NIDIS began developing its drought early
warning system in the Missouri Basin, it was fortunate to develop in such a
collaborative landscape and have the ability to assess drought in the context
of how we also anticipate floods. The work that NIDIS is conducting in the
Missouri Basin is more comprehensive as a result and hopefully can add to
how we anticipate and respond to both droughts and floods in the future.

Acknowledgments

The authors would like to thank Kevin Grode from the US Army Corps
of Engineers (USACE), Northwestern Division, for providing the Missouri
River Basin map and the mainstem runoff data. They would also like to
thank him and Mike Swenson (also from the USACE Northwestern Division)
for their helpful comments and review of the manuscript.

References

Bartels, A. J., and M. Spencer. 2012. The great flood of 2011. *Nebraska Life Magazine*,
    May–June. http://www.nebraskalife.com/The-Great-Flood-of-2011/(accessed
    December 5, 2016).
Hobbins, M., A. Wood, D. McEvoy, et al. 2016. The Evaporative Demand Drought
    Index: Part I – Linking drought evolution to variations in evaporative demand.
    Journal of Hydrometeorology 17(6):1745–1761.
Hoerling, M., J. Eischeid, A. Kumar, et al. 2014. Causes and predictability of the 2012
    Great Plains drought. *Bulletin of the American Meteorological Society* 95(2):269–282.
Hoerling, M., J. Eischeid, and R. Webb. 2013. *Climate Assessment Report: Understanding
    and explaining climate extremes in the Missouri River Basin associated with the
    2011 flooding.* http://www.esrl.noaa.gov/psd/csi/factsheets/pdf/noaa-mrb-
    climate-assessment-report.pdf (accessed January 23, 2017).
Lawson, M. L. 2009. *Dammed Indians revisited: The continuing history of the Pick-Sloan
    plan and the Missouri River Sioux.* Pierre, SD: South Dakota State Historical
    Society Press.
Livneh, B., M. Hoerling, A. Badger, and J. Eischeid. 2016. *Climate assessment report:
    Causes for hydrologic extremes in the Upper Missouri River Basin.* https://www.
    esrl.noaa.gov/psd/csi/factsheets/pdf/mrb-climate-assessment-report-
    hydroextremes_2016.pdf (accessed January 23, 2017).
NDMC (National Drought Mitigation Center). 2014. *From too much to too little: How
    the central U.S. Drought of 2012 evolved out of one of the most devastating floods
    on record in 2011.* http://drought.unl.edu/Portals/0/docs/CentralUSDrought
    Assessment2012.pdf (accessed on January 10, 2017).

NWS (National Weather Service). 2012. *Service assessment: The Missouri/Souris River floods of May–August 2011*. Kansas City, Missouri. http://www.weather.gov/media/publications/assessments/Missouri_floods11.pdf.

Otkin, J. A., M. C. Anderson, C. Hain, et al. 2016. Assessing the evolution of soil moisture and vegetation conditions during the 2012 United States flash drought. *Agricultural and Forest Meteorology* 218–219:230–242.

Otkin, J. A., M. Shafer, M. Svoboda, et al. 2015. Facilitating the use of drought early warning information through interactions with agricultural stakeholders. *Bulletin of the American Meteorological Society* 96:1073–1078.

Pegion, K., and R. Webb. 2014. *Seasonal precipitation forecasts over the Missouri River Basin: An assessment of operational and experimental forecast system skill and reliability*. http://www.esrl.noaa.gov/psd/csi/factsheets/pdf/noaa-mrb-fcst-skill-assessment-report.pdf.

Svoboda, M., D. LeCompte, M. Hayes, et al. 2002. The drought monitor. *Bulletin of the American Meteorological Society* 83:1181–1190.

USACE (U.S. Army Corps of Engineers). 2012a. *Post 2011 flood event analysis of Missouri River mainstem flood control storage*. US Army Corps of Engineers, Northwestern Division, Omaha, NE.

USACE (U.S. Army Corps of Engineers). 2012b. *Review of the Regulation of the Missouri River Mainstem Reservoir System During the Flood of 2011*. U.S. Army Corps of Engineers, Northwestern Division, Omaha, NE.

USACE (U.S. Army Corps of Engineers). National Oceanic and Atmospheric Administration and the Natural Resources Conservation Service, et al. 2013. *Upper Missouri River Basin Snow Sampling and Instrumentation Recommendations*. Kansas City, Missouri.

16

Managing Drought in Urban Centers: Lessons from Australia

Joanne Chong, Heather Cooley, Mary Ann Dickinson,
Andrea Turner, and Stuart White

CONTENTS

16.1 Introduction to Australia's Millennium Drought

Australia is the world's driest inhabited continent. Many regions have highly variable climates and are prone to severe multiyear droughts. From around 1997 to 2012, however, Australia endured the "Millennium Drought," which affected a larger area of Australia, and in many locations it lasted far longer than any previous drought on record. Figure 16.1 illustrates the pattern of precipitation deficiency in Australia between November 2001 and October 2009. Falling reservoir levels and persistently low precipitation rates fueled concerns that major urban centers, including many capital cities, would face severe water shortages and, in some cases, concerns that they might run out of water.

Ultimately, because of a comprehensive drought response effort, Australian cities did not run out of water. This chapter draws on the experiences of a range of stakeholders including water utilities, government agencies, businesses, and communities across Australia to examine how this was achieved, as well as how these efforts could have been improved. In collaboration with the Alliance for Water Efficiency and the Pacific Institute, the Institute for Sustainable Futures at the University of Technology Sydney examined how Australia's experiences could be applied to California during its most severe

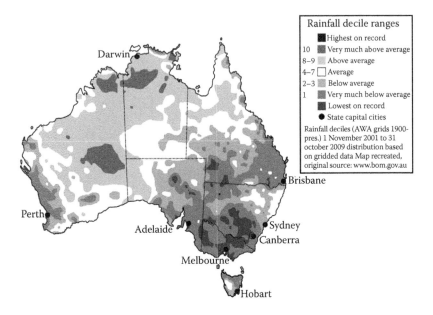

FIGURE 16.1
Most Australian capital cities experienced significantly lower rainfall than average during the Millennium Drought. Rainfall deciles (AWA grids 1900-present). 1 November 2001 to 31 October 2009. Distribution based on gridded data. Map created, original source Australian Bureau of Meteorology (www.bom.gov.au).

drought on record (Turner et al. 2016). The lessons learned from Australia's millennium drought can also inform drought preparedness and response efforts in urban centers around the world.

16.2 Leverage Opportunity but Beware of Politicized Panic

The Millennium Drought provided an opportunity to leverage community concern and political will for change and innovations in the way that urban water systems were managed and planned. Falling reservoir levels and growing concern about climate change brought about a realization that the country was too vulnerable to drought because of high water usage and reliance on rain-dependent water sources. These concerns highlighted the need to adopt comprehensive demand management programs and diversify water sources. In response, Australian cities and towns implemented world-leading approaches and ideas to build more resilient and sustainable water systems.

Major, comprehensive demand management programs were an essential element of drought response efforts across Australia. Large-scale conservation and efficiency programs were implemented quickly by utilities and governments across drought-affected states, building on decades of industry experience in demand management. Some of the programs implemented included:

- "Do-it-yourself" water savings kits and subsidized home water audits, leak repairs, and installation of water-efficient fixtures, conducted by authorized plumbers
- Showerhead exchange, toilet replacement, and washing machine rebate programs
- Information and product rebates for outdoor water savings, such as swimming pool covers, irrigation systems, rain sensors, and faucet timers
- Targeted information, support, and incentives (both rewards and punitive measures) for the highest residential water users
- Business water efficiency management plans and water saving actions plans, in some cases mandatory for high water users, developed with the support of utilities or governments (including for audits and other technical advice)

These investments, in combination with restrictions on water use, proved to be highly cost-effective ways to reduce reservoir depletion, delay or eliminate the need for major expenditures on new supply and treatment infrastructure, and prevent cities from running out of water (Chong and White 2017b). In many cities, large structural and behavioral shifts in water demand were achieved. For example, in South East Queensland, residential water demand fell by 60 percent to 33 gpcd (125 lcd) and has only increased to around 45 gpcd (170 lcd) since then (Turner et al. 2016).

The Millennium Drought also presented opportunities for new policy and management approaches to urban water planning. For the first time, governments considered "real options planning," an approach pioneered in the finance industry and predicated on the principle that the value of an investment lies in its *readiness* to be implemented, *when and if required*. For example, Sydney Water incorporated this approach by being "ready to construct" a desalination plant should dam levels drop below a specified trigger level, which was calculated through stochastic modeling of rainfall, water demand, and the amount of time needed to bring the desalination plant online (Metropolitan Water Directorate 2006). This planning approach allows greater flexibility for investment in large capital items by making the expenditure "staged" and modular, and it allows the option to curtail completion of a plant if conditions change.

In addition to these innovations, the Millennium Drought also spurred crisis-driven, politicized decisions that set aside the extensive planning undertaken by government agencies and utilities. These included the New South Wales government's decision to construct the desalination plant regardless of dam levels and prior to the trigger point established through real options planning. The government persisted in releasing tenders to construct even after the drought broke. Another example is the controversial decision to construct the Traveston Dam in South East Queensland (Chong and White 2007)—a decision subsequently overturned. In several locations, crisis-driven decisions resulted in costly, and in some cases energy-intensive, investments that were ultimately not used.

16.3 Partnerships and Collaboration

Strong partnerships, knowledge sharing, and coordination between organizations—states, agencies, utilities, researchers, and industries—contributed to success during the Millennium Drought. Governments and utilities invested heavily in partnerships that leveraged and deepened the networks and collaborations between stakeholder groups, and these partnerships were fundamental to the successful design and implementation of water savings efforts. These collaborations also signaled the "we're all in it together" ethos for conserving water and helped to foster greater public support.

Governments and utilities formed crucial partnerships with businesses that used water, as well as with those that manufactured and supplied water-using devices and provided services to help customers manage their water use. Most water efficiency programs were funded and led by utilities and state governments, but industry associations and trade groups (and their members) participated extensively in their design and implementation. Additionally, the Western Australian utility, WA Water Corporation, and the Western Australian government engaged irrigation and landscaping businesses in programs to build capacity and provide accreditation schemes.

Government agencies and utilities also formed interdepartmental drought response teams to coordinate state and regional efforts. For example, in the state of New South Wales, the Water CEOs group was chaired by the head of the Cabinet Office and comprised the heads of all water-related agencies and utilities. Likewise, in Melbourne, the Drought Coordinating Committee, which comprised members from across utilities and governments, was instrumental in coordinating drought actions. The committee has subsequently reconvened to review and revise the approach to drought planning and response.

The Millennium Drought also encouraged significant sharing of information and experiences across jurisdictions and cities. For example, in Perth and Melbourne, detailed surveys and analyses of how people used water were shared among utilities, spawning a new era of detailed sector- and end use-based forecasting of water demand that was used to design water efficiency programs and conduct long-term planning. Similarly, agencies in

South East Queensland drew on the long-term experience of Sydney Water to quickly design and launch large-scale water efficiency programs.

Additionally, water companies, utility industry associations, and state governments commissioned extensive research to inform their real-time responses to drought. By building industry experience, knowledge, and networks, Australian cities ensured they were well-positioned to deal with the drought, train a new generation of industry professionals, and plan for future climate uncertainty. However, since the end of the drought, the focus has shifted away from water efficiency programs, posing ongoing challenges in terms of maintaining expertise and knowledge throughout the industry.

16.4 Community Engagement

Communication and public engagement on drought conditions and water savings programs were instrumental to the success of water savings initiatives. Water efficiency marketing and media campaigns were also effective in fostering support and empowering the community to take action. The strategies used in these campaigns included:

- Linking water-use restrictions to information about the availability of incentives, rebates, and other water savings initiatives
- Applying clear and consistent messages to focus community support on achieving a common goal, such as the Target 140 campaign in South East Queensland and the Target 155 campaign in Victoria, which reduced household water use to 37 gpcd (140 lcd) and 41 gpcd (155 lcd), respectively
- Communicating directly with high water users through a direct mail-out containing a survey and links to water saving offers, along with additional follow-up if there was no response
- Promoting case studies of businesses that participated in programs to save water

Successful community engagement means effective listening as well as speaking. Decision-making during drought involves trade-offs—and it is important to invite the community to provide their input on these trade-offs. This helps to ensure that decisions reflect community preferences and in turn engender support for the decisions made. For example, in Western Australia, a robust and comprehensive community engagement process on water security was undertaken in 2003, including a citizens' forum held at Parliament House that was addressed by the State Premier.

There were, however, many missed opportunities to implement best-practice public engagement efforts during the drought. In most states, decisions about investments, policy choices, water use trade-offs, and levels of service were made centrally. Occasionally, this was done in consultation with industry representative organizations, but in many cases, it did not involve direct engagement with members of the broader community. Governments did not take advantage of the level of innovation that Australia had demonstrated in deploying robust forms of community engagement.

16.5 Implementing Supply- and Demand-Side Options

During the Millennium Drought, Australian cities faced a significant issue that confronts all water-scarce urban regions: the tension between investing in demand management programs and investing in water supply infrastructure. Both options contribute to water security, but measures to improve water efficiency are in many cases among the most cost-effective options. Established approaches to planning built on integrated resource planning (Fane et al. 2011) provide guidance on how to analyze all types of options, but the application of these approaches has not been universal, including during the millennium drought.

Several low-cost measures on both the demand and supply sides were pursued during the drought, such as increased water efficiency, accessing dead storage, and intercatchment transfers. However, it was apparent during this period, and it has become even more apparent since then, that regulatory and institutional settings provide imperfect signals for utilities and governments to invest in the most efficient, sustainable, cost-effective, and resilient portfolio of options, regardless of whether they are supply or demand measures.

Australian cities are largely served by regulated government-owned monopolies, and the conventional regulatory and institutional settings do not, on the whole, provide signals for efficient investment in both supply and demand. Water efficiency programs incur operating costs, and they can reduce water demand and hence revenue. In contrast, supply measures involve mainly capital costs. Conventional regulatory settings encourage utilities to minimize operating expenditures and to set prices designed to earn a rate of return on capital. Price regulation that allows cost pass-through of capital expenditure and a return on capital favors supply options over demand management measures (White et al. 2008a, 2008b).

Regulatory arrangements can allow for revenue neutrality, and they can also be used to ensure that the costs of investments in water efficiency programs are passed on to the customer. Australia, at the time of the millennium drought, had a mixture of regulatory arrangements for utilities; some had resolved this tension by allowing a price pass-through and others had not.

Most utilities continue to face this tension between conserving the resource and reducing long-term costs for the customer on the one hand and satisfying regulatory and state government drivers on the other. Indeed, the situation has worsened in Australia since the drought ended, because the large over-supply can have the effect of providing utilities with an incentive to encourage greater water use, or at least a disincentive to improving water efficiency.

A further barrier is that urban water options are often analyzed on an individual project scale in terms of costs and water security (and other) benefits, rather than on a system scale. For example, an individual precinct reuse system, if considered in isolation to the rest of the system, might be assessed as contributing only in a limited way to water security. However, an integrated planning approach that considers a network of multiple decentralized systems might prove more resilient and cost-effective, when avoided costs of the centralized system are taken into account. Furthermore, across jurisdictions there is generally no clear process that allows for the livability, physical, and health benefits of climate-resilient water provision to be considered in planning. Consequently, opportunities to design systems to cost-effectively irrigate green spaces and trees, and hence provide amenity and cooling long term as well as during drought, are likely to be forgone.

16.6 Conclusions and Applications Beyond Australia

Several lessons can be drawn from Australia's Millennium Drought that can inform drought preparedness and response efforts in urban centers around the world. First, a fundamental conclusion is that responding to drought requires the implementation of *both* supply- and demand-side options. There is significant industry knowledge in Australia and elsewhere on how to evaluate the full range of costs and benefits of water security options, and how to design and implement portfolios using a real-options, readiness approach. However, in Australia as elsewhere, the existing regulatory and institutional settings can often be biased toward supply-side investments. Shifting policy goals toward water services and resilience can require a complex and lengthy process of reform, but this shift is essential for developing more sustainable water systems.

Second, a considerable number of low-cost options for reducing water demand in urban areas exist without affecting the services and benefits that water provides. Worldwide, and in different institutional settings, there is a wealth of expertise on implementing demand management programs *before* drought. These long-term measures increase resilience *during* drought by slowing down the rate of depletion of available water resources. Evidence from Australia and elsewhere indicates there is often significant potential for savings inside and outside homes (Chong and White 2016,

2017b; Turner et al. 2016), as well as in commercial, industrial, and institutional settings. For example, a recent study by Heberger et al. (2014) found that repairing leaks, installing the most efficient appliances and fixtures, and replacing lawns and other water-intensive landscaping with plants requiring less water could reduce urban water use in California by 30 to 60 percent, saving 3.6 to 6.4 cubic kilometers per year. Designing and implementing such water efficiency programs relies on good data and robust monitoring and evaluation, critical for estimating the potential water savings.

Third, an effective supply-side strategy should integrate modular, scalable, diverse, and innovative technology options. Although advances have been made in seasonal forecasting, the length and severity of the droughts we will experience in the future are uncertain. A rapid and progressive approach to developing supply infrastructure, and especially contract design, can avoid commitments to unnecessary, costly expenditures.

Fourth, communication and transparency are paramount to garnering public support and participation in responding to drought. Collaborating with and involving the community builds consumer confidence, trust, and the energy to act. Additionally, collaboration will help to develop leaders within industry and the community, and they will in turn promote water savings. Ultimately, however, if the public is to trust the decisions of governments and utilities, then community voices and values must be integrated into planning processes, and the resulting plans, which are essentially developed under a social contract between governments, utilities, industries, and residents, must be implemented.

References

BOM. 2015. *Recent rainfall, drought and southern Australia's long-term rainfall decline.* Bureau of Meteorology, Australian Government. http://www.bom.gov.au/climate/updates/articles/a010-southern-rainfall-decline.shtml (accessed January 22, 2017).

Chong, J., and S. White. 2007. Decisions for the urban drought: Paternalism or participation? In *2007 ANZSEE Conference, Re-Inventing Sustainability: A Climate for Change.* Australia and New Zealand Society for Ecological Economics. http://cfsites1.uts.edu.au/find/isf/publications/chongwhite2007paternalismorparticipation.pdf. Accessed January 22, 2017.

Chong, J., and S. White. 2016. *Drought response and supply—Demand management planning: Iloilo City, Philippines.* Report prepared by the Institute for Sustainable Futures, University of Technology Sydney for USAID: Be Secure.

Chong, J., and S. White. 2017a. *São Paulo metropolitan region roadmap for urban water security.* Report prepared for Secretaria de Saneamento e Recursos Hídricos, Government of São Paulo, Brazil.

Chong, J., and S. White. 2017b. Urban water reform in Australia—Pathways to sustainability. In *Decision making in water resources policy and management: An Australian perspective,* eds. B. Hart and J. Doolan. Amsterdam: Elsevier, pp. 85–96.

Fane, S., A. Turner, J. Fyfe, et al. 2011. *Integrated resource planning for urban water—Resource papers.* Waterlines Report Series No. 41. Prepared by the Institute for Sustainable Futures, University of Technology Sydney. Canberra: National Water Commission.

Heberger, M., H. Cooley, and P. Gleick. 2014. *Urban water conservation and efficiency potential in California.* Issue Brief 14-05-D. Oakland, CA: Pacific Institute.

Metropolitan Water Directorate. 2006. *2006 Metropolitan water plan executive summary.* Government of NSW. https://www.metrowater.nsw.gov.au/sites/default/files/publication-documents/mwp_exec_summary.pdf (accessed January 22, 2017).

Turner, A., S. White, J. Chong, M. Dickinson, H. Cooley, and K. Donnelly. 2016. *Managing drought: Learning from Australia.* Prepared by the Alliance for Water Efficiency, the Institute for Sustainable Futures, University of Technology Sydney and the Pacific Institute for the Metropolitan Water District of Southern California, the San Francisco Public Utilities Commission and the Water Research Foundation. Institute for Sustainable Futures, New South Wales, Australia.

White, S., S. Fane, D. Giurco, and A. Turner. 2008a. Putting the economics in its place: Decision-making in an uncertain environment. In *Deliberative ecological economics*, eds. C. Zografos and R. Howarth, pp. 80–106. New Delhi: Oxford University Press.

White, S., K. Noble, and J. Chong. 2008b. Reform, risk and reality: Challenges and opportunities for Australian urban water management. *The Australian Economic Review* 41(4), 428–434.

17

Managing Drought and Water Scarcity in Federal Political Systems

Dustin E. Garrick, Lucia De Stefano, and Daniel Connell

CONTENTS

17.1 Introduction

During the summer of 2015, droughts and water shortages affected federal countries ranging from Australia, Brazil, and Canada to the United States, South Africa, and India. Drought involves coordination challenges in federal political systems where national and subnational governments each play a critical role. By blurring key roles and responsibilities, droughts create stress tests for transboundary water governance, requiring intergovernmental coordination between states (known as horizontal coordination) and multilevel coordination between states and national governments (known as vertical coordination). These governance challenges increase the importance of conflict resolution and other institutional mechanisms to share risks and enhance resilience to severe, sustained drought events.

According to the Forum of Federations, 25 countries have a federal political system, including many of the oldest federations, Australia, Canada, and the United States, with large geographic territories facing diverse challenges related to drought. Democratization has also brought federalism to countries with a long history of centralized governance, such as Spain and Ethiopia, where droughts are a recurrent feature of the hydroclimatology. Federations, therefore, vary in their policy approaches and institutional structures, particularly in the level of centralization of key governance tasks, and how this varies according to the duration, intensity, and severity of droughts.

In this context, sharing knowledge, experiences, and best practices developed across a spectrum of federal countries is a powerful way to build understanding and capacity to address present and future challenges posed by droughts and other extreme climate events. This chapter aims to advance our understanding of the factors and institutions influencing cooperation, conflict resolution, and capacity to adapt to drought and water scarcity in federations.

To do so, we develop case studies of drought management in federal political systems by addressing the following questions:

- What is the (recent) history of droughts in the country and its major river basins?
- What are the major sources of tensions and disputes between states, between states and the national government, and between the different interests within the basin?
- What are the institutional mechanisms available for responding to these tensions and coordination challenges?
- What are the lessons learned about barriers, enabling conditions, and strategies for cooperation and conflict resolution during drought?

Australia, Spain, and the United States are used to illustrate the diversity of approaches to address the coordination challenges arising during droughts. The three countries are prone to droughts. However, the countries vary in their federal system of water governance. On one end of the spectrum, Spain has a relatively centralized approach to water governance and drought planning. On the other end of the spectrum, the United States has a relatively decentralized system of water governance and drought planning concentrated at the state level, until more recent efforts were undertaken to strengthen national programs. Australia represents a mixed approach involving strong state and national roles. Therefore, this group of cases offers insights about the challenges and responses associated with droughts in federal political systems.

17.2 Australia

17.2.1 History of Major Droughts

In the Murray–Darling Basin (MDB) since the late nineteenth century, droughts have been drivers of institutional change in the way water is shared and managed. The severe drought of the late nineteenth and early twentieth century was a powerful spur for federation in 1901, particularly for South Australia at the end of the River Murray system. Eventually it resulted in the River Murray Waters Agreement, ratified by the three southern state governments (New South Wales, Victoria, and South Australia) and the national government, through identical parallel legislation in 1914/1915.

The drought of the early 1980s provided the stimulus for the MDB Agreement, which reflected increasing concern about the need for a whole-of-catchment perspective to take account of development pressures (Helman 2009). These were highlighted by the 1995 Water Audit, which revealed that because of the growth of extractions, upstream drought conditions at the Murray Mouth had increased in frequency from 5 to 63 percent of years.

The drought of the early 2000s was the most intense recorded, and it led to the national government takeover of MDB policy through its Water Act in 2007/ 2008. Research conducted during this period predicted that unregulated development pressures combined with climate change would result in further drastic reductions in flows. Climate predictions are for greater rainfall variability within a long-term drying trend. The need for coordinated holistic management to minimize and share increasing costs is now widely accepted. The main source of disagreement is about the underlying priorities that should shape the management framework (Connell 2007, 2011).

17.2.2 Tensions

Federal political systems provide a greater diversity of options for political action than do unitary systems. Discussions about federalism usually focus on the dynamics between the states and the national government. However, much of the action within federal systems reflects tensions between states and the behavior of stakeholders moving between the levels of government, depending on how they perceive their interests and opportunities.

For example, the place and role of South Australia is central to the history of cross-border water sharing in the Murray system. For South

Australia at the end of the system, the very existence of the state depends on access to flows in the River Murray, so it has persistently used its leverage within the national government to ensure that its interests are taken into account. It is no accident that since federation a disproportionate number of the national government ministers responsible for policy affecting the River Murray have come from that state. South Australia has long been a strong advocate of total catchment management in its various forms, arguing that the environmental health of the lower lakes and estuary should be the yardstick for effective river management. This has intensified conflicts with the upper states as development pressures have increased throughout the catchment.

The growing focus on whole-of-catchment management has fueled debates about priorities. Traditionally, promoting irrigation along the Murray corridor and "drought proofing" the towns and cities of South Australia have been the major goals. The institutional reforms of the 2000s were aimed at more comprehensive management designed to take account of a greater range of stakeholders and the need for long-term sustainability. However, the irrigation sector has worked very effectively to protect its interests. The original intention was that water markets would operate across more than the irrigation sector, but this proved difficult to achieve. Ambitious and well-funded national government plans to restore environmental conditions by water purchases from willing sellers at market price have increasingly been stifled by the irrigation sector operating at state and national level. In the same way, attempts to improve the water security of the major cities of Adelaide and Melbourne by water purchases in the MDB have been blocked by opposition from irrigation-based communities. Instead, both cities have built multibillion-dollar desalination plants even though the water needed is only a very small percentage of the volumes currently being diverted for agriculture in the MDB.

17.2.3 Institutional Mechanisms

The River Murray Waters Agreement in 1914/1915 was based on a water-sharing agreement whose key elements are still the core of the intergovernmental agreement about water sharing in the MDB (Connell 2007). It required the two upriver states, New South Wales and Victoria, to provide a defined quantity at the South Australian border except in time of drought, when each of the three states is entitled to a third of whatever is in Lake Hume, the strategic storage near the top of the catchment. This worked fairly well for a number of decades, but since the 1980s there has been an intense political struggle between the states and different stakeholders over attempts to expand it to take greater account of environmental and urban interests outside the agricultural sector, which uses about 95 percent of all water extracted from the system.

In recent decades, the national government, driven by international trends in river management and research, economic pressures, and growing public concern about environmental decline, has led attempts to define sustainable management in ways that can be operationalized. The area of greatest success has been in the introduction of water markets across the three states. During the intense drought of the 2000s, water trading within and between states played a major role in maintaining economic productivity in the agricultural sector at near predrought levels. The development of a cross-border water market was a significant achievement given strong opposition from the state governments of New South Wales and Victoria (South Australia has long believed that a water market will result in more water moving to that state) and the existence of many different types of water entitlements before the integration process began.

The original intent was that the water reform program would be implemented within a working river framework that would provide security for economic development and restore the MDB as an environmentally attractive and ecologically functional system able to weather the future challenges of climate change. But despite the national government's Water Act of 2007/2008, based on a requirement for sustainable diversion limits across the entire MDB, it can be argued that the irrigation sector has been effective in eroding the water recovery targets and the mechanisms for achieving them. One of the most significant of these retreats has been in the area of monitoring and auditing with the abolition of the National Water Commission, responsible for reporting on progress on water reform, and the Sustainable Rivers Audit, which was to provide information about long-term environmental trends.

17.2.4 Lessons

Perspectives about what lessons should be learned from the history of water management in the MDB depend on the learner. The irrigation sector has learned how to be very effective in molding the forces of change even when the times seem hostile. Urban centers such as Adelaide and Melbourne have learned to depend on themselves—hence the shift to large-scale investment in desalination. Advocates of the reforms foreshadowed in the National Water Initiative of 2004 probably need to think more about what is needed to promote basin-wide consciousness and political support for sustainable management. The case for effective monitoring and auditing appears central. Without the information that was provided by the National Water Commission and the Sustainable River Audit, public policy debates will not achieve traction. There should also be more thinking about how to engage the public. One option would be to devolve management of much of the environmental water currently tightly managed by the national government to elected community regional organizations working within strong reporting frameworks. This would be messy, but it would provide members of the public with reason to get engaged. Water management left to the experts will continue as it has in the past.

17.3 Spain

17.3.1 History of Droughts

During the past 30 years, prolonged droughts affecting large portions of Spain's territory have occurred every decade (1980–1983, 1990–1994, and 2005–2008). Droughts have proved to be catalysts for legal reforms and investments in water infrastructure, as they often revealed weaknesses in the water management system that were tackled either during or shortly after the end of the dry spell. For instance, the first major reform to the 1985 Water Act was passed in 1999, after a major drought (1990–1994) that resulted in significant economic losses and large-scale water supply restrictions. Between 2005 and 2008, a new drought in several regions of Spain had less severe impacts relative to the one of the 1990s, partly because it was less intense and partly because several actions had been taken to avoid severe restrictions and environmental problems experienced during the previous drought (Estrela and Vargas 2012).

Droughts are also clear windows of opportunity to spur the implementation of solutions already present in Spain's water debate before the dry period. For instance, water markets were introduced in the 1999 reform of the Water Act but had been rarely implemented until the 2005–2008 drought. Moreover, that drought period served to "test" interbasin water trade—which was explicitly banned from the 1999 law reform (Hernández-Mora and Del Moral 2015) but had strong support from some farmer lobbies and urban supply actors. Similarly, in 2006 the government passed a large program for the modernization of irrigation systems in the whole country. This program had been on the policy agenda since the early 1990s and was passed using a fast-track approval process using drought as a justification (Urquijo et al. 2015).

With the advent of democracy in the 1970s, Spain adopted a decentralized political system in which regions have competences over many policy domains (e.g., education, health, and environment protection). The 1978 Constitution also established that water had to be managed by river basin authorities linked to the central government for the basins shared by two or more regions, whereas for intraregional basins (i.e., those located within a single region), water had to be managed by regional water authorities. In this context, in each basin, droughts are managed mainly by the corresponding water authorities, often with strong support from the central government when investments or special legal provisions are needed.

17.3.2 Tensions and Cooperation Challenges

During droughts, most of the tensions occur among water users, whose water rights are regulated by a complex water rights system that establishes prioritization of uses in case of drought and is framed within river basin management plans. Tensions among users can spill over into interregional relations when regional government advocates for the interests of its constituency (e.g., to avoid water restrictions or to support the need for investments in new water infrastructure). Most of the interregional tensions, however, occur over competences over water resources and water allocation. In particular, water allocation as defined in the river basin management plans (revised and approved every 6 years), the construction and operating of water infrastructure, and water transfers are significant sources of tensions and disputes between regions and between regions and the central government. This can be explained by the fact that Spain has a strong regulatory framework for water management, and those issues and instruments create the foundation for any decision taken to manage drought. Thus, regions engaged in bitter disputes over the approval of the National Hydrological Plan in 2001, as it included the construction of a long list of new water infrastructure and a major water transfer from the Ebro basin in the northeastern part of the country to several regions along the Mediterranean coast. Moreover, since the 2000s there have been several judicial cases in the Supreme and Constitutional Courts where regions have sued one another in order to gain greater control over water resources development in their territories and to increase their competences over water resources planning and management (for an overview, see López-Gunn and De Stefano 2014; Moral Ituarte and Hernández-Mora Zapata 2016).

17.3.3 Institutional Mechanisms and Adaptation Options

The 2001 Law of the National Hydrological Plan mandated the elaboration of specific drought management plans for all the river basins and for towns with more than 20,000 inhabitants. These plans include early warning and drought monitoring indicators that facilitate the declaration of different levels of drought alert. They also specify sets of measures that could or should be implemented, depending on the level of alert. Most of the river basin drought management plans were approved in 2007 and they are currently under revision (Estrela and Sancho 2016), whereas drought management plans for urban supply have been progressively approved for the largest towns.

The existence of these plans and the strong national regulatory framework create an important foundation for drought management. River basins are the overarching spatial domain where water resources are shared, while demand and supply are managed according to smaller systems, the so-called water exploitation systems. Thus, regional interests and disputes during drought are somehow diluted by the fact that water is distributed and shared mainly based on hydrological and hydraulic criteria rather than based on boundaries between regions. During drought, there is usually a quite good level of cooperation among actors in each basin, facilitated by the existence of participatory bodies where consumptive water users and the competent authorities periodically meet to approve water deliveries and, more recently, by the creation of a drought commission summoned by the river water authority to discuss drought-related decisions with consumptive water users and other stakeholders.

During droughts, the central government can issue royal decrees that serve to approve urgent investments and legal changes to address "exceptional situations" (Article 58, Water Act). This can include temporarily increasing the powers of the river basin authority in order to facilitate temporal reallocation of water volumes or authorizing the drilling of emergency wells to supplement ordinary water supply. These royal decrees, however, have been used by the central government also as a way of bypassing the regular approval procedure and approving highly disputed investments or legal reforms that do not provide immediate relief from drought (Urquijo et al. 2015).

In terms of mechanisms to manage interregional relations over water planning and management, formal venues are mainly hosted within the river basin authority and have proved to be venues for voting decisions rather than for actual discussion, negotiation, and consensus building. Negotiations prior to the voting sessions are mainly bilateral between the central government and each region and usually occur in informal forums.

17.3.4 Lessons Learned

Focusing specifically on droughts, Spain's experience confirms the importance of having good monitoring systems in place and defining clear indicators and thresholds to declare different levels of drought alert. This type of system creates a common platform of information that serves as a reference for all the actors, thus contributing to reducing uncertainty and disputes among both governmental and nongovernmental stakeholders.

As mentioned earlier, during droughts the central government can pass special decrees to temporarily increase the power of the river basin authority and to expedite water-related investment decisions or even legal reforms. These legal instruments in some cases are needed to address immediate needs, but in other cases, fast-track procedures for decision-making are used by the central government to assuage tensions and maintain social peace at the expense of taxpayers. Moreover, investments and legal changes approved through special decrees are not subject to the same level of debate and public scrutiny as regular decisions, which reduces accountability and public participation options.

When considering consensus building over water planning and water allocation, the experience of Spain showcases that in a politically decentralized system, subnational governments have the capacity of significantly hindering decision-making processes when they do not feel their interests are sufficiently taken into account in the official water-related, decision-making venues. In this context, an important lesson about interregional relations relates to the role of courts in conflict management: the history of recent interregional conflicts shows that judicial rulings tend to create winners and losers, which leads to entrenchment of disputes and postponement of negotiated solutions.

As mentioned earlier, official venues are mainly for voting decisions that have already been discussed and negotiated beforehand. On one side, this can create power imbalances, as some actors remain excluded from bilateral, informal discussions. On the other side, it underscores that mutual trust, fluid exchange of information, and strong collaboration at the technical level among different levels of government are all key ingredients to build consensus around decisions that will be sanctioned in the official venues.

17.4 United States

17.4.1 History of Droughts

Droughts in the United States range from localized events and seasonal deficits to severe droughts that affect regional or continental areas over one or many years (Cook et al. 2013; Diodato et al. 2007; Overpeck 2013). Drought is a defining feature of the intermountain western region of the United States, including the iconic events such as the Dust Bowl (1930s) and post-World War II (1950s) droughts that prompted a range of water infrastructure and management responses. In recent years, population growth in humid and semiarid areas alike has increased the exposure to drought events in California, Georgia and the Southeast, Hawaii, the Southwest, the Pacific Northwest, and Texas. In this context, some regions are vulnerable to substantial impacts even for drought events of moderate intensity, duration, and severity. Together with the continental scale droughts in 2012–2013, these events have made drought an issue of increasing national interest and concern, prompting coordination of monitoring, planning, and other drought management actions (Folger and Cody 2015).

River basins in the western United States have experienced sustained droughts and will be the primary focus here, highlighting the experience of the Colorado and Rio Grande/Bravo basins, two international rivers shared by the United States and Mexico. Here the primary focus will be the US portion of each basin.* The Colorado River Basin drains almost 700,000 km²

---

* The interstate dynamics within Mexico and Mexico's delivery obligations to the United States in the Rio Bravo are beyond the scope of this analysis.

with its territory covering parts of nine states (seven in the United States, two in Mexico). The basin's hydroclimatology has been marked by sustained droughts, including the period from the late 1940s through the 1950s, which was the drought of reference for modeling and planning purposes until recently. Since 2000, the Colorado River Basin has experienced a dry period without precedent in the observed record, although reconstructions of the paleoclimate indicate droughts of longer duration and intensity (Udall and Overpeck 2017; Woodhouse et al. 2006).

The Rio Grande/Bravo Basin drains approximately 450,000 km^2 (excluding endorheic zones), including almost equal territory in the United States and Mexico. Both the United States (upper Rio Grande) and Mexico (Rio Conchos) contain important tributaries shared by multiple states within each country. Like the Colorado River, the basin has experienced sustained droughts in the 1950s and since 2000. However, the history of drought in the Rio Conchos and Upper Rio Grande has not been correlated per the paleoclimate record, which means that droughts can affect one tributary without impacting the other (Woodhouse et al. 2012).

17.4.2 Tensions and Cooperation

The legal and institutional framework for water allocation in the western United States creates the basis for tensions during droughts by establishing a "zero-sum" game, where some water users have highly reliable water rights and others lose access to water completely. Water allocation in the western United States is governed by the principle of prior appropriation and beneficial use, known colloquially as "first in time, first in right" and "use it or lose it," respectively. During drought periods, the first to establish and maintain a beneficial use is the last to lose access. This principle applies at the level of water users and their associations (irrigation districts and municipal utilities). The Colorado River and Rio Grande illustrate the tensions and cooperation associated with transboundary management of droughts in the western US system of water allocation. In the US portion of both the Colorado and Rio Grande/Bravo, interstate water apportionment agreements exist to share water based on principles of equitable use between states. Each agreement defines allocation rules requiring the delivery of volumes of water from upstream to downstream states, creating coordination challenges during droughts.

In the Colorado River, drought conditions and competition for water have placed intergovernmental water agreements under pressure, exacerbating tensions stemming from the structural imbalance between demand and supply (i.e., the overcommitment of the river's annual renewable runoff). The 1922 Colorado River Compact and a series of additional laws, rules, court cases, and operational criteria constitute the Law of the River, apportioning water between the states sharing the river on the US side. Drought and shortage conditions were not addressed by this institutional framework, creating uncertainty about the operational criteria used to manage reservoirs and share shortages between states until recently, as discussed below. Tensions between states within the US portion of the river basin have produced a legacy of intense conflict over interstate water allocation matters, including the landmark 1963 Supreme Court case, Arizona v. California, which clarified and confirmed prior interstate accords. Despite the concerns and fears that severe sustained drought would trigger interstate conflict, the dry period since 2000 has been marked by unprecedented cooperation, culminating in a series of agreements and institutional mechanisms for building resilience to drought and water scarcity, including shortage sharing agreements that include Mexico and ongoing efforts to negotiate a drought contingency plan that would address severe shortages.

In the Rio Grande/Bravo, drought has exacerbated pressures related to urbanization and endangered species issues in the US portion and has caused both interstate and international tensions with Mexico. A complex set of intergovernmental agreements divide water between states in the United States (1938 Rio Grande Compact) and between countries (1906 Convention and 1944 Water Treaty). In the United States, drought since 2000 has heightened dependence on and conflict over groundwater pumping, with several interstate coordination challenges because of effects of groundwater pumping in Colorado on New Mexico's water supplies and effects of groundwater pumping in Southeast New Mexico on the reliability of water deliveries to Texas under the Rio Grande Compact. This dispute has prompted a court case between Texas and New Mexico that is currently before the Supreme Court.

17.4.3 Institutional Mechanisms

Institutional responses to drought in the Colorado and Rio Grande rivers have required coordination mechanisms to facilitate cooperation and manage conflicts between the states. In the Colorado River Basin, the primacy of the states in water allocation gives them a central role in drought management. However, federally constructed and managed reservoirs require interstate coordination during drought to share shortages and build flexibility. The 2001 Interim Surplus Guidelines and 2007 Shortage Sharing Guidelines have established rules for managing the river's two main reservoirs—Lakes Powell and Mead—to address water supply variability, working within the rules of the National Environmental Policy Act and the requirement for formal rulemaking processes. In 2012, the Bureau of Reclamation (a federal agency) and seven basin states completed a basin study under the 2009 Secure Water Act to assess water supply–demand imbalances under future climate change scenarios to support planning for water variability (Reclamation 2012). The 2007 Record of Decision for the shortage sharing rules highlight the rationale and principles guiding these institutional adaptations:

> During the public process, a unique and remarkable consensus emerged
> in the basin among stakeholders including the Governor's representatives

of the seven Colorado River Basin States (Basin States). This consensus had a number of common themes: encourage conservation, plan for shortages, implement closer coordination of operations of Lake Powell and Lake Mead, preserve flexibility to deal with further challenges such as climate change and deepening drought, implement operational rules for a long—but not permanent—period in order to gain valuable operating experience, and continue to have the federal government facilitate—but not dictate—informed decision-making in the Basin. (US Department of Interior 2007, Colorado River Interim Guidelines for Lower Basin Shortages and Coordinated Operations for Lake Powell and Lake Mead)

The Upper Rio Grande in the United States has had less progress in interstate and multilevel coordination of drought adaptation (Garrick et al. 2016). The 1938 Compact governing water allocation between Colorado, New Mexico, and Texas arguably provides a more flexible and adaptive framework for drought adaptation because it shares water based on proportions of available supplies (rather than fixed volumes in the Lower Colorado River Basin) and allows the accrual of short-term debts and credits to buffer supply variability. However, groundwater is not addressed by the Compact, and this omission has created legal uncertainty and hydrological impacts that impede coordinated efforts. Drought has exacerbated tensions related to groundwater pumping and endangered fish species, requiring institutional mechanisms to resolve conflicts between states and between states and the federal government. Groundwater pumping in the Elephant Butte region led to changes in the Bureau of Reclamation's operating rules. A negotiated agreement between El Paso and irrigation districts in New Mexico and Texas attempted to resolve complaints from Texas about groundwater pumping in New Mexico but was invalidated by the New Mexico attorney general, prompting the ongoing Supreme Court dispute. Although a range of operational and informal mechanisms is available to facilitate coordination among stakeholders, the progress toward an institutional framework for interstate drought adaptation experienced in the Colorado River Basin has proved elusive.

17.4.4 Lessons Learned

The Colorado and Rio Grande basins illustrate the range of coordination challenges and responses associated with drought adaptation in the western United States. First, state control over water allocation and federal roles in reservoir construction and management create coordination challenges related to sharing water among states during droughts and operating reservoirs. Second, interim rules and integrated data, modeling, and planning systems facilitate cooperation and learning among states, including the identification of positive sum or win–win options that boost system reliability. Third, the exclusion of groundwater, or of key stakeholders, can create a vicious cycle of legal uncertainty and conflict that weakens capacity to adapt and fuels costly conflicts. Finally, the national government plays an important facilitation role through funding resources, infrastructure, and monitoring networks to provide information and incentives for joint management. In both the Colorado and Rio Grande, the threat of federal action has been a powerful stimulus for cooperation or conflict resolution during droughts.

17.5 Conclusion

17.5.1 Tensions

All three countries experienced tensions between states and across levels of governance during droughts, illustrating the different coordination challenges caused by drought. State governments promote the interests of their constituents, defending water rights within their territory at the possible expense of regional and basin-wide interests. Water users petition their state or regional governments to defend their interests in interstate forums. For example, regional governments of Castilla-La Mancha and Murcia, the regions located at the two extremes of the Tagus-Segura aqueduct, voice concerns and interests from their water users when transferrable water volumes are being negotiated during droughts. Similar dynamics unfold in Australia with the states of New South Wales and South Australia defending their upstream and downstream interests during droughts, respectively. New Mexico shows the potential for a state to become internally divided because of interstate commitments. The 1938 Rio Grande Compact delivers water from New Mexico to Texas at Elephant Butte Reservoir in New Mexico, upstream of the Texas border; groundwater use by New Mexico farmers in this region has diminished the surface water deliveries from New Mexico to Texas. As a result, the New Mexico farmers between Elephant Butte Reservoir and the Texas border find themselves at odds with their own state government, which is bound by its legal obligations to pass water to Texas. This illustrates how coordination challenges also have a vertical dimension—that is, the coordination between state governments and national governments during droughts.

17.5.2 Institutional Mechanisms and Adaptation Options

While all three countries have basin-level institutions, states hold different levels of authority and responsibilities over water allocation, which shapes the intensity of interstate disputes and conflicts between the states and national governments over decisions related to drought. This affects the ability to make regional and national level decisions related to drought management and water allocation. In Spain, water is allocated by a river basin management plan approved at the national level, not per regional boundaries. On the one hand, national control over allocation has led to coordinated drought planning and management. On the other hand, it limits the ability of regions to dispute water allocation with other regions. In the United States, states' rights over water allocation have constrained basin-wide governance and limited capacity for coordinated drought planning and management; recent initiatives have sought to address these limitations through a national drought resilience action plan and interagency working groups aimed at coordinating roles and capacities. Australia's approach to water allocation and drought management takes a middle path, recognizing states' rights over water allocation within the framework established by basin-wide planning. This has led to tensions in two directions: between states and between states and the Commonwealth government.

17.5 3 Lessons for the Future

The comparison of the three countries provides insights about the role of the central government and the resolution of conflicts between states. All three countries used financial resources from the central government to mitigate the impacts of droughts and facilitate cooperation between users and states. For example, the 2009 Secure Water Act in the United States has provided an infusion of funding to share costs with states to conduct water supply and demand studies. The central government also plays a key role in building and coordinating capacity in an integrated way. All three countries have used national programs in developing or coordinating information and monitoring systems to equip lower levels of government with the data needed to forecast, monitor, and manage droughts. This can come in many forms, depending on the context, including data collection and standards or guidelines for the development and application of drought indicators tailored to local and regional conditions. Finally, droughts present stress tests and expose ambiguities in the institutional framework governing water planning and allocation; adaptation requires multiple formal and informal venues for states and national governments to sort out conflicts and facilitate cooperative efforts that will only become more important in the face of climate change and hot droughts.

References

Connell, D. 2007. *Water Politics in the Murray–Darling Basin.* Annandale, NSW: Federation Press.

Connell, D. 2011. Water reform and the federal system in the Murray–Darling Basin. *Water Resources Management* 25(15):3993–4003.

Cook, B. I., J. E. Smerdon, R. Seager, and E. R. Cook. 2013. Pan-continental droughts in North America over the last millennium. *Journal of Climate* 27(1):383–397.

Diodato, D. M., D. A. Wilhite, and D. I. Nelson. 2007. Managing drought in the United States: A roadmap for science and public policy. *Eos, Transactions American Geophysical Union* 88(9):109.

Estrela, T., and T. A. Sancho. 2016. Drought management policies in Spain and the European Union: From traditional emergency actions to Drought Management Plans. *Water Policy* 18(S2):153–176.

Estrela, T., and E. Vargas. 2012. Drought management plans in the European Union. *The case of Spain. Water resources management* 26(6): 1537–1553.

Folger, P., and B. A. Cody. 2015. *Drought in the United States: Causes and Current Understanding*. Congressional Research Service Report 7-5700. Washington, DC: CRS.

Garrick, D., E. Schlager, and S. Villamayor-Tomas. 2016. Governing an international transboundary river: Opportunism, safeguards and drought adaptation in the Rio Grande. *Publius* 46(2):170–198.

Helman, P. 2009. *Droughts in the Murray–Darling Basin since European Settlement*. Griffith Centre for Coastal Management Research Report No. 100. Published by Murray–Darling Basin Authority, Canberra.

Hernández-Mora, N., and L. Del Moral. 2015. Developing markets for water reallocation: Revisiting the experience of Spanish water mercantilización. *Geoforum* 62:143–155.

López-Gunn, E., and L. De Stefano. 2014. Between a rock and a hard place: Redefining water security under decentralization in Spain. In *Federal Rivers: Managing Water in Multi-Layered Political Systems*, eds. D. Garrick, G. R. M. Anderson, D. Connell, and J. Pittock, p. 158. Cheltenham, Gloucestershire: Edward Elgar Publishing.

Moral Ituarte, L. D., and N. Hernández-Mora Zapata. 2016. Nuevos debates sobre escalas en política de aguas. Estado, cuencas hidrográficas y comunidades autónomas en España. *Ciudad y Territorio Estudios Territoriales* 48(190):563–583.

Overpeck, J. T. 2013. Climate science: The challenge of hot drought. *Nature* 503:350–351.

Reclamation (U.S. Bureau of Reclamation). 2012. *Colorado River Basin Water Supply and Demand Study*. Washington, DC: US Department of Interior.

Udall, B., and J. Overpeck. 2017. The 21st century Colorado River hot drought and implications for the future. *Water Resources Research* 53(3):2404–2418. doi:10.1002/2016WR019638.

Urquijo, J., L. De Stefano, and A. La Calle. 2015. Drought and exceptional laws in SpaIn: The official water discourse. *International Environmental Agreements: Politics, Law and Economics* 15(3):273–292.

U.S. Department of Interior. 2007. *Colorado River Interim Guidelines for Lower Basin Shortages and Coordinated Operations for Lake Powell and Lake Mead*. US Department of Interior, Washington, DC. https://www.usbr.gov/lc/region/programs/strategies/RecordofDecision.pdf. Accessed July 3, 2017.

Woodhouse, C. A., S. T. Gray, and D. M. Meko. 2006. Updated streamflow reconstructions for the Upper Colorado River basin. *Water Resources Research* 42(5):W05415. doi:10.1029/2005WR004455.

Woodhouse, C. A., and D. W. Stahle. 2012. Rio Grande and Rio Conchos water supply variability over the past 500 years. *Climate Research* 51(2):147.

18

Drought Risk Management: Needs and Experiences in Europe

Jürgen V. Vogt, Paulo Barbosa, Carmelo Cammalleri,
Hugo Carrão, and Christophe Lavaysse

CONTENTS

18.1 Introduction

The European Drought Observatory (EDO) was developed as a response to the need to better understand, monitor, and forecast the interlinked phenomena of water scarcity and drought (WS&D) in Europe and to provide input for the development of evidence-based policies in the field. A first attempt to address the WS&D problem in the European Union (EU) was included in the European Water Framework Directive (WFD 2000), which requires drought management plans to be developed in all river basin districts prone to prolonged droughts. However, this requires a clear definition of a "prolonged drought" and adequate monitoring and assessment systems. In 2007, the European Commission (EC) published a specific Communication to the European Parliament and the Council, *Addressing the challenge of water scarcity and droughts in the European Union* (European Commission 2007). This communication explicitly asks for the development of EDO and acknowledges its use for the enhancement of the knowledge of the issue. It further underlines the fact that efficient alert systems are an essential dimension of risk management and that an early warning system will, therefore, follow suit to improve the drought preparedness of the relevant authorities. It details the need for a system that "will integrate relevant data and research results, drought monitoring, detection and forecasting on different spatial scales, from local and regional activities to continental overview at EU level, and will make it possible to evaluate future events" (European Commission 2007, p. 9).

This communication and the general lack of harmonized drought information at the European level led the European Commission Joint Research Centre (JRC) to start the development of such a system in close collaboration with the EU member states, the European Environment Agency, Eurostat, and representatives from the electricity and water industries. EDO targets efficient methods to monitor and forecast meteorological, agricultural, and hydrological droughts at European scale and at the same time to benchmark the developed methods with national to subnational information systems. It is a distributed system, where data and indicators are handled at each spatial scale by the responsible authorities (stakeholders) and visualized through Web mapping services. This requires calculation of a suite of core indicators according to defined standards at all scales. With increasing detail, additional locally important indicators can be added by the responsible authorities. While JRC handles data and computes indicators at the continental level (so-called awareness-raising indicators), national, regional, and river basin authorities add more detailed information for their area of interest. As detail increases, indicators become more relevant for day-to-day water management. EDO can be accessed through JRC's web portal at http://edo.jrc.ec.europa.eu/ and at the same time it serves as the European node in the first prototype of a distributed global drought information system (GDIS) hosted by NOAA at https://www.drought.gov/gdm/ and developed as part of the Group on Earth Observation (GEO) work plan.

EDO provides a suite of drought indicators at different spatial and temporal scales, including 10-daily and monthly updated maps on the occurrence and evolution of drought events, as well as a 7-day forecast of soil moisture. Medium- to long-term forecasting is under development using probabilistic ensemble methods. Currently, EDO includes meteorological indicators (e.g., Standardized Precipitation Index [SPI] and temperature), soil moisture (output of a distributed hydrological model), vegetation condition (based on satellite-derived measurements of the photosynthetic activity of the vegetation), and river low flows. At the more detailed levels it includes, for example, indicators on groundwater levels and trends (France) and warning levels for water management in irrigated and non-irrigated areas (Ebro river basin).

The variety of indicators proved useful for the expert user and, in case of severe drought events, for the production of drought reports by the JRC drought team. The information content, including the drought reports, is well received by the stakeholder community, as shown by the number of web accesses and downloads. For the policymakers and high-level managers, however, this level of detail proved to be too complicated. They require synthetic high-level combined indicators, showing different alert levels, to be used for awareness raising as well as for policy and decision-making. Such combined indicators need to be developed by sector (e.g., for agriculture, public water supply, energy production, and waterborne transport).

The development of the first Combined Drought Indicator (CDI) for agriculture, therefore, was a major breakthrough in providing information on the drought propagation within the hydrological cycle, that is from the rainfall deficit to a deficit in soil moisture and the resulting impacts on the vegetation cover. The CDI provides easy-to-understand sector-specific information for decision makers in the form of alert levels. Like the North American Drought Monitor (NADM; see Chapters 7 and 19), EDO provides reports of exceptional drought events, albeit not in a regular manner. More recently, EDO has been extended to the global level in order to provide information to the Emergency Response Coordination Centre (ERCC) of the EC, which supports and coordinates a wide range of prevention and preparedness activities in the area of natural and man-made disasters. This extended system, called the Global Drought Observatory (GDO; http://edo.jrc.ec.europa.eu/gdo) adds the component of risk and impact assessment. A first drought risk assessment for food security has been implemented and a Likelihood of Drought Impact (LDI) indicator has been developed that serves as a high-level alert indicator combining the hazard with exposure and vulnerability to evaluate the evolving drought risk for that sector.

In the following sections, we provide more detail on various aspects of the EDO and GDO systems. In Section 18.2, we discuss the core indicators and the approach to forecasting drought events. Section 18.3 then details the current approach implemented for drought risk assessment at the global level, and Section 18.4 provides some information on the setup of the GDO. Section 18.5 provides conclusions and an outlook.

18.2 Drought Monitoring and Forecasting

The large array of sectors impacted by drought, as well as the spatio-temporal variability in its traits, suggests the need for a variety of indicators to cover the most common drought types: meteorological, agricultural, and hydrological (see Chapter 1). Following this classification of drought, three sets of indices are used in both EDO and GDO to capture the nature of the drought phenomenon; these indices will be discussed separately in the next sections.

18.2.1 Meteorological Drought Monitoring

Shortage in precipitation drives most drought events. This is why the SPI (McKee et al. 1993) is one of the key indicators for meteorological drought monitoring, as highlighted by the World Meteorological Organization (WMO 2006). The computation of SPI is based on an equiprobability transformation of the probability of observed precipitation into standardized z-score values. Within our systems, SPI is computed on different accumulation periods of n-month (SPI-n, with n = 1, 3, 6, 12, 24, 48 months) by using a reference period of 30 years (1981–2010). In the first step, the accumulated precipitation data are fitted using the gamma probability density function; successively, the fitted cumulative distribution function (*cdf*) is converted into standardized normal variable values through the standardized normal *cdf* with null mean and unity variance.

For EDO, SPIs are computed from daily rainfall records at SYNOP (surface synoptic observations) stations from the JRC-MARS (http://mars.jrc.ec.europa.eu/) database, and then interpolated to a 0.25 degree grid by blending those maps with 1.0 degree resolution SPI maps derived from the Global Precipitation Climatology Centre (GPCC) monthly precipitation dataset (http://gpcp.dwd.de). GPCC data are specifically useful for regions with an insufficient number of stations with long-term precipitation records. For the global system, precipitation data currently used only are from the GPCC dataset only, resulting in a resolution of 1.0 degrees.

18.2.2 Agricultural Drought Monitoring

Focusing on the effect of drought on agricultural or naturally vegetated lands, a monitoring system can be based on either hydrological quantities affected by water shortage and related to growth and yield (i.e., soil moisture, actual evapotranspiration) or on indicators of the biomass amount or ecosystem productivity (i.e., vegetation indices, leaf area index, fraction of Absorbed Photosynthetically Active Radiation [fAPAR]) (Mishra and Singh 2010).

Soil moisture (θ) is seen as one of the most suitable variables to monitor and quantify the impact of water shortage on plants. More specifically, drought events are commonly detected by means of soil moisture anomalies (deviation from the climatology) computed as z-score values (e.g., Anderson et al. 2012) for a given aggregation period (e.g., dekad or month):

$$Z_{i,k} = \frac{\theta_{i,k} - \mu_i}{\sigma_i} \tag{18.1}$$

where $\theta_{i,k}$ is the soil moisture for the *i*-th aggregation period at the *k*-th year, and μ_i and σ_i are the long-term average and standard deviation for the *i*-th aggregation period, respectively. The use of z-scores is suitable to detect soil moisture conditions that are drier than usual according to a past climatology, which can be considered a good indicator of the occurrence of agricultural drought.

Often, hydrological model outputs are used to spatially reconstruct the temporal dynamic of soil moisture over a certain region. In EDO, the root zone soil moisture outputs (in terms of soil water suction, pF) of the LISFLOOD model (de Roo et al. 2000) are used to obtain dekadal (three approximately 10-day periods per month) anomalies over Europe on a 5-km grid. Near-real time runs of LISFLOOD from the European Flood Awareness System (EFAS, Thielen et al. 2009) are used for that purpose. At global scale, we are testing to combine different sources of soil moisture, including outputs from LISFLOOD and thermal and passive/active microwave remote sensing data for a merged global product at 0.1 degree resolution (Cammalleri and Vogt 2017a).

Applications over large areas have highlighted how under some circumstances the simple anomalies can be insufficient to detect negative effects on the plant cover, mainly in areas characterized by high water content values (i.e., low or null water deficit). For this reason, Cammalleri et al. (2016a) have developed a soil moisture-based Drought Severity Index (DSI) that accounts for both the rarity of a soil moisture status (derived from the z-score) and the actual magnitude of the vegetation water deficit. This drought index is computed as a geometric mean of two indicators:

$$DSI = \sqrt{p \cdot d} \qquad (18.2)$$

where the term d represents a soil moisture-derived water deficit index and p represents a dryness probability index, the latter being related to the probability that d is drier than the mode of the reference climatology. The use of the geometric mean allows having a high value of DSI only when both d and p are high, and hence the soil moisture status is both rare and stressing for plants.

The plots in Figure 18.1 exemplify how d is directly derived from θ by means of the s-shaped water stress curve proposed by van Genuchten (1987) and p is computed after fitting a beta probability distribution function (*pdf*) to the climatological data (Gupta and Nadarajah 2004). It is worth noting that d is > 0 only when θ is greater than the critical value for which water stress starts to occur (Seneviratne et al. 2010) and p is > 0 only if d is significantly higher than the mode of the *pdf*.

An alternative approach for agricultural drought monitoring is to directly observe the variation in vegetation growth or greenness to detect areas affected by drought events. In this context, remote sensing-derived vegetation indices are very useful tools for such analysis over large areas (e.g., Ghulam et al. 2007; Peters et al. 2002). Among the quantities that can be derived from space, the absorbed photosynthetically active radiation (fAPAR) has been widely identified as a suitable proxy of the greenness and health status of

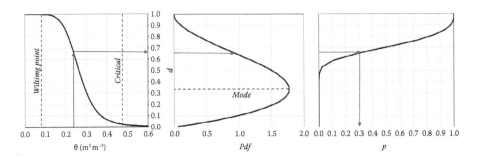

FIGURE 18.1
Schematic representation of the procedure to compute d and p factors. On the left plot θ is converted in d through a water stress curve, in the central panel d is compared against the climatology to evaluate if it is drier than the mode, and on the right panel d is converted into p by accounting for the *pdf* of the climatology. More details can be found in Cammalleri et al. (2016a).

vegetation, thanks to its central role in both plant primary productivity and carbon dioxide absorption (Gobron et al. 2005a). The observed sensitivity of fAPAR to vegetation stress has suggested its use in drought monitoring (Gobron et al. 2005b), particularly using anomalies. Both EDO and GDO systems use long records (starting in 2001) of fAPAR maps derived from the Terra satellite MODerate-resolution Imaging Spectroradiometer (MODIS) 8-day standard product (MOD15A2); these maps are quality checked to ensure the use of only high-quality data, filtered and averaged to dekadal time scale, and used as input of Equation 18.1 to derive z-score values. Further studies tried to combine the analysis of the vegetation phenological cycle and fAPAR anomalies to improve the accuracy of drought detection (Cammalleri et al. 2016b).

fAPAR anomalies can also be related to a variety of other stress factors (e.g., heat and pests); hence, further information on water stress needs to be used to associate recorded anomalies with drought. Following these considerations, Sepulcre-Cantó et al. (2012) developed the CDI to account for the cascade process from a shortage in precipitation to yield reduction through a soil moisture deficit. The authors investigated the relationship between three types of indices: (1) the n-month accumulation standardized precipitation index (SPI-n), (2) the soil moisture anomalies in terms of soil suction (pF), and (3) the fAPAR anomalies.

Figure 18.2 highlights the conceptual framework that constitutes the CDI; a watch status is issued when a significant precipitation deficit is observed (e.g., SPI-3 or SPI-1 < −1), which is then converted into a warning and then into an alert when a significant soil moisture deficit and negative fAPAR anomalies are observed as well. Two recovery classes are also added to track down the return to normal conditions of rainfall and vegetation status, respectively.

An example of the CDI output from EDO is shown in Figure 18.3.

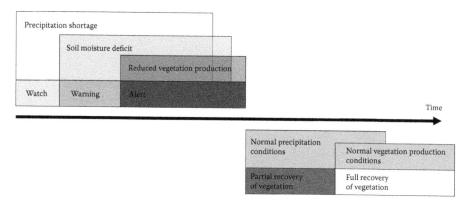

FIGURE 18.2
Schematic representation of the stages of the idealized agricultural drought cause–effect relation-ship that inform the concept of the CDI and the associated warning levels that are outputs of the CDI. (Adapted from Sepulcre-Cantò et al., *Natural Hazards and Earth System Sciences* 12:3519–3531, 2012.)

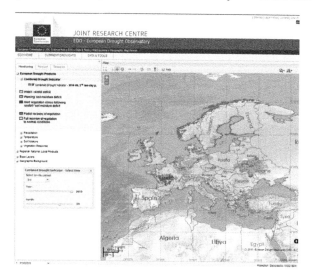

FIGURE 18.3
The European Drought Observatory (EDO). Example of the combined drought indicator (CDI) for September 21–30, 2016.

18.2.3 Hydrological Drought Monitoring

Monitoring hydrological droughts in EDO is based on the capability to cap-ture the dynamic nature of this specific class of drought events. A low-flow index focusing on streamflow data was developed by Cammalleri et al. (2017b) to evaluate when the total deficit of water discharge is below a certain threshold computed on a climatological time series of data.

As schematically represented in Figure 18.4, a daily low-flow threshold is computed as a percentile (i.e., 95th) of the historical dataset.

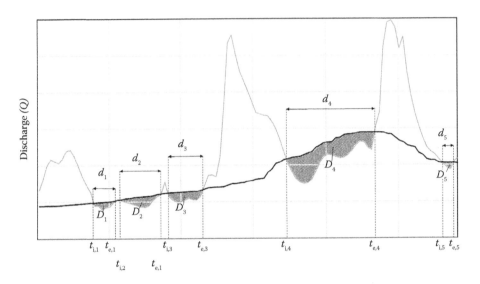

FIGURE 18.4
Schematic representation of a sequence of low-flow events. Actual discharge values are shown as a gray line and the low-flow threshold as a black line. (From Cammalleri, C., et al., *Hydrolog. Sci. J.*, 62(3):346–358, 2017b.)

The periods with discharge continuously below the threshold are considered low-flow events with a specific deficit (D) value that has a frequency characterized by means of an exponential distribution:

$$F(D_i; \lambda) = 1 - e^{-\lambda D_i} \text{ with } D_i \geq 0 \qquad (18.3)$$

The value of $F(D)$ is used as a measure of the severity of the hydrological drought for that river section. In order to minimize the impact of small events, as well as of close dependent events, both pooling and small-event removal procedures are applied (Cammalleri et al. 2017b).

In the operational EDO system, the discharge data for the low-flow index computation are derived from the LISFLOOD runs, which are also used for the computation of the soil moisture-based indices. These data are mapped on a 5-km regular grid, even if the low-flow index is computed only on the river cells with a drainage area of at least 1,000 km². Finally, the resulting grid is mapped on the catchment characterization and modeling (CCM) river network for Europe, a detailed vector-based and fully connected hydrological network that allows for a structured analysis of up- and down-stream relationships (Vogt 2007a, 2007b).

18.2.4 Forecasting

Currently, different long-term forecast products are proposed and tested. They are based on ensemble forecasts of standardized precipitation indices (e.g., SPI-3 to SPI-12) with lead times reaching from 1 to 7 months. Note that for lead times that are shorter than the accumulation periods, the SPIs are calculated using a mix of observed precipitation during the period before the starting date and the forecasts thereafter (Dutra et al. 2014). This method allows seamless forecasts of SPI with 1 to 12 months lead time.

Two information types are derived from the ensemble system. The first one is associated with the probabilistic forecast. It is defined as the number of members that are associated with an SPI lower than –1 (drought) or larger than 1 (flood), or normal conditions (between –1 and 1). If there is a consistency between the members (e.g., more than 50 percent) to forecast a specific anomaly or normal conditions, then the information is reported. If none of these three specific conditions emerged predominantly (i.e., large uncertainties of the members), the SPI forecasted is considered not significant and no conclusion can be provided. This method does not quantify the intensity of a drought but the consistency and the spread of the ensemble can be related to the uncertainties of the forecasts.

The second information type is the ensemble mean of the SPI. In case of consistency within the ensemble members, the mean values, provided for the same accumulation periods and lead times as previously discussed, are reported and provide information on the strength of a forecasted event. Because of the use of the ensemble mean, this method tends to underestimate the intensities of observed events but allows distinguishing abnormal from extreme abnormal (dry or wet) conditions.

The predictability of the forecasts at such long lead times is obviously a big challenge, and there are only few papers in the literature that deal with this kind of assessment. Dutra et al. (2013) have assessed the prediction capabilities of the ECMWF Seasonal S4 model for different basins in Africa. Even if their model presents better scores than the climatology, we expect a positive effect of the underlying data construction, with observations added for the shorter lead times that will influence the results. In Lavaysse et al. (2015), the scores of the SPI forecasted have been quantified using different ensemble systems over Europe. At a monthly time scale (e.g., for SPI-1 with 1-month lead time), it has been shown that 40 percent of all droughts (defined as an SPI lower than –1) are correctly forecasted 1 month in advance. This score may not seem large, but according to the challenge and the difficulty to forecast precipitation at those lead times (one month of cumulated precipitation) and given the strict definition of an event, this score is clearly above the climatology (16 percent) and provides a first state-of-the-art predictability of these events. To increase the predictability scores of droughts, some studies suggest working with variables that have more persistency, such as the soil moisture (Sheffield et al. 2014). But to be comparable, the score related to the climatology also should be provided.

Ongoing works at JRC are testing the use of atmospherical predictors to forecast extreme precipitation anomalies. For instance, in Lavaysse et al. (2016), it has been shown that the occurrence of weather regimes (WRs) that classify atmospherical circulation patterns in predetermined anomaly patterns could improve the forecast scores in Europe. Over certain regions, such as Scandinavia, the precipitation anomaly patterns are strongly connected to blocking situations. These situations are generally better forecasted in the atmospheric models than the precipitation anomalies. It has been shown that using a simple best-correlation attribution of WR occurrences and precipitation anomalies, the forecast of WRs could improve the probability of the detection of droughts (i.e., SPI lower than –1) by more than 20 percent. Indeed, over Scandinavia in winter, 65 percent of the events are correctly forecasted 1 month in advance (comparing to about 40 percent using the forecasted precipitation). Obviously, this region and this season are well known to be strongly connected to the synoptic circulation; elsewhere or during other seasons, the forecasted precipitation could be better than this alternative method. To provide the best forecast product possible, an assessment of each forecast method has been made over Europe, and these results allow identification of the most accurate forecast for each region and each season. According to these past evaluations, the best forecast products can be chosen per region or grid cell in order to provide more robust information.

18.3 Drought Risk Assessment at the Global Level

Drought risk assessment is an essential component of any comprehensive drought management plan. Assessing risk is crucial to identify relief, coping, and management responses that will reduce drought damage to society. In this context, the JRC has developed the Likelihood of Drought Impact (LDI) to support the drought risk management activities of the European Union (EU) Emergency Response Coordination Centre at the Directorate General for Humanitarian Aid and Civil Protection (ECHO). The LDI is a categorized metric that expresses the probability of harmful consequences or potential losses resulting from interactions between drought hazard (i.e., the possible occurrence of drought events), drought exposure (i.e., the total population and its livelihoods and assets in an area in which drought events may occur), and drought vulnerability (i.e., the propensity of exposed elements to suffer adverse effects when impacted by a drought event). The determinants of the LDI can be schematized in the following mathematical form:

$$LDI = f \text{ (hazard, exposure, vulnerability)} \qquad (18.4)$$

The scores of the LDI are currently expressed on a scale of three categories of impact that need to be interpreted by the user: low—potential for drought establishment; medium—potential for drought affecting sector activities; high—potential for development of a drought emergency.

Given the conceptual relationship presented in Equation 18.4, no hazard or no exposure will result in an LDI that is null (as proposed by Hayes et al. 2004). It is important to note that the proposed categorical scale of likelihood is not a measure of absolute losses or the magnitude of actual damage to human population and its assets or the environment, but it is suitable for informing decisions on preparedness actions and on responses to potential impacts.

18.3.1 Exposure

To assess the impacts of drought hazard, the first step is to inventory and analyze the environment that can be damaged (Di Mauro 2014). In general, exposure data identify the different types of physical entities that are on the ground, including built-up assets, infrastructures, agricultural land, and people, to cite but a few (Peduzzi et al. 2009). Drought exposure is very different from that of other hazard types. First, unlike earthquakes, floods, or tsunamis, which occur along generally well-defined fault lines, river valleys, or coastlines, drought can impact extended areas and can occur in most parts of the world (even in wet and humid regions), with the exception of desert regions, where it does not have meaning (Dai 2011; Goddard et al. 2003). Second, drought develops slowly, resulting from a prolonged period (from weeks to years) of precipitation below average or expected value at a particular location (Dracup et al. 1980; Wilhite and Glantz 1985). Therefore, droughts have an impact on different water use sectors as a function of the timing, duration, and amount of a precipitation deficit. For example, the immediate impacts of short-term (i.e., a few weeks) droughts might be a fall in crop production, poor pasture growth, or a decline in fodder supplies from crop residues for livestock farming. Prolonged water shortages (i.e., several months or years) may lead to effects such as lower earnings from agriculture, reduced energy production (e.g., reduced hydropower production, reduced cooling capacities for nuclear plants), problems in public water supply (both quantity and quality), reduced inland water transport, problems for tourism, job losses, food insecurity, and human casualties (Downing and Bakker 2000; Mishra and Singh 2010).

To address the diversity of drought impacts, we compute exposure by means of a nonparametric and noncompensatory Data Envelopment Analysis (DEA) (Cook et al. 2014; Lovell and Pastor 1999), as recently proposed by Carrão et al. (2016). This approach to drought exposure is multivariate and takes into account the spatial distribution of human population and numerous physical assets (proxy indicators) characterizing agriculture and primary sector activities, namely: crop areas (agricultural drought), livestock (agricultural drought), industrial/domestic water use (hydrological drought), and human population (socioeconomic drought). In the DEA methodology, the exposure of each region to drought is relative and is determined by a normalized multivariate statistical distance to the most exposed region.

Currently, drought exposure is computed on the basis of four spatially explicit geographic layers that completely cover the global land surface, namely: global agricultural lands in the year 2000 (Ramankutty et al. 2008); gridded population of the world, version 4 (GPWv4) (Balk et al. 2006; Deichmann et al. 2001; Tobler et al. 1997); gridded livestock of the world (GLW) v2.0 (Robinson et al. 2014); and baseline water stress (BWS) (Gassert et al. 2014a, 2014b).

18.3.2 Vulnerability

Since the location, severity, and frequency of water deficits (the hazard) cannot be deterministically assessed and exposure is dynamic because of economic and population changes, interventions to reduce drought impacts may have to focus on reducing vulnerability of human and natural systems. For estimating the LDI, we take vulnerability to drought into account and adopt the framework proposed by UNISDR (2004): a reflection of the state of the individual and collective social, economic, and infrastructural factors of a region at hand. Social vulnerability is linked to the level of well-being of individuals, communities, and society; economic vulnerability is highly dependent upon the economic status of individuals, communities, and nations; and infrastructural vulnerability comprises the basic infrastructures needed to support the production of goods and sustainability of livelihoods (Scoones 1998).

Vulnerability to drought is computed as a two-step composite model, as recently proposed by Carrão et al. (2016). It derives from the aggregation of proxy indicators representing the economic (Econ), social (Soc), and infrastructural (Infr) factors of vulnerability at each geographic location, similar to the drought vulnerability index (DVI) (Naumann et al. 2014). Each factor is characterized by a set of proxy indicators (e.g., GDP per capita, government effectiveness, and percentage of retained renewable water) selected from, for example, the World Bank and the Food and Agriculture Organization, in accordance with the work of Naumann et al. (2014) and substantiated by the vulnerability studies of Scoones (1998), Brooks et al. (2005), and Alkire and Santos (2014). In the first step, indicators for each factor are combined using a DEA model, similar to drought exposure. In the second step, individual factors resulting from independent DEA models are arithmetically aggregated into a composite model of drought vulnerability (dv), as follows:

$$dv = (Soc + Econ + Infr) / 3 \qquad (18.5)$$

18.3.3 Likelihood of Drought Impact (LDI)

Currently, the LDI is computed and updated every 8 days. For its computation, we take into account the CDI, as a dynamic layer of drought hazard (as presented in Section 18.2.2), and the structural layers of exposure and vulnerability (as presented in Sections 18.3.1 and 18.3.2). The original three CDI classes have been extended to five classes ranging from very low to very high.

Exposure and vulnerability to drought are updated on a yearly basis, as they are derived from structural information, such as population distribution and socioeconomic indicators that do not change significantly for shorter periods of time. Once estimated for a specific year, these numerical and continuous determinants of drought risk are converted into nine categories of intensity (Figure 18.5), namely: exceptionally low, extremely low, very low, low, medium, high, very high, extremely high, and exceptionally high. Categories are computed independently for each determinant according to percentiles of their empirical probability distribution for the whole world.

To compute the LDI, we first merge the exposure and vulnerability layers into a structural layer of spatial propensity to damage (Figure 18.6a) with five intensity classes (very low, low, medium, high, and very high), defined according to the matrix shown in Table 18.1.

The global map of propensity to damage is then combined with the CDI to derive the final LDI categories (low, medium high), according to the matrix presented in Table 18.2. Since drought exposure and vulnerability are updated only once per year, the 8-day LDI is driven by short-term changes in the values of the CDI. In Figure 18.6b, we show an LDI map of the world for the 8-day time interval covering the period October 8–15, 2015.

18.4 Supporting the Global Activities of the European Emergency Response Coordination Centre (ERCC)

In order to satisfy the needs of the ERCC, an operational unit working on a 24/7 basis for the coordination of EU responses to natural and technological disasters in the world, a drought monitoring system capable of providing both high-level alert information and detailed indicator information was developed. The creation of semiautomatic analysis reports to help duty officers quickly extract relevant information and provide electronic and printed documents during meetings of experts and decision makers was included as a specific feature. If more detailed information is required, the JRC drought team, together with the ERCC analytical team, produces targeted analytical reports for the case at hand.

As a first approach to a high-level alert indicator, the LDI targeted to food insecurity has been implemented (see Section 18.3). Besides the map, the system provides the duty officer with a hierarchical list of affected countries and quick links to further country information (e.g., population, GDP, area affected, and people affected by each LDI class). By simply clicking in an administrative unit (mostly units at subcountry level), the duty officer enables the system to generate an on-the-fly report providing the most important information for the selected unit. It includes, for example, statistics on the areal extent of different alert levels, the number of people affected, and the land use and land cover types in the different LDI classes (i.e., low, medium,

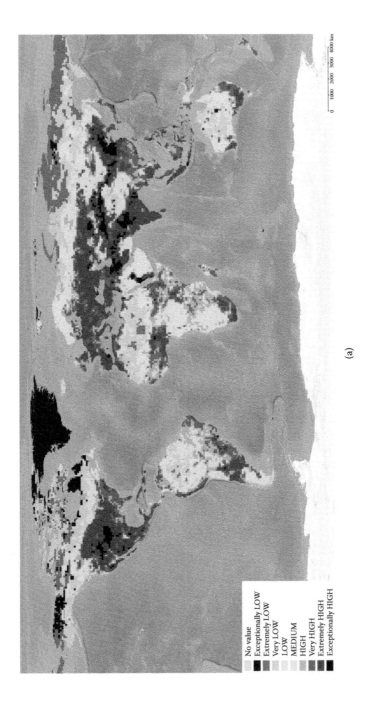

(a)

FIGURE 18.5
Global maps of drought exposure (a) and vulnerability (b).

(Continued)

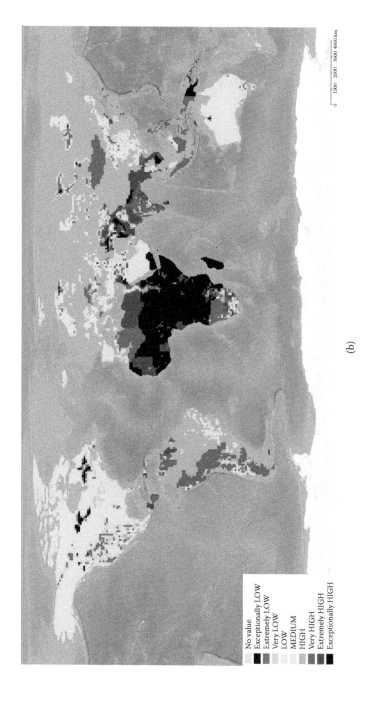

FIGURE 18.5 (Continued)

Global maps of drought exposure (a) and vulnerability (b).

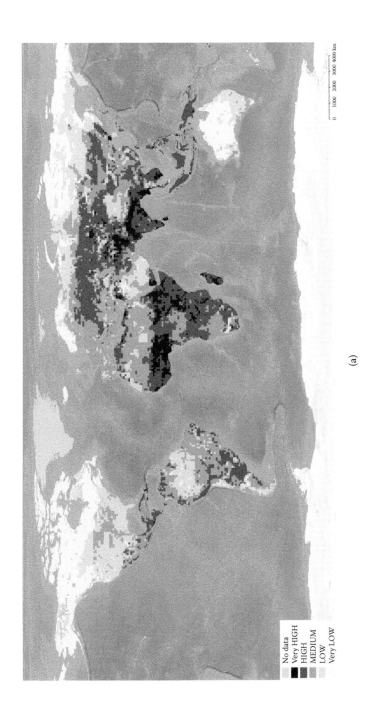

(a)

FIGURE 18.6
Global map of propensity to damage resulting from combining the discrete classes of exposure and vulnerability (a) and of LDI for
October 8–15, 2015 (b).

(Continued)

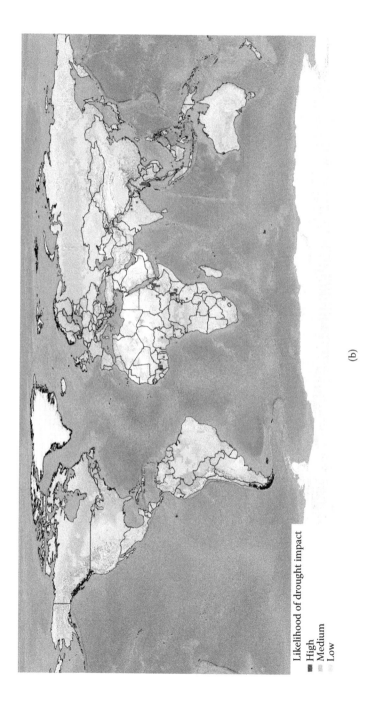

(b)

FIGURE 18.6 (Continued)
Global map of propensity to damage resulting from combining the discrete classes of exposure and vulnerability (a) and of LDI for October 8–15, 2015 (b).

TABLE 18.1

Matrix Combining the Discrete Classes of Exposure and Vulnerability into Categories of Propensity to Damage

	Vulnerability								
Exposure	Exc. LOW	Ext. LOW	Very LOW	LOW	MEDIUM	HIGH	Very HIGH	Ext. HIGH	Exc. HIGH
Exc. LOW	0	0	0	0	0	0	0	0	0
Ext. LOW	VL	VL	VL	VL	VL	VL	L	L	L
Very LOW	VL	VL	VL	VL	VL	L	L	M	M
LOW	VL	VL	VL	VL	L	L	M	M	M
MEDIUM	VL	VL	VL	L	L	M	M	H	H
HIGH	VL	VL	L	L	M	M	H	H	H
Very HIGH	VL	L	L	L	M	H	H	H	VH
Ext. HIGH	L	L	L	M	M	H	H	VH	VH
Exc. HIGH	L	L	M	M	M	H	VH	VH	VH

TABLE 18.2

Matrix Combining the Discrete Classes of the CDI and "Propensity to Damage" into LDI Categories

	Propensity to Damage					
CDI	Zero	Very LOW	LOW	MEDIUM	HIGH	Very HIGH
Zero	0	0	0	0	0	0
Very LOW	0	L	L	L	L	L
LOW	0	L	L	L	L	M
MEDIUM	L[a]	L	L	M	M	M
HIGH	L[a]	L	L	M	M	H
Very HIGH	L[a]	L	M	M	H	H

Note: L - low; M - medium; H - high likelihood of drought impact.

[a] *Although the system identifies no "propensity" for these cases, we decided to release a warning to protect for potential impacts not identified due to an underestimation of input exposure layers.*

and high likelihood of impact). Currently, the system information is updated at 8-day intervals.

A snapshot of the GDO system is shown in Figure 18.7, presenting the top-level map of the LDI, alongside the hierarchical list of all affected countries visible in the map. When the user zooms into the map, this list is automatically updated, showing only countries visible on the map. An example of part of an on-the-fly generated report is shown in Figure 18.8.

For the more experienced user, the system provides access to all underlying indicators through an expandable menu on the left-hand side of the map. It gives, for example, access to the meteorological, soil moisture, and vegetation indicators and allows customizing the geographical information shown

FIGURE 18.7
The Global Drought Observatory (GDO). Example of the Likelihood of Drought Impact (LDI) indicator for October 8–15, 2015.

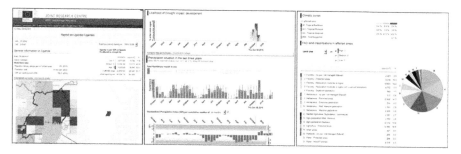

FIGURE 18.8
Example of an on-the-fly generated report for Uganda for October 8–15, 2015. The online version allows interactive selection of different graphs and statistics.

on the map. In the future, forecasting information will be added. During the El Niño event of 2015–2016, the system proved its capability to capture most of the severe impacts noted throughout the globe and helped both the ERCC staff and the JRC drought team to provide useful reports and targeted information to decision makers and field officers alike. As the system is still under development, more specific indicators (e.g., on temperatures and heatwaves) will be added and the spatial and temporal resolution will be improved as new data become available. Finally, LDIs for different economic and environmental sectors will be added.

18.5 Conclusions and Outlook

This chapter discussed examples of the calculation of a suite of scientifically based drought indicators and their use in online information systems (i.e., map servers and analysis tools), which provide accurate and up-to-date information on the occurrence and evolution of droughts in Europe and globally to decision makers at different levels. These tools proved useful for raising the awareness of the drought problem and for steering proactive as well as emergency measures in the case of emerging droughts. They are used in an operational context by decision makers throughout the EU and the European Commission. The general public has shown considerable interest as well.

The key challenge in reducing drought risk is to move from the prevailing reactive approach, fighting the highly diverse drought impacts, to a proactive society that is resilient and adapted to the risk of drought (i.e., the adoption and implementation of proactive risk management) (see Chapters 1 through 4). This requires practitioners, policymakers, and scientists to use a consistent set of drought definitions and characteristics, as well as the availability of adequate monitoring and early warning systems that provide information not only on the natural hazard but also on the risk or likelihood of impacts in different economic and public sectors (e.g., public water supply, food security, energy production, transport, and health) and the environment (ecosystems). Such targeted information can be used for the implementation of drought management plans and in the coordination of the deployment of humanitarian aid and emergency responses by civil protection mechanisms. In addition, current, but also future, societal exposure, and context-specific vulnerability should be identified to eventually assess the evolving drought risk. Knowing all these aspects, drought risk can be successfully managed.

References

Alkire, S., and M. Santos. 2014. Measuring acute poverty in the developing world: Robustness and scope of the multidimensional poverty index. *World Development* 59:251–274.

Anderson, W. B., B. F. Zaitchik, C. R. Hain, M. C. Anderson, M. T. Yilmaz, J. Mecikalski, and L. Schultz. 2012. Towards an integrated soil moisture drought monitor for East Africa. *Hydrology and Earth System Sciences* 16:2893–2913.

Balk, D., U. Deichmann, G. Yetman, F. Pozzi, S. Hay, and A. Nelson. 2006. Determining global population distribution: Methods, applications and data. *Advances in Parasitology* 62:119–156.

Brooks, N., W. N. Adger, and P. M. Kelly. 2005. The determinants of vulnerability and adaptive capacity at the national level and the implications for adaptation. *Global Environmental Change* 15:151–163.

Cammalleri, C., F. Micale, and J. Vogt. 2016a. A novel soil moisture-based drought severity index (DSI) combining water deficit magnitude and frequency. *Hydrological Processes* 30:289–301.

Cammalleri, C., F. Micale, and J. Vogt. 2016b. Analysing the fAPAR dynamic over Europe using MODIS data. In *Bionature 2016: The Seventh International Conference on Bioenvironment, Biodiversity and Renewable Energies*, ed. B. Gersbeck-Schierholz, 17–21. Red Hook, NY: Curran Associates.

Cammalleri, C., and J. V. Vogt. 2017a. Combining modelled and remote sensing soil moisture anomalies for an operational global drought monitoring. *Geophysical Research Abstracts* 19:EGU2017-7527.

Cammalleri, C., J. V. Vogt, and P. Salamon. 2017b. Development of an operational low-flow index for hydrological drought monitoring over Europe. *Hydrological Sciences Journal* 62(3):346–358. doi:10.1080/02626667.2016.1240869.

Carrão, H., G. Naumann, and P. Barbosa. 2016. Mapping global patterns of drought risk: An empirical framework based on sub-national estimates of hazard, exposure and vulnerability. *Global Environmental Change* 39:108–124.

Cook, W. D., K. Tone, and J. Zhu. 2014. Data envelopment analysis: Prior to choosing a model. *Omega* 44:1–4.

Dai, A., 2011. Drought under global warming: A review. *Wiley Interdisciplinary Reviews: Climate Change* 2:45–65.

de Roo, A., C. Wesseling, and W. van Deusen. 2000. Physically based river basin modelling within a GIS: The LISFLOOD model. *Hydrological Processes* 14:1981–1992.

Deichmann, U., D. Balk, and G. Yetman. 2001. *Transforming population data for interdisciplinary usages: From census to grid*. Technical Report. NASA Socioeconomic Data and Applications Center (SEDAC), CIESIN. Columbia University, Palisades, NY.

Di Mauro, M. 2014. Quantifying risk before disasters occur: Hazard information for probabilistic risk assessment. *WMO Bulletin* 63:36–41.

Downing, T., and Bakker, K. 2000. Drought discourse and vulnerability. In *Drought: A Global Assessment*, ed. D. A. Wilhite, 213–230. Natural Hazards and Disasters Series. London: Routledge.

Dracup, J. A., K. S. Lee, and E. G. Paulson Jr. 1980. On the definition of droughts. *Water Resources Research* 16:297–302.

Dutra, E., L. Magnusson, F. Wetterhall, H.L. Cloke, G. Balsamo, S. Boussetta, and F. Pappenberger. 2013. The 2010–2011 drought in the Horn of Africa in ECMWF reanalysis and seasonal forecast products. *International Journal of Climatology* 33(7):1720–1729.

Dutra, E., F. Wetterhall, F. Di Giuseppe, G. Naumann, P. Barbosa, J.V. Vogt, W. Pozzi, and F. Pappenberger. 2014. Global meteorological drought—Part 1: Probabilistic monitoring. *Hydrology and Earth System Sciences* 18:2657–2667.

European Commission. 2007. *Communication from the Commission to the European Parliament and Council: Addressing the challenge of water scarcity and droughts in the European Union*. COM(2007) 414 final. http://eur-lex.europa.eu/legal-content/EN/TXT/PDF/?uri=CELEX:52007DC0414&qid=1498125347173&from=EN. Accessed June 22, 2017.

Gassert, F., M. Landis, M. Luck, P. Reig, and T. Shiao. 2014a. Aqueduct global maps 2.1. Working Paper. https://www.wri.org/sites/default/files/Aqueduct_Global_Maps_2.1.pdf. Accessed June 22, 2017.

Gassert, F., M. Luck, M. Landis, P. Reig, and T. Shiao. 2014b. *Aqueduct global maps 2.1: Constructing decision-relevant global water risk indicators.* Working Paper. http://www.wri.org/sites/default/files/Aqueduct_Global_Maps_2.1-Constructing_Decicion-Relevant_Global_Water_Risk_Indicators_final_0.pdf. Accessed June 22, 2017.

Ghulam, A., Z.-L. Li, Q. Qin, and Q. Tong. 2007. Exploration of the spectral space based on vegetation index and albedo for surface drought estimation. *Journal of Applied Remote Sensing* 1:13529–13512.

Gobron, N., B. Pinty, F. Mélin, M. Taberner, M. M. Verstraete, A. Belward, T. Lavergne, and J.-L. Widlowski. 2005b. The state of vegetation in Europe following the 2003 drought. *International Journal of Remote Sensing* 26(9):2013–2020.

Gobron, N., M. M. Verstraete, B. Pinty, O. Aussedat, and M. Taberner. 2005a. Potential long time series fAPAR products for assessing and monitoring land surfaces changes. In Proceedings of the 1st International Conference on Remote Sensing and Geoinformation Processing in the Assessment and Monitoring of Land Degradation and Desertification (Trier, Germany, September 7–9), ed. A. Röder, and J. Hill, 53–59. European Commission, Brussels.

Goddard, S., S. K. Harms, S. E. Reichenbach, T. Tadesse, and W. J. Waltman. 2003. Geospatial decision support for drought risk management. *Communications of the ACM* 46:35–37.

Gupta, A. K., and S. Nadarajah. 2004. *Handbook of Beta Distribution and its Applications.* Boca Raton, FL: CRC.

Hayes, M., O. Wilhelmi, and C. Knutson. 2004. Reducing drought risk: Bridging theory and practice. *Natural Hazards Review* 5:106–113.

Lavaysse, C., J. V. Vogt, and F. Pappenberger. 2015. Early warning of drought in Europe using the monthly ensemble system from ECMWF. *Hydrology and Earth System Sciences* 19(7):3273–3286.

Lavaysse, C., J. V. Vogt, A. Toreti, J. Thielen, and G. Masato. 2016. A combined method for forecasting Extreme Monthly Precipitation in Europe. 23th Conference on Probability and Statistics in Atmospheric Sciences, AMS, New Orleans, 9–14 January, 2016. https://ams.confex.com/ams/96Annual/webprogram/Paper289076.html. Accessed June 22, 2017.

Lovell, C., and J. T. Pastor. 1999. Radial DEA models without inputs or without outputs. *European Journal of Operational Research* 118(1):46–51.

McKee, T. B., N. J. Doesken, and J. Kleist. 1993. The relationship of drought frequency and duration to time scales. In Proceedings of the 8th Conference of Applied Climatology, 17–22 January, 1993. Anaheim, California, 179–184. American Meteorological Society, Boston, MA.

Mishra, A. K., and V. P. Singh. 2010. A review of drought concepts. *Journal of Hydrology* 391:202–216.

Naumann, G., P. Barbosa, L. Garrote, A. Iglesias, and J. Vogt. 2014. Exploring drought vulnerability in Africa: An indicator based analysis to be used in early warning systems. *Hydrology and Earth System Sciences* 18:1591–1604.

Peduzzi, P., H. Dao, C. Herold, and F. Mouton. 2009. Assessing global exposure and vulnerability towards natural hazards: The disaster risk index. *Natural Hazards and Earth System Sciences* 9:1149–1159.

Peters, A. J., E. A. Walter-Shea, L. Ji, A. Vina, M. Hayes, and M. D. Svoboda. 2002. Drought monitoring with NDVI-based standardized vegetation index. *Photogrammetric Engineering & Remote Sensing* 68:71–75.

Ramankutty, N., A. T. Evan, C. Monfreda, and J. A. Foley. 2008. Farming the planet: 1. Geographic distribution of global agricultural lands in the year 2000. *Global Biogeochemical Cycles* 22:GB1003.

Robinson, T. P., G. R. W. Wint, G. Conchedda, T.P. van Boeckel, V. Ercoli, E. Palamara, G. Cinardi, L. D'Aietti, S.I. Hay, and M. Gilbert. 2014. Mapping the global distribution of livestock. *PLoS One* 9:e96084.

Scoones, I. 1998. Sustainable rural livelihoods: A framework for analysis. Working Paper 72. Institute of Development Studies (IDS), Brighton, UK.

Seneviratne, S. I., T. Corti, E. L. Davin, M. Hirschi, E.B. Jaeger, I. Lehner, B. Orlowsky, and A.J. Teuling. 2010. Investigating soil moisture–climate interactions in a changing climate: A review. *Earth-Science Reviews* 99:125–161.

Sepulcre-Cantó, G., S. Horion, A. Singleton, H. Carrão, and J. V. Vogt. 2012. Development of a combined drought indicator to detect agricultural drought in Europe. *Natural Hazards and Earth System Sciences* 12:3519–3531.

Sheffield, J., E. F. Wood, N. Chaney, K. Guan, S. Sadri, X. Yuan, L. Olang, A. Amani, A. Ali, S. Demuth, and L. Ogallo. 2014. A drought monitoring and forecasting system for sub-Sahara African water resources and food security. *Bulletin of the American Meteorological Society* 95(6):861–882.

Thielen, J., J. Bartholmes, M.-H. Ramos, and A. de Roo. 2009. The European flood alert system—Part 1: Concept and development. *Hydrology and Earth System Sciences* 13:125–140.

Tobler, W., U. Deichmann, J. Gottsegen, and K. Maloy. 1997. World population in a grid of spherical quadrilaterals. *International Journal of Population Geography* 3:203–225.

UNISDR. 2004. *Living with Risk: A Global Review of Disaster Reduction Initiatives.* Review Volume 1. Geneva, Switzerland: United Nations International Strategy for Disaster Reduction.

van Genuchten, M. T. 1987. *A Numerical Model for Water and Solute Movement in and Below the Root Zone.* Research Report No 121. U.S. Salinity Laboratory, USDA ARS, Riverside, CA.

Vogt, J. V., P. Soille, R. Colombo, M. L. Paracchini, and A. de Jager. 2007b. Development of a pan-European river and catchment database. In *Digital Terrain Modelling*, eds. R. Peckham and G. Jordan, 121–144. Berlin: Springer. https://link.springer. com/chapter/10.1007%2F978-3-540-36731-4_6. Accessed June 22, 2017.

Vogt, J. V., P. Soille, A. de Jager, E. Rimaviciute, W. Mehl, S. Foisneau, K. Bodis, J. Dusart, M.L. Paracchini, P. Haastrup, and C. Bamps. 2007a. *A Pan-European River and Catchment Database.* JRC Reference Report EUR 22920 EN, Luxembourg. http:// ccm.jrc.ec.europa.eu/documents/CCM2-Report_EUR-22920-EN_2007_STD. pdf. Accessed June 22, 2017.

WFD (Water Framework Directive). 2000. *Directive 2000/60/EC of the European Parliament and of the Council of 23 October 2000 Establishing a Framework for Community Action in the Field of Water Policy.* Official Journal of the European Communities, Luxembourg. http://eur-lex.europa.eu/resource.html?uri=cellar:5c835afb-2ec6-4577-bdf8-756d3d694eeb.0004.02/DOC_1&format=pdf. Accessed June 22, 2017.

Wilhite, D. A., and M. H. Glantz. 1985. Understanding the drought phenomenon: The role of definitions. *Water International* 10:111–120.

WMO (World Meteorological Organization). 2006. *Drought Monitoring and Early Warning: Concepts, Progress and Future Challenges.* WMO No. 1006. Geneva, Switzerland: WMO.

19

National Drought Policy in Mexico: A Paradigm Change from Reactive to Proactive Management

Mario López Pérez, Felipe I. Arreguín Cortés, and Oscar F. Ibáñez

CONTENTS

19.1 Introduction

The persistent drought that most of Mexico experienced from the last months of 2010 until 2013 exposed the limitations of dealing with the phenomenon using a reactive approach. Some early warning signs were provided through the Mexican drought monitor, and unprecedented state and local government efforts were coordinated to help the affected communities. Despite these efforts, most of the infrastructure investments were not in place fast enough to help with the emergency, although mitigation programs that included temporary jobs, water and food for poor communities, and insurance coverage for agricultural and farm losses, did provide some relief.

Droughts have significant impacts on almost any social activity, depending upon the vulnerability of the affected region, community, or country. Agriculture is usually the first economic sector affected by lack of water, especially areas without irrigation infrastructure. Even under normal conditions, these areas do not have a reliable source of water and therefore cannot plan on agricultural activities, including subsistence agriculture. When droughts occur, peasants are left with no means of sustenance. This situation can lead to forced migration.

Through its impacts on water quantity and quality, drought may also affect the health of vulnerable groups such as children and the elderly. In urban areas with a deficit of potable water, drought exacerbates unhealthy living conditions, which may lead to fatalities.

The environment is also deeply affected by lack of water, since every ecosystem needs water to be sustained. Forestry is affected not only because of dry conditions but also because drought augments the risk of fires and plagues, which end up affecting the animals and plants that live in the forest or derive some benefit from it.

Drought impacts may lead to social and political unrest in different sectors of the community, and since reactive approaches seldom help to solve water crises, a vicious circle of anger, opposition, and unrest makes it even harder to effectively allocate available water among these sectors.

Tourism, recreational activities, and other economic activities that demand large quantities of water are also affected as a result of reduced water levels in lakes, dams, and rivers.

19.1.1 Drought in Mexican History

Drought has been a significant natural hazard over most regions of Mexico for centuries. Current research suggests that it may have been partly responsible for the demise and fall of the ancient Toltec, Mayan, and Teotihuacan cultures (Desastres y Sociedad 1993; Gill 2008; Kennett et al. 2012). Even though systematic and extensive data on droughts are not available for the region in the pre-Hispanic period (1500 BC–AD 1521), the impacts have been documented to some degree and expressed in ancient codes. Once the central and northern territories of Mexico had been inhabited by Spaniards moving north during the colonial period (1521–1821), better records were kept, and drought events and effects were documented in these regions of the country.

Extensive drought in 1785 and 1786 affected most of the populated territory of New Spain, and its economic effects lasted more than two decades after the episode. These impacts can be linked to another drought episode in the early nineteenth century (1808–1810) (SARH 1980). These two droughts had a devastating effect, particularly on agriculture in the Bajio region, a region that sparked the Independence movement. Some historians note that the bad economic conditions associated with the drought served as a catalyst for the civil uprising that was associated with the Independence movement (Brading 2008; Tutino 1990).

After Mexico's independence from Spain, a series of regional conflicts, civil rebellions, and disputes made record-keeping difficult, and data collection was scattered and random. However, the effects of at least 10 droughts were documented in different regions in Mexico from 1821 to 1875. After the civil war ended, the Porfirian era (1876–1911) provided some stability, and governmental information collection was enhanced. The foundation of the Central Meteorological Observatory in 1877 (the founding institution that later became the National Meteorological Service) led to the systematic recording of meteorological data (CONAGUA 2012a).

From this latter period, regional records of food and water deficits, cattle deaths and famine, migration, and social conflict reveal the economic impacts of drought on agriculture and farming. In Mexico, 29 drought episodes have been noted. The impacts on the northern lands prompted water users to ask the government to build different types of irrigation and water storage infrastructure in Sinaloa, Sonora, and Baja California, and along the Rio Bravo. Since that time, we can find records of the first mitigation measures, such as moving cattle herds to nondrought affected regions in order to avoid massive die-offs (Contreras Servin 2005).

During the twentieth century, a more detailed record of droughts and their effects was available in Mexico. Some of those droughts were part of events that affected several regions of the world, such as in 1951 in Europe, Asia, and Oceania; 1956 in Europe, Asia, and America; and 1972 in Oceania, Asia, and America. Out of 38 droughts recorded in Mexico from 1911 to 1977, 17 can be linked to world events.

For Mexico, at least 20 drought episodes qualified as severe droughts that affected economic production and people's livelihoods. However, the drought years of 1925, 1935, 1957, 1960, 1962, 1969, and 1977 were considered to be extremely severe. In particular, the 1960, 1962, and 1969 droughts created a crisis in agriculture that spilled over into the rest of the Mexican economy and society (Castorena and Florescano 1980).

The 1956–1957 drought that mainly affected the northern border states of Tamaulipas, Chihuahua, Coahuila, and Sonora, and extended to Sinaloa, Durango, Zacatecas, Colima, Aguascalientes, and Oaxaca had significant social effects, causing unemployment and migration (Cerano Paredes et al. 2011). The government reacted with some infrastructure works, but it was not enough to stop migration.

In 1969, the drought affected 20 percent of the nonirrigated land, forcing the government to create a plan to fight drought that included temporary jobs for farmers and instructions for insurance coverage on unpaid bank loans. Every event was treated as a disaster that had to be attended to by the government, which reacted to minimize the impact of the phenomena. Infrastructure works were undertaken after the fact, generating some marginal economic activity in the affected zones. During this period, concerns about environmental impacts were not included in governmental strategies to minimize drought effects.

During the last decade of the twentieth century, Chihuahua and Sonora experienced extreme droughts, with devastating effects in those states and adjacent areas. These droughts did not affect the rest of the country (Núñez-López et al. 2007). Some irrigation districts suffered because so much water stored in dams was used during the drought that the dams were almost empty by 1994. It is clear that proactive drought management protocols were not in place at the time (Mussali and Ibáñez 2012).

Several observations can be made from this brief review of drought events in Mexico. First, there was a lack of data for the whole territory, since vast areas were uninhabited until the colonial period, when the north slowly began to grow in population. It was not until the creation of the National Meteorological Service that reliable and systematic records of events were available. These records increased in detail (including social and economic effects) during the twentieth century as new governmental institutions were created to manage water.

Second, despite the sparse information available, specialists have determined that these droughts were major drivers of the demise of the ancient pre-Hispanic cultures in Mexico. Droughts and their social and economic impacts during the colonial period are considered key factors in subsequent political conflicts, mainly the war for independence that began at the end of a drought in the states of the Bajio region, where most of the insurrection took place.

By the end of the nineteenth century, after the conclusion of the war for independence and several other civil conflicts, Mexico had gained some stability. The available data from that era show the impacts of drought on agricultural activities; because the majority of the population at the time was agrarian, the effects of these droughts rippled through the entire society.

Third, the government's reliance on a reactive approach to addressing the drought hazard can be highlighted, given that drought is a normal feature of the climate for vast regions of the country and sometimes affects most of the country. Some mitigation and adaptation measures were developed and used during various droughts, but decisions were made on a case-by-case basis, without a national policy that could be used to systematically improve the efficiency of the programs by incorporating measures to address drought in a proactive manner.

19.1.2 The 2010–2013 Drought and the Reactive Approach

From 2010 through mid-2013, an extreme drought extended over 90 percent of Mexico, with different severity levels. It was considered the largest drought on record since the beginning of the historical climate record in Mexico, and it represented an enormous challenge to the Mexican government, which was barely recovering from the economic world crisis of 2008–2009. The effects of the drought were felt mainly in the rural communities and areas without irrigation, but it also affected several economic activities associated with agriculture and cattle ranching.

In the northern Mexican states, the drought lasted longer. For the states of Durango and Aguascalientes, it was the worst drought on record. For Guanajuato and Zacatecas, it was the second driest year on record. For Coahuila and Baja California Sur, it represented the third driest year in their climatic record. For Nuevo León, it was the fourth driest year and the fifth driest year for the state of Chihuahua. For other states such as Sonora, it was the nineteenth driest year on record (Table 19.1).

The impacts of the drought were mainly on migration, unemployment, and lack of food and water in small rural and isolated communities, depending upon the different levels of drought through the period. In some parts of the country, drought was mild, and in some southern regions of the country, the phenomena lasted only for a few months. Still, the drought mobilized political and governmental actors throughout the country.

The Mexican drought monitor, in place since 2003, provided advanced warning of the drought in 2010, but not as part of a national strategy or nationwide accepted tool to help address drought. The Mexican drought monitor was developed when the governments of Canada, the United States, and Mexico agreed to share information and models to prepare the North American drought monitor, and its main purpose was to provide the country collaboration for the latter monitor.

Several high-level meetings were held within the Mexican government to prepare for the 2011 drought, but at that time, it was hard to predict either duration or extent over the country, and the approach was reactive.

TABLE 19.1

Precipitation Anomalies for Northern States in Mexico

State	Precipitation Anomaly (%)	Precipitation January–December 2011	Average Precipitation January–December 2011 (mm)	January–December 1941–2011
Durango	−51.1	245.7	502.4	1°+ SECO
Aguascalientes	−44.4	257.8	463.8	1°+ SECO
Zacatecas	−38.5	378.6	615.4	2°+ SECO
Coahuila	−47.5	176.1	335.4	3°+ SECO
Baja California Sur	−60.4	70.5	178.2	3°+ SECO
Chihuahua	−40.2	259.2	433.1	5°+ SECO
Nuevo León	−39.0	374.7	614.1	4°+ SECO
Sonora	−14.8	359.7	422.4	19°+ SECO
Colima	54.3	1367.8	886.4	1°+ HÚMEDO
Chiapas	20.2	2373.5	1975.1	4°+ HÚMEDO
Quintana Roo	16.8	1473.5	1261.3	9°+ HÚMEDO

Source: CONAGUA, *Estadísticas del agua en México,* Edición 2011, Comisión Nacional del Agua, México, DF, pp. 9, 50, 62–64, 2011.
Note: HÚMEDO, wettest; SECO, driest.

Figure 19.1 includes available records since 2003 from the Mexican drought monitor (CONAGUA 2017).

During 2011, the federal government allotted resources for the second half of the year for mitigation measures including infrastructure construction, and financial and insurance coverage support for affected areas. Some temporary jobs were offered and restrictions for water allotments in irrigation districts were in place, as well as recommendations for crops with lower water consumption.

An intense political fight was evolving at the same time that the drought was affecting the country, and the states and legislators did not know how to react. Legislators had to approve the 2012 budget, and it did not include provisions to attend to the crisis. Less than a month later, some legislators were demanding extra money from the federal government to address the emergency.

The only policy in place to deal specifically with drought involved two mitigation funding mechanisms: the national fund for natural disasters (FONDEN), which had specific rules that included the assessment of drought effects during December at the beginning of the dry season, and the

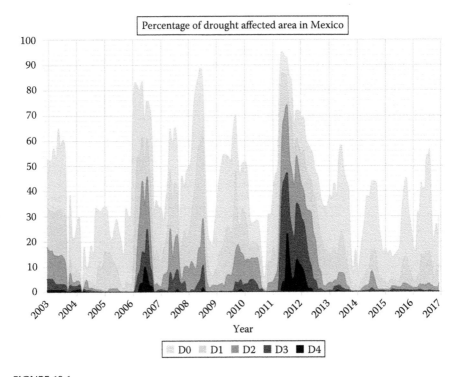

FIGURE 19.1

Percentage of drought-affected area in Mexico, 2003–2017. (From CONAGUA, *Servicio Meteorológico Nacional: Monitor de sequía en México*, 2017. http://smn.cna.gob.mx/es/climatologia/monitor-de-sequia/monitor-de-sequia-en-mexico.)

attention component for natural disasters for agriculture and fishing sectors (CADENA). Both of these funding mechanisms depend on predetermined levels from the historical analysis of long data records and statistical methods such as Streamflow Drought Index (SDI) and the Standard Precipitation Index (SPI).

Given the extent and impact of the ongoing drought, the federal government implemented a presidential decree in January 2012 that organized several programs within the federal government from different ministries to deal with the 2011 drought effects. A two-tier approach was defined: one set of measures aimed at protecting production and infrastructure, and another for humanitarian help for families and communities affected by the different levels of drought.

The production and infrastructure component included temporary employment in affected areas, insurance protection for lost crops and livestock deaths, maintenance of production capabilities, financial support for economic activities in affected areas, and sound and sustainable use of water. The humanitarian component included water and food for affected communities and family income protection.

Regardless of the collaboration between state, municipal, and federal governments, and the coordination of at least five ministries and three other federal offices from the federal government (Secretariat of Communications and Transport [SCT]; Secretariat of the Environment and Natural Resources [SEMARNAT]; Secretariat of Agriculture, Livestock, Fisheries and Food [SAGARPA]; Secretariat of Social Development [SEDESOL]; Secretariat of Finance and Public Credit [SHCP]; National Water Commission [CONAGUA]; Trust Funds for Rural Development [FIRA]), the approach was undoubtedly reactive, and showed severe limitations. Therefore, new ideas began to be considered to tackle future events more proactively.

19.1.3 The International Context

Mexico initiated collaboration with Canada and the United States through CONAGUA back in 2002 to create the North American drought monitor. Since then, several collaboration activities have been undertaken with international organizations, specifically the World Meteorological Organization (WMO) for training and skills development for the Meteorological National Service (SMN), and CONAGUA's technical areas dealing with extreme events (flooding and drought).

During the Conference of the Parties (COP) of the United Nations Framework Convention on Climate Change held in Cancun, Mexico, in 2010, the Mexican delegation proposed that water adaptation measures be included as part of the working agenda. This was a significant step in the paradigm change from mitigation to adaptation strategies, considering that climate change impacts are felt first during extreme events associated with water. The same proposal was made at different international forums (CONAGUA 2012b).

CONAGUA personnel participated at an experts meeting in 2011, *Towards a Compendium on National Drought Policy*, convened by WMO, George Mason University, the Environmental Science and Technology Center, the National Drought Mitigation Center (NDMC), and the United States Department of Agriculture (Sivakumar et al. 2011). The 2010–2013 drought was an event that impacted not only Mexico but several countries at different levels. At this meeting, experiences were shared, and the need for a paradigm reform of national policies was a major conclusion.

The experiences of Spain, the United States, Australia, India, and China were used in developing prevention and mitigation strategies for drought in Mexico. Special attention was placed on the California and Colorado experiences as well as the policies of some public water systems in Texas (City of San Antonio 2014; Colorado Water Conservation Board and National Integrated Drought Information System 2012; Sivakumar et al. 2011; WMO and UNCCD 2013).

While Mexico was designing the new policy toward drought, the High-level Meeting on National Drought Policy (HMNDP) was held in Geneva in March 2013. The Mexican delegation participated and approved the documents emanating from that meeting, and they were used as a reference in the ongoing efforts to define the Mexican National Drought Policy (HMNDP 2013).

These efforts attracted the attention of the World Bank, WMO, Turkey, and Brazil and other Latin American countries, and Mexican experts attended several regional and international workshops and offered technical assistance.

Overall, the alignment of the severe drought in Mexico and the international efforts in which Mexico actively participated in national drought policy and water adaptation measures regarding climate change created a timely synergy for an interactive process incorporating local and international experiences.

19.2 Theoretical Framework Considerations

Considerations of paradigm shifts are to be taken very seriously because of the lengthy and complex process involved in such endeavors (Kuhn 1962). Because governmental affairs are characterized by inertial and incremental adaptations, adopting a paradigm shift is even more complicated than just a conceptual redefinition in academic terms. The long history of governments reacting to drought events instead of preparing for them and the relatively recent use of concepts involving risk management, reduction of vulnerability, adaptation measures, and prevention as part of drought policies worldwide set the stage for the level of difficulty involved in shifting from crisis management to risk management (HMNDP 2013).

The theoretical framework to analyze the creation of this public policy may be considered postpositivist according to the description in Torgerson (1986). There must be a concerted effort among governmental officials, professional experts, and academics to develop the core of the policy, including a strong technical background and experience of dealing *in situ* with real drought events, but also with a definitive aim to generate public participation in the creation of initiatives to be used at the basin level.

Another theoretical perspective that better resonates with water policy efforts is provided by Schmandt (1998), who states that complex problems like drought need to be assessed using the most current expertise, and solutions must utilize the initial assessment and engage stakeholders. In Mexico's case, local representatives and water users, as well as federal authorities are stakeholders, and they were included in defining the proactive mitigation measures considered for each case.

The principles considered the backbone of the Mexican drought policy corresponds to these theoretical considerations and are used to define the main elements of the drought policy. These are the preparedness or proactive approach, which drives the policy shift from a reactive perspective toward proactive management. This aspect of the process entails monitoring and information outreach to provide authorities and stakeholders with information to implement previously defined mitigation measures (after certain thresholds are reached) that are aimed at coping with less water. This also involves vulnerability baseline definitions to evaluate the risk at different drought intensity levels according to the ranges of the Mexican and North American drought monitors.

Decentralization is a second basic goal of the policy that involves local stakeholders, since the problems can be prevented better at the affected community level. Nevertheless, institutional as well as citizen training and empowerment is required since the current top-down approach usually leaves communities without the proper preparedness and resources to handle drought.

Governance is a third element of the process, with a basin council structure as the core part of the governance. The goal of this element is to use institutional capacity to strengthen governance required for drought preparedness and mitigation as well as to build resilience into the process to mature and sustain it beyond national and state political systems. Involving local universities and supporting technical decisions of stakeholders and authorities provides the basis for public participation and the strong governance needed to reduce vulnerabilities.

A fourth element of this process is training and research. The paradigm shift and decentralization requires understanding and proper training of the new concepts involved in risk management, and the lack of baseline information on vulnerability, drought impacts, and extreme event forecasting and understanding underscores the need for research.

The fifth element of this process is gradualism and evaluation. This concept involves building a transitional process between paradigms, breaking away from inertial attitudes, and developing and evaluating performance indicators for continuous improvement.

The sixth element of this process is institutional coordination. One of the obvious problems of reactive policies compared to preparedness approaches involves lack of institutional coordination. A vast number of governmental ministries, offices, and programs are needed to respond to droughts, which are complex in nature, and structural shifts can only be implemented through good and systematic coordination.

The process involves two major elements developed in accordance with the principles listed above. These elements address current drought situations and the transition from reactive institutions and rules to the new mechanisms designed for the new paradigm:

- Elaboration of drought prevention and mitigation programs (PMPMS) in every basin council in the country
- Mitigation measures to face ongoing drought emergencies

The paradigm shift involves a policy shift while multiple governmental and cultural practices exist. As Kuhn (1962) clearly explains, there is a transitional period where both paradigms coexist. The key to moving from one paradigm to a new one is to use the existing institutional framework to create the new institutions and rules that will regulate drought policy in the future. This will be attained by identifying the old institutions that need to be replaced and carefully managing change through training and capacity development, both institutional and personal.

19.3 Mexican National Drought Policy

In January 2013, at the beginning of the new federal administration, the president of Mexico instructed the National Water Commission (CONAGUA) to implement the national program against drought (PRONACOSE). The decision was made to design the national drought policy (NDP), which moved from a reactive to a proactive approach. At the same time, the effects of the 2010–2013 drought were still being felt in some northern states, and the existing policies and rules of operation for drought emergency programs were under implementation. Therefore, a program with two major components, the drought prevention and mitigation programs (PMPMS) and the Interministerial Commission (IC) for the Attention of Drought and Flooding was designed and implemented.

The design and development of the NDP was an interactive task. Through the implementation of PRONACOSE, most of the elements fell in place along the way, with multiple modifications. Other adjustments were made to adapt international experiences to the national institutional framework and regional characteristics. A simplified diagram of the NDP is presented in Figure 19.2.

19.3.1 PRONACOSE

The main objectives of PRONACOSE are to develop and implement the PMPMS and address drought events at the watershed level. Other objectives are to develop local institutional capacity while simultaneously coordinating and implementing drought mitigation activities as needed and to promote drought research and build a historical archive.

PRONACOSE's mission is to develop the basis for the paradigm shift on drought response from reaction to proactive actions (based on risk management) while also attending to current drought impacts. The vision of the program is to guarantee planning and implementation of drought measures,

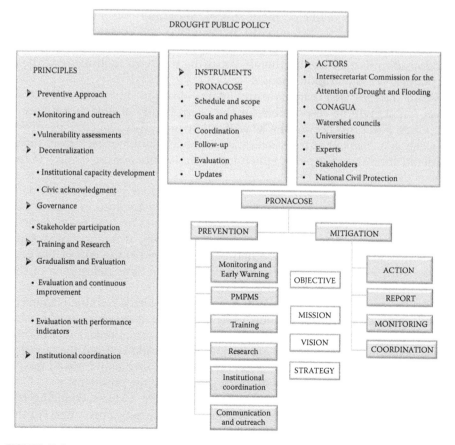

FIGURE 19.2
National drought policy diagram.

involving public participation in the definition of actions to reduce vulnerability as a pillar for the Mexican strategy on climate change adaptation (expressed in the Climate Change General Law and the National Waters Law, and linked with the National Civil Protection Service activities).

The program envisions that every watershed council will have a PMPMS and periodical evaluation and updates involving members' participation to improve the programs, with interinstitutional coordination instruments at the national level.

The strategy to implement the policy consists of gradually decentralizing drought response, involving stakeholders through the watershed councils and also pairing the watershed councils with teams of academic experts from local universities. The goal is to develop local institutional capacity and begin to change the old reactive top-down approach to addressing drought.

The PMPMS are the main instruments to carry out the strategy, and the goal is that once the basin councils adopt and implement these instruments, they will move to delineate more specific programs for cities and irrigation districts. The PMPMS need to be evaluated and updated regularly to assure that the first updated programs will be institutionalized before the federal administration ends, thereby ensuring they will continue regardless of political administration change.

While the PMPMS are in place and institutional capacity exists at the local level, the IC will be responsible for the coordination of federal mitigation activities. After this occurs, the IC will follow up and evaluate the progress. Another key element of the strategy includes constant feedback from international entities and experts, which may help to enhance the quality of the policy developments.

The process of developing the NDP began immediately after the presidential order calling for its creation. Some of the people involved in responding to the ongoing drought were asked to assemble ideas on how to implement the paradigm shift while coping with the existing drought. Some legal guidelines for drought response had been approved at the end of the previous administration, and these were used for the basic definitions of the phenomena (Diario Oficial de la Federación 2012).

The technical staff of CONAGUA and the watershed council contacts were convened to prepare the new policy approach. Academic experts were invited to develop training courses and to analyze national and international experiences on drought management that could be used for capacity building. At the same time, the legal framework to create and install the IC was prepared in conjunction with the legal office of the president.

Issues concerning the early warning system were addressed during the first technical discussions, and it was decided to have two approaches. For the new policy and the ongoing mitigation efforts, the Mexican drought monitor was to be used to activate alarms and define the drought severity stages. In addition, the SPI and SDI were kept to run the mitigation programs FONDEN, which will provide funding to address emergency response and damages for water utilities derived from drought periods, and CADENA, which will cover cattle and agriculture damages.

Mexico's participation in the HMNDP in the early design stages was essential in order to connect with evolving international efforts and to continue interaction at subsequent forums, where the Mexican experience was also considered by other international experts and governmental representatives. Mexico became a member of both the advisory and management committees of the integrated drought management program (IDMP), and the Mexican case was part of the *National Drought Management Policy Guidelines – A Template for Action* published by the IDMP (2014). CONAGUA was also invited to by the World Bank and WMO to present the PRONACOSE experience at meetings in Brazil, Central America, and Turkey. In 2015, the Mexican Water Technology Institute became part of the same IDMP committees due to its strong collaboration with CONAGUA in the design and implementation of the PRONACOSE.

Several training courses were held with CONAGUA technical personnel from the 26 watershed councils and academics from 12 universities throughout the country. Some of these meetings were face to face, while others were through video conferences. The aim was to agree on the basic premises contained in the guidelines. Participation and counseling by national and international experts was part of the process at this early stage.

In addition to the CONAGUA guidelines used by the universities and watershed councils to develop the PMPMS, a detailed supervision tool was developed to help the councils comply with the ambitious process and schedule for developing the programs. A website was developed that included online materials, reports, information, and multimedia tools for everyone to use: www.pronacose.gob.mx.

The PMPMS content for the main 26 Mexican basins was defined in the guidelines (CONAGUA 2013) to have a minimum standardization format. It included the following:

1. Abstract
2. Presentation
3. Watershed characterization
4. Task force definition within the watershed council to coordinate and follow up on the elaboration of the PMPMS
5. Objectives definition
6. Drought history and assessment of drought impacts
7. Vulnerability assessment
8. Mitigation and response strategies
9. Drought phases
10. Triggers and measures' objectives

11. Specific program with measures for every drought phase

12. Implementation

13. Monitoring

14. Conclusions

15. Annexes

The process for developing the PMPMS in every basin council should include the following steps, to be accomplished in about 9 months. It is noteworthy that some of the steps correspond to sections of the report: Training workshop to launch the PRONACOSE; a letter of intention from the participating institutions; contact with the technical director of the corresponding basin organization; organization of the directive task force (GTD); work plan and organizational chart; a general report on drought history and impacts in the basin; characterization of the basin; report on the basin's vulnerability; mitigation and responses expected for drought management; phases and characterization of associated indicators; a detailed program for every phase; first version of PMPMS; agreements with stakeholders in at least three meetings; final version of PMPMS; and implementation.

Stakeholders participating in the 26 basin councils were involved in the process of defining measures to be implemented at different stages of drought in accordance with predefined indicators. The initial proposals were presented to them by university experts in consultation with technical personnel from CONAGUA. The university experts facilitated the appropriation and implementation of the PMPMS by the basin council. A highly interactive process was undertaken that involved knowledge of and discussions about the alternatives and some negotiation based on the measures that the stakeholders were willing to apply in their basin.

The principles of decentralization, governance, training, gradualism, and institutional coordination are all considered within the process of building the PMPMS with the purpose of reorienting drought management policy. The first version of PMPMS is intended to be a good approximation (i.e., a tentative first draft) of what is expected to be an accepted, adjusted, and viable program. The process is designed to be a gradual transition to achieve the policy shift.

19.3.2 Interministerial Commission for the Attention of Drought and Flooding (IC)

The creation of the IC, which also helped to organize and coordinate efforts for the ongoing drought, had important implications for policy implementation. The fact that it was convened at the beginning of a new federal administration was extremely important, since it helped to generate a different approach to the policy shift.

The Commission was installed in April 2013 and included 14 secretariats and federal offices. The IC was chaired by the Secretariat of Environment and Natural Resources (SEMARNAT). Other members of the IC included the Secretariat of the Interior (SEGOB); Secretariat of National Defense (SEDENA); Secretariat of the Navy (SEMAR); Secretariat of Finance and Public Credit (SHCP); Secretariat of Social Development (SEDESOL); Secretariat of Energy (SENER); Secretariat of Economy (SE); Secretariat of Agriculture, Livestock, Fisheries and Food (SAGARPA); Secretariat of Communications and Transportation (SCT); Secretariat of Health (SALUD); Secretariat of Rural, Territorial and Urban Development (SEDATU); Electricity Federal Commission (CFE); and the National Water Commission (CONAGUA) as the executive secretariat.

One of the first tasks of the commission was to install a committee to review all the federal programs that might have an impact on drought. This exercise led to the identification of 114 programs, and 19 of those were reviewed together with the two funding protocols from SEGOB to address drought impacts: FONDEN and Preventive Trusteeship (FIPREDEN). They were examined in detail to determine if they needed to be reorganized or modified to avoid duplication and enhance efficiency of federal actions to handle drought (OMM 2013). Modification of the programs' rules of operation is the next step to align them with the new approach (CONAGUA 2014).

Another responsibility of the IC was to create a technical experts' committee to provide counseling for and evaluation of the PRONACOSE activities and to discuss and propose research needed to strengthen understanding of drought. Also, all the federal secretariats agreed to provide research funding for selected topics. The agenda was approved in early 2015.

The IC also served as the coordination tool for all the mitigation activities that the federal government implemented during the last part of the long drought.

19.3.3 Policy Instruments

The implementation of PRONACOSE requires a specific schedule and scope, goals and phases, coordination, follow-up mechanisms, evaluation, and updates. The IC is the coordination instrument. The experts committee will help evaluate the programs reviewed by every watershed council. The goal is to update the PMPMS at least once after their first implementation.

The main goal of the program is to implement a policy shift conceived by the federal administration. This includes the elaboration of the first version of the PMPMS and its continued evaluation and improvement. Another key element is to make sure that the PRONACOSE becomes part of the national civil protection system (SNPC), particularly for the early warning system. This system is already in place for flooding, but the distinct characteristics of drought events make it harder to call for emergency preparedness meetings and protocols, as is done for flooding events. Currently, the SNPC works for mitigation activities and emergency response when drought impacts are affecting communities, and the challenge is to utilize the early warning system at different stages to initiate preparedness and reduce vulnerability.

During the first phase of the PRONACOSE implementation, which will last 1 year, the first version of the 26 basin PMPMS will be completed, the IC will be in full operation, and the basic training and agreements with watershed council members will be completed. The second phase, which spans a 2-year period, includes the elaboration of the first PMPMS for two cities (water utilities) and one irrigation district in each basin, the research agenda definition and development of vulnerability evaluation criteria, a media campaign to present and publicize information about the PMPMS, and the beginning of interaction with the SNPC to implement early warning protocols for the different basins.

The third and fourth phases are designed to evaluate and update the PMPMS and develop more PMPMS at the water utilities and irrigation districts level, as well as to coordinate with the SNPC to integrate Mexico's National Atlas of Risks with drought information on vulnerabilities and protocols to be utilized by the IC. The fifth and sixth phases are intended for evaluation of the NDP, implementation of revised PMPMS, and institutional adjustment of federal, state, and municipal governments' programs to be aligned with the new policy.

The implementation of the PRONACOSE is in its third year. By the middle of the second phase, all PMPMS had been completed, and most of the PMPMS intended for at least two water utilities per basin were finished. Also, all of the institutional coordination and advisory mechanisms are working, including the IC, the committee to review the budget alignment of the federal programs to the PRONACOSE criteria. More than 20 federal programs out of 102 related to drought-addressing activities are aligned to the priorities established in the PMPMS. Also, the experts committee defined the research agenda, which was approved by the IC with consensus about the topics to receive funding from several secretariats, and the early monitoring system has been implemented.

19.4 Findings and Preliminary Conclusions

The principles described above to develop the Mexican National Drought Policy helped to advance the implementation process. The ambitious goal of having a policy shift seems to be advancing, consistently breaking some of the inertial attitudes at the basin councils and the federal government. However, the final test will be implementing the PMPMS before and during a drought event so they can be verified and enhanced.

The inclusive effort aimed at decentralizing the program has strengthened governance, and the emphasis on training has improved the capacity building needed for decentralization. Research created an incentive for the universities and experts to participate in and embrace the program, and the IC enabled the different areas of the federal government to gradually assume the policy shift.

Some challenges that need to be carefully considered appeared during the process of defining and implementing the policy. It may be possible to tackle some challenges through mild adjustments, while others might not be solved in the near future. But it is important to identify these challenges and look for alternative ways to respond to these challenges.

Four main drivers for change can be identified in the process. The 2010–2013 drought event was the main source of political momentum and the first driver. The existence of a long period of drought presented the opportunity for a strong political response and deployment of several programs and institutional instruments to manage the crisis.

The second driver emerged during the emergency response to the drought. This response exposed the limitations of a reactive approach, leading CONAGUA technical personnel to consider a paradigm shift to adaptation and proactive measures so that the Mexican government would be better equipped to deal with climate change.

Mexican participation in expert meetings convened by WMO and others to address the concept of risk management as the cornerstone for national drought policy constitutes the third driver.

The final driver was that the change in federal administration presented the political opportunity for a new approach. The directive given by the president of Mexico streamlined the process and helped to quickly implement the PRONACOSE and the IC for the Attention of Drought and Flooding while the ongoing drought event persisted through the first half of 2013.

There are challenges too, such as the rainy season in the second half of 2013 and the confluence of hurricanes Ingrid and Manuel that hit Mexico from the Atlantic and the Pacific in September 2013, which completely shifted political attention and priorities. Nevertheless, the previously existing momentum helped officials finish the 26 PMPMS later that year, although the cessation of drought impeded the implementation of the PMPMS measures, stalling their evaluation and potential adjustment.

The situation quite clearly resembled the famous hydro-illogical cycle diagram proposed by Wilhite (2011) (Figure 4.2), where it does not matter how severe a drought event is; once the rain comes, things go back to a "business as usual" attitude. The institutional design of the IC for the Attention of Drought and *Flooding* (emphasis added), which was influenced by an unexpected favorable circumstance (because of the hydrologic conditions in Mexico in 2013 and 2014), helped to consolidate the institutional coordination mechanism for drought.

In terms of opportunities, perhaps the least developed effort was publicizing information, since it has been restricted mainly to members and representatives that participate in the basin councils. Outreach efforts need to be improved so that the public will want to participate in the measures defined by the PMPMS. One way to solve this weakness would be to incorporate early monitoring systems and protocols into the SNPC, which provides timely information for media broadcasts to the public.

Another way to improve education on drought issues, using early monitoring systems, and reducing vulnerability would be to involve the Ministry of Public Education, which currently is not part of the IC. In any case, educational programs might be included gradually as part of the basic educational curriculum.

Publicizing national drought management policy and preparedness plans, building public awareness and consensus, and developing educational programs for all ages and stakeholder groups are included as important steps in the recommendations published by the IDMP (WMO and GWP 2014).

The modification of FONDEN, FIPREDEN (Natural Disaster Preventive Trust Fund), and CADENA is one of the main challenges that remain since these programs were crafted with the presumption of a predefined amount of money available for response. It is difficult to adjust them for preparedness or structural changes that will reduce vulnerability and reduce the need for reconstruction or mitigation funds. They are the institutional remnants of the old policy approach and they need to be kept functioning while a different set of rules is designed for the new policy. The link with the National Crusade Against Hunger represents a window of opportunity to reach less-developed communities, with a joint goal of reducing their vulnerability.

Another challenge that remains unfulfilled is homogenization of criteria for vulnerability assessment. Currently, several methods are used, and the debate about pros and cons of each method is very much alive. It will be important to reach some consensus about it, considering the importance of evaluation and improvement. Another huge challenge is to allocate federal resources to ensure a permanent linkage of drought vulnerability and probability maps as well as implementation and permanent updating of the PMPMS.

Risk needs to be calculated based on quantifiable impacts. Thus, having databases and models that will help us estimate how much money and how many resources and lives are saved by using this approach is paramount, rather than continuing to react to drought through crisis management.

Other limitations for the new policy are more difficult to overcome because of their structural character. For example, the level of complexity of the basin boundaries, which do not correspond to states' political boundaries, makes it difficult to define priorities and budgets. Local authorities may be reluctant to embrace risk management because reacting to emergencies results in greater political gains.

The extraordinary speed of the implementation of PRONACOSE, including the completion of PMPMS for the entire country, the construction of a basic institutional capacity at the basin council level, the installation of the IC and related coordinating committees, and the existence of an operational early warning system, makes the Mexican experience an important case study for other nations seeking to build a national drought policy.

The sudden end to the long drought in Mexico led to a halt in the implementation of the newly developed measures contained in the PMPMS, which will need to be improved once they are put to the test under real drought circumstances. On the other hand, this pause gives officials time to review and discuss vulnerability assessment methodologies, accomplish specific PMPMS for about 50 cities throughout the country, and define topics and funding for further research on drought-related issues.

Finally, CONAGUA's interactions with drought policy workshops, conferences, and forums will help them evaluate and improve their own policy using experiences from other countries.

References

Brading, D. A. 2008. *Miners and Merchants in Bourbon México 1763–1810*. Cambridge Latin American Studies (Book 10). Cambridge: Cambridge University Press.

Castorena, G., and E. Florescano. 1980. *Análisis Histórico de las Sequías en México*, Comisión del Plan Nacional Hidráulico, Mexico DF.

Cerano Paredes, J., J. Villanueva Díaz, R. Valdez Cepeda, E. Cornejo Oviedo, I. Sánchez Cohen, and V. Constante García. 2011. Variabilidad histórica de la precipitación reconstruida con anillos de árboles para el sureste de Coahuila. *Revista Mexicana de Ciencias Ambientales*, 2(4):31–45.

City of San Antonio. 2014. *City of San Antonio Drought Operations Plan*. City of San Antonio Office of Sustainability. http://www.sanantonio.gov/sustainability/Environment/DroughtOperationsPlan.aspx (accessed December 28, 2014)

Colorado Water Conservation Board and National Integrated Drought Information System. 2012. *Summary Report: Colorado Drought Tournament*. http://cwcbweblink.state.co.us/WebLink/ElectronicFile.aspx?docid=168487&searchid=ea4295dc-9c5a-4c7e-8416-f3c4770e9720&dbid=0 (accessed December 28, 2014)

CONAGUA. 2011. *Estadísticas del agua en México*. Edición 2011. México, DF: Comisión Nacional del Agua, pp. 9, 50, 62–64.

CONAGUA. 2012a. *Servicio Meteorológico Nacional: 135 años de historia en México*. México DF: SEMARNAT.

CONAGUA. 2012b. *Recuento de la Cooperación Internacional de la CONAGUA 2009–2012*. México DF: CONAGUA.

CONAGUA. 2013. *Guía para la formulación de Programas de Medidas Preventivas de Mitigación de la Sequía*. Jiutepec: Convenio con Instituto Mexicano de Tecnología del Agua/Coordinación de Hidrología.

CONAGUA. 2014. *Propuesta de adecuación de las reglas de operación de los programas federales vinculados al PRONACOSE*. Enrique Aguilar. Informe OMM/PREMIA No. 233 Edición 2014. Comisión Nacional del Agua, Mexico DF.

CONAGUA. 2017. *Servicio Meteorológico Nacional: Monitor de sequía en México*. http:// smn.cna.gob.mx/es/climatologia/monitor-de-sequia/monitor-de-sequia-en-mexico (accessed January 28, 2017)

Contreras Servin, C. 2005. *Las sequías en México durante el siglo XIX*. Investigaciones Geográficas. no.56. México abr. http://www.scielo.org.mx/scielo.php?pid= S0188-46112005000100008&script=sci_arttext (accessed January 2, 2015)

Desastres y Sociedad. 1993. *Revista semestral de la red de estudios sociales en prevención de desastres en América latina/1993/No.1*, Revista Desastres y Sociedad, N° 1.Red de Estudios Sociales en Prevenciónde Desastres en América Latina y el Caribe. Panamá.

Diario Oficial de la Federación. 2012. Lineamientos que establecen criterios y mecanismos para emitir acuerdos de carácter general en situaciones de emergencia por la ocurrencia de sequía, así como las medidas preventivas y de mitigación, que podrán implementar los usuarios de las aguas nacionales para lograr un uso eficiente del agua durante sequía. http://dof.gob.mx/nota_detalle.php?codigo =5278695&fecha=22/11/2012 (accessed January 02, 2015).

Gill, R. B. 2008. *Las grandes sequías mayas. Agua, vida y muerte*. 1ª ed. en español, Trad. M. O. Arruti y Hernández, Fondo de Cultura Económica. Mexico DF.

HMNDP. 2013. High-level Meeting on National Drought Policy. Ginebra, International Conference Center (CICG), 11–15 March. National Drought Management Policy (policy document). http://www.wmo.int/pages/prog/ wcp/drought/hmndp/documents/PolicyDocumentRev_12-2013_En.pdf (accessed May 01, 2015)

Kennett, D. J., S. F. M. Breitenbach, V. V. Aquino, Y. Asmerom, J. Awe, J. U. L. Baldini, P. Bartlein, et al. 2012. Development and disintegration of Maya political systems in response to climate change. *Science* 338(9):788–791.

Kuhn, T. 1962. *The structure of scientific revolutions*. Chicago, IL: University of Chicago Press.

Mussali, R., and O. Ibáñez. 2012. *"Más de 2 décadas de historia" de la CONAGUA*.

Núñez-López, D., C. A. Muñoz-Robles, V. M. Reyes-Gómez, I. Velasco-Velasco, and H. Gadsden-Esparza. 2007. Caracterización de la sequía a diversas escalas de tiempo en Chihuahua, México. *Agrociencia* 41:253–262.

OMM (Organización Meteorológica Mundial) 2013. *Análisis de Programas Federales en Sinergia con las Acciones del PRONACOSE Orientadas hacia la Prevención (Enrique Aguilar) Actividad Giaba 02/2013*. Informe OMM/PREMIA no. 229, Mexico DF.

SARH. 1980. *Análisis histórico de las sequías en México*. México, DF: Secretaría de Agricultura y Recursos Hidráulicos.

Schmandt, J. 1998. Civic science. *Science Communication* 20(1):62–69.

Sivakumar, M. V. K., R. P. Motha, D. A. Wilhite, D. A, and J. Qu. 2011. Towards a compendium on national drought policy: Proceedings of an expert meeting. http:// www.wamis.org/agm/pubs/agm12/agm12.pdf (accessed January 5, 2015).

Torgerson, D. 1986. Between knowledge and politics: Three faces of policy analysis. *Policy Sciences* 19(1):33–59.

Tutino, J. 1990. *De la insurrección a la revolución en México. Las bases sociales de la violencia agraria*. México: Ed. ERA, pp. 1750–1940.

Wilhite, D. A. 2011. Addressing impacts and societal vulnerability. In *Towards a Compendium on National Drought Policy: Proceedings of an Expert Meeting*, ed. M. V. K. Sivakumar, R. P. Motha, D. A. Wilhite, and J. Qu, pp. 13–22. http:// www.wamis.org/agm/pubs/agm12/agm12.pdf (accessed December 29, 2014).

WMO (World Meteorological Organization) and (GWP) Global Water Partnership. 2014. National drought management policy guidelines: A template for action (D.A. Wilhite). Integrated Drought Management Programme (IDMP) Tools and Guidelines Series 1. WMO, Geneva, Switzerland.

WMO (World Meteorological Organization) and UNCCD (United Nations Convention to Combat Desertification). 2013. Science document: Best practices on national drought management policy. High Level Meeting on National Drought Policy (HMNDP) CICG, Geneva 11–15 March 2013.

20

Drought Risk Management in the Caribbean Community: Early Warning Information and Other Risk Reduction Considerations

Adrian Trotman, Antonio Joyette, Cedric Van Meerbeeck,
Roche Mahon, Shelly-Ann Cox, Neisha Cave, and David Farrell

CONTENTS

20.1 Introduction: The Caribbean Context of Drought

The Caribbean Community (CARICOM) is a political and economic grouping consisting of 20 countries—15 member states and 5 associate members stretching from Belize (in Central America) across the Antilles (including The Bahamas) to Guyana and Suriname on the South American continent. All CARICOM member states are recognized as small island developing states (SIDS) or SIDS associate members.

Climate-related hazards are the most frequently occurring natural hazards in the Caribbean. The region's acute vulnerability to climate-related hazards (including strong winds, storm surge, flooding, and drought) is reflected in loss of life, economic and financial losses, and damage to the environment. The drought hazard itself is anticipated to result in multifarious impacts on the Caribbean region. Already, the history of droughts in the Caribbean has revealed that impacts are socially wide-reaching, economically substantial, and diverse across sectors (Farrell et al. 2010; CIMH and FAO 2016), and that the region is plagued by inadequate risk management. As such, drought poses a significant challenge to the region's sustainable development. Current climate change predictions for the region indicate that the frequency and intensity of drought will increase in the future (CIMH and FAO 2016). As a result, addressing drought represents a critical aspect of the region's adaptation to climate change. Since the devastating drought of 2009–2010, significant progress has been made in monitoring, forecasting, and mitigating the impacts of drought in the region, such that by the 2014–2016 event, the region was better prepared. While significant progress has been made through a range of initiatives that will be discussed in this chapter, it is recognized that more work needs to be done at community, national, and regional levels in areas such as (1) the development and implementation of drought policy and plans, (2) forecasting, early warning, and integrative decision-support systems, and (3) stakeholder communication systems.

Seven of the world's top 36 water-stressed countries are in the Caribbean (WRI 2013). Island states such as Barbados, Antigua and Barbuda, and St. Kitts and Nevis, with less than 1,000 m^3 freshwater resources per capita, are deemed water-scarce (CIMH and FAO 2016). Within non-water-scarce countries, local communities and cities may be chronically water-scarce, especially under water-stressed conditions. Water scarcity on Caribbean islands is increasing because of the expansion of the tourism industry, population growth (although at slowing rates in several states), urbanization, increasing societal affluence, ineffective water management practices and strategies, and declining water quality due to anthropogenic activities and climatic factors—including changing spatio-temporal climate patterns that will likely lead to increased occurrences of drought.

During the past decades, the Caribbean has experienced several drought events (CIMH and FAO 2016), including the two most recent events in

2009–2010 and 2014–2016. Changing climate in the Caribbean is expected to further exacerbate the impacts from drought (Farrell et al. 2007; Hughes et al. 2010; Joyette et al. 2015; Mumby et al. 2014; Pulwarty et al. 2010), with annual losses of US$3.8 million by 2080 (Toba 2009). This would be attributed to declines in rainfall, particularly in the wet season (Angeles et al. 2007; IPCC 2013; Taylor et al. 2012), and increasing temperature and associated increases in evaporation (Dai 2011, 2013) that are projected for the future.

Until the late 1800s, managing the impacts of drought focused primarily on preserving crown and estate wealth by decreasing the losses to plantation crops and livestock (Cundall 1927). From the late 1800s until recent decades, the focus of drought management shifted toward greater consideration of local societal needs. For example, authorities in the region commenced construction of water storage infrastructure, developed new sources of water and implemented water resources protection strategies and legislation, expanded and enhanced distribution networks, and organized water rationing during drought to ensure adequate supplies of water (Cramer 1938; Lindin 1973; MPDE 2001). At the turn of the twenty-first century, managing the response to drought became the responsibility of emergency management officials as drought was recognized as a national disaster (Maybank et al. 1995). The response to drought largely centered on potable water management through public alerts, encouraging public water conservation, and systematic water rationing. National disaster agencies charged with managing the response to drought are often overextended and overburdened with limited national coordination and linkages, and exiguous national policies and planning. Where national strategies exist, implementation can, at times, be a serious cause for concern.

20.2 Nature of Caribbean Rainfall

20.2.1 Characteristics, Seasonal Patterns, and Trends

The location and diverse topography of the Caribbean influence the amount of annual rainfall and its pattern. At least 70–80 percent of the region's rainfall is realized during the wet season (Enfield and Alfaro 1999), with high variability in the onset, duration, and quantum of rainfall during wet and dry seasons. From The Bahamas and Belize in the west to Trinidad and Tobago in the southeast, the wet season begins around May or June and ends around November or December. The remainder of the year largely represents the dry season. North of around 18°N, wet season rainfall exhibits a bimodal peak interposed by a distinct drier episode, colloquially referred to as a "mid-summer drought" (Gamble et al. 2008). With respect to monthly rainfall totals, much of the region shows a primary maximum in the latter half of the wet season—September to November. This is true in terms of the chance and the intensity of rainfall on any given day (Figure 20.1).

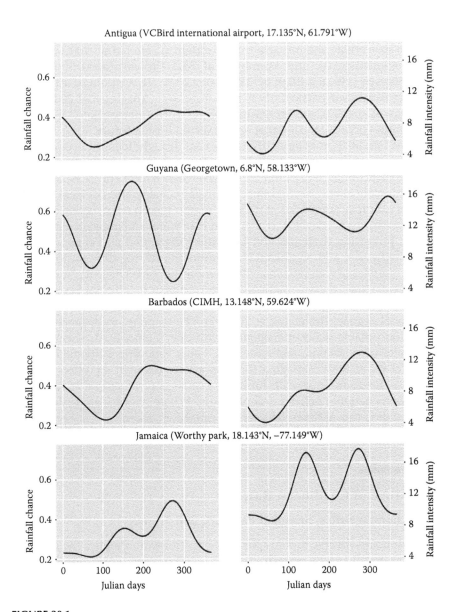

FIGURE 20.1

Smoothed rainfall seasonality in four select stations from CARICOM countries. The left column depicts the chance of a wet day (i.e., a calendar day with >0.85 mm), whereas the right column depicts the average rainfall intensity on a wet day for each Julian day of the year. The remainder of the year, the Caribbean dry season, shows a rainfall deficit, particularly when compared to evaporative losses. Rainfall declines through December continues until late February/early March, the peak of the dry season. Rainfall totals between April and May tend to be highly variable from year to year, with May being among the wettest months of the year in some years and virtually dry, with relatively high potential evaporation rates in other years. These dry months of May extend the impacts of the dry season.

In northern Guyana and Suriname, however, two wet and two dry seasons are experienced per year.

Variability in the intensity and timing of the events mentioned above often results from the El Niño southern oscillation (ENSO) and the gradient in SST between the Pacific and Atlantic (Enfield and Alfaro 1999; Giannini et al. 2000, 2001; Taylor et al. 2002, 2011;) Stephenson et al. 2008, the North Atlantic oscillation (Charlery et al. 2006) and North Atlantic high pressure cell (Gamble et al. 2008), decadal fluctuations (Taylor et al. 2002), the Madden Julian oscillation (MJO) (Martin and Schumacher 2011), and the Caribbean low-level jet (Cook and Vizy 2010; Taylor et al. 2012). Increasing attention is also given to the role of the Saharan air layer in reducing Caribbean rainfall (Prospero and Lamb 2003; Prospero and Nees 1986; Rodriguez 2013). However, the ENSO is likely the single most important factor in the interannual rainfall variability in the Caribbean.

20.3 Impacts from Recent Drought Events—the Case for Enhancing Drought Management in the Caribbean

Because of the limited water resources on most CARICOM SIDS and the fact that all the land area of most CARICOM SIDS is simultaneously impacted by drought, the impacts of drought can be severe, with all socioeconomic sectors being impacted directly by a lack of water or through complex intersectoral interactions.

20.3.1 Declines in Rainfall

For most of the Caribbean, rainfall during the latter part of the 2009 rainy season was below normal and was followed by a drier-than-normal dry season (Farrell et al. 2010). The authors also reported that Caribbean states experienced rainfall totals in the lowest 10 percent, or record lows, between September 2009 and May 2010.

Between May 2010 and early 2014, the Caribbean experienced normal to above-normal rainfall most months. But by the end of 2014, some Caribbean countries (in particular, Jamaica and Antigua) were reporting significant impacts due to the onset of drought conditions. The year 2015 was the driest on record at rainfall stations in many Caribbean islands, including Antigua, Tobago, Barbados, Jamaica, and Saint Lucia (Stephenson et al. 2016). Relief from the drought conditions came in late August/September 2015 at some locations, but by late November, drought conditions were reestablished (CariCOF 2016a). Though some relief finally came to much of the region in June or July 2016, a severely dry August, with record low rainfall in some Caribbean territories, including at three stations in Barbados (CariCOF 2016b), raised fresh concerns.

20.3.2 Impacts on Sectors

Impacts from the 2009–2010 event ranged from reduction in crop yields, low reservoir levels and reduced streamflows, significant increases in the number of bushfires and acreage burned, to a significant number of landslides on overexposed slopes with the return of the rains (Farrell et al. 2010; see also Table 20.1). Farrell et al. (2010) further noted that in St. Vincent and the Grenadines, hydropower production fell by 50 percent in the first quarter of 2010. The climate and health assessment prepared for the Government of Dominica and published in 2016 (Government of Dominica 2016) cited poor storage and treatment of water to mitigate drought as contributors to the proliferation of *Aedes aegypti* mosquitoes, the vector culpable for the transmission of dengue, chikungunya, and Zika across and within sectors. Similar circumstances led to gastrointestinal diseases in Jamaica (Table 20.1). Refer to Farrell et al. (2010) for more detailed impacts of the 2009–2010 event.

The impacts of the 2014–2016 episode were quite similar, with extreme results in some cases. By September 2014, reservoirs in Antigua were empty

TABLE 20.1
Socioeconomic Impacts Due to Drought During 2014 to 2016

Socioeconomic Sector	2014–2016 Impacts
Agriculture	• Decrease in agricultural production reported in Anguilla, Antigua and Barbuda, Barbados, Belize, Dominica, Haiti, Jamaica, Saint Kitts and Nevis, Saint Lucia, Trinidad and Tobago • Reports of increased food prices in Haiti • Increase in plant pests and diseases in Antigua and Barbuda • Losses of livestock in Jamaica • Reports of an increase in destructive bushfires in Jamaica, Saint Kitts and Nevis, and Trinidad and Tobago
Water	• Water shortages reported in Antigua and Barbuda, Barbados, Grenada, Guyana, Saint Kitts and Nevis, Saint Vincent and the Grenadines, Trinidad and Tobago (forcing water rationing) • Potworks Dam in Antigua was only 10 percent full by the end of 2014, and by the end of 2015, consumption of desalinated water was greater than 90 percent, compared with the normal of 60 percent
Energy	• Energy produced by hydroelectric plants down by 15 percent in Jamaica
Health	• Gastroenteritis in Barbados as a result of improper water storage practices • Hot spell warning issued in Trinidad with an advisory for implications to human health
Tourism	• Tourism industry crippled by water crisis in Tobago, with many hotel cancellations due to water shortages

Sources: The Anguillian 2015, Government of Belize 2016, Nation News 2016, Caribbean 360 2014, 2015, 2016; CIMH Caribbean Drought Bulletins; http://rcc.cimh.edu.bb/climate-bulletins/drought-bulletin/, Jamaica Observer 2014a, 2014b; Jamaica Observer 2015.

and water availability in Jamaica was a significant issue. In Georgetown, Guyana's Shelter Belt location, 21,000 persons were affected by limited potable water, and the island of St. Kitts introduced island-wide water rationing for the first time in its history during 2015 (personal communication, Dennison Paul, Acting General Manager of the St. Kitts Water Services Department [WSD], 2016).

In the agriculture sector in 2015, crop production in Belize (inclusive of sugar cane and citrus fruits) declined, with export losses estimated in the millions of dollars. The drought conditions also forced the cancellation of the annual Mango Array & Tropical Fruit Festival in British Virgin Islands (personal communication, Mr. Bevin Braithwaite, Chief Agricultural Officer, British Virgin Islands). Barbados Water Authority (BWA) indicated that increases in irrigation contributed to severe water shortages in some communities across the island (BWA 2015).

20.4 Drought Early Warning Information in the Caribbean

20.4.1 Drought Early Warning Prior to 2009

Chen et al. (2005) noted that monitoring agricultural drought in the Caribbean was historically a case of comparing monthly and annual rainfall totals to their respective averages and monitoring biological indicators in the field, related to agricultural production. This perspective changed following, arguably, the strongest El Nino on record in 1997–1998. As a consequence of the significant impacts suffered by the region from the event and the region's lack of effective early warning and preparedness, the Caribbean and international partners formed the Caribbean Climate Outlook Forum (later referred to as the CariCOF) tasked with preparing 3-month precipitation outlooks for the region, indicating the probability of above- and below-normal and normal rainfall for the period and therefore an indication of drought potential.

Today, climate monitoring and forecasting has evolved beyond this, with a routine suite of tools and products geared toward decision-support and updated regularly. In addition, notable emphasis has been placed on drought monitoring and forecasting by the establishment of the Caribbean Drought and Precipitation Monitoring Network (CDPMN).

20.4.2 Establishing the CDPMN

In January 2009, CIMH launched the CDPMN (CIMH and FAO 2016; Trotman et al. 2009) aimed at monitoring drought (and excessive precipitation) and delivering prognostic climate information at both the national and regional scales. The operationalized system immediately showed its value by providing CARICOM governments with situation analyses and advice beginning in January 2010 (CIMH and FAO 2016). The standardized precipitation index (SPI) (McKee et al. 1993), recommended by WMO for monitoring meteorological drought (Hayes et al. 2011), is integrated in the CDPMN. Deciles (Gibbs and Maher 1967) were also integrated into the CDPMN where, along with the SPI, they were subjectively coupled with the Caribbean precipitation outlook to provide critical information and advice to regional governments in 2010.

20.4.3 Drought Early Warning Information and Products

CIMH plays a leading and critical role in the region in the production and delivery of several climate monitoring and forecasting information products as part of its recently designated role as the WMO Regional Climate Centre for the Caribbean (http://rcc.cimh.edu.bb). Several of these products and associated tools address drought early warning.

20.4.4 Regional Drought Monitoring in CARICOM

Initial regional rainfall (drought) monitoring maps were produced in April 2009 to highlight the precipitation status at the end of March 2009 on four time scales (1-, 3-, 6-, and 12-month) using both SPI and deciles (CIMH and FAO 2016; Farrell et al. 2010; Trotman et al. 2009). The different time intervals are established to reflect the fact that the duration of drought can have different sector impacts (WMO and GWP 2016). Experiences in the Caribbean have illustrated that because of the socioeconomic importance of rain-fed agriculture, the provision of drought information on 3-month time scales mitigates risks in the agricultural sector. Similarly, given the small watersheds and aquifers in the region, and the strong seasonal variability for potable water due to key economic sectors such as tourism, that provision of drought information on timescales of 6 months may be optimal. An example of the 3-month SPI is shown in Figure 20.2.

Insufficient drought impact reporting and methods for robust impact assessment have limited the region's understanding and awareness of the impacts of drought on the performance of social and economic sectors and ultimately the region's socioeconomic development. Building climate-resilient societies that mitigate such phenomena as drought requires collecting such information to enable the development and implementation of robust planning and decision-support systems.

To address this lack of information, which is increasingly being requested, CIMH created the Caribbean climate impacts database (CID) (http://cid.cimh.edu.bb)—an open-source geospatial inventory that archives, among other things, sector-based impacts from various climate phenomena, including drought. The CID also captures (1) planning and response mechanisms used in the disaster risk management sector in the form of standard operating procedures (SOPs) and (2) rainfall impacts via a

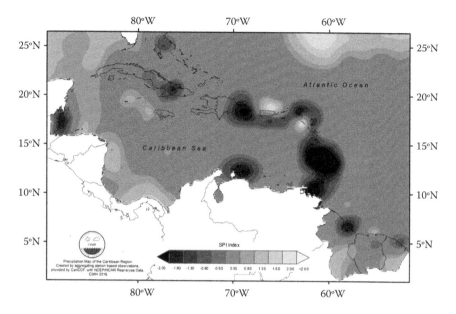

FIGURE 20.2
SPI map for October 2009 to March 2010. (Courtesy of CIMH.)

reporter website that captures real-time submission of impacts that supplement drought monitoring.

20.4.5 Regional Drought Forecasting in CARICOM

A statistical downscaling climate prediction software package, Mason (2011) has been used to provide seasonal tercile rainfall forecasts since 2012. A factor limiting the effective utilization of tercile-based forecast information as a prediction component in drought early warning is that precipitation forecasts are expressed in probabilities of normal, above-normal, and below-normal rainfall, which may be difficult to effectively integrate in sector-based decision-making processes. Although these forecasts can provide an initial indication of pending periods of unusual dryness or wetness, they can neither distinguish unusual from extreme rainfall totals nor ensure sufficient certainty on an extreme outcome, such as rainfall deficits serious enough to cause severe drought.

The first regional drought forecasting system that extended the capabilities of the CPDMN and utilized the CPT was presented at the Wet Season CariCOF held in Jamaica in May 2014 (http://rcc.cimh.edu.bb/long-range-forecasts/caricof-climate-outlooks/). Using this system, forecasts are updated each month for moving 6-month periods (Figure 20.3a), as well as for 12-month periods (Figure 20.3b) ending at the end of the wet season or dry season that the CariCOF calls the *hydrological year*. Through this

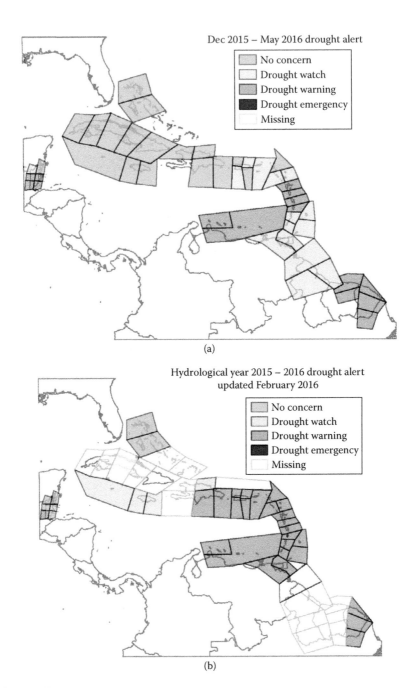

FIGURE 20.3
(a) Drought forecast covering the 6-month period December 2015 to May 2016 and (b) the drought forecast for the hydrological year (June 2015 to May 2016) released at the end of February 2015. The forecasts indicate the impact that is likely to be felt from any deficits in rainfall for the time periods considered.

approach, which adds observed precipitation totals at the start of the period to more uncertain, probabilistic precipitation forecasts for the remainder of the period, the CariCOF drought alert outlooks provide more confidence than the probabilistic precipitation forecasts themselves. This approach enables the identification of 88 percent of observed rainfall deficit levels consistent with severe long-term hydrological drought by the end of the dry season, for example, as early as November (i.e., with a 6-month lead time).

The forecast information included in the CariCOF drought outlooks contributes to a seasonal decision-support system based on alert levels that are tied to increasing probabilities of crossing specific SPI thresholds—with specific but different SPI thresholds being established for the wet and dry seasons. This approach is used to construct drought alert maps with actions corresponding to each alert level following extensive sectoral stakeholder engagement.

Communicating early warning information from the operational drought monitors and forecasts to sectoral stakeholders occurs through packaged products disseminated via the monthly Caribbean drought bulletin (http://rcc.cimh.edu.bb/climate-bulletins/drought-bulletin/) and discussed at the CariCOF.

20.5 Drought Early Warning Supporting Risk Reduction in the Caribbean—Policymaking and Planning

The assessment of Farrell et al. (2010), following the watershed drought of 2009–2010, uncovered several important capacity issues, including limited national capacity in key areas such as (1) early warning, (2) systemic problems within countries that limit information sharing between key stakeholder institutions, (3) inadequate policies and plans, and (4) limited finances to implement and sustain key activities. Further, early warning information should be built into any policies and plans so that any actions taken are relative to the ongoing and expected duration and severity of dryness.

20.5.1 Enabling Environment for Drought Risk Management

Early warning information is recognized by the *Regional Comprehensive Disaster Management (CDM) Strategy and Programming Framework 2014–2024* (implemented by the Caribbean Disaster Emergency Management Agency, CDEMA) and the *Implementation Plan for the Regional Framework for Achieving Development Resilient to Climate Change* (led by the Caribbean Community Climate Change Centre, CCCCC) as critical to the effective management of climate-related disasters and adaptation to climate variability, extremes, and change. Further, drought management was an area of focus at the 53rd Special Meeting of the Council for Trade and Economic Development (COTED) in February 2015, where CIMH was called on to enhance early warning systems for drought in CARICOM. However, while guiding frameworks existed at the regional level, there were major gaps at the national level.

20.5.2 Drought Risk Management: Capacity Building at the National Level

Two core recommendations emerging from Farrell et al. (2010) were that (1) regional institutions include in their activities the development of early warning systems and appropriate indicators to support member states in their planning and adaptation strategies, and (2) Caribbean countries implement appropriate multisector working groups to ensure that each sector is familiar with the various sensitivities and needs of others to ensure timely and effective decision-making. The Caribbean, through the material support of the international donor community (Government of Brazil, USAID) as well as the technical and coordination support of international, regional, and subregional organizations such as CIMH, CDEMA, OECS, FAO, and National Drought Mitigation Center (NDMC) at the University of Nebraska–Lincoln, has since embarked on three strategic initiatives to address these gaps:

1. Phase 1—CARICOM/Brazil/FAO Cooperation Program on Disaster Risk Reduction (DRR)
2. Phase 2—USAID-funded OECS Reduce Risk to Human & Natural Assets Resulting from Climate Change (RRACC) project
3. Phase 3—USAID-funded Building Regional Climate Capacity in the Caribbean program

Capacity building has focused on three main areas: (1) addressing gaps in NMHS drought early warning capabilities; (2) building the capacity of sectoral practitioners, decision makers, and policymakers in the Caribbean to use drought monitoring and outlook products; and (3) enhancing national frameworks, policies and plans, and terms of reference (TOR) to more effectively manage drought risk and, in many cases, commence the development of drought plans that integrate drought early warning information.

Phase 1: CARICOM/Brazil/FAO Cooperation on Disaster Risk Reduction (DRR)

In addition to the provision of extensive training in drought monitoring and planning, a major output of this inaugural initiative that commenced in 2012 was consensus building around the development of a draft national drought management framework for the Caribbean (Figure 20.4). This framework

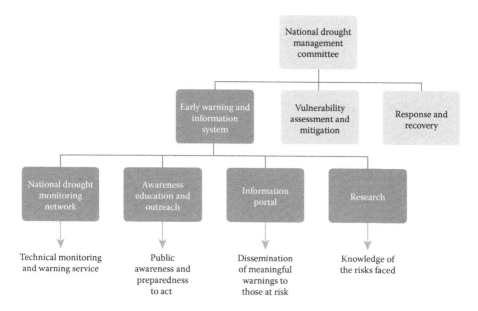

FIGURE 20.4
Proposed framework for national drought management in the Caribbean.

focused on building drought early warning information systems (DEWIS) through networks and working groups, whose activities would be managed by a national drought management committee that would also be responsible for other aspects of disaster management related to drought (i.e., vulnerability assessment and mitigation, and recovery and response). This framework is in line with international thinking regarding the four main components of early warning information systems as proposed by UNISDR (2009). During this phase, St. Lucia developed the TORs for the Flood and Drought Mitigation Committee that was ratified by its government. Draft TORs were also developed for national drought monitoring networks for Grenada and Jamaica.

Phase 2: USAID-funded OECS Reduce Risk to Human and Natural Assets Resulting from Climate Change (RRACC) Project

Drought conditions between 2014 and 2015 in the eastern Caribbean emphasized the urgency of continuing to advance efforts commenced under the CARICOM/Brazil program. The highlight of this phase was the policy and plan development process carried out through a series of "writeshops" where the writing, editing, and enhancing of national documents were clear objectives. The ultimate goal was to support member states in amending existing national hazard management plans to include mitigation and response to the drought hazard or to developing new plans, including DEWIS plans (Table 20.2).

TABLE 20.2

Phase 2 Drought Writeshops Output and Progress for Select OECS Member States

Objectives	Country	
• To help advance drought preparedness • To build upon drought monitoring and forecasting initiatives • To develop capacity to interpret and apply products • To further advance draft implementation plans for national DEWISs in the context of a national drought management framework	Saint Lucia	1. Enhanced the Saint Lucia disaster management policy framework 2. Updated the water and sewage company's water management plan for drought conditions to better reflect drought considerations 3. Further refined the roles and responsibilities outlined in the existing flood and drought management committee's TOR
	Grenada	1. Assessment of the drought management framework 2. Advanced Grenada flood and drought early warning and information systems draft TOR commenced under Phase 1 3. Developed a Grenada plan of action
	Antigua and Barbuda	1. Developed draft institutional and legislative framework review document 2. Developed draft Antigua and Barbuda national drought management committee TOR
	St. Kitts and Nevis	1. Developed draft assessment of national and sectoral drought policies and plans document 2. Initiated work on a new draft water services drought management plan 3. Developed draft drought management committee TOR

Phase 3: USAID-funded BRCCC Program

The BRCCC Program continues the writeshop process to advance the suite of revised and new documents. This will be done by continuing and completing the draft documents for the four select countries in Table 20.2, so that these documents can be considered for ratification by the cabinets of the national governments.

20.6 Building on the Foundation: Advancing Drought Management in CARICOM through Early Warning and Other Risk Reduction Options

The 2009–2010 event challenged the region to give greater consideration to drought as a disaster that has to be managed more strategically, even at the level of CARICOM heads of government, particularly in light of climate change projections (Farrell et al. 2010). The most recent event in 2014–2016 illustrated that, despite the severity and extensive nature of the drought, the region was better prepared, having had early warning information and some built capacity since the 2009–2010 event. However, the event has also indicated that there is still work to be done.

With policy and plan development deemed necessary for CARICOM states, one of the immediate priorities is the completion of the suite of policies and plans for the four states in Table 20.2. This may motivate other member states to follow, prompted by recent impacts of the 2014–2016 event. The CIMH has also encouraged member states to develop drought plans specific to their sectors. Though many CARICOM states have some level of action plans for water, they mainly support management of potable water resources. Also, the region, with the assistance of the FAO, has sought to develop national agriculture disaster risk management (ADRM) plans (Roberts 2013), with plans being ratified in Guyana, Belize, and St. Lucia (personal communication from Dr. Lystra Fletcher-Paul, Caribbean Sub-regional Director of the FAO). Most of the draft and ratified ADRM plans lacked comprehensive, if any, actions to mitigate drought. Since then, mitigation measures have been elaborated for the Caribbean agriculture sector (CIMH and FAO 2016), and a drought template to enhance adaptation in the agriculture sector is being pursued under CARICOM's agriculture policy program (APP).

Since 2015, CIMH has been working toward developing sectoral early warning information systems across climate timescales (EWISACTs) aimed at providing user-oriented climate early warning information tailored to specific user needs. The development of sectoral EWISACTs supports the regional implementation of the global framework for climate services (GFCS) (WMO 2011). In the Caribbean, six climate-sensitive sectors have been prioritized, namely the agriculture, water, energy, health, disaster risk management, and tourism sectors. The region has formalized an approach to the codesign, codevelopment, and codelivery of climate early warning information, including for drought, for Caribbean sectors in the form of a multisectoral partnership between lead regional technical organizations in these six climate-sensitive sectors and the CIMH (as the climate services provider)—the Consortium of Regional Sectoral EWISACTs Partners (Figure 20.5). It is anticipated that (1) the consortium and its activities will filter down to the national level, enhancing climate (including drought) early warning

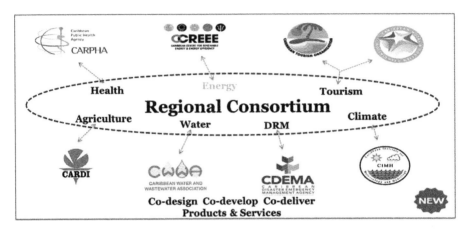

FIGURE 20.5
Consortium of regional sectoral EWISACTs partners.

and its uptake, and (2) the consortium framework and structure will be replicated at the national level. Supported and encouraged by this process, and led by CIMH, the region is also keen to advance the work of the CDPMN and the CariCOF that provides early warning information.

The work of the consortium also helps to close current gaps in physical and social research because this affects the nature, quality, and usefulness of early warning products (e.g., improving spatial and temporal resolution, communicating climate science in targeted, user friendly language). Further, stakeholders have raised the absence of funding/budget for addressing drought as a critical barrier that hampers action. Mobilizing consistent funding to apply to proactive efforts that seek to build capacity for drought risk management even under nondrought conditions remains key.

Since 2009, the CARICOM has enhanced its response to drought risk management. The enabling environments being established at the regional and national levels are set to enhance management with resilience-building as well as the reduction of impacts and losses from this hazard in the near future. Based on the preliminary evidence, it has been suggested that a systematic and sustained effort should be attained. The path forward should be centered on implementing a stakeholder-driven agenda that continues to address gaps in four key areas: (1) improving the provision and usefulness of scientific drought-related products, (2) connecting drought early warning information to national and sectoral decision-making tools/systems, (3) strengthening the enabling environment in national and sectoral systems, and (4) sustainability. With these priority areas, the region stands to benefit, particularly with drought-sensitive sectors that would be more productive and efficient.

References

Angeles ME, Gonzalez JE, Erickson DJ, Hernandez JL. 2007. Predictions of future climate change in the Caribbean region using global general circulation models. *Int J Climatol* 27: 555–569.

Anguillian. 2015. *The way forward for agriculture in Anguilla.* http://theanguillian. com/2015/06/the-way-forward-for-agriculture-in-anguilla (accessed December 20, 2016).

Barbados Water Authority (BWA). 2015. *Reasons for reduced water.* supply. http://barbadoswaterauthority.com/?p=3163 (accessed December 27, 2016).

Caribbean360. 2014. *Antigua Reservoirs remain empty after a year of severe drought conditions.* http://www.caribbean360.com/news/ (accessed December 20, 2016).

Caribbean360. 2015. *British Virgin Islands on drought watch with dramatic drop in rainfall.* http://www.caribbean360.com/news/british-virgin-islands-on-drought-watch-with-dramatic-drop-in-rainfall (accessed December 19, 2016).

Caribbean360. 2016. *Water rationing resumes in St. Kitts and Nevis.* http://www. caribbean360.com/news/water-rationing-resumes-in-st-kitts-and-nevis-after-below-normal-rainfall (accessed August 18, 2016).

CariCOF. 2016a. Caribbean climate outlook newsletter. March to August. CIMH, Bridgetown, Barbados.

CariCOF. 2016b. Caribbean climate outlook newsletter. October to December. CIMH, Bridgetown, Barbados.

Charlery J, Nurse L, Whitehall K. 2006. Exploring the relationship between the North Atlantic oscillation and rainfall patterns in Barbados. *Int J Climatol* 26: 819–827.

Chen AA, Falloon T, Taylor M. 2005. Monitoring agricultural drought in the West Indies. *In Monitoring and Predicting Agricultural Drought: A Global Study.* eds. VK. Boken, AP. Cracknell, RL. Heathcote, pp. 144–155. New York: Oxford University Press.

CIMH and FAO. 2016. *Drought characteristics and management in the Caribbean.* 2–3. FAO. http://www.fao.org/3/a-i5695e.pdf (accessed December 27, 2016).

Cook KH, Vizy EK. 2010. Hydrodynamics of the Caribbean low-level jet and its relationship to precipitation. Special U.S. CLIVAR drought collection. *J Clim* 23: 1477–1494.

Cramer LW. 1938. *Annual report of the Governor of the Virgin Islands.* Washington, DC: United States Government Printing Office.

Cundall F. 1927. *Chronological outlines of Jamaica history, 1492–1926.* Kingston, Jamaica: Government Printing Office.

Dai A. 2011. Drought under global warming: A review. *Clim Change* 2(1): 45–65. doi:10.1002/wcc.81.

Dai A. 2013. Increasing drought under global warming in observations and models. *Nat Clim Change* 3(1): 52–58. doi:http://www.nature.com/nclimate/journal/v3/n1/abs/nclimate1633.html#supplementary-information.

Enfield DB, Alfaro EJ. 1999. The dependence of Caribbean rainfall on the interaction of the tropical Atlantic and Pacific oceans. *J Clim* 12: 2093–2103.

Farrell D, Trotman A, Cox C. 2010. Drought early warning and risk reduction: A case study of the drought of 2009–2010. http://www.preventionweb.net/english/hyogo/gar/2011/en/bgdocs/Farrell_et_al_2010.pdf (accessed December 27, 2016).

Farrell DA, Caesar K, Whitehall K. 2007. *Climate Change and its Impact on Natural Risk Reduction Practices, Preparedness and Mitigation Programmes in the Caribbean.* Elements for Life, Tudor Rose, London, UK.

Gamble DW, Parnell DB, Curtis S. 2008. Spatial variability of the Caribbean mid-summer drought and relation to north Atlantic high circulation. *Int J Climatol* 28(3):343–350. doi:10.1002/joc.1600.

Giannini A, Chang JCH, Cane MA, Kushnir Y, Seager R. 2001. The ENSO teleconnection to the tropical Atlantic ocean: Contributions of the remote and local SSTs to rainfall variability in the tropical Americas. *J Clim* 14: 4530–4544.

Giannini A, Kushnir Y, Cane MA. 2000. Interannual variability of Caribbean rainfall, ENSO and the Atlantic Ocean. *J Clim* 13: 297–310.

Gibbs WJ, Maher JV. 1967. *Rainfall deciles as drought indicators.* Bureau of Meteorology Bulletin No. 48, Melbourne, Australia.

Government of Belize. 2016. *Belize Drought Assessment.* Belize City: Ministry of Agriculture.

Government of Dominica. 2016. *Assessment of Climate Change and Health Vulnerability and Adaptation in Dominica).* Ministry of Health, Roseau, Dominica.

Hayes M, Svoboda M, Wall N, Widhalm M. 2011. The Lincoln Declaration on Drought Indices: Universal meteorological drought index recommended. *Bull Am Meteorolog Soc* 92(4): 485–488.

Hughes TP, Graham NAJ, Jackson JBC, Mumby PJ, Steneck RS. 2010. Rising to the challenge of sustaining coral reef resilience. *Trends Ecol Evol* 25: 633–642.

IPCC, 2013: Climate Change 2013: The Physical Science Basis. Contribution of Working Group I to the Fifth Assessment Report of the Intergovernmental Panel on Climate Change [Stocker, T.F., D. Qin, G.-K. Plattner, M. Tignor, S.K. Allen, J. Boschung, A. Nauels, Y. Xia, V. Bex and P.M. Midgley (eds.)]. Cambridge University Press, Cambridge, United Kingdom and New York, NY, USA, 1535 pp, doi:10.1017/CBO9781107415324.

Jamaica Observer. 2014a. *Drought cuts hydro energy output 15%.* http://www.jamaicaobserver.com/business/Drought-cuts-hydro-energy-output-15-17343276 (accessed December 20, 2016).

Jamaica Observer. 2014b. *Worst of the drought is yet to come, Pickersgill warns.* http://www.jamaicaobserver.com/news/Worst-of-the-drought-is-yet-to-come--Pickersgill-warns-_17211082 (accessed December 20, 2016).

Joyette ART, Nurse LA, Pulwarty RS. 2015. Disaster risk insurance and catastrophe models in risk-prone small Caribbean islands. *Disasters* 39(3): 467–492. doi:10.1111/disa.12118.

Lindin HJ. 1973. "Drought in the Caribbean." *The Tribune* 5: 7.

Martin ER, Schumacher C. 2011. Modulation of Caribbean precipitation by the Madden–Julian oscillation. *J Clim* 24: 813–824. http://journals.ametsoc.org/doi/10.1175/2010JCLI3773.1

Mason SJ. 2011. Seasonal forecasting using the Climate Predictability Tool (CPT). *Proceedings of the 36th NOAA Annual Climate Diagnostics and Prediction Workshop,* Fort Worth, Texas, TX. pp. 180–182. http://www.nws.noaa.gov/ost/climate/STIP/36CDPW/36cdpw-smason.pdf (accessed December 27, 2016).

Maybank J, Bonsai B, Jones K, Lawford R, O'Brien EG, Ripley EA, Wheaton E. 1995. Drought as a natural disaster. *Atmosphere-Ocean* 33(2): 195–222. doi:10.1080/07055900.1995.9649532.

McKee TB, Doesken NJ, Kleist J. 1993. The relationship of drought frequency and duration to time scales. *Preprints, 8th Conference on Applied Climatology*, pp. 179–184. American Meteorological Society, Boston, MA.

MPDE. 2001. *State of the environment report 2000*. GEO Barbados. Barbados: Ministry of Physical Development and Environment.

Mumby PJ, Flower J, Chollett I, Box S, Bozec Y, et al., 2014. *Towards Reef Resilience and Sustainable Livelihoods. A Handbook for Caribbean Coral Reef Managers.* Exeter, England: University of Exeter.

NationNews. 2016, Drought drying up crop. *Nationnews*, January 24. http://www.nationnews.com/nationnews/news/77038/drought-drying-crop (accessed December 20, 2016).

Prospero JM, Lamb PJ. 2003. African droughts and dust transport to the Caribbean: Climate change implications. *Science* 302(5647): 1024–1027.

Prospero JM, Nees RT. 1986. Impact of the North African drought and El Niño on mineral dust in the Barbados trade winds. *Nature* 320: 735–738. doi:10.1038/320735a0.

Pulwarty R, Nurse L, Trotz U. 2010. Caribbean islands in a changing climate. *Environment: Science and Policy for Sustainable Development* 52(6): 16–27.

Roberts D. 2013. *Status of disaster risk management. Plans for floods, hurricanes and drought in the agriculture sector: A Caribbean perspective.* Barbados: FAO Subregional Office. http://www.fao.org/docrep/018/i3341e/i3341e.pdf (accessed December 27, 2016).

Rodriguez A. 2013. *African dust clouds worry Caribbean scientists.* August 27. http://phys.org/news/2013-08-african-clouds-caribbean-scientists.html. (accessed December 27, 2016).

Stephenson TS, Chen AA, Taylor MA. 2008. Toward the development of prediction models for the primary Caribbean dry season. *Theor Appl Climatol* 92(1–2): 87–101. doi: 10.1007/s00704-007-0308-2.

Stephenson TS, Taylor MA, Trotman AR, et al. 2016. Caribbean. *Bull Am Meteorolog Soc* 97(8): S181–182.

Taylor MA, Enfield DB, Chen AA. 2002. Influence of tropical Atlantic versus the tropical Pacific on Caribbean rainfall. *J Geophys Res* 107(C9): 3127. doi:10.1029/2001/JC001097.

Taylor MA, Whyte FS, Stephenson TS, Campbell JD. 2012. Why dry? *Int J Climatol* 33(3): 784–792. doi:10.1002/joc.3461.

Taylor MA, Stephenson TS, Owino A, Chen AA, Campbell JD. 2011. Tropical gradient influences on Caribbean rainfall. *J Geophys Res* 116: D00Q08. doi:10.1029/2010JD015580.

Toba N. 2009. *Potential economic impacts of climate change in the Caribbean community. LCR Sustainable Development.* Working Paper No. 32, 35–47. Washington, DC: World Bank.

Trotman AR, Moore A, Stoute S. 2009. The Caribbean drought and precipitation monitoring network: The concept and its progress. In *Climate Sense*, pp. 122–125. Tudor-Rose. Leicester, United Kingdom.

UNISDR 2009. Terminology on Disaster Risk Reduction. United Nations Office for Disaster Risk Reduction. Geneva, Switzerland. 30pp.

WMO. 2011. *Climate knowledge for action: A global framework for climate services—Empowering the most vulnerable.* The Report of the High-level TaskForce for the Global Framework for Climate Services. Geneva: World Meteorological Organization.   https://www.wmo.int/gfcs/sites/default/files/FAQ/HLT/HLT_FAQ_en.pdf (accessed December 30, 2016).

WMO and GWP. 2016. *Handbook of drought indicators and indices.* eds. M. Svoboda, BA. Fuchs. Integrated Drought Management Programme (IDMP), Integrated Drought Management Tools and Guidelines Series 2. Geneva.

WRI. 2013. *World's 36 most water-stressed countries.* World Resources Institute, December 12. http://www.wri.org/blog/2013/12/world%E2%80%99s-36-most-water-stressed-countries (accessed December 27, 2016).

21

Facilitating a Proactive Drought Management and Policy Shift: Recent Lessons from Northeast Brazil

Nathan L. Engle, Erwin De Nys, and Antonio Rocha Magalhães

CONTENTS

21.1 Droughts and Their Management in Brazil

Brazilians have a long history of living with harsh conditions in the Northeast.[*] The majority of the Northeast is characterized as the *sertão*, or the semiarid region that is defined by its long, almost rainless, dry season of several months. The people of Northeast Brazil have managed over the years to cope with these conditions, including through the introduction of water infrastructure projects and the advent of institutions responsible for planning the socioeconomic development of the region. The improvements in supply expansion to address water needs and support to farmers have helped the region progress over the decades. However, when extreme droughts hit the Northeast, structural solutions, while necessary, are often insufficient to withstand these multiyear periods of below-average rainfall.

Since 2012, and continuing through early 2017, the semiarid Northeast has been suffering through an intensive prolonged drought. Reservoirs are at historically low levels, and even if rainfall improves throughout 2017, hydraulic systems will require several additional years to recover to

---

[*] The Northeast is a very large area of 1,561,177 km² that consists of nine states: Maranhão, Piauí, Ceará, Rio Grande do Norte, Paraíba, Pernambuco, Alagoas, Sergipe, and Bahia (from north to south).

full capacity. The systems that have remained resilient over the last 5 years will not remain operational for an additional dry year, leading to the likely collapse of water supply in small cities and increases in already severe water rationing in state capital cities across the Northeast (eight of the nine capital cities have been historically buffered from the direct impacts of drought because of their location along the more rain-abundant Atlantic coast in the Northeast region). This has threatened the ability of society to maintain adequate drinking water supplies and water for other uses, such as irrigation, hydropower, industrial production, and environmental goods and services. The impacts of prolonged droughts are often concentrated in the rural poor communities living in this semiarid region. Ultimately, these impacts threaten the considerable gains in terms of economic, social, and human development that the region has experienced in the past several decades and place many communities at risk of slipping back into extreme poverty. Figure 21.1 shows the extent of the drought in one Northeast state, Ceará.

Brazil, like many nations, has invested heavily in emergency actions to mitigate the economic losses from prolonged periods of droughts as they unfold. Examples include but are not limited to increased emergency lines of credit, renegotiation of agricultural debts, expansion of social support programs such as Bolsa Estiagem and Garantia-Safra (cash transfer programs to poor families and farmers), and Operação Carro Pipa (water truck deliveries of emergency drinking water to rural communities).

Gaining access to many of these programs and resources relies on the municipalities declaring a situation of emergency or a state of public calamity, loosely defined as an intense and serious shift of the normal conditions

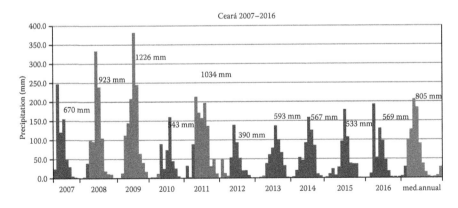

FIGURE 21.1
Monthly average rainfall and rainfall distribution (January–December) for Ceará state in Northeast Brazil, from 2007 to late 2016. Wet years are depicted in blue, dry years in red. The average year in terms of distribution and amount is located at the far right of the figure in gray. (Courtesy of FUNCEME, Fortaleza, Brazil.)

that affects the locale's response capacity. The state and federal governments then verify and provide access to drought-emergency resources and programs. However, this declaration and assistance process does not involve a systematic procedure for objectively defining droughts and what should constitute an emergency or public calamity situation. Without a specific set of scientifically informed indicators or criteria on which to base the declaration, drought management has been historically reactive to the emergency situation occurring on the ground, and subsequent relief measures are often slow, inefficiently targeted, and subject to political capture and corruption.

The recent drought in the Northeast spurred an intense debate within the country to improve drought policy and management. In recognition of both the need to move away from the crisis management of droughts and the opportunity presented by the current drought and water scarcity situations to make lasting progress, the Ministry of National Integration (MI) in 2013 requested analytical, advisory, and convening services from the World Bank (Bank) to help it in its endeavor to shift its traditional crisis management of droughts to a more prepared and risk-based management approach.* As a result of MI's request, the Bank developed the drought preparedness and climate resilience program (Program) to assist in this endeavor, which was implemented between 2013 and 2016.

The main objective of the Program was to help stakeholders in Brazil (both at the national and state levels, and more specifically in the Northeast region) develop and institutionalize proactive approaches to drought events, with an ancillary benefit of developing tools, frameworks, processes, and exchange platforms from which other countries and sectors/regions could learn and eventually foster innovation around this topic.

This chapter describes the recent advances in drought policy and management in Brazil that were supported by the Program over the past 3½ years, the main results, and the anticipated next steps. A more complete account of these efforts, as told through the various perspectives of some of the key stakeholders involved, is detailed in De Nys et al. (2016a) (English version) and De Nys et al. (2016b) (Portuguese version).

---

* This request paralleled activities that were occurring on the international stage for improving drought resilience, most notably the High-level Meeting on National Drought Policy (HMNDP), in Geneva, Switzerland, in March 2013. At the HMNDP, Brazil declared its commitment to discuss and debate how to design, coordinate, and integrate comprehensive policy on drought planning and management in order to reduce impacts and increase resilience to future droughts and climate change. In December 2013 (in partnership with this program), Brazil and the MI also hosted a follow-up international workshop for the HMNDP process, which gathered over one dozen countries to build capacity for developing national drought policies across the Latin America and Caribbean region. By the end of 2013, MI's endeavor to shift the drought paradigm in Brazil was well underway.

21.2 Shifting the Paradigm along the Three Pillars Framework

The Bank and its partners structured the Program around a framework since referred to as the *three pillars of drought preparedness*. This framework, illustrated in Figure 21.2, consists of (1) monitoring and forecasting/early warning, (2) vulnerability/resilience and impact assessment, and (3) mitigation and response planning and measures.

The Program was designed as two mutually reinforcing tracks. Track 1, or the "drought policy track," sought to support a national/regional and state dialogue and framework on drought preparedness policy.

Track 2, or the "Northeast pilot track," endeavored to implement a Northeast Brazil regional program to demonstrate tangible tools and strategies for proactive drought management through the design and development of both a Northeast drought monitor (Monitor) and operational drought preparedness plans across selected case studies. Five preparedness or contingency planning exercises were developed through dialogue and requests among different sectors and at different scales of decision-making: two urban water utility case studies in the Fortaleza Metropolitan Region and the Agreste Region of Pernambuco, respectively; two plans (a basin-wide drought preparedness plan and a multiple-use plan for a small açude, or water storage reservoir, called the Cruzeta Reservoir) in the Piranhas-Açu River Basin, which is shared between the states of Paraíba and Rio Grande do Norte; and a plan for rain-fed agriculture at the level of a municipality in central Ceará state, the Piquet Carneiro municipality.

These tracks were bridged by an econometric assessment of the drought in the Northeast, which explored linkages with the Monitor and various federal and state data sources to identify impacts and costs of the current Northeast drought.

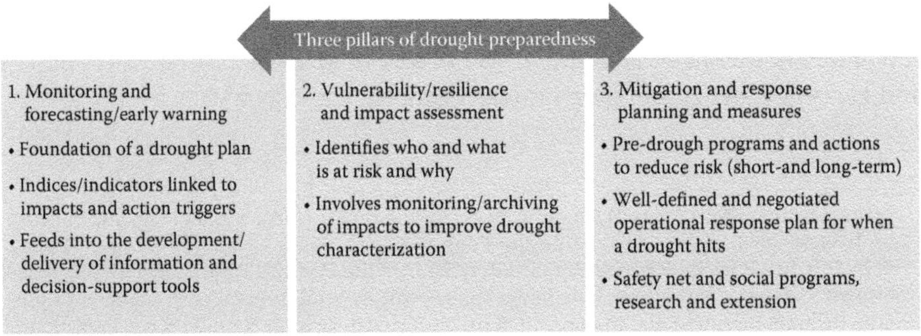

Three pillars of drought preparedness

1. Monitoring and forecasting/early warning	2. Vulnerability/resilience and impact assessment	3. Mitigation and response planning and measures
• Foundation of a drought plan	• Identifies who and what is at risk and why	• Pre-drough programs and actions to reduce risk (short-and long-term)
• Indices/indicators linked to impacts and action triggers	• Involves monitoring/archiving of impacts to improve drought characterization	• Well-defined and negotiated operational response plan for when a drought hits
• Feeds into the development/delivery of information and decision-support tools		• Safety net and social programs, research and extension

FIGURE 21.2
The three pillars of drought preparedness that served as the guiding framework for this Program to support a paradigm shift away from reactive crisis management and toward more proactive approaches to drought events.

21.3 Key Advancements in Drought Policy and Management

The Monitor quickly developed into the anchor for the broader technical and institutional upgrades being sought by MI. In its most visible form, the Monitor is a monthly map that describes the current state of drought across the Northeast, determined by several meteorological/hydrological indicators (e.g., standardized precipitation index and standardized evapotranspiration index). The indicators are weighted to produce a composite five-stage drought severity index (on a scale of S0–S4, where S means *seca*, or drought), and thus add nuance, objectivity, and consistency to the definition of drought in Northeast Brazil. The categories are defined as the percentile of recurrence across the index of weighted indicators: S0 (30th percentile, going into and coming out of drought); S1 (20th percentile, moderate drought); S2 (10th percentile, severe drought); S3 (5th percentile, extreme drought); and S4 (2nd percentile, exceptional drought). The occurrence of an S4 drought, therefore, represents a 1 in 50 year event. Figure 21.3 provides a recent example of the monthly Monitor map.

The Monitor is inspired by the efforts of the US Drought Monitor and similar efforts in Mexico, and as such, it has benefited from close collaboration and training with the individuals and institutions in these countries responsible for their respective drought monitoring efforts. Similar to how it operates in these countries, the Monitor is far more than a map for the stakeholders in Brazil. Rather, it is an organizational construct of people, institutions, and processes, which are as important as the map itself. Its production has taken considerable collaboration and behavioral changes and now involves close coordination between senior-level technical specialists from institutions of the nine Northeast states and several federal entities. Three states (i.e., Ceará, Pernambuco, and Bahia) are playing the role of authors of the map, and all nine states are involved as validators to ensure the map is accurately depicting the drought conditions in their respective areas, while also helping to refine and constantly improve the characterization of drought throughout the region. Each month, the authors take turns leading a 2-week-long process of gathering and processing information from a newly established data integration and sharing process that was facilitated among the states and federal government via the Program, and subsequently drawing the map using a geographical information system. This process also includes revisions, discussions, and data exchanges to validate and improve the map before it is published.

The Monitor was officially launched in March 2016, and it has been operational and available to the public since then. One of the key federal partners, the National Water Agency (ANA), leads the coordination among the federal and state institutions, plays the role of Executive Secretary, and hosts the website (http://monitordesecas.ana.gov.br/).

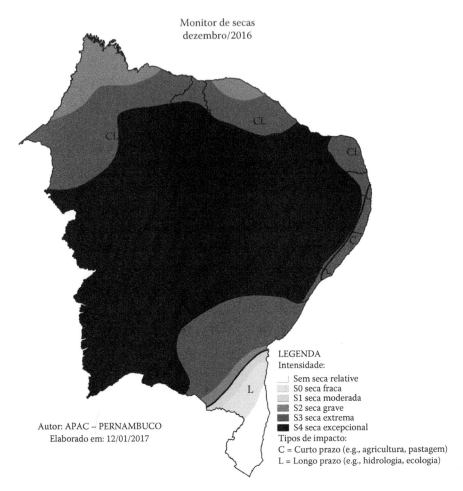

FIGURE 21.3
Drought map for December 2016 produced by the Northeast Drought Monitor.

Along with the Monitor, the drought plans endeavored to make the elements of drought preparedness tangible to decision makers and demonstrate the paradigm shift toward proactive drought management. The plans all characterize drought impacts and vulnerabilities, key institutional actors, planning measures for mitigating drought risk, and emergency responses. As such, the teams and partners designing and implementing the plans attempted to consolidate the plans along the three pillars framework, and, to the extent possible, begin to make the links between the Monitor and its categorization of drought across S0–S4 with context-specific policy and management actions triggered by these categories in the drought preparedness plans. Whereas some of the plans were unable to define policy and management actions triggered as the drought progresses to higher stages (e.g., S0–S4), such as in Piquet Carneiro (the rain-fed agriculture plan), others, such as the two urban plans, formulated a range of actions to be triggered across these various stages of drought.

Some of the plans have already become operational in the communities for which they were designed, the intention being that they will be used to guide decisions as the next drought unfolds and also to help guide longer-term investments to address underlying vulnerabilities and mitigate future drought risks. Moreover, these concrete examples of drought preparedness plans are helping to drive the conversation among federal and state governments on how to scale up these planning exercises across the Northeast.

None of the plans are able to pull information directly from the Monitor to inform the policy actions/triggers, because in its early stages the Monitor does not yet have the breadth and granularity of indicators to warrant a direct link between it and the plans. However, most of the plans adhere to the new S0–S4 categorization of drought severity and intend to use this categorization to feed back into the Monitor to inform its characterization of drought (e.g., the reservoir levels in the urban plan associated with S0–S4 will help the Monitor define drought severity in those areas). All of the plans highlight a need for continued iteration, and in these future updates to the plans, to strengthen links with the Monitor.

The Program produced several analytical products to explain the socio-economic, institutional, technical, political, and social aspects of Brazilian drought policies. These included a rapid impacts and cost analysis (carried out from September 2014 to January 2015), which provided a qualitative and quantitative evaluation of the drought across multiple sectors of the Northeast, identified the key actors for supporting the institutionalization of the second pillar, and demonstrated a methodology that these actors can replicate for evaluating the impacts of droughts and drought responses in the future. The Program also helped to define a set of principal action items for advancing and institutionalizing a national drought policy and program, which informed the initial discussions around a national seminar process in late 2013 and produced a multicountry comparative drought policy study to identify lessons and good practices from several other drought-prone nations (i.e., Australia, Mexico, Spain, and the United States) (Cadaval Martins et al. 2015).

21.4 Lessons Learned and Next Steps

The three pillars framework introduced through this Program has resonated strongly within Brazil. MI, ANA, and many of the Northeast state partners have elected to pursue future drought policies and strategies along this framework, particularly strengthening the second and third pillars (vulnerability/risk/impact assessment and mitigation and response planning and measures, respectively). Overall, there has been an improvement in awareness of drought definitions and declaration and response processes, and partners are confident that recent advancements will bring more permanence/institutionalization to drought preparedness planning and management, but this will not be truly visible until the next drought hits.

While the Monitor and drought preparedness plans have served as important technical and tangible underpinnings, the most significant progress made over the past 3 to 4 years has been the institutional advancements and the change in mentality and approach to droughts. The network of institutions, people, and processes across the Northeast states that are now committed to the Monitor and the broader drought paradigm shift (as reflected by the robust authors and validators processes and networks) is indicative of these fundamental advancements. Despite these achievements, other agencies and ministries (e.g., those responsible for implementing policies and actions) still need to be more involved with the Monitor.

The drought plans all represent significant strides in local/state collaboration and agreements toward improving resilience in these respective communities, as well as the building of capacity and buy-in for the broader drought paradigm shift. The various meetings, planning workshops, and other capacity-building activities that were supported by the Program will help institutionalize drought preparedness planning. One very positive sign is that several states have already started to integrate drought preparedness concepts into statewide planning and policy decisions, as informed by the efforts of the Program. This includes discussions of a state drought plan for Ceará that will incorporate the Monitor and preparedness planning, as well as a development policy loan from the Bank to the state of Sergipe, which has drought policy preparedness as one of its requirements for disbursements.

Despite these achievements, there is still work to do to solidify the paradigm shift. Moving beyond the strong support of MI and ANA (as well as the three authoring states of Ceará, Pernambuco, and Bahia), other federal and state partners still need to become acquainted with the new approach to drought policy and management, and the benefits of implementing proactive measures. Since the official launch of the Monitor, the process has sought to maintain and continue to build higher levels of trustworthiness and relevance. This has been challenging with the current transitions in politics and leadership, particularly the replacement of champions at the highest levels of MI. The drought preparedness plans will also face challenges in remaining relevant during these political changes. And perhaps the most significant test for both the Monitor and the plans will be whether they can maintain relevance and support once the current drought eventually begins to subside.

To overcome these challenges, Brazil needs to focus on further integrating institutional processes associated with drought preparation and response, including by making drought committees, which are typically *ad hoc* intergovernmental emergency planning mechanisms put together in the middle of the crisis, more permanent institutional coordinating bodies. Moving these away from *ad hoc* response mechanisms and toward deliberative decision-making forums through which leading agencies can regularly discuss long-term and short-term programs, policies, and approaches will help to institutionalize the drought policy paradigm shift. It will also provide a vehicle for implementing policy actions based on the Monitor, as well as developing and coordinating between various drought preparedness plans.

A proactive approach to managing droughts will take continued communication efforts across government and society. It is imperative for the main partners within the state and federal governments to make clear two key arguments for those policy officials not yet familiar with the benefits of drought preparedness: (1) it reflects good management of public expenditures because it saves money and reduces hardship associated with drought impacts, and (2) it is a way to minimize political losses throughout the duration of their term as the current droughts persist, or in the wake of a new drought.

It is also important for Brazil to embark upon efforts to develop and institutionalize pillars two and three of the three pillars framework, and to do this by continuing a national drought policy dialogue and building from the strong foundation of the Monitor and drought preparedness plans. Establishing the Monitor, which will in itself still require tremendous effort and commitment moving forward, was an important first step. However, to fully realize the benefits of drought preparedness and to achieve the desired paradigm shift, strengthening vulnerability and risk/impact assessment and decision-making through mitigation and response planning and measures is critical.

There is also a need and an opportunity for the country to adapt to the achievements in the Northeast over the past few years in other areas of the country, particularly the Southeast. The recent water scarcity crisis in and around greater São Paulo catapulted drought into the spotlight as an issue of national concern and debate. Internationally, the world is now looking closely at how Brazil is addressing drought preparedness.

Finally, it will be important for governing bodies in Brazil to situate the drought preparedness efforts within medium- and long-term development and water security objectives, including how to capitalize on the progress of the Program to more broadly build climate resilience. For example, Brazil is making major investments in water infrastructure such as irrigation schemes, reservoirs, and strategic water transfers between river basins in the Northeast, including the integration project of the São Francisco River (PISF). The PISF will soon bring considerable water flows to the Northeast and therefore increase water systems' resilience to climate variability and boost development in the region. Important challenges are thus on the horizon with respect to how to develop planning and management strategies for maintaining longer-term resilience and how to harmonize such processes against future droughts and water shortages, energy supply and demand, ecosystem goods and services, rain-fed and irrigated agriculture needs, and regional economic development policies and programs.

At present, the discrepancy between the design assumption of the water allocation planning and the operational reality of needing to manage greater uncertainty with climate change impacts and demand increases leaves little margin to maneuver and is often a main driver of water conflict. Analyzing, documenting, and understanding the key vulnerabilities across these integrated sectors and projects will help facilitate adaptation to the hydrological effects of climate change, particularly increasing droughts and water shortages. The ability to mitigate the impacts of future droughts is, therefore, inherently tied to regional economic development, ecosystems recovery and management, and water and energy investments, among other factors. Using the momentum from the drought preparedness efforts to date will help Brazil to begin making these important connections.

References

Cadaval Martins, J., N. L. Engle, and E. De Nys. 2015. Evaluating national drought policies. A comparative analysis of Australia, Brazil, Mexico, Spain and the United States. Centro de Gestão e Estudos Estratégicos (CGEE). *Parcerias Estratégicas* 20(41):57–88.

De Nys, E., N. L. Engle, and A. R. Magalhães. 2016a. *Drought in Brazil: Proactive Management and Policy.* Boca Raton, FL: CRC Press.

De Nys, E., N. L. Engle, and A. R. Magalhães. 2016b. *Secas no Brasil: Política e Gestão Proativas.* Brasilia: Centro de Gestão e Estudos Estratégicos (CGEE) e Banco Mundial.

22

Droughts and Drought Management in the Czech Republic in a Changing Climate

Miroslav Trnka, Rudolf Brázdil, Adam Vizina,
Petr Dobrovolný, Jiří Mikšovský, Petr Štěpánek,
Petr Hlavinka, Ladislava Řezníčková, and Zdeněk Žalud

CONTENTS

22.1 Introduction

Droughts have important negative impacts on human society and many of its key activities. Several studies warned of a growing risk of droughts in central Europe in past decades (see Brázdil et al. 2015a; Trnka et al. 2016 for discussions of these studies). The problem of droughts and their negative impacts may become more severe in the context of future climate change because of enhanced anthropogenic forcings (Eitzinger et al. 2013; Trnka et al. 2011, 2014).

22.2 Past and Present Droughts

Different data can be used to characterize past droughts over the territory of the Czech lands (known as the Czech Republic since 1993). Based on meteorological measurements for the 1961–2014 period (chosen because it is best covered by meteorological stations), we noted a trend for increased drought occurrence at most stations. These tendencies have been documented through a series of drought indices (Figure 22.1) as well as in estimated soil moisture anomalies (Figure 22.2). Particularly during the period from April to June, a marked increase in the number of days with insufficient water content in the main rooting zone was identified. This increase can be attributed mainly to rising air temperatures, global radiation, and vapor pressure deficits, combined with little change in total precipitation (Trnka et al. 2015). The progressive decrease of water reserves in the soil in May and June, which are the critical months for agricultural and forest production, is alarming. As Figure 22.2 illustrates, these changes are widespread and well pronounced. The increased depletion of soil moisture reserves accumulated over winter months that has been observed during the April to June period also explains the increased variability of soil moisture content in the summer months (Trnka et al. 2015). As a result of the generally lower soil water at the end of June, the July–September soil moisture becomes more dependent on summer rainfall, which is highly variable. Therefore, interannual variability of the soil moisture content has increased as well.

Applying data from several secular meteorological stations, Czech temperature and precipitation series can be used to calculate a monthly series of several drought indices since 1805, namely the Standardized Precipitation Evapotranspiration Index for 1 and 12 months (SPEI-1 and SPEI-12), Palmer Z-index (Z-index), and Palmer Drought Severity Index (PDSI). All these indices showed a significant trend toward increased dryness in spring. Indices representing the long-term anomaly of water balance (SPEI-12 and PDSI)

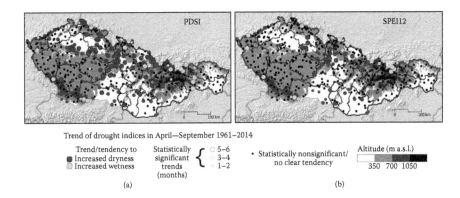

FIGURE 22.1
Number of months within the summer half-year (April–September) with positive (light gray–toward more wet conditions) and negative (dark gray–toward more dry conditions) significant trends ($\alpha = 0.05$) over the territory of the Czech and Slovak republics and northern Austria for (a) self-calibrated PDSI and (b) 12-month scSPEI. Evaluation of trends/tendencies was carried out individually for each month in the 1961–2014 period.

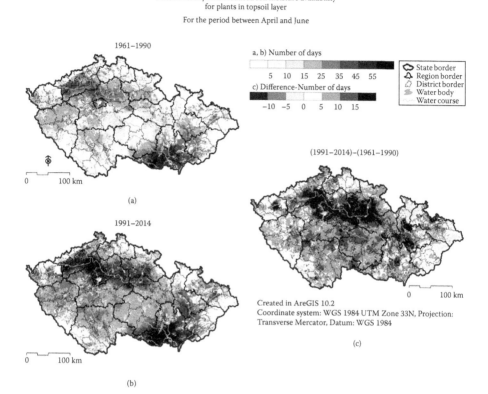

Number of days with reduced soil moisture availability
for plants in topsoil layer

For the period between April and June

1961–1990

a, b) Number of days

5 10 15 25 35 45 55

c) Difference-Number of days

−10 −5 0 5 10 15

State border
Region border
District border
Water body
Water course

(1991–2014)–(1961–1990)

0   100 km

(a)

1991–2014

Created in AreGIS 10.2
Coordinate system: WGS 1984 UTM Zone 33N, Projection:
Transverse Mercator, Datum: WGS 1984

(c)

0   100 km

0   100 km

(b)

FIGURE 22.2
Number of days with reduced soil moisture availability in 0–0.4 m topsoil layer in April–June,
the critical period for vegetation, over the territory of the Czech Republic. Mean number of
days for the (a) 1961–1990 and (b) 1991–2014 periods is presented together with the (c) difference
between the periods.

showed the same tendency for the whole year and summer, and to a lesser
extent for autumn. No conclusive drying trends were identified during the
last two centuries for winter, and in some regions a trend toward increased
wetness was found (Brázdil et al. 2015a). The drying trends in the April–June
period have been driven particularly by major temperature increases, leading to higher potential evapotranspiration.

Instrumental observations before 1800 are scarce, but knowledge of
droughts for the preinstrumental period can be obtained from documentary evidence (Brázdil et al. 2005, 2010). Extensive documentary evidence
related to droughts and their impacts in the Czech lands allowed analysis of the occurrence and severity of dry episodes on an annual timescale.

This information can be combined with drought indices calculated for the instrumental period to create long-term decadal frequencies of droughts from 1501 to 2012 (Brázdil et al. 2013). Despite great interdecadal variability (Figure 22.3a), the highest frequency of years with dry episodes during 50-year periods occurred between 1951 and 2000 (26 years), followed by 1751–1800 (25), 1701–1750 (24), and 1801–1850 (24). The lowest rate of dry years was recorded for 1651–1700 (16) and 1551–1600 (19). More detailed evidence of long-term drought fluctuations in the Czech lands can be obtained from a series of four drought indices (SPI, SPEI, Z-index, and PDSI) derived on the seasonal, half-year, and annual levels from documentary and instrumental data for the 1501–2015 period (Brázdil et al. 2016a). As shown in Figure 22.3b and c, fluctuations in annual SPEI-12 and PDSI and demonstrates great interannual and interdecadal variability.

The available documentary data also provide convincing evidence about several extraordinary episodes of droughts, such as those in 1534, 1536, 1540 (classified as a year of unprecedented European heat and drought by Wetter et al. 2014), 1590, 1616, 1718, 1719, 1726, 1746, and 1790. Their list can be extended using instrumental records for droughts in 1808, 1809, 1811, 1826, 1834, 1842, 1863, 1868, 1904, 1911, 1917, 1921, 1947, 1953(–1954), 1959, 1992, 2000, 2003, 2007, 2012, and 2015. Reported dry episodes had significant impacts on the daily life of the population and, in many cases, led to significant increases in food prices, followed by the adoption of various emergency measures. The broad extent of various impacts, including economic, social, and political, was documented by Brázdil et al. (2016b) in an analysis of the disastrous drought in 1947, which also had a broader European context.

Overall results of multiple studies (e.g., Brázdil et al. 2015a, 2016a) conclusively show that despite relatively strong variability in drought frequency, a trend toward increasing drought intensity could be identified in recent decades (see Figure 22.3), which is also closely linked with changed frequency of the drought conductive circulation types over central Europe (e.g., Brázdil et al. 2015a; Trnka et al. 2009).

Because documentary data from the Czech lands before 1500 are rather sporadic, some potential for drought reconstruction is offered by tree-ring width (TRW) data, represented by annually resolved oak TRW chronology, covering 761–2010 (Dobrovolný et al. 2015). Despite the complicated relationship of TRW to drought in the central European scale, minimal TRW values may identify the occurrence of dry seasons. The existing chronology showed greater frequency of years with minimal increments of wood (sign of a growth depression potentially caused by drought) at the end of the ninth century, the turn of the twelfth and thirteenth centuries, the mid-seventeenth century, and the beginning of the nineteenth century. Conversely, a smaller number of years with low growth was typical for the end of the eleventh century, the second half of the fourteenth century, and the first half of the eighteenth century.

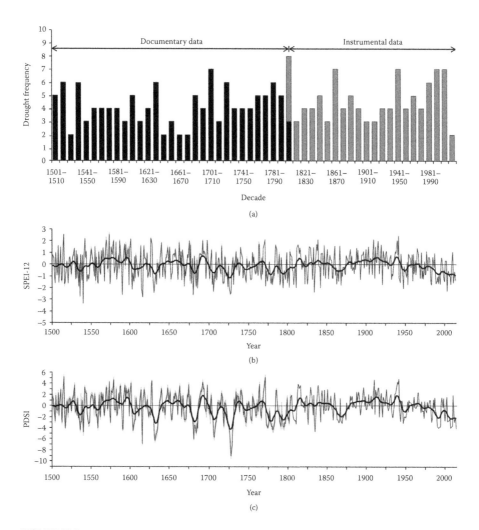

FIGURE 22.3

Long-term fluctuations of droughts in the Czech lands since 1501 based on documentary and instrumental data: (a) decadal frequencies of years with detected drought episodes, (b) annual SPEI-12, and (c) annual PDSI. Series in (b) and (c) are smoothed by Gaussian filter for 20 years. (Compiled from Brázdil et al. 2013, 2016a.)

22.3 Synoptic and Anthropogenic Factors of Droughts

Droughts in the Czech lands are related to the prevailing anticyclonic character of the weather with lack of precipitation and above-mean temperatures, enhancing the severity of droughts by intensified evapotranspiration. Analysis of the synoptic conditions favoring dry episodes from April to

September in the 1850–2010 period showed the connection of droughts with ridges of high pressure over central Europe extending from the area of the Azores High. In some cases, ridges extending from anticyclones located east or southeast of the Czech Republic may occur. Further supporting situations are related directly to isolated anticyclones over central Europe. On the other hand, low air pressure during drought episodes is typical for the area of the North Atlantic, Scandinavia, and the eastern Mediterranean (Brázdil et al. 2015a).

The severity of drought and its impacts can be influenced by human activities in the landscape, particularly the ability to retain water on the land. Changes in land use play an important role, affecting both overall and regional impacts on the amount of water in the landscape and its runoff. This effect was particularly significant in the floodplains of major rivers that were straightened by imposing regulations to achieve faster runoff of water. In order to obtain new arable land or building areas, numerous land reclamation projects took place that affected waterlogged zones and dramatically reduced the size of natural floodplain forests. At the same time, significant volumes of water have been retained in large water reservoirs (dams), which greatly reduce the occurrence of low flows on the main rivers.

22.4 Climate Forcings of Droughts

Various series of external or internal climate forcings were investigated with regard to their possible effect on Czech droughts (Brázdil et al. 2015a, 2015b). The method of multiple linear regression applied to Czech areal and station series of drought indices proved the influence of several large-scale climate-forcing factors (Figure 22.4). Statistically significant relationships were identified that were associated with anthropogenic factors, such as the temperature increase driven by increasing concentrations of greenhouse gases that contribute to lower calculated drought indices because of the impact of higher temperatures on evaporative demand (SPEI, Z-index, and PDSI). Significant links were also found with the North Atlantic oscillation (NAO). Although the NAO is distinctly seasonally dependent (i.e., in winter, the positive phase of NAO increases values of drought indices), for other seasons and the year as a whole, the opposite tendency is true. Volcanic factors show a slight tendency to drier patterns in periods following major volcanic eruptions, but significant relationships are small. There are also some indications of the influence of the El Niño southern oscillation, though only statistically significant for some indices and locations. Other climate forcings, specifically solar activity and the Atlantic multidecadal oscillation, do not show a significant relation to Czech droughts. It is, therefore, obvious that significant changes in the hydrological balance over the Czech territory, which have been demonstrated

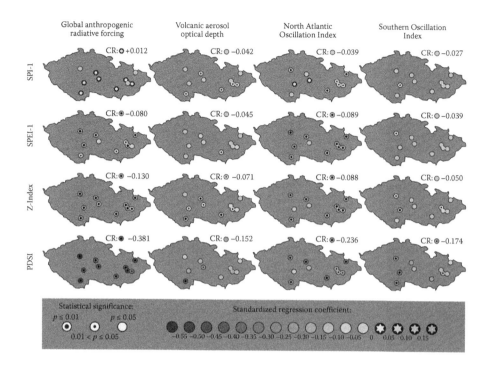

FIGURE 22.4
Standardized regression coefficients obtained by the multiple linear regression between series characterizing selected external and internal climate forcings (columns) and drought indices (rows), over the 1883–2010 period. Results for 10 individual Czech stations are shown along with those for Czech mean areal drought indices (marked CR). Coefficients significantly different from zero are indicated by a central dot (moving-block bootstrap was applied for statistical significance evaluation).

by lower soil moisture content in some seasons, can be attributed to climate change. Projected global temperature increases could thus have significant negative consequences for drought occurrence. However, it can also be noted that the presented statistical analysis alone does not explicitly confirm a causal nature of the respective links, especially in the case of long-term trends.

22.5 Future Droughts

Although drought trends are obvious (Figures 22.1 and 22.2) and clearly attributable to anthropogenic activities, it is important to estimate potential changes in the length, frequency, and intensity of the future droughts. To evaluate the impact of climate change on the values of the selected indicators, we tested the change in the number of days with drought stress in the topsoil layer during the period from April to June. For each 500 m grid, the weather data were modified based on the expected climate change conditions for the region. To be able to assess the development of conditions during the 2021–2040 period, we modified 1981–2010 daily weather data using a delta approach and five global circulation models. These models were selected as representations of mean values (IPSL–model of Institute of Pierre Simon Laplace, France) and to best capture the variability of expected changes in precipitation and temperature (BNU–Beijing Normal University, China; MRI–Meteorological Research Institute, Japan; CNMR–National Centre for Meteorological Research, France; and HadGEM–Hadley Centre Global Environment Model, UK). The five selected GCMs represent variability of 40 circulation models available in the CMIP5 database (Taylor et al. 2012). The RCP 4.5 (representative concentration pathway) greenhouse gas concentration trajectory and a climatic sensitivity of 3.0 K were used. As Figure 22.5 illustrates, the drought risk in the near future will not remain stable, and all five GCMs (Figure 22.5b through f) show a marked increase in the number of days with drought stress in the topsoil compared to the baseline period. The area with mean occurrence of a lack of water longer than 1 month is about 11.4 percent under the baseline climate but increases to 18–27 percent, with a fairly significant area with a water shortage on average lasting 55 days or more. Such changes would mean profound increases in the overall drought hazard. In the southeast region of the Czech Republic,

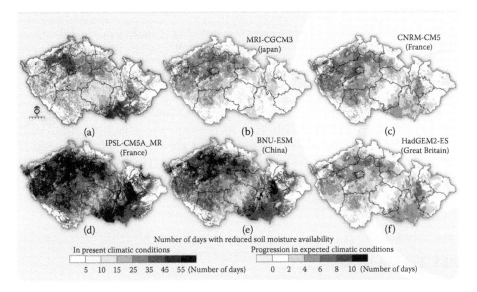

FIGURE 22.5
Number of days with reduced soil moisture for the 1981–2015 (baseline) period and the change in comparison with the baseline estimated using five representative GCMs for the 2021–2040 period and RCP 4.5. (a) Present climatic conditions: 1981–2015 (b–f) Expected climatic conditions: 2021–2040, RCP 4.5, 5 global circulation models (GCM).

the expansion of the highest hazard area occurs in a northward direction, while in the west the expansion covers the Elbe river lowland. Both areas are presently considered the most fertile regions in the country. An additional factor of concern is the occurrence of drought spots across the entire country, with the only exception being the northeast region. The increased incidence of drought at these sites is driven primarily by a lower soil water-holding capacity. These results indicate that hazard levels are not static and are likely to change in the future. In addition, this dynamic (i.e., hazard levels in relation to climatic change) must be considered when areas most at risk are defined. What is surprising, however, is the magnitude of the predicted changes that could occur over such a short time frame in the near future. The probability of extreme drought increases considerably under predicted future climate conditions, and these changes may occur more quickly than previously anticipated. This finding is of great concern and suggests the urgency of improving drought resilience.

22.6 Improving Drought Preparedness

Technological advances in agriculture have significantly increased production levels, and the effects of drought episodes should not directly threaten the country's food security under a stable climate condition. However, in the last 20 years, a disturbing trend toward increased sensitivity of food production to the occurrence of drought has been identified. The situation in the forested areas was found to be even more dramatic because of the risk of forest fires, which together with higher incidence of vegetation stress (frequently caused by drought) should lead to changes in traditional management principles. Despite the fact that drought events are not as likely as other events to draw the attention of the media and have been left out of public debates for a long time, citizens and municipalities in general are aware of the risks and the need for adaptation measures to be implemented. This shows that the problem of drought and its effects might have (or might be perceived to have) a greater impact on society than the media coverage would suggest, which may lead to increased media coverage.

The scientific evidence presented above points out the fact that Czech society has had and will continue to have to deal with episodes of droughts in all parts of the country. Analysis of historical data shows evidence of exceptional drought episodes with the capability to seriously harm the agrarian economy. Despite technological advancements, agriculture continues to be dependent on rainfall as its source for water. Major droughts in more recent years (i.e., 2000, 2003, 2007, 2011–2012, and 2014–2015) were catalysts for changes in the attitudes of policymakers. The adoption of a comprehensive drought policy is urgently needed since the frequency and intensity of drought episodes is the highest in regions with low availability of ground-water resources (Figure 22.6). Previously, this common problem was alleviated by storing water in the higher parts of the river catchments in reservoirs constructed during the twentieth century.

Although drinking water and energy requirements will still be met in most situations in the next 20–30 years, manageable water resources for irrigation are not sufficient to cover expected increases in water demand. Agricultural water consumption is much lower than in neighboring countries because irrigation systems are underused (and underdeveloped). There are several reasons for this, the main one being generally favorable climatic conditions with most precipitation occurring during summer months and water from winter months accumulating in the soil. As the changing climate erodes reliability of soil water and precipitation as sufficient sources of water for crops (especially those of higher value like hops, grapes, or vegetables), the demand for irrigated agriculture will increase. However, if present irrigation systems (covering less than 4 percent of arable land) were to run at full capacity, the water amount required to cover the crop needs would surpass the resources in some catchments, particularly during the dry years. Because of changes in climatic conditions during the period from 2021 to 2040, we expect that an increase in water use of up to 33 percent (compared with the 1981–2010 baseline period) will be necessary to maintain the same cropping systems. Future climate can allow profitable

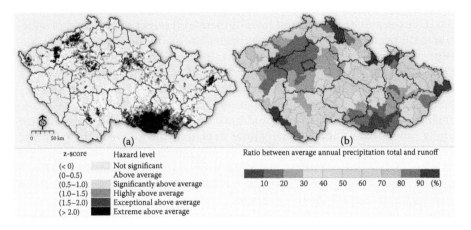

FIGURE 22.6
(a) Areas considered to be most at risk from the occurrence of drought based on the occurrence of days with drought stress in the first (April–June) and second (July–September) half of the summer, half-year and proportion of soils with extremely low soil water-holding capacity. Z-scores of all three parameters were averaged per cadastral unit. (b) Ratio between annual runoff and precipitation expressing proportion of rainfall that on average ends in the streams and rivers. Third-order catchments are represented (larger catchments are divided in smaller parts). Both maps represent the 1981–2014 period.

irrigated agriculture but will require an extension or increase of irrigated areas. Although a significant increase of the amount of irrigated land is theoretically possible, it would require large infrastructure investments, and consensus on this investment has not been achieved by policymakers and the general public.

Figure 22.7 points to an even more pressing issue that is likely to arise in the near future. Ongoing climate change will likely cause a significant decrease of overall water resources in the eight principal catchments of the Czech Republic in the next 20 years. Four out of five simulation runs based on a representative set of global circulation models signal a major drop in the amount of potentially available water (i.e., sum of annual discharge after the current water withdrawals have been subtracted). In the case of a 10-year drought, all iterations lead to a significantly reduced amount of available water compared to 1981–2010. This is especially true for the southeast and northwest parts of the country, where the occurrence of droughts is likely to increase in the next 20 years (Figure 22.5).

Measures focusing on reducing the vulnerability of the territory to drought must be considered across the country, but consideration must also be given to other hydrometeorological risks. The ongoing global climate change in central Europe will lead not only to the already mentioned increase in the frequency and severity of droughts but also to the increased frequency of other hydrometeorological extremes such as floods and flash floods, or heat waves (Stocker et al. 2013). Therefore, it is necessary to consider adaptation and mitigation measures that deal with the increasing risk of droughts *and* floods. Minimizing vulnerability to these hazards requires that adaptation and mitigation measures for each hazard be evaluated together, as these measures may not be mutually appropriate.

To deal with the negative climatic trends and growing exposure of the area to major drought events, it is important to strengthen the institutional capacity of the governing bodies (especially the regional and central government authorities) and prepare specific strategies for dealing with drought proactively as well as responding more effectively during drought. During a drought episode, little can be done except to respond to the crisis or emergency to limit damages. Therefore, it is necessary to work proactively and systematically with individual businesses, communities, and drought-affected sectors to increase their resilience to drought episodes. Current policies that have been implemented or proposed in the legislative process include:

- Systematic support to improve retention capacity of the soil and landscape as a whole
- Optimizing the crop structure and crop/cultivar diversification, including appropriate utilization of soil tillage and other agricultural technology

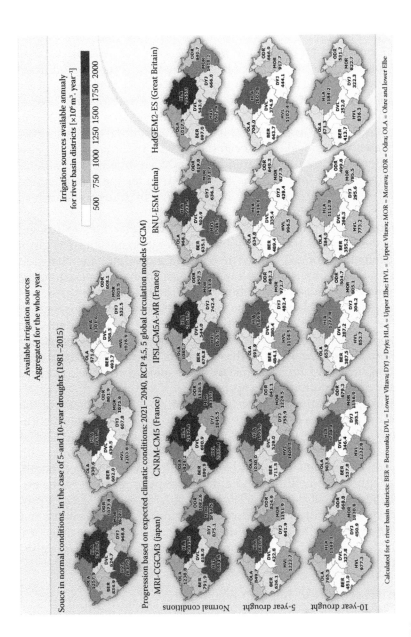

FIGURE 22.7
Potentially available water resources represented as discharge through the closing profile of the given catchment area per year after subtracting existing water uses. Top three maps represent baseline conditions (1981–2015) for "normal" year, and 5- and 10-year drought. The series of 15 maps represent estimates for the 2021–2040 period for five representative global circulation models.

- A focus on the selection of resilient tree species and forest types for drought and other known climate risks

- The early detection of drought onset, with the will to take appropriate and timely actions, and an emphasis on minimizing economic, social, and environmental losses, including building awareness and understanding of those sectors most affected

Delayed response to drought at all levels of management may result in the multiplication of potential damage. Some other policies currently under discussion include:

- Increasing manageable water resources (new dams, ponds, or underground water reservoirs) of various sizes

- Raising awareness among the population to promote and support individual responsibility for improving drought resilience (e.g., the economic use of water resources and the use of various individual water storage systems)

- Systematic preparation for the economic consequences of drought episodes during "good" years (e.g., creation of a fund to cover uninsurable risks for farmers, which would top-up premium payments by farmers from the public resources, i.e., the state would match the premium payments by a private company, and provide support during drought only to those actively participating in the funding scheme, and discontinuation of *ad hoc* interventions during droughts)

- Preparation of specific, direct, and useful drought plans with clearly defined competences and regular updating

22.7 Role of Drought Monitoring and Forecast

In all cases, it is important to improve spatial and temporal information about the current state of the drought and forecast likely development of the given drought event. Especially in the case of agriculture, the existence of early and mid-term seasonal forecasts is crucial in order for this sector to adopt appropriate management strategies to minimize the impacts of drought. The application of information that can be provided by early warning systems, including long-term forecasts, can also be useful for many other sectors (e.g., energy, transportation, forestry, and tourism and recreation) that are discussed in other chapters of this book. For this reason, a specialized portal (http://www.drought.cz) has been created. It summarizes the current status of drought using:

- A combination of ground observations and high-resolution soil moisture modeling that provides daily information on the drought levels at 500×500 m resolution (Figure 22.8a)
- Remotely sensed vegetation status data (250 m resolution) that can be used to assess soil moisture deficit impact on field crops, permanent or perennial cultures (vineyards and orchards), and forests (Figure 22.8b) as well as soil moisture estimated throughout microwave radar that provides an additional method of soil moisture status assessment
- Near-real-time drought impact reporting by a network of farmers that report on the soil moisture content but in particular based on observed drought impacts at the farm level for a given week (Figure 22.8c)

At the present time, close to 300 respondents are actively participating in providing information on the drought status and drought impacts at their farms and forests, with more than 120 of these respondents reporting each week. There is currently an effort to increase the number of

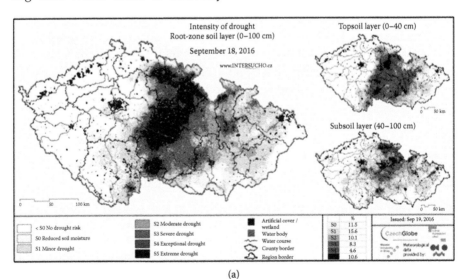

(a)

FIGURE 22.8

(a) Soil moisture content based intensity of drought for September 19, 2016; (b) corresponding map of vegetation status on October 1 based on enhanced vegetation index anomaly from 2000–2015 values (Terra-MODIS satellite); and (c) estimated impacts of drought on the main crops and soil moisture content as provided by farmers for week of September 19 (a map is published with a 1-week lag). *(Continued)*

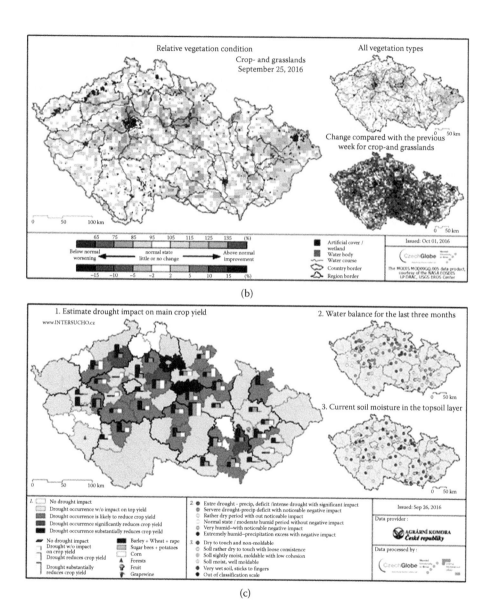

FIGURE 22.8 (Continued)
(a) Soil moisture content based intensity of drought for September 19, 2016; (b) corresponding map of vegetation status on October 1 based on enhanced vegetation index anomaly from 2000 to 2015 values (Terra-MODIS satellite); and (c) estimated impacts of drought on the main crops and soil moisture content as provided by farmers for week of September 19 (a map is published with a 1-week lag).

regularly reporting respondents to more than 250 to achieve proper spatial representation. However, for efficient management of water resources during drought events, the drought status information needs to be supplemented by drought forecasts. This is being done with an ensemble of five numerical weather prediction models for up to 10 days (Figure 22.9a) and a probabilistic forecast up to 2 months (Figure 22.9b). This information is available to all users on a daily basis, free of charge, and is highly valued by users. This system was used by 45,411 users in 2016, with more than 250,000 page views, which represents more than double the number of users in the very dry year of 2015 (20,614 users with 130,021 page views). It also confirms the high interest of users in drought forecasting products.

22.8 Summary and Outlook

There is no doubt that the risk of droughts and their impacts will continue to require systematic attention by researchers and policymakers alike in the coming years. This will be the case not only in the Czech Republic but throughout central Europe. Collaboration on this topic with neighboring countries is both economical and necessary. For example, the networked monitoring and warning system for agricultural droughts in central Europe (http://www.drought.cz) currently serves both the Czech and Slovak republics. This is a no-cost/no-project mutual collaboration between the Global Change Research Institute of the Czech Academy of Sciences in Brno (CzechGlobe) and the Slovak Hydrometeorological Institute in Bratislava. The Slovak partner provides meteorological data and expertise and communicates information to Slovakian stakeholders, while CzechGlobe continues to run the monitoring system itself as well as the preparation of current status and forecast maps. This mutual collaboration in sharing know-how and data provides tangible benefits to both sides and can serve as a suitable model for other regions. Plans exist to share the methodology and know-how between institutions across the region (Figure 22.10), including the status of vegetation (provided by CzechGlobe) and a Soil Water Index (provided by the Technical University in Vienna via Copernicus). It would be ideal in the near future to develop comprehensive systems to provide overviews of meteorological, agricultural, and hydrological drought status and forecasts, and their impacts for the region. Close collaboration is required not only in the field of drought monitoring but also in the preparation and implementation of drought plans. The exchange of experience is not only economical and efficient but necessary as well, since any measures implemented to respond to a drought event can have consequences for all countries in the region.

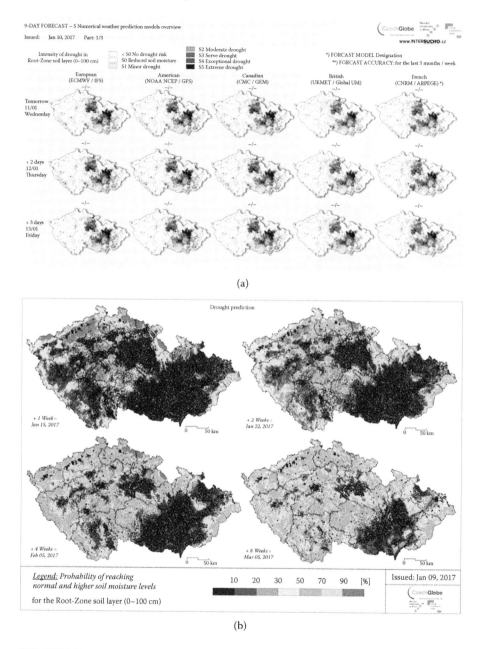

(a)

(b)

FIGURE 22.9

(a) Example of the drought forecast using ensemble of five numerical weather prediction models. Forecast was issued on January 10, 2017. The forecast is presently available for n+9 days. (b) Probabilistic forecast of normal or higher than normal soil moisture levels for 1, 2, 4, and 8 weeks, issued on January 9, 2017.

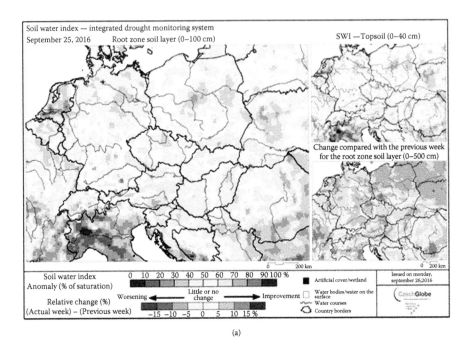

(a)

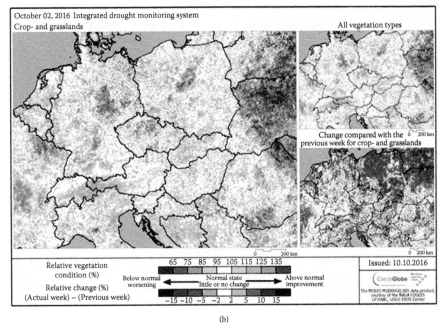

(b)

FIGURE 22.10
(a) Soil water index for September 25, 2016 and (b) corresponding two maps of vegetation status on October 2 based on Enhanced Vegetation Index anomaly from 2000 to 2015 values (Terra-MODIS satellite).

Acknowledgments

The presented results were obtained with support from Ministry of Education, Youth and Sports of the Czech Republic within the National Sustainability Program I (NPU I), grant number LO1415, and Academy of Sciences Strategy 21. The drought forecasting was supported by National Agency for Agricultural Research of the Czech Republic, Project No. QJ1610072 "System for monitoring and forecast of impacts of agricultural drought." MT, RB, PD, JM, PŠ, and LŘ were supported by Czech Science Foundation for the project no. 17-10026S. We acknowledge a broad team of authors not named in this chapter, who contributed to the book of droughts in the Czech Republic (see Brázdil et al. 2015a).

References

Brázdil, R., P. Dobrovolný, J. Luterbacher, et al. 2010. European climate of the past 500 years: New challenges for historical climatology. *Climatic Change* 101:7–40. doi:10.1007/s10584-009-9783-z.

Brázdil, R., P. Dobrovolný, M. Trnka, et al. 2013. Droughts in the Czech Lands, 1090–2012 AD. *Climate of the Past* 9:1985–2002. doi:10.5194/cp-9-1985-2013.

Brázdil, R., P. Dobrovolný, M. Trnka, et al. 2016a. Documentary and instrumental-based drought indices for the Czech Lands back to AD 1501. *Climate Research* 70:103–117. doi:10.3354/cr01380.

Brázdil, R., C. Pfister, H. Wanner, H. von Storch, and J. Luterbacher. 2005. Historical climatology in Europe—The state of the art. *Climatic Change* 70:363–430.

Brázdil, R., P. Raška, M. Trnka, et al. 2016b. The Central European drought of 1947: Causes and consequences, with particular reference to the Czech Lands. *Climate Research* 70:161–178. doi:10.3354/cr01387.

Brázdil, R., M. Trnka, J. Mikšovský, L. Řezníčková, and P. Dobrovolný. 2015b. Spring-summer droughts in the Czech Land in 1805–2012 and their forcings. *International Journal of Climatology* 35:1405–1421. doi:10.1002/joc.4065.

Brázdil, R., M. Trnka, L. Řezníčková, et al. 2015a. *Sucho v českých zemích: minulost, současnost, budoucnost (Drought in the Czech Lands: Past, Present, Future)*. Centrum výzkumu globální změny Akademie věd České republiky, v.v.i., Brno, Czech Republic.

Dobrovolný, P., M. Rybníček, T. Kolář, R. Brázdil, M. Trnka, and U. Büntgen. 2015. A tree-ring perspective on temporal changes in the frequency and intensity of hydroclimatic extremes in the territory of the Czech Republic since 761 AD. *Climate of the Past* 11:1453–1466. doi:10.5194/cp-11-1453-2015.

Eitzinger, J., M. Trnka, D. Semerádová, et al. 2013. Regional climate change impacts on agricultural crop production in Central and Eastern Europe—Hotspots, regional differences and common trends. *Journal of Agricultural Science* 151:787–812.

Stocker, T. F., D. Qin, G.-K. Plattner, et al. (eds.). 2013. *Climate Change 2013: The Physical Science Basis*. Contribution of Working Group I to the Fifth Assessment Report of the Intergovernmental Panel on Climate Change. Cambridge: Cambridge University Press.

Taylor, K. E., R. J. Stouffer, and G. A. Meehl. 2012. An overview of CMIP5 and the experiment design. *Bulletin of the American Meteorological Society* 93:485–498.

Trnka, M., J. Balek, P. Štěpánek, et al. 2016. Drought trends over part of Central Europe between 1961 and 2014. *Climate Research* 70:143–160.

Trnka, M., M. Dubrovsky, M. Svoboda, et al. 2009. Developing a regional drought climatology for the Czech Republic. *International Journal of Climatology* 29:863–883.

Trnka, M., P. Hlavinka, and M. A. Semenov. 2015. Adaptation options for wheat in Europe will be limited by increased adverse weather events under climate change. *Journal of Royal Society Interface* 12:20150721.

Trnka, M., J. E. Olesen, K. C. Kersebaum, et al. 2011. Agroclimatic conditions in Europe under climate change. *Global Change Biology* 17:2298–2318.

Trnka, M., R. P. Rötter, M. Ruiz-Ramos, et al. 2014. Adverse weather conditions for European wheat production will become more frequent with climate change. *Nature Climate Change* 4:637–643.

Wetter, O., C. Pfister, J. P. Werner, et al. 2014. The year-long unprecedented European heat and drought of 1540—A worst case. *Climatic Change* 125:349–363. doi:10.1007/s10584-014-1184-2.

23

Drought Planning and Management in the Iberian Peninsula

Rodrigo Maia and Sergio M. Vicente-Serrano

CONTENTS

23.1 Introduction

Drought is one of the most damaging natural hazards in the Iberian Peninsula (IP), causing varied socioeconomic and environmental impacts. Largely because of semiarid climatic characteristics and intensifying water use, the southern Europe region, which includes the IP, has been historically highly vulnerable to droughts. Precipitation variability and drought occurrence are two common characteristics of the climate in the IP (Martín-Vide 1994). Important climate differences exist across the IP (Font-Tullot 1988), with droughts affecting both humid and dry regions (Gil and Morales 2001; Vicente-Serrano 2006a, 2013). The main negative impacts of this phenomena are found in the regions with an annual average precipitation below 600 mm (Vicente-Serrano 2007).

The perception of the impacts of droughts in the IP has noticeably changed in the past few decades (Pita 1989). This is a consequence of the changes that transformed a rural society to a dominant urban society in which the main economic activities are not related to the primary sector (agriculture and livestock). The main impacts of droughts before the 1960s were mostly recorded in dry cultivated areas, causing frequent famine episodes in agrarian societies. This was also a primary impact of the strong drought episodes that affected the IP in the 1940s.

At present, the weight of the primary sector in the Iberian economy is much lower than some decades ago. In addition, different adaptation measures (e.g., dams, irrigated lands, agricultural insurance, and pasture insurance) have reduced the vulnerability of the agricultural sector to drought, although droughts are still causing significant crop failures in dryland agricultural areas (Austin et al. 1998; Molinero 2001; Páscoa et al. 2017). These measures, together with urban growth and the high importance of the tourism sector, have caused an important change in drought perception by society (Morales et al. 2000). Currently, that perception mostly considers drought as a hydrological hazard, and societal alerts are usually related to decreases in reservoir water levels, which may trigger supply problems in irrigated lands and urban areas. In addition, there is a growing interest in the possible environmental impacts of droughts, which are difficult to separate from the historical use of the territory and the current land management. In the past few decades, drought events have been occurring with some frequency (Vicente-Serrano 2006a), affecting large areas and causing significant impacts for various economic activities, especially agriculture and livestock (Maia et al. 2015). For example, the drought that affected the IP between 1992 and 1995 caused water restrictions for 12 million inhabitants and caused about EUR 3,500 million in losses. Also, a decrease of EUR 1,200–1,800 million in crop production was recorded during this period. The driest year recorded in the instrumental record of the IP was 2005; impacts included decreased agricultural production, a high frequency of forest fires (mostly in Portugal) (Gouveia et al. 2012), and a noticeable decrease of hydropower production. Hydropower production in 2005 was the lowest on record since 1965 (Jerez et al. 2013). In 2012, the last severe drought that affected the IP, a large increase in the surface area affected by wildfires (the greatest areal extent since 1994) occurred in Spain.

This chapter analyzes current drought planning and management in the IP, framing it in terms of the climatic characteristics of the region and the European Union (EU) water policy, with which both Portugal and Spain have agreed to comply (taking into account institutional differences, commonalities, and cooperation agreements between the two Iberian countries). In this context, the characterization of climate and hydrological drought patterns is discussed (Section 23.2), followed by the current state of drought planning and management in Portugal and Spain, and compliance with European policy (Sections 23.3 and 23.4). Section 23.5

suggests some potential actions to improve drought risk management at the Iberian level, including some recommended drought management policy best practices.

23.2 Characterization of Droughts in Iberian Peninsula

Different studies have analyzed the occurrence of droughts in the IP over the centuries using dendrochronological (Tejedor et al. 2015), documentary (Martín-Vide and Barriendos 1995; Vicente-Serrano and Cuadrat 2007a), and geological (Corella et al. 2016) records. These studies show a high incidence of droughts, providing further evidence that drought is a general climate feature of the region. For instance, Domínguez-Castro et al. (2012) analyzed drought occurrence using ecclesiastical records (pro-pluvia rogations) obtained from 17 archives during the eighteenth and nineteenth centuries. They not only found that droughts were more frequent in some decades (e.g., 1750s, 1780s, and 1820s) but also showed important spatial differences in the occurrence and severity of drought events. The consequences of droughts in the preindustrial period, with an economy based on agriculture and livestock, were devastating, with frequent famine and mortality episodes associated with dry conditions (Cuadrat et al. 2016).

The high frequency of droughts is also identified in the instrumental records available from the second half of the nineteenth century. A regional series for the entire IP based on the 12-month Standardized Precipitation Index (SPI) since 1901 shows different major episodes (Figure 23.1). The decades beginning in 1910, 1920, and 1930 showed low-severity drought periods. In contrast, after 1940 the variability of the SPI increased noticeably.

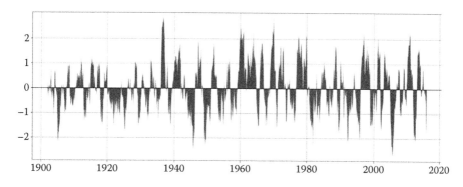

FIGURE 23.1
Evolution of the 12-month standardized precipitation index for the entire Iberian Peninsula between 1901 and 2015.

The decades beginning in 1940 and 1950 were dominantly dry and the 1960s and 1970s were dominantly humid. Two of the most extreme drought events since the beginning of the twentieth century were in 1945–1946 and 1949–1950. After 1980 there is a dominance of dry periods in the series. Thus, 1981–1984 and 1992–1995 were two periods affected by long and severe droughts, and the driest year of the series was recorded in 2005.

Various studies have shown that spatial variability of droughts in the IP can be very important, even at the regional scale (Vicente-Serrano and Cuadrat 2007b; Vicente-Serrano et al. 2004; Vicente-Serrano and López-Moreno 2006). It is not uncommon to find a region in drought conditions while other areas of the IP show normal or even humid conditions. Thus, few historical episodes have affected more than 75 percent of the total surface area of the IP (Vicente-Serrano 2006a). Using climate drought indices and multivariate techniques, it is possible to identify some homogeneous regions in which the temporal variability of the droughts has been similar. In the IP, six main regions that demonstrated a particular temporal evolution in the drought occurrence were identified (Vicente-Serrano 2006b).

The strong spatial variability of droughts in the IP is mostly a consequence of the different atmospheric circulation patterns that control the precipitation of this region (Rodríguez-Puebla et al. 1998). Atmospheric circulation anomalies during winter and spring usually determine water availability conditions some months in advance (Lorenzo-Lacruz et al. 2011; Vicente-Serrano et al. 2016). The North Atlantic oscillation (NAO) is the atmospheric mechanism that mostly controls the interannual variability of precipitation during the cold season in the IP (Trigo et al. 2002), causing important agricultural, hydrological, and socioeconomic impacts (Vicente-Serrano and Trigo 2011). Severe drought periods as observed in the 1980s and 1990s have been associated with positive phases of the NAO. Nevertheless, the problem is much more complex, because although large areas of the IP show stronger precipitation by the NAO, other regions show the influence of other atmospheric circulation patterns (Martín-Vide and López-Bustins 2006; Vicente-Serrano et al. 2009). Even so, an explanation of the main drought episodes is not possible with simple approaches based on atmospheric circulation indices since atmospheric conditions that trigger a drought episode may strongly vary from one episode to another (García-Herrera et al. 2007; Trigo et al. 2013).

Possible trends in drought frequency and severity worldwide and in Europe have recently been discussed (Trenberth et al. 2014). Studies based on precipitation data have suggested a reinforcement of drought conditions in the Mediterranean area (Hoerling et al. 2012). Nevertheless, the strong spatial variability of droughts makes it difficult to establish a general pattern of drought trends. The SPI series since the beginning of the twentieth century shows a significant increase in the severity and duration of droughts in the southwest and northeast IP (Vicente-Serrano 2013). In other areas (e.g., northwest, southeast, and central regions), the trend is toward lower magnitude and duration of the drought episodes. As a consequence of the strong temporal variability of the index, the average SPI series for the IP from 1901 to 2015, pictured in Figure 23.1, does not show a significant trend toward more negative SPI values. Nevertheless, studies based on objective hydrological drought metrics (e.g., streamflows) show a noticeable increase in the frequency and severity of drought events in the past decades. Lorenzo-Lacruz et al. (2013a) analyzed the evolution of hydrological droughts in the IP between 1945 and 2005 and showed a clear increment of hydrological droughts in most of the basins of the south and east IP. This is also observed, although with less severity, in other basins of the northeast and the northwest.

It is difficult to establish a distinction between climatic and anthropogenic drivers of hydrological drought severity. Current management and use of water has affected the duration and severity of hydrological droughts. Thus, water management significantly alters the relationship between climatic and hydrological droughts, both in the magnitude of the relationship and in the timescale of the response (López-Moreno et al. 2013; Lorenzo-Lacruz et al. 2013b). Water management may also alter drought duration and severity. López-moreno et al. (2009) showed how the development of large reservoirs on the Spanish-Portuguese boundary has resulted in an intensification of the water scarcity episodes downstream from the dams in comparison to upstream sectors. In addition, increased water demands by irrigation may cause an important accentuation of the severity of the hydrological droughts downstream from the reservoirs and irrigation polygons. Recently, Vicente-Serrano et al. (2017) analyzed the evolution of climatic and hydrological drought events in the Segre basin, a highly regulated basin in northeast Spain, and showed that intensification of irrigated lands has increased the severity of hydrological drought events downstream in comparison to the observed evolution of the climate droughts.

In any case, the recent warming processes identified in the IP have caused a noticeable increase of the atmospheric evaporative demand (AED), with an average of 25 mm decade^{-1} since 1960 (Vicente-Serrano et al. 2014a). Evidence suggests that this increase has reinforced drought severity in past decades. Vicente-Serrano et al. (2014b) analyzed the possible impact of the AED on drought severity comparing two drought indices: the SPI, based on precipitation, and the Standardized Precipitation Evapotranspiration Index (SPEI), based on the climatic balance between precipitation and the AED (Vicente-Serrano et al. 2010). The results showed a clear influence of the increased AED on the climatic drought severity and also on the current availability of water resources. Thus, although precipitation trends are mostly not significant in the IP, a reinforcement of hydrological drought conditions has been identified in the majority of the Iberian basins, including natural nonregulated basins, which are not influenced by water regulation and consumption, and mostly respond to climate variability (Figure 23.2).

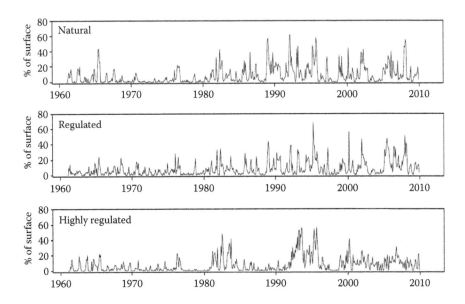

FIGURE 23.2
Percentage of surface area affected by streamflow drought from 1961 to 2009, based on the standardized streamflow index for the natural, regulated, and highly regulated basins of the entire Iberian Peninsula.

23.3 Setting the Scene: European and Iberian Drought Planning and Management Context

The Water Framework Directive (WFD), published in 2000, set a new perspective for European water policy and management, considering a river basin level approach and the development of river basin management plans (RBMPs) to protect European water bodies. The WFD identifies the need to mitigate the effects of droughts, highlighting the possibility of complementing RBMPs through special programs and management plans to deal with specific water issues (e.g., drought management plans) (EC 2000). In compliance with the EU WFD, Spain and Portugal approved in 2016 the second (6 years) cycle of the RBMPs of most of their river basin districts (RBDs), namely the ones corresponding to (each country's part of) the shared transboundary RBDs (four—one with two rivers).

In 2007, a Communication from the Commission to the European Parliament and the Council on Water Scarcity and Drought (WS&D) framed a set of policy options to address and mitigate water scarcity and drought within the EU (EC 2007a). These options are aimed at the improvement of drought risk management, by fostering development of drought management plans, drought early warning systems, and a European drought observatory (EC 2007a; Estrela and Vargas 2012). Furthermore, according to EC (2007a), drought management plans (DMPs) should include (1) indicators and drought level thresholds, (2) measures to be taken accordingly to each drought level, and (3) a clear organizational framework for drought management (EC 2007b).

The two IP countries are aware of the importance of drought management and have been developing efforts to cope with drought issues in the WFD and EU drought policy context since 2005, when a severe drought event occurred. Drought planning and management in the IP depends on cooperation and interaction between Portugal and Spain, as the five shared river basins (Minho, Lima, Douro, Tejo, and Guadiana) correspond to 45 percent of IP territory, in what may be considered a singular case regarding transboundary management. The specificity of this case calls for joint river basin management agreements (Pulwarty and Maia 2015). That need is particularly significant for Portugal, since 64 percent of its territory corresponds to the shared river basins. In fact, the downstream location of Portuguese parts of the shared river basins makes the country extremely vulnerable to Spain's water uses, flow regimes, and sediment transport (Do Ó 2011; López-Moreno et al. 2009; Pulwarty and Maia 2015). The bilateral cooperation on the shared river basins management is framed by the Albufeira Convention, signed in 1998 (DR 1999).

23.3.1 The Albufeira Convention

Since the nineteenth century, Portugal and Spain have been establishing partnerships and treaties, mainly to define boundaries and uses of the rivers' bordering stretches. The Convention on Cooperation for Portuguese–Spanish River Basins Protection and Sustainable Use, usually referred as the Albufeira Convention and active since 2000, was framed under the WFD principles and was the first to address all the shared rivers, at the river basin scale.

The Albufeira Convention defines the framework of bilateral cooperation for sustainable water management of the shared water resources, within a river basin (DR 1999). It promotes coordination on specific bilateral issues, such as flow regime, droughts, and emergency situations (Maia 2011). The Convention states that Portugal and Spain shall "coordinate actions to prevent and control drought and water scarcity situations" and should "undertake joint studies of drought and water scarcity."

One of the most important achievements under the convention, undertaken by the Commission for Convention Development and Appliance (CADC), was the revision of the provisory minimum flow regime (MFR) established by the convention. Those values must be guaranteed at some control sections, in nonexceptional years (defined mostly based on values of referenced precipitation monitoring stations), as described by Maia (2008). However, that regime may be revised to take into account the environmental flow regime established by the RBMPs. In fact, currently, Spanish RBMPs have larger minimum flow values than those established by the Albufeira Convention; Portuguese RBMPs have yet to establish these values.

23.4 Current Drought Planning and Management in the Iberian Peninsula

Both the Albufeira Convention and the WFD highlight the need for coordination and cooperation on transboundary river basin management between Spain and Portugal (Maia 2009). Concerning drought, the Albufeira Convention states that both countries must enhance drought prevention and control coordination by the definition of common criteria for exceptions to the MFR and for drought risk management, and by the establishment of monitoring points and indicators, trigger values, and measures to be applied in drought situations (Maia 2009). Nevertheless, up to now, the two countries are far from having accomplished or even agreed on a common or coordinated drought management framework, in line with a recommendation by the European Commission to foster transboundary cooperation (EC 2007b). Even the establishment of a common and homogeneous drought indicator system, which was to have been developed by the CADC, has still not been achieved. In fact, concerning drought management and planning, the Iberian countries have never reached a common understanding, with the two countries currently in different stages (Maia 2011). A more detailed description of the status of drought management (institutional framework, planning, and monitoring) within each Iberian country (Portugal and Spain) is presented below. At the end of this chapter, an overall comparison of drought management is provided for these two countries.

23.4.1 Drought Institutional Framework

In Spain, the current legal framework for managing water resources is the Spanish Water Act and the WFD (transposed into Spanish law in 2003) (Stefano et al. 2015). The WFD was transposed into Portuguese law in 2005. Figure 23.3 illustrates Spanish and Portuguese RBDs (Portuguese "Regiões Hidrográficas," RH; Spanish "Demarcaciones Hidrográficas").

In terms of institutional framework, Spain's water management is organized in a multilevel structure, divided between the central government, autonomous communities (which are defined by the Spanish Constitution as the regional governments), and river basin district administrations (RBDAs) (Sánchez-Martínez et al. 2012). The Spanish Water Act established two river

FIGURE 23.3
Spanish and Portuguese river basin districts (representation of Portuguese islands [Madeira and Azores archipelagos] RBDs is missing).

basin types: (1) intraregional basins, in which boundaries lie within a single autonomous community and (2) interregional basins, whose boundaries encompass more than one autonomous community and/or are transboundary (such as the ones shared with Portugal). Intraregional basins are managed by the regional government of the autonomous community, through a hydraulic administration (Administración Hidráulica). The latter are managed by river basin management agencies (Confederaciones Hidrográficas). The national drought policy is defined by central government (Ministry of Environment). The RBDAs are responsible for drought planning and operational management. When a drought situation is declared, a Permanent Drought Commission is formed under the approval of a royal decree by the government. The Commission is composed of representatives of the administrations and stakeholders, with the role of managing water resources systems, in the basin area where the drought situation is declared (Estrela and Sancho 2016).

The situation is different in Portugal. The Portuguese Environmental Agency (Agência Portuguesa do Ambiente, APA), as the National Water Authority, represents the state in water issues, having the responsibility for water planning and management. That authority and responsibility was transferred to APA in 2012, with the integration of the former RBDAs (five in the IP mainland Portuguese territory) as decentralized services at the regional level (APA 2016). According to the Portuguese Water Law, the National Water Authority declares drought situations and manages, together with other relevant organizations, the application of drought mitigation measures. Nevertheless, following the 2012 drought event, the Commission for the Prevention and Monitoring of Effects of Drought and Climate Change (in Portuguese, CSAC) and a technical working group under it were created as permanent bodies responsible for preventing, monitoring, and following drought conditions and climate change impacts, as well as for providing and assessing risk measures to mitigate drought effects (DR 2012). The working group, coordinated by the Office of Planning, Policy and General Administration (GPP), is composed of 20 entities in various areas (such as meteorology, agriculture, natural conservation, food, territory, finance). The National Water Authority (APA) is one of these entities.

23.4.2 Drought Management

Drought plans are very important since they create a framework for a risk management approach. These plans are developed to "provide tools to inform decision making, define tasks and responsibilities for all drought management, and identify a host of mitigation and response actions" (Stefano et al. 2015). Because each drought event can evolve in unpredictable ways and affect different areas, some *ad hoc* mechanisms must be arranged by governmental agencies and sectoral institutions to respond to evolving drought conditions. In the following section, a description of the developmental frame of drought management in Spain and Portugal, based on DMPs' implementation, is presented. Both Portuguese and Spanish DMPs provide a framework for action in drought situations, showing some common tools to trigger actions (drought indicators), measures associated with each drought situation, and the main entities and roles in drought management.

Spain

The National Hydrological Plan Act established in 2001 stated that DMPs should be developed by the RBDAs. Following the severe drought episode affecting the IP in 2005, Spain developed a guidance document for DMPs. As a consequence, DMPs were developed for all Spanish RBDs and approved in 2007 by ministerial order, being considered as specific plans of RBMPs, according to the Hydrological Planning Royal Decree (RD 907/2007). The National Hydrological Plan also stated that public administrations responsible for urban water supply systems with more than 20,000 inhabitants should have a contingency plan for drought situations (BOE 2001).

Spain was a pioneer in the development of hydrological drought plans at the basin scale, which are based on accurate real-time monitoring information on the hydrological conditions of the basin. The hydrological drought plans are developed at the basin level by the different water management agencies (e.g., the Ebro basin drought plan can be obtained at http://www.chebro.es/contenido.streamFichero.do?idBinario=5889). These drought plans are independent for each basin since they are adapted to the specific basin characteristics, but all use the same concept for drought monitoring (detailed below).

Spanish DMPs include a drought diagnosis (i.e., drought indicators and thresholds), a program of related measures (of different types, associated with each drought status), and a management and follow-up system (with an organizational framework to deal with drought in each river basin district) (CHG 2007; Maia 2009). The 2007 DMPs are currently being revised and adapted to the second cycle RBMPs (adopted at the beginning of 2016) and are expected to be completed by the end of 2017 (BOE 2016).

Spain does not currently have an active drought early warning system (DEWS); however, the drought indicators to be defined in the revised DMP are expected to include seasonal meteorological forecasts, provided by AEMET, the Spanish national meteorological entity. Currently, the contingency plans are only developed in some large cities, such as Madrid, Barcelona, Seville, and Malaga (MAPAMA 2016a).

Portugal

No DMPs have been developed to date at the RBD scale. A proposal for a national Plan for Prevention, Monitoring and Contingency for Drought Situations (hereafter referred to as PT DMP) was made public in 2015, but it has not yet been officially approved (APA 2016). As illustrated by the severe drought of 2004/2005, drought situations have been managed in a reactive way, mostly based on the activity of a drought commission established by the government in cases of severe drought, operating at the national level. During the next severe drought period, in 2012, the Commission (CSAC, as referred to above) was established as a permanent body together with a technical advisor working group (DR 2012). The PT DMP proposal (developed by the working group) considers the establishment of drought indicators and drought levels and sets some specific measures that are associated with those levels. However, at the moment, those features may not be fully implemented because of the DMP's provisional character. A pilot version of a DEWS was developed by the former Portuguese Water Authority (INAG) and was presented to the public in 2011, but it has not yet been implemented (Maia 2011).

The DMP proposal provides information about drought prevention, monitoring, and contingencies. This plan includes (1) some preventive structural and nonstructural measures, (2) the variables to be used for drought indicator development, (3) the periodicity of drought monitoring, and (4) a proposal for action during a drought event (such as the entities involved and the measures associated with each drought alert). The DMP proposal also describes the terms of public disclosure, including the frequency of publication of monitoring results by the entities involved. Furthermore, the proposal calls for all public water supply and irrigation supply management entities to prepare a contingency plan for drought situations.

23.5 Drought Monitoring Systems

Drought monitoring systems are key to tracking and mapping drought spatial extent and intensity throughout time, establishing links between drought status and adequate actions to be taken, and analyzing the effectiveness of the measures implemented.

23.5.1 Spain

Spain has a countrywide drought monitoring system managed by the Ministry for Agriculture, Food and the Environment (MAGRAMA). This system is based on information from several hydrological variables, such as reservoir storage, groundwater piezometric levels, streamflow, reservoir inflows, and precipitation. These variables are measured at several locations throughout the river basins and are weighted to obtain an integrated indicator, or a concept definition specific to Spain, on the basis of a national indicator system (Estrela and Vargas 2012). That indicator (Indice de Estado—Index of Status) is meant to represent the hydrological status in each river basin. For example, if the water uses of a region depend on the water stored in a reservoir, the drought index is based on the reservoir level; if water uses depend on groundwater, the piezometric levels of the aquifer are the data used to calculate the index. The index is obtained according to the percentiles of each key variable for each hydrological system. The values obtained for the current situation are compared to the drought thresholds, resulting in the establishment of a corresponding drought alert level. The standardized values of the indicators (ranging from 0 to 1) provide the basin drought status levels, classified as *normal*, *prealert*, *alert*, and *emergency*. A normal situation corresponds to a hydrological condition better than the average conditions, defined by historical evaluation. The other drought status levels (prealert, alert, and emergency) correspond to hydrological conditions with values below the average (CHG 2007).

Each of the drought alert levels is associated with specific management actions, which are specified in the DMPs, developed for each RBD. For example, in a prealert situation, the RBA encourages farmers to revise their cropping plans to consider the available water resources (Estrela and Vargas 2012; Stefano et al. 2015).

The RBDAs periodically send follow-up information to the Water Directorate of MAGRAMA, which compiles the data from every RBDA to produce a common dataset and reports on drought information, conditions, and measures in the Spanish territory. This information (with maps, graphs, and statistics) is presented by the National Drought Observatory, for the entire Spanish territory, usually on a monthly basis (MAPAMA 2016b) (Figure 23.4).

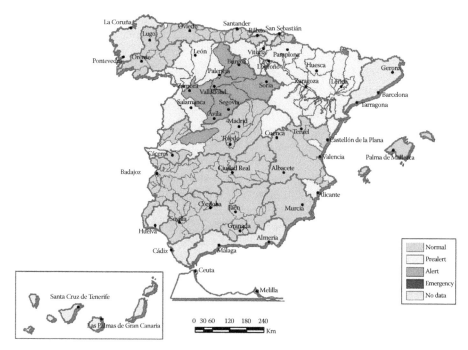

FIGURE 23.4
Spatial distribution of the hydrological status index in June 2012, corresponding with the last severe drought episode that affected the IP

In parallel, there is some available information on real-time drought conditions in Spain. The Spanish State Meteorological Agency, AEMET, produces national maps of the monthly information about meteorological drought status using the SPI. This index is calculated based on precipitation records (AEMET 2016).

Agricultural drought monitoring in Spain follows two approaches. The first approach involves a drought monitoring system for pasturelands based on anomalies recorded from the time series of MODIS images (http://www.mapama.gob.es/es/enesa/lineas_de_seguros/seguros_ganaderos/perdida_pastos.asp). According to the vegetation indices calculated from these images, a threshold is fixed to determine if pasturelands of each region show a significant decrease of leaf production and net primary production associated with drought. Hence, based on the remote sensing system, and in accordance with the evolution of expected losses in productivity, economic compensation is given to the farmers that have pasture insurance. The second approach is based on agricultural monitoring developed by the national company for agricultural insurance (AGROSEGURO, http://agroseguro.es/). High spatial density field surveys are carried out to determine crop conditions in the different phenological stages. If crop failure is declared in an area, an evaluation is done to determine the expected losses in each agricultural use and the corresponding insurance compensation to the farmers.

23.5.2 Portugal

Following the 1995 drought, the Drought Vigilance and Alert Program (PVAS—Programa de Alerta e Vigilância de Secas) was developed (between 1995 and 1998) by INAG, the National Water Institute, an institution that was incorporated in the Portuguese Environmental Agency (APA) in 2012, which became the national Portuguese Water Authority. The PVAS was used to monitor the 2004–2005 and 2012 drought events. The PT DMP proposal calls for the use of PVAS, including some modifications such as the reinforcement of the piezometer monitoring network. Through PVAS, it will be possible to identify long-duration drought and water scarcity events and impacts, based on information on common hydrometeorological variables: precipitation, river flows and water storage in reservoirs, and aquifers. From that information, an evaluation is carried out in four periods of the hydrological year (end of January, March, May, and September), in order to assess exceptional circumstances due to drought.

Monthly reports on common meteorological information (prepared by the Portuguese Meteorological Institute, IPMA) that may be relevant for assessment of drought conditions, such as SPI and PDSI (Palmer drought severity index), are made public at the GPP (coordinator of CSAC's working group) website. The PT DMP proposal also foresees monitoring (in normal and drought conditions) of rain-fed agriculture and extensive livestock production indicators.

According to the PT DMP proposal, the monitored variables are the basis to assess agrometeorological and hydrological drought status, which in turn is associated with the four levels for drought conditions: normal, prealert, alert, and emergency. Each drought level may be related to certain drought measures. For example, the alert level may be related to the implementation of mitigation measures, such as restrictions in water uses (APA 2016). It should be noted that DMP PT overall monitoring procedures (action plan) are not yet fully implemented.

Currently, in Portugal, some rain-fed agricultural and extensive livestock production activities are monitored, and the results are made available (by the SIMA and RICA systems—in Portuguese, respectively, Sistema de Informação de Mercados Agrícolas and Rede de Informação de Contabilidades Agrícolas). According to the PT DMP proposal, the information obtained from these current monitoring systems could be used to evaluate drought impacts. This kind of information was used by CSAC's advisory technical working group when evaluating the drought situation during the 2014/2015 meteorological (prealert) drought event.

23.6 Overall Comparison between Portugal and Spain

Portugal and Spain have a long and common history of severe drought events, which pose several challenges. Those events highlight the importance of adapting water resources planning and management to face the corresponding impacts and water use limitations. The two countries follow a common frame and approaches on drought planning and management, but their implementation differs. One difference relates to the approach and time for the development of DMPs. While Spain developed DMPs in 2007 for the Spanish RBDs, Portugal is expected to implement a national DMP, following a proposal prepared in 2015 but not yet formally approved. In fact, in some Spanish RBDs (e.g., Douro and Guadiana), a review and/or adjustment of some aspects of the corresponding DMPs (namely, the indicators' assessment and/or evaluation) was already conducted in the scope of the first and second RBMPs approval. A complete revision of Spanish DMPs is being undertaken and is expected to be completed by the end of 2017.

In terms of institutional frameworks, namely concerning RBDs governance, the countries exhibit some differences in the broad range of water resources management. In Portugal, the water resources administration concentrates on a national institution (APA), which is the national water authority; in Spain, this structure is shared by central government, autonomous communities, and the RBDAs. This difference in governance structure is also reflected in terms of drought management. The different spatial scale approach may also be observed at the operational stage. That can be illustrated by the role of regional, local, and water users' organizations on drought management. In Portugal, where the recently (2012) established drought management bodies have a nationwide monitoring area, the role of local/regional entities when a drought situation is declared remains unclear, denoting a lack of a drought operational framework. In Spain, the RBDAs play an important role in drought management provision, and drought management actions are taken at all water user levels. Indeed, Spanish DMPs define the role of each drought organization/stakeholder as far as the drought event develops.

Spain has fully implemented a drought monitoring system that provides information on the level of drought in different regions (zonas de explotación) of all Spanish river basins. In Portugal, the PT DMP's proposed drought monitoring system remains to be fully implemented. As in the case of Spain, the system is expected to provide information on a similar number of levels of drought, taking into consideration the evolution of hydrological variables.

Regarding drought early warning systems, neither Spain nor Portugal has a fully operational DEWS at regional and/or national scales, as recommended in the 2007 European guidelines (EC 2007a). Hence, there is a clear need to develop those systems in both countries. In Portugal, a pilot version for a DEWS was developed by the former Portuguese Water Authority (INAG), but it is not yet operational (Maia 2011). Nevertheless, the Portuguese National Water Plan has established a goal to implement a DEWS by 2021 (DR 2016). In Spain, efforts are evolving to develop a DEWSs for the Spanish river basin, in order to enable the medium-term prediction of droughts in addition to the current capability to confirm the presence of drought. Furthermore, within the DMPs revision, Spanish drought indicators will be defined in order to include seasonal meteorological forecasts, provided by AEMET.

Table 23.1 provides a brief comparison of the main topics associated with drought planning and management in Portugal and Spain.

The evaluation of drought planning and management in both countries makes it clear that a common or coordinated drought monitoring and management is still far away. To achieve this, there is a need to first adjust the indicators, thresholds, and alert levels between both countries, as agreed on and planned under CADC work. In fact, those adjustments must be suited to transboundary agreements, notably the Albufeira Convention, as described in Maia (2009).

Under the terms of the Albufeira Convention, Spain is to guarantee MFRs at the upstream border of the shared rivers, except in exceptional years—this corresponds to the accumulated precipitation (in referenced monitoring stations), in a defined period, being lower than a preestablished minimum. That means that the MFRs (except for Guadiana, where the storage volume on predefined reservoirs is also taken into account) are only dependent on precipitation values from predefined precipitation stations. The drought monitoring system used in Spain, defined in the DMPs, includes several hydrological indicators as noted before, but not the pluviometric indicators defined within the Albufeira Convention in some river basin systems (e.g., Douro and Tejo); if those indicators are included, as in the case of the Guadiana RBD, they are mixed with other pluviometric indicators (Maia 2009).

Figure 23.5 enables a comparison of the drought alert levels obtained for the same period when considering the pluviometric indicators of the Albufeira Convention (Figure 23.5a) or the indicators used in the system defined in the DMPs (Figure 23.5), in this case from the Douro river basin (CHD 2016).

As Figure 23.5 shows, different results and correspondent classifications are obtained for both situations. For example, alert and emergency situations are more frequent using the indicators established by the Albufeira Convention (CHD 2016; Maia 2009). This highlights the necessity to reach a concrete homogenization of indicators between the two countries and their previous agreements. Thereafter, within the second RBMP of Spanish Douro, a revision of the indicators to be used was performed, and it considered the inclusion of the Albufeira Convention precipitation indicators in the Spanish indicator system for that RBD (CHD 2016).

TABLE 23.1

Drought Monitoring and Management Comparison between Portugal and Spain as of March 2017

	Spain	Portugal
Drought management plan	Adopted, since 2007, for each RBD A revision of DMPs is expected in 2017	A provisional national plan was proposed in 2015, but has not yet been officially approved
Responsible bodies	MAGRAMA; RBD administrations; drought permanent commissions	Commission for the Prevention and Monitoring of Effects of Drought and Climate Change (CSAC), and technical working group
Drought monitoring systems	Meteorological monitoring based on SPI index, at national scale, by AEMET national drought indicator system: Information provided at river basin scale, based on: • Water storage in reservoirs • Groundwater storage • River flows • Precipitation Use of index of status Four drought levels based on hydrological monitoring (normal, prealert, alert, emergency)	Meteorological monitoring, at national scale, based on SPI and PDSI, by IPMA PVAS*: Information provided at national scale, considering: • Water storage in reservoirs • Groundwater storage • River flows • Precipitation No specific integrated index Four drought levels based on the hydrological monitoring (normal, prealert, alert, emergency) (*currently not fully implemented)
Drought conditions reports	*At RBD level:* Monthly** reports by the RBD administrations *At national level:* Monthly** reports by the National Drought Observatory (** or more frequently, if drought conditions justify)	*At national level:* Monthly meteorological reports produced by IPMA and presented by the CSAC's advisory working group. In accordance with the anticipated PT DMP, upon the aggravation of drought conditions, the CSAC's advisory working group reports shall be produced on a more frequent basis
Drought measures	Prevention and contingency measures presented in the DMPs for each drought level, applied during drought events	Measures associated with each drought level are defined by the DMP (to approve)

(Continued)

TABLE 23.1 *(Continued)*

Drought Monitoring and Management Comparison between Portugal and Spain as of March 2017

	Spain	Portugal
Contingency plans	Systems with more than 20,000 inhabitants should have contingency plans; currently, those are only developed in some large cities	PT DMP foresees that all public water supply and irrigation supply management entities shall prepare a contingency plan
Drought early warning system (DEWS)	No DEWS are active, at global RBDs level; the drought indicators to be defined in the 2017 DMPs revision are expected to include seasonal meteorological forecasts, provided by AEMET	The Portuguese National Water Plan establishes the goal to implement a drought early warning system by 2021

23.7 Conclusions/Future Perspectives

Drought planning and management in the IP is framed under EU common water policy requirements (WFD) and specific policy option recommendations on WS&D. Although the spatial variability of droughts in the IP is significant, homogenous regions with similar temporal variability of droughts can be identified. As noted earlier, drought planning and management in the IP is mostly dependent on cooperation between Portugal and Spain, as the two countries five shared river basins area corresponds to 45 percent of IP territory. The required drought policy harmonization shall be in line with existing river basin management bilateral agreements. Although the two countries have been cooperating and working toward a declared joint management of the shared RBDs, they are currently at different stages concerning drought planning and management, with no agreement on a common or coordinated drought management framework.

In fact, Spain's drought management policy is based on a more proactive and planned approach whereas Portugal's drought policy still reflects a crisis management approach, with no DMP approved yet. In order to definitively shift to an effective and anticipated (since 2012) drought proactive management, Portugal will likely need to redefine the operational institutional framework for drought management (involving institutions that are currently under different ministries than in 2012) and the role of and/or coordination with the National Water Authority. It is also aimed and expected that both countries will contribute to and/or foster best practices on drought

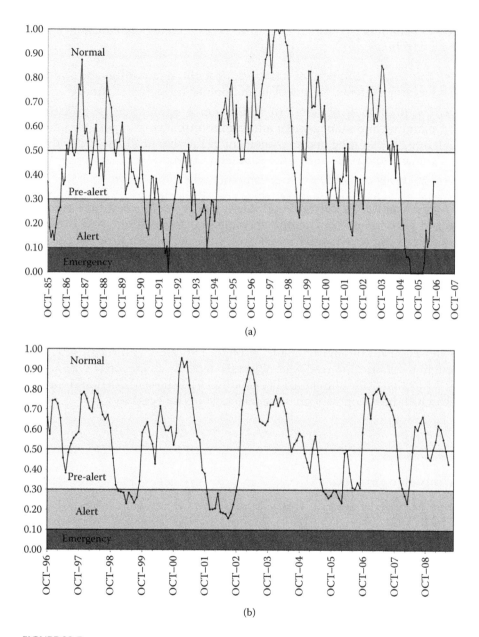

(a)

(b)

FIGURE 23.5

Temporal evolution of the index of status and alert levels on Douro river basin considering (a) precipitation indicators established in the Albufeira Convention and (b) global operational indicators of the river basin. (Adapted from CHD, Actualización del Plan Especial de Sequías (Drought Management Plan Actualization), Anejo 13.1 del Plan Hidrológico de la Parte Española de la Demarcación Hidrográfica del Duero (2015–2021), 2016.)

management policy (WMO 2012) framed on EU and regional bilateral agreements, as follows:

- The implementation of effective drought monitoring and early warning systems, to anticipate and predict drought occurrence
- The promotion of standard and common approaches in drought planning and management aimed at coordinated and possible joint drought planning and management of the shared RBDs, under the frame of the WFD and of the Albufeira Convention
- The development of a sound common indicators system

In addition, although the current drought management measures have contributed to reduced vulnerability and impacts in agriculture and livestock sectors, and improved water management during critical drought periods, it is necessary to reinforce DMPs and indicators corresponding to environmental droughts. Environmental drought impacts have grown in the IP over the past two decades, in association with climatic warming processes over the entire Mediterranean. Droughts have affected forest growth and caused forest decline in large areas (Camarero et al. 2015; Carnicer et al. 2011) and contributed to land degradation processes (Vicente-Serrano et al. 2012). Nevertheless, the development of environmental drought indicators is difficult since the drought resistance and resilience of forests depends on the forest type and is also a function of the climate aridity conditions (Pasho et al. 2011), and further research and development on these relationships is necessary.

Acknowledgments

Sergio M. Vicente-Serrano was supported by the research projects PCIN-2015-220, CGL2014-52135-C03-01 financed by the Spanish Commission of Science and Technology and FEDER, and IMDROFLOOD financed by the Water Works 2014 cofunded call of the European Commission.

References

AEMET. 2016. *Meterological drought vigilance.* http://www.aemet.es/es/servicioscli maticos/vigilancia_clima/vigilancia_sequia. Accessed January 2017.
APA. 2016. Plano de Gestão das Regiões Hidrográficas (2° ciclo), 2ª parte, Guadiana (2nd Cycle Portuguese Guadiana River Basin Management Plan, 2nd Part). https://www.apambiente.pt/_zdata/Politicas/Agua/PlaneamentoeGestao/ PGRH/2016-2021/PTRH7/PGRH7_Parte2.pdf. Accessed January 2017.

Austin, R. B., C. Cantero-Martínez, J. L. Arrúe, E. Playán, and P. Cano-Marcellán. 1998. Yield-rainfall relationships in cereal cropping systems in the Ebro river valley of Spain. *European Journal of Agronomy* 8: 239–248.

BOE. 2001. Boletín Oficial del Estado (Official State Journal) 161. Law 10/2001, pp. 24228–24250. Spanish National Hydrological Plan. https://www.boe.es/boe/dias/2001/07/06/pdfs/A24228-24250.pdf. Accessed January 2017.

BOE. 2016. Boletín Oficial del Estado (Official State Journal) 16. Decree-Law 1/2016, pp. 2972–4301. https://www.boe.es/boe/dias/2016/01/19/pdfs/BOE-A-2016-439.pdf. Accessed January 2017.

Camarero, J. J., A. Gazol, G. Sangüesa-Barreda, J. Oliva, and S. M. Vicente-Serrano. 2015. To die or not to die: Early warnings of tree dieback in response to a severe drought. *Journal of Ecology* 103: 44–57.

Carnicer, J., M. Coll, M. Ninyerola, et al. 2011. Widespread crown condition decline, food web disruption, and amplified tree mortality with increased climate change-type drought. *Proceedings of the National Academy of Sciences of the United States of America* 108(4): 1474–1478.

CHD. 2016. Actualización del Plan Especial de Sequías (Drought Management Plan Actualization). Anejo 13.1 del Plan Hidrológico de la Parte Española de la Demarcación Hidrográfica del Duero (2015–2021).

CHG. 2007. Plan Especial de Sequía de la Confederación Hidrográfica del Guadiana (Drought Management Plan for Guadiana River Basin). http://www.chguadiana.es/corps/chguadiana/data/resources/file/sequia/Plan_Especial_de_Sequias.pdf. Accessed January 2017.

Corella, J. P., B. L. Valero-Garcés, S. M. Vicente-Serrano, A. Brauer, and G. Benito. 2016. Three millennia of heavy rainfalls in Western Mediterranean: Frequency, seasonality and atmospheric drivers. *Scientific Reports* 6: 38206. doi:10.1038/srep38206.

Cuadrat, J. M., F. J. Alfaro, E. Tejedor, R. Serrano-Notivoli, M. Barriendos, and M. A. Saz. 2016. La sequía de mediadios del siglo XVII en el valle del Ebro. Características climáticas e impacto social del evento. In *Paisaje, cultura territorial y vivencia de la geografía*, eds. J. F. Vera-Rebollo, J. Olcina Cantos, and M. Hernández, pp. 923–934. Alicante, Spain: University of Alicante.

Domínguez-Castro, F., P. Ribera, R. García-Herrera, J. M. Cuadrat, and J. M. Moreno. 2012. Assessing extreme droughts in Spain during 1750–1850 from rogation ceremonies. *Climate of the Past* 8(2): 705–722.

Do Ó, A. 2011. Gestão Transfronteiriça do Risco de Seca na Bacia do Guadiana: Análise Comparativa das Estruturas Nacionais de Planeamento. VII Congreso Ibérico sobre Gestión y Planificación del Agua "Rios Ibéricos +10. Mirando al futuro tras 10 años de DMA". Talavera de la Reina. February. http://www.fnca.eu/images/documentos/VII%20C.IBERICO/Comunicaciones/A8/02-DoO.pdf. Accessed January 2017.

DR. 1999. Diário da República (Portuguese Government Official Journal) 191/1999. Convention on Co-operation for the Protection and Sustainable Use of Portuguese-Spanish River Basins (in Portuguese), pp. 5410–5430. http://www.cadc-albufeira.eu/pt/documentos/. Accessed January 2017.

DR. 2012. Diário da República (Portuguese Government Official Journal) 62/2012. Ministers Council Resolution 37/2012. Drought Commission and Workgroup. http://dre.pt/util/getpdf.asp?s=diad&serie=1&iddr=2012.62&iddip=20120603. Accessed January 2017.

DR. 2016. Diário da República (Portuguese Government Official Journal) 215/2016. Decree-Law 76/2016. National Water Plan. https://dre.pt/application/conteudo/75701996. Accessed January 2017.

EC. 2000. Directive 2000/60/EC of the European Parliament and of the Council of 23 October 2000 establishing a framework for Community action in the field of water policy. Official Journal of the European Communities, L 327, 22.12.2000, p. 1. http://eur-lex.europa.eu/resource.html?uri=cellar:5c835afb-2ec6-4577-bdf8-756d3d694eeb.0004.02/DOC_1&format=PDF. Accessed January 2017.

EC. 2007a. Communication from the Commission to the European Parliament and the Council—Addressing the challenge of water scarcity and droughts in the European Union. *COM/2007/414 final.* http://eur-lex.europa.eu/legal-content/EN/TXT/PDF/?uri=CELEX:52007DC0414&from=EN. Accessed January 2017.

EC. 2007b. Drought Management Plan Report. Including Agricultural, Drought Indicators and Climate Change Aspects. *Water Scarcity and Droughts Expert Network.* Luxemburg: European Commission. http://ec.europa.eu/environment/water/quantity/pdf/dmp_report.pdf. Accessed January 2017.

Estrela, T., and T. Sancho. 2016. Drought management policies in Spain and the European Union: From traditional emergency actions to Drought Management Plans. *Water Policy* 19(3). doi:10.2166/wp.2016.018.

Estrela, T., and E. Vargas. 2012. Drought management plans in the European Union. The case of Spain. *Water Resources Management* 26: 1537. doi:10.1007/s11269-011-9971-2.

Font-Tullot, I. 1988. *Historia del clima en España.* Madrid: Instituto Nacional de Meteorología.

García-Herrera, R., D. Paredes, R. M. Trigo, D. Barriopedro, and M. A. Mendes. 2007. The outstanding 2004/05 drought in the Iberian Peninsula: Associated atmospheric circulation. *Journal of Hydrometeorology* 8(3): 483–498.

Gil, A., and A. Morales. 2001. *Causas y consecuencias de las sequías en España.* Interniversitario de Geografía. Universidad de Alicante, ISBN: 84-7908-600-9. Spain.

Gouveia, C. M., A. Bastos, R. M. Trigo, and C. C. Dacamara. 2012. Drought impacts on vegetation in the pre- and post-fire events over Iberian Peninsula. *Natural Hazards and Earth System Sciences* 12(10): 3123–3137.

Hoerling, M., J. Eischeid, J. Perlwitz, X. Quan, T. Zhang, and P. Pegion. 2012. On the increased frequency of Mediterranean drought. *Journal of Climate* 25: 2146–2161.

Jerez, S., R. M. Trigo, S. M. Vicente-Serrano, et al. 2013. The impact of the North Atlantic Oscillation on the renewable energy resources in south-western Europe. *Journal of Applied Meteorology and Climatology* 52: 2204–2225.

López-Moreno, J. I., S. M. Vicente-Serrano, S. Beguería, J. M. García-Ruiz, M. M. Portela, and A. B. Almeida. 2009. Downstream propagation of hydrological droughts in highly regulated transboundary rivers: The case of the Tagus River between Spain and Portugal. *Water Resources Research* 45: W02405. doi:10.1029/2008WR007198.

López-Moreno, J. I., S. M. Vicente-Serrano, J. Zabalza, et al. 2013. Hydrological response to climate variability at different time scales: A study in the Ebro basin. *Journal of Hydrology* 477: 175–188.

Lorenzo-Lacruz, J., E. Moran-Tejeda, S. Vicente-Serrano, and J. López-Moreno. 2013a. Streamflow droughts in the Iberian Peninsula between 1945 and 2005: Spatial and temporal patterns. *Hydrology and Earth System Sciences* 17: 119–134. doi:10.5194/hess-17-119-2013.   ∙

Lorenzo-Lacruz, J., S. M. Vicente-Serrano, J. C. González-Hidalgo, J. I. López-Moreno, and N. Cortesi. 2013b. Hydrological drought response to meteorological drought at various time scales in the Iberian Peninsula. *Climate Research* 58: 117–131.

Lorenzo-Lacruz, J., S. M. Vicente-Serrano, J. I. López-Moreno, J. C. González-Hidalgo, and E. Morán-Tejeda. 2011. The response of Iberian rivers to the North Atlantic Oscillation. *Hydrology and Earth System Sciences* 15: 2581–2597.

Maia, R. 2008. The EU Water Framework Directive implementation in the Iberian Context. 13th IWRA World Water Congress. Proceedings, Protection and Restoration of the Environment VIII, 1–4 September 2008, Montpellier, France.

Maia, R. 2009. Drought Management in the Iberian Peninsula Shared River Basins: Developments and Perspectives. EWRA 7th International Conference on Water Resources Conservation and Risk Reduction Under Climatic Instability, Cyprus.

Maia, R. 2011. Drought Planning and Management on an European Transboundary Context: The Case of the Iberian Peninsula [online]. In: *Proceedings of the 34th World Congress of the International Association for Hydro-Environment Research and Engineering: 33rd Hydrology and Water Resources Symposium and 10th Conference on Hydraulics in Water Engineering.* Valentine, EM., Apelt, CJ., Ball, J., Chanson, H., Cox, R., Ettema, R., Kuczera, G., Lambert, M., Melville, BW., Sargison, JE (eds.). A.C.T.: Engineers, Barton, Australia, pp. 567–574.

Maia, R., E. Vivas, R. Serralheiro, and M. Carvalho. 2015. Socioeconomic evaluation of drought effects. Main principles and application to Guadiana and Algarve case studies. *Water Resources Management* 29(2): 575–588. doi:10.1007/s11269-014-0883-9.

MAPAMA. 2016a. *Emergency plans webpage.* Spanish Drought National Observatory Website. http://www.mapama.gob.es/es/agua/temas/observatorio-nacional-de-la-sequia/planificacion-gestion-sequias/observatorio_nacional_sequia_3_3_planes_emergencia.aspx. Accessed January 2017.

MAPAMA. 2016b. *Monitoring reports and maps.* Spanish Drought National Observatory Website. http://www.mapama.gob.es/es/agua/temas/observatorio-nacional-de-la-sequia/informes-mapas-seguimiento/default.aspx. Accessed January 2017.

Martín-Vide, J. 1994. Diez características de la pluviometría española decisivas en el control de la demanda y el uso del agua. *Boletín de la AG* 18: 9–16.

Martín-Vide, J., and M. Barriendos. 1995. The use of rogation ceremony records in climatic reconstruction: A case study from Catalonia (Spain). *Climatic Change* 30: 201–221.

Martín-Vide, J., and J. A. López-Bustins. 2006. The Western Mediterranean oscillation and rainfall in the Iberian Peninsula. *International Journal of Climatology* 26: 1455–1475.

Molinero, F. 2001. Consecuencias agrarias de las sequías en Castilla y León. In *Causas y consecuencias de las sequías en España*, eds. A. Gil Olcina and A. Morales Gil, pp. 261–276. Alicante, Spain: University of Alicante, Interuniversity Institute of Geography.

Morales, A., J. Olcina, and A. M. Rico. 2000. Diferentes persepciones de la sequía en España: Adaptación, catastrofismo e intentos de corrección. *Investigaciones Geográficas* 23: 5–46.

Páscoa, P., C. M. Gouveia, A. Russo, and R. M. Trigo. 2017. The role of drought on wheat yield interannual variability in the Iberian Peninsula from 1929 to 2012. *International Journal of Biometeorology* 61(3): 439–451.

Pasho, E., J. J. Camarero, M. de Luis, and S. M. Vicente-Serrano. 2011. Spatial variability in large-scale and regional atmospheric drivers of Pinus halepensis growth in eastern Spain. *Agricultural and Forest Meteorology* 151: 1106–1119.

Pita, M. F. 1989. La sequía como desastre natural. Su incidencia en el ámbito español. Norba 6(7): 31–61.

Pulwarty, R. S., and R. Maia. 2015. Adaptation challenges in complex rivers around the world: The Guadiana and the Colorado Basins. *Water Resources Management* 29(2): 273–293. doi:10.1007/s11269-014-0885-7.

Rodríguez-Puebla, C., A. H. Encinas, S. Nieto, and J. Garmendia. 1998. Spatial and temporal patterns of annual precipitation variability over the Iberian Peninsula. *International Journal of Climatology* 18: 299–316.

Sánchez-Martínez, M. -T., M. Salas-Velasco, and N. Rodríguez-Ferrero. 2012. Who manages Spain's water resources? The political and administrative division of water management. *International Journal of Water Resources Development* 28(1): 27–42. doi:10.1080/07900627.2012.640610.

Stefano, L., B. Willaarts, H. Cooley, and A. Garrido. 2015. International Workshop on Drought Management in Spain and California: Lessons Learned. *Rosenberg International Forum*. Santander, Spain: Botín Foundation.

Tejedor, E., M. de Luis, J. M. Cuadrat, J. Esper, and M. Á. Saz. 2015. Tree-ring-based drought reconstruction in the Iberian Range (east of Spain) since 1694. *International Journal of Biometeorology* 60(3): 361–372.

Trenberth, K. E., A. Dai, G. van der Schrier, et al. 2014. Global warming and changes in drought: Expectations, observations and uncertainties. *Nature Climate Change* 4: 17–22.

Trigo, R. M., J. Añel, D. Barriopedro, et al. 2013. The record winter drought of 2011–12 in the Iberian Peninsula. *Bulletin of the American Meteorological Society* 94(9): S41–S45.

Trigo, R. M., T. J. Osborn, and J. M. Corte-Real. 2002. The North Atlantic Oscillation influence on Europe: Climate impacts and associated physical mechanisms. *Climate Research* 20(1): 9–17.

Vicente-Serrano, S. M. 2006a. Spatial and temporal analysis of droughts in the Iberian Peninsula (1910–2000). *Hydrological Sciences Journal* 51(1): 83–97.

Vicente-Serrano, S. M. 2006b. Differences in spatial patterns of drought on different time scales: An analysis of the Iberian Peninsula. *Water Resources Management* 20: 37–60.

Vicente-Serrano, S. M. 2007. Evaluating the impact of drought using remote sensing in a Mediterranean, semi-arid region. *Natural Hazards* 40: 173–208.

Vicente-Serrano, S. M. 2013. Spatial and temporal evolution of precipitation droughts in Spain in the last century. *Adverse weather in Spain, Madrid, WCRP Spanish Committee*, pp. 283–296.

Vicente-Serrano, S. M., C. Azorin-Molina, A. Sanchez-Lorenzo, et al. 2014a. Reference evapotranspiration variability and trends in Spain, 1961–2011. *Global and Planetary Change* 121: 26–40.

Vicente-Serrano, S. M., S. Beguería, J. I. López-Moreno, A. M. El Kenawy, and M. Angulo-Martínez. 2009. Daily atmospheric circulation events and extreme precipitation risk in Northeast Spain: The role of the North Atlantic Oscillation, Western Mediterranean Oscillation, and Mediterranean Oscillation. *Journal of Geophysical Research-Atmosphere* 114: D08106. doi:10.1029/2008JD011492.

Vicente-Serrano, S. M., S. Beguería, and J. I. López-Moreno. 2010. A multi-scalar drought index sensitive to global warming: The Standardized Precipitation Evapotranspiration Index—SPEI. *Journal of Climate* 23: 1696–1718.

Vicente-Serrano, S. M., and J. M. Cuadrat. 2007a. North Atlantic oscillation control of droughts in Northeast of Spain: Evaluation since A.D. 1600. *Climatic Change* 85: 357–379.

Vicente-Serrano, S. M., and J. M. Cuadrat-Prats. 2007b. Trends in drought intensity and variability in the middle Ebro valley (NE Spain) during the second half of the twentieth century. *Theoretical and Applied Climatology* 88: 247–258.

Vicente-Serrano, S. M., R. García-Herrera, D. Barriopedro, et al. 2016. The Westerly index as complementary indicator of the North Atlantic Oscillation in explaining drought variability across Europe. *Climate Dynamics* 47: 845–863.

Vicente-Serrano, S. M., J. C. González-Hidalgo, M. de Luis, and J. Raventós. 2004. Spatial and temporal patterns of droughts in the Mediterranean area: The Valencia region (East-Spain). *Climate Research* 26: 5–15.

Vicente-Serrano, S. M., and J. I. López-Moreno. 2006. The influence of atmospheric circulation at different spatial scales on winter drought variability through a semiarid climatic gradient in north east Spain. *International Journal of Climatology* 26: 1427–1456.

Vicente-Serrano, S. M., J. I. Lopez–Moreno, S. Beguería, et al. 2014b. Evidence of increasing drought severity caused by temperature rise in southern Europe. *Environmental Research Letters* 9: 044001. doi:10.1088/1748-9326/9/4/044001.

Vicente-Serrano, S. M., and R. Trigo, eds. 2011. *Hydrological, socioeconomic and ecological impacts of the North Atlantic Oscillation in the Mediterranean region.* Advances in Global Research (AGLO) series. New York: Springer-Verlag.

Vicente-Serrano, S. M., J. Zabalza-Martínez, G. Borràs, et al. 2017. Extreme hydrological events and the influence of reservoirs in a highly regulated river basin of northeastern Spain. *Journal of Hydrology: Regional Studies* 12: 13–32.

Vicente-Serrano, S. M., A. Zouber, T. Lasanta, and Y. Pueyo. 2012. Dryness is accelerating degradation of vulnerable shrublands in semiarid Mediterranean environments. *Ecological Monographs* 82: 407–428.

WMO. 2012. Science Document: Best Practices on National Drought Management Policy. World Meteorological Organization. High Level Meeting on National Drought Policy (HMNDP). Geneva: CICG.

24

Establishing the Queensland Drought Mitigation Centre

Roger C. Stone

CONTENTS

24.1 Introduction

The large State of Queensland, Australia, has the highest levels of naturally occurring year-to-year rainfall variability in the world (Love 2005; Nicholls et al. 1997). Yet, drought in Australia and Queensland had long been regarded by policymakers mostly as an aberration to an otherwise long-term "normal" climate pattern (Botterill and Wilhite 2005; Wilhite 2005). Indeed, from the time of European settlement, Australian governments "responded to the concept of drought being a natural disaster through various Commonwealth-State *natural disaster relief arrangements* which treated drought in a similar manner to disasters such as tropical cyclones or floods" (Botterill and Wilhite 2005).

However, major droughts keep reoccurring and losses to the national economy, especially rural economies, keep increasing in Australia. The 1963–1968 drought saw a 40 percent drop in Australia's wheat harvest and a decrease in farm income of between AUD$300 million and AUD$600 million. The 1982/1983 drought resulted in a loss of AUD$3 billion to the Australian economy. The 1991–1995 drought period resulted in a loss of AUD$5 billion to the Australian economy, with AUD$590 million provided directly to farmers and rural business through drought relief payments. The major drought periods mentioned here were all driven by the well-recognizable (and predictable, within a risk management framework) El Niño phenomenon in the central and eastern equatorial Pacific.

Meanwhile, by the 1990s, it was becoming increasingly untenable to support the currently existing drought policy in the light of ongoing and improved understanding of Australia's climate patterns, especially its high natural levels of year-to-year climatic variability. In the 10–20 years leading up to 1992, considerable research activity was also to provide fundamental insight into the potential predictability of much of Australia's and Queensland's rainfall variability, at least on a seasonal basis, and thus provide the basis for improved preparedness for Queensland's crippling droughts (Stone 2014).

Drought research for a profitable and sustainable rural sector had become wide-ranging, and by 1992 included whole farm management systems that integrated climate prediction, technical, biological, and financial information; control strategies for weeds and pests; socioeconomic factors; and the needs of rural communities and farm families in times of stress (Australia's National Drought Policy, 1992, in White et al. 1993, 2005).

Yet, a key aspect remained on how to most effectively harness these climatic and agricultural research outcomes, together with provision and development of the ongoing climatic and whole farm systems research (including farm supply chain research) that has been active in Australia and globally, into such a preparedness program (Everingham et al. 2003; Meinke and Stone 2005; Stone and Meinke 2005). It is further suggested that a major technological advancement, in addition to the improved understanding of key climatic mechanisms over this period in Queensland and Australia, has been the realization of the key value of the use of crop simulation modeling (such as the agricultural production systems simulation model [APSIM]) to aid *planning* of improved preparedness for agricultural drought purposes utilizing management decisions such as more appropriate planting dates and fertilizer applications (Keating et al. 2003). A further associated technological advancement has been the capability to provide prerun APSIM outputs as a form of decision support to aid in the preparedness for extremes in climate variability, including likely low rainfall periods that can be coupled with very low antecedent soil moisture conditions (Cox et al. 2004).

Mainly through interaction and facilitation provided by such initiatives as the integrated drought management program (IDMP; see Chapter 3), a collaborative program between the Global Water Partnership (GWP) and the World Meteorological Organization (WMO) in Geneva, it was also realized that much can be learned by combining global knowledge of successes and problems associated with similar such initiatives in drought preparedness. IDMP brings together many world experts and program managers in drought mitigation and includes those from the US National Drought Mitigation Center (NDMC), which aims to help institutions develop and implement measures to reduce societal vulnerability to drought, stressing preparedness and risk management rather than crisis management. Indeed, Pulwarty (2011) recognized that no single nation or organization can afford to tackle all of the hurdles involved in creating a drought early warning system in its entirety and stressed the need for international and national linkages.

Thus, there is a need to build initiatives that, in turn, recognize the need for global clearinghouses in understanding the complexity of drought research and planning, and the need to work with and across other key agencies in order to fully tackle the many facets of effective drought preparedness and planning. It has been through recognizing this need for global collaboration in drought research and to harness local preparedness research, development, and planning that the Queensland Drought Mitigation Centre (QDMC) came into being in 2017.

24.2 The Queensland Drought Mitigation Centre

The key facilitator in the QDMC has been the Queensland Government, which recently announced the rural assistance and drought package totaling AUD$77.9 million over 5 years. This package includes the drought assistance package of AUD$44.9 million for 2016–2017 and the drought and climate adaptation program (DCAP), of which the QDMC is a key component, of AUD$3.5 million for 2016–2017. However, should rural and associated industry and community support be forthcoming during 2016–2017, then there is a high probability of the DCAP initiative being extended a further 4 years. The entire package comprises: the focused grazing best management program (BMP), which includes whole farm economic planning and incorporates components to help graziers (generally called *producers* in Australia) be better prepared for all risks, but including drought as a major component; government department research and development programs to help Queensland agriculture adapt to future climate scenarios; and the QDMC, which incorporates drought and climate research components, including research into named-peril agricultural insurance programs (details listed below).

The QDMC has three nodes, with the core component located in Toowoomba at the University of Southern Queensland within the International Centre for Applied Climate Sciences (ICACS). The QDMC will be a center of excellence for research (including monitoring, prediction, and early warning), communication, and capacity-building in areas relating to the mitigation of the impacts of drought on agriculture and the wider community. The QDMC is bringing together and building on the internationally recognized expertise in seasonal climate forecasting, climate risk management, and decision-support systems that exists within ICACS and the state government departments of Agriculture and Fisheries (DAF) and Science, Technology and Innovation (DSITI). Linkages and collaborations with national and international research and development agencies will be critical to ensure all new relevant science and systems research innovations, globally, are properly captured by the needs of QDMC and DCAP. These collaborative agencies may include the NDMC, the US national integrated drought information system, the Australian Bureau of Meteorology (especially its research division in seasonal and decadal climate forecasting research), the UK Met Office (especially its Hadley Centre for Climate Research and its work on decadal prediction), the Australian CSIRO, key Queensland and Australian Universities, the New South Wales Department of Primary Industries, the WMO (especially its Technical Commission for Agricultural Meteorology and its associated IDMP and the key program areas in drought and food security), and, potentially, the Drought Management Centre for Southeastern Europe and the European Drought Centre.

In addition, the QDMC will provide a focus for attracting in-kind and cash investments in drought preparedness from participating agencies and research, development, and extension funders. In essence, the budget requested for the QDMC provides seed funding for additional investment by other key interested agencies, especially those closely connected to farmers, graziers, and agricultural industry.

The QDMC has the following approach to improving drought preparedness in farming systems based on the following principles:

1. Use of the latest advances in climate science and seasonal climate forecasting to provide an enhanced understanding of drought risk and drought mitigation options, as well as climate risk management more generally

2. A risk management approach that recognizes the short- and long-term production and financial risks induced by climatic variability, in particular drought

3. A focus on supporting the tactical and strategic decisions that farm business managers make across short- and long-term timeframes in managing their businesses

4. A recognition that making the most of good seasons is a key component of managing climatic variability and improving drought preparedness

In total, the QDMC's work program will initially comprise the projects shown in Table 24.1.

24.3 Summary

Developing improved drought preparedness programs and processes in the State of Queensland has entailed a long process that originated

TABLE 24.1

Listing of Formal Initial QDMC Projects for 2016/17 as Part of the Overall DCAP Program Funded by the Queensland Government (Missing Numbered Projects Are Those in Other Fields and Not Part of the Overall QDMC Project List)

DCAP2 Improving seasonal climate forecasts (QBO, STR, etc. included in conjunction with ENSO)

DCAP3 Improve the ability of forecasts to predict multiyear drought—integrating the UKMO DePreSys model or similar into decadal forecasts (collaboration with BoM/UKMO)

DCAP5 Regional climate change adaptation for agricultural industries

DCAP6 Producing enhanced/named-peril crop insurance systems (collaboration with major international reinsurance agencies)

DCAP7.1 Developing products for use in drought monitoring: drought index application

DCAP7.2 Developing products for use in drought monitoring: improved crop yield and production forecasts (integrating seasonal forecasts with a multicrop modeling approach)

DCAP9 Developing and customizing decision support tools (e.g., "GRAZe-ON," "Droughtplan," "Rainman," "ClimateARM")

DCAP13 Revamping the successful *Managing for Climate* strong user engagement workshops across the state

DCAP14 Crop production modeling under climate change and regional adaptation

DCAP15 Assessing the economic value of improved climate risk management strategies through the application of seasonal climate forecasts for key agricultural industries in Queensland

---

in a more thorough understanding of the climatic mechanisms responsible for the extremely high levels of year-to-year rainfall variability in Queensland in the 1990s. Coupled with the improved understanding of the causes of this climate variability—and hence offering the possibility of predicting and preparing for high rainfall variability and drought— has been the development of many crop and pasture simulation modeling systems and associated decision-support systems. Nevertheless, drought preparedness has remained a relatively slow and incomplete task over the past 20–30 years, despite the fact that so much progress had been made in Queensland in climate and farming systems research over that period. A remarkable breakthrough recently occurred in 2016 with the Queensland Government now utilizing these scientific and decision-support system advances to incorporate the QDMC. Through a series of subprojects, which range from further improving seasonal climate and decadal forecasting to innovative named-peril farm insurance systems and reintroducing previously successful *Managing for Climate* farmer workshops, it is envisaged that Queensland rural producers and others along the agricultural supply chain will now be better able to further enhance their capability to manage for the many droughts and protracted droughts that will occur in Queensland in the future.

References

Botterill, L. C., and D. A. Wilhite. 2005. *From Disaster Response to Risk Management: Australia's National Drought Policy.* Dordrecht: Springer.

Cox, H., G. L. Hammer, G. McLean, and C. King. 2004. National Whopper Cropper— Risk management discussion support software. In *New directions for a Diverse Planet: Proceedings of the 4th International Crop Science Congress*, eds. T. Fischer, N. Turner, J. Angus, et al. Brisbane, Australia, September 26–October 1, The Regional Institute Ltd, Gosford, Australia, 2250. http://www.cropscience.org.au/icsc2004/poster/4/1/1/402_cox.htm#TopOfPage (accessed June 30, 2017).

Everingham, Y. L., R. C. Muchow, R. C. Stone, and D. H. Coomans. 2003. Using Southern Oscillation Index phases to forecast sugarcane yields: A case study for northeastern Australia. *International Journal of Climatology* 23(10):1211–1218.

Keating, B. A., P. S. Carberry, G. L. Hammer, et al. 2003. An overview of APSIM, a model designed for farming systems simulation. *European Journal of Agronomy* 18(3–4):267–288.

Love, G. 2005. Impacts of climate variability on regional Australia. *ABARE Outlook Conference Proceedings (National Agricultural and Resources Outlook Conference Proceedings)*, Australian Government Department of Agriculture, Fisheries and Forestry, Canberra, Climate Session Papers, 10–19. http://pandora.nla.gov.au/pan/45562/20050622-0000/PC13021.pdf (accessed March, 2017).

Meinke, H., and R. C. Stone. 2005. Seasonal and inter-annual climate forecasting: The new tool for increasing preparedness to climate variability and change in agricultural planning and operations. *Climatic Change* 70:221–253.

Nicholls, N., W. Drosdowsky, and B. Lavery. 1997. Australian rainfall variability and change. *Weather* 52:66–72.

Pulwarty, R. S. 2011. Drought information systems: Improving international and national linkages. *Drought Mitigation Center Faculty Publications.* Paper 59. http://digitalcommons.unl.edu/droughtfacpub (accessed 17 February 2017).

Stone, R. C. 2014. Constructing a framework for national drought policy: The way forward—The way Australia developed and implemented the national drought policy. *Weather and Climate Extremes* 3:117–125. http://dx.doi.org/10.1016/j.wace.2014.02.001.

Stone, R. C., and H. Meinke. 2005. Operational seasonal forecasting of crop performance. *Philosophical Transactions of the Royal Society B* 360:2109–2124.

White, D. H., L. C. Botterill, and B. O'Meagher. 2005. At the intersection of science and politics: Defining exceptional drought. In *From Disaster Response to Risk Management: Australia's National Drought Policy*, eds. L. C. Botterill and D. A. Wilhite, 99–111. Dordrecht: Springer.

White, D. H., D. Collins, and S. M. Howden. 1993. Drought in Australia: Prediction, monitoring, management, and policy. In *Drought Assessment, Management and Planning: Theory and Case Studies*, eds. D. A. Wilhite, 213–236. Dordrecht: Kluwer Academic Publishers.

Wilhite, D. A. 2005. Drought policy and preparedness: The Australian experience in an international context. In *From Disaster Response to Risk Management: Australia's National Drought Policy*, eds. L. C. Botterill and D. A. Wilhite, 157–176. Dordrecht: Springer.

Section V

Integration and Conclusions

25

Drought and Water Crises: Lessons Drawn, Some Lessons Learned, and the Road Ahead

Donald A. Wilhite and Roger S. Pulwarty

CONTENTS

25.1 Introduction

By the middle of the 21st century more than half of the planet will be living in areas of water stress where supply cannot sustainably meet demand (The Economist, Nov. 5, 2016).

Despite the fact that drought is an inevitable feature of climate for nearly all climatic regimes, progress on drought preparedness has been extremely slow. Historically, the approach taken by most nations has been reactive, responding to drought in a crisis management mode. Many nations now feel a growing sense of urgency to move forward with a more proactive, risk-based drought management approach (Wilhite 2000; Wilhite et al. 2014) as highlighted in Parts I and II of this book, Chapters 1 through 5, and in several of the case studies in Part IV. Certainly, the widespread occurrence of this insidious natural hazard in recent years has contributed to the sense of urgency. But, drought occurs in many parts of the world and affects portions of many countries, both developing and developed, on an annual basis. For example, the average area affected by severe and extreme drought in the United States each year is about 14 percent. This figure has been as high as 65 percent (1934 and 2012) and has hovered in the 35–40 percent range for most years since 2000. (Thus, global estimates of drought occurrence trends can mask significant regional and local changes.) So, does the widespread occurrence of drought in the United States over the last 10 years explain the emergence of several national and state initiatives centered on drought monitoring and preparedness, given that events of this magnitude have not motivated policy makers to act in the past? To some degree, we would say, yes. Widespread, severe, and multiyear droughts in other countries or regions in recent years (e.g., Australia, Brazil, the Greater Horn of Africa, and Mexico) have also been instrumental in moving the conversation on improving drought monitoring and preparedness forward. However, our experience would suggest that this is only one of the factors contributing to the increased attention to drought risk management in many drought-prone countries.

Climate change and the potential threat of an increase in frequency and severity of extreme events is also a contributing factor. In the first edition of this book, we concluded that climate change was probably not playing a significant role in this trend because most policymakers have difficulty thinking beyond their term of office or the next election, and many businesses were unwilling to look beyond their next quarterly report. At this writing, we suggest that it is now more likely that climate change is playing a significant role in driving decisions for countries, communities, and water managers to invest time and resources into drought risk management, specifically directed toward the development of more comprehensive drought monitoring, early warning and information systems, vulnerability assessments, preparedness planning, and national drought policies. This increased attention to improving drought management is likely because more countries are now experiencing noticeable changes in their climate along with an increase in the frequency and severity of drought. For regions that have experienced either a downward trend of annual precipitation or a higher frequency of drought events (perhaps multiyear in length), or both, the potential threat of climate change now seems more real. In addition, a series of activities and actions have also stimulated greater interest in and progress on drought risk management. For example, the convening of the High-level Meeting on National Drought Policy (HMNDP) in 2013 with 87 countries in attendance (see Chapter 2) and a number of follow-on activities to that conference (e.g., the launching of the Integrated Drought Management Programme [IDMP]— see Chapter 3), the regional workshops on building capacity for national drought policies, new initiatives in the EU (see Chapter 18), and the 2016 African drought conference (http://allafrica.com/stories/201608180693. html) have helped to spread the concepts and garner support for drought risk management and national drought policies.

Decision-making under uncertainty is onerous but unavoidable. Some policymakers and resource managers remain of the opinion that climate change projections are in error, preferring to presume that there will not be a change in the climate state and that extreme climatic events such as drought will not change in frequency or severity. Little consideration is given to the real possibility that projected changes in climate may be too conservative, or underestimate the degree of change in the frequency and severity of extreme events for some locations. In some areas, drying due to climate change is overlaid on the periodic droughts those areas have always experienced. As illustrated in this volume, the occurrence of extreme climate variability, including drought, coupled with high temperatures and other atmospheric factors and land surface conditions, can result in events that exceed climate model projections.

In addition to the factors mentioned above, it is our opinion that the growing interest in drought preparedness is also associated with the documented increase in social, economic, and environmental vulnerability, as exemplified by the increase in the magnitude and complexity of impacts. Although global figures for the trends in economic losses associated with drought do not exist, a report from the UN Development Programme (UNDP Bureau of Crisis Prevention and Recovery 2004) indicates that annual losses associated with natural disasters increased from US$75.5 billion in the 1960s to nearly US$660 billion in the 1990s. Losses resulting from drought likely follow a similar trend, but actual impacts, including economic loss numbers, are not well known. These figures for natural disasters, and especially drought, are likely significantly underestimated because of the inexact reporting or insufficiency of the data. Loss estimates do not include social and environmental costs over time and secondary and higher order impacts such as on hydropower. This increase has been observed in both developing and developed countries, although the types of impacts differ markedly in most cases, as illustrated by numerous authors in this book. Most estimates of drought-related losses exclude indirect losses—livelihoods, informal economies, intangible losses including ecosystem services, quality of life, and cultural impacts (Pulwarty and Verdin 2013).

With respect to drought, how can we define vulnerability? It is usually expressed as the degree of a society's capacity to anticipate, cope with, resist, or adapt to, and recover from the impact of a natural hazard. Capacity on paper does not always translate to capability on the ground. Urbanization is placing more pressure on limited water supplies and overwhelming the capacity of many water supply systems to deliver that water to users, including agriculture, especially during periods of peak demand. More sophisticated technology decreases our vulnerability to drought in some instances while increasing it in others. Greater awareness of our environmental and ecosystem services and the need to preserve and restore environmental quality is placing increased pressure on all of us to be better stewards of our physical and biological resources. Environmental degradation such as desertification is reducing the biological productivity of many landscapes and increasing vulnerability to drought events. All of these factors emphasize that our vulnerability to drought is dynamic and must be evaluated periodically. The recurrence of a drought today of equal or similar magnitude to one experienced several decades ago will likely follow different impact trajectories and result in far greater economic, social, and environmental losses and conflicts between water users.

25.2 Moving from Crisis to Risk Management: Changing the Paradigm

In 1986, an international symposium and workshop was organized at the University of Nebraska that focused on the principal aspects of drought, ranging from prediction, early warning and impact assessment to response, planning, and policy. The goal of this meeting was to review and assess our current knowledge of drought and determine research and information needs to improve national and international capacity to cope with drought (Wilhite and Easterling 1987). Reflecting on this meeting today, 30 years later, and its outcomes, it would seem that the symposium may represent the beginning of the movement to a new paradigm for drought management—one focusing on reducing societal vulnerability to drought through a more proactive approach.

Following directly on this framing, the National Drought Policy Commission (NDPC 2000) noted that drought risk management should:

1. Favor preparedness over insurance, insurance over relief, and incentives over regulation

2. Set research priorities based on the potential of the research results to reduce drought impacts

3. Coordinate the delivery of federal services through cooperation and collaboration with nonfederal entities

In addition, the European Union Water Framework Directive (EU 2000, Chapter 18 of this volume) in addressing water scarcity and drought identifies "improving drought risk management" as one of its seven key policy options.

As previously stated, progress on shifting the paradigm for drought management has been slow. Clearly, it has taken time for the policy and the practitioner communities to become more aware of the diverse and escalating impacts of drought, their complexities, and the ineffectiveness of the reactive postimpact or crisis management approach. The factors explaining the slow emergence of this new paradigm are many, but it is clear that it has emerged today in many countries and in many international organizations dealing with disaster management and development issues. Drought-related crises, such as those that recently occurred in the Horn of Africa and elsewhere, reveal major concerns regarding detailed examination of the root causes of the lack of early action (Verver 2011). A more proactive, risk-based approach to drought management must rely on a strong science component. It also must occur at the interstices of science and policy—a particularly uncomfortable place for many scientists.

While some have argued (see Stakhiv et al. 2016) that the entire development of water infrastructure in regions such as the western United States,

Egypt, and elsewhere illustrates our capabilities in effectively managing drought risk, we posit that in most such cases (e.g., the western United States), the reality is that this infrastructure evolved over a century or more and relied on instream water originating in wetter climates upstream that was technically in surplus (i.e., not yet fully exploited) since full development had not yet taken place. Most such systems now find themselves to be closed. As illustrated by the Colorado River, understanding how vulnerability is shifting is central to identifying future management pathways and reform options (Kenney et al. 2010). In addition, this infrastructure is also now aging and in need of retrofitting and repair.

Figure 4.1 illustrates the cycle of disaster management, depicting the interconnectedness or linkages between crisis and risk management. The traditional crisis management approach has been largely ineffective, and there are many examples of how this approach has increased vulnerability to drought because of individuals' (i.e., disaster victims) greater reliance on the emergency response programs of government and donor organizations. Drought relief or assistance, for example, often rewards the poor resource manager who has not planned for drought, whereas the better resource manager who has employed appropriate mitigation measures is not eligible for this assistance. Thus, drought relief is often a disincentive for improved resource management. Should government reward good stewardship of natural resources and planning or unsustainable resource management? Unfortunately, most nations have been following the latter approach for decades because of political and other pressures associated with crises and the lack of preparation. Redirecting this institutional inertia to a new paradigm offers considerable challenges for the science and policy communities.

As has been underscored many times by the contributors to this book, reducing future drought risk requires a more proactive approach, one that emphasizes preparedness planning and the development of appropriate mitigation actions and programs, including improved drought monitoring through the development of comprehensive early warning and information delivery systems. The contrasting characteristics of crisis versus risk management are illustrated in Table 25.1. However, this approach has to be multithematic and multisectoral because of the complexities of associated impacts and their interlinkages. Risk management favorably complements the crisis management part of the disaster management cycle such that in time one would expect the magnitude of impacts (whether economic, social, or environmental) to diminish. However, the natural tendency has been for society to revert to a position of apathy once the threat accompanying a disaster subsides (i.e., the proverbial hydro-illogical cycle; see Figure 4.2 in Chapter 4).

This raises an important point that has been addressed by many authors throughout this book: what constitutes a crisis? Crises are inextricably tied to decision-making. The Merriam-Webster, 1977 dictionary gives the following definitions of crisis: the decisive moment (as in a literary play); an unstable or crucial time or state of affairs whose outcome will make a decisive difference

TABLE 25.1

Crisis Versus Risk Management: Characteristics, Costs, and Benefits

Crisis Management	Risk Management
Expensive • Costs + costs of inaction • Repeats past mistakes	Investment • Short-term—EWS, building networks, collaborations, institutional capacity • Long-term—structural adjustments, policy shifts
Postimpact • Drought relief/emergency assistance	Preimpact • Risk assessments, mitigation
Rewards poor resource management	Identifies and addresses the root causes of vulnerability
Treats the symptoms of vulnerability (i.e., impacts)	Promotes improved stewardship of natural resources
Increases vulnerability, reliance on government and donors	Reduces vulnerability, builds self-reliance, reduces need for government and donor interventions
	Assists climate change adaptation

for better or for worse. The word *crisis* is taken from the Greek *krisis*, which literally means "decision." A crisis may be said to be occurring if a change or the cumulative impacts of changes in the external or internal environment generates a threat to basic values or desired outcomes and results in a high probability of involvement in conflict (legal, military, or otherwise), and there is a finite time for a response to the external value threat. A crisis is not yet a catastrophe; it is a turning point. Crisis situations can be ameliorated if different levels of decision makers perceive critical conditions to exist and if a change of the situation is possible for the actors. Thus, informed decision-making is key to effective mitigation of crises conditions and the proactive reduction of risk to acceptable levels. Being proactive about hazard management brings into play the need for decision support tools to inform vulnerability reduction strategies, including improved capacity to use information about impending events.

A key decision support tool for crisis mitigation is embedded within the concept of early warning. Early warning systems are more than scientific and technical instruments for forecasting hazards and issuing alerts. They should be understood as scientifically credible, authoritative, and accessible information systems that integrate information about and coming from areas of risk that facilitate decision-making (formal and informal) in a way that empower vulnerable sectors and social groups to mitigate potential losses and damages from impending hazard events (Maskrey 1997; NIDIS 2007).

Natural hazard risk information (let alone vulnerability reduction strategies) is rarely if ever considered in development and economic policymaking. Crisis scenarios can let us view risk reduction as much from the window of opportunity provided by acting before disaster happens as from the other smaller, darker pane window following a disaster. Given the slow onset and persistent nature of drought, mitigating potential impacts, in theory and in practice, must be recast as an integral part of development planning and implemented at national, regional, and local levels. Impact assessment methodologies should reveal not only why vulnerability exists (who and what is at risk and why) but also the investments (economic and social) that, if chosen, will reduce vulnerability or risk to locally acceptable levels. Studies of the natural and social context of drought should include an assessment of impediments to flows of knowledge and identify appropriate information entry points into policies and practices that would otherwise give rise to crisis situations. Issues of sustainable development, water scarcity, transboundary water conflicts, environmental degradation and protection, and climate change are contributing to the debate on what outcomes are being valued in the management of water.

25.3 Emerging Issues

Since the US Dust Bowl years of the 1930s, a great deal of reliable knowledge on reactive and anticipatory approaches to drought hazards and disasters has been derived. However, in an increasingly interconnected and rapidly changing world, several areas of concern are emerging. We highlight the following five areas of concern and opportunity:

1. *Uncertainties associated with a changing climate and its manifestation at regional and local levels*

   There is a strong need to approach climate model outputs far more critically than at present, especially for impact assessment to support adaptation at the local level. Multiyear droughts are, at present, not addressed by any forecast system, including the increasing role of evaporative demand. Many hotspots that show fragility in the face of climate change also exhibit soil moisture and soil quality reduction combined with reduced adaptive capacity. Scenario planning (based on past, present, and projected events) may provide better understanding of whether and how best to use probabilistic information with past data and cumulative risks across climate timescales. Central to all of the above is a sustained network of high-quality monitoring systems.

2. *The complex pathways of drought impacts: water-energy-food nexus*

   The United Nations (FAO 2014) describes the water-energy-food nexus as follows:

   "Water, energy, and food are inextricably linked. Water is an input for producing agricultural goods in the fields and along the entire agro-food supply chain." Agriculture is currently the largest user of water at the global level, accounting for 70 percent of total withdrawal. The food production and supply chain accounts for about 30 percent of total global energy consumption. Energy is required to produce and distribute water and food, to pump water from groundwater or surface water sources, to power irrigation systems, and to process, store, and transport agricultural goods. Global demand for energy is expected to increase by 400 percent by 2050. In areas where hydropower plays a significant role in national energy supply, such as Brazil and Zambia, blackouts and jumps in energy prices have occurred during extended periods of drought. Similarly, in 2014 as a result of low flows, the Glen Canyon Dam (Colorado River) had to purchase US$60 million of thermal power to offset market demands in the US Southwest, the fastest growing region in the country. Since the 1990s, average increases in the yields of maize, rice, and wheat at global levels have begun to level off at just about 1 percent per annum (FAO 2017). There are many synergies and trade-offs between water, energy use, and food production. Increasing irrigation might promote food or biofuel production, but it can also reduce river flows and hydropower potential through increased overall water withdrawals and, thus, jeopardize food security. In most cases, each component has been studied and managed individually, without consideration of the trade-offs, cultural similarities (and differences), interactions, and complementarity for jointly ensuring water, energy, and food security.

3. *The costs of drought impacts and the benefits of action and costs of inaction*

The major assumption behind proactive action around drought is that present or upfront actions and investments can produce significant future benefits. Support for such claims has been difficult to document. However, a US study found that each dollar spent in three federal natural hazard mitigation grant programs (the Hazard Mitigation Grant Program, Project *Impact*, and the Flood Mitigation Assistance Program) saves society an average of US$4 in future avoided losses (Godschalck et al. 2009). This conclusion is derived from the fact that between 1993 and 2003 the US Federal Government spent US$3.5 billion on mitigation while saving society US$14 billion in estimated losses (Mittler et al. 2009). No such study exists for drought. However, in this volume, Gerber and Mirzabaev (Chapter 5) have begun to make some headway in assessing benefits of action and the costs of inaction. In the area of drought and other hazards, much more work needs to be done to realize what has been called the "triple dividend of resilience" (Tanner et al. 2015). The dividends provide for three types of public and private benefits that make a business case for proactive disaster risk investments and also a narrative for reconciling short- and long-term objectives. These benefits include:

a. Avoiding losses when disasters strike

b. Stimulating economic activity thanks to reduced disaster risk

c. Development cobenefits, or uses, of a specific disaster risk management investment

4. *The role of technology, efficiency … and policy*

Since 1980, water use in the United States has returned to 1970 levels of use. During this period, the US population increased by 33 percent (Rogers 1993). This transformation illustrates the cumulative effectiveness of behavioral and efficiency changes. However, the major drivers were national policies reducing average annual demand and freshwater withdrawals in the United States (Rogers 1993), demonstrating the value of enabling legislation and regulation in leading to conservation measures. These included the Clean Water Act (1972), National Environmental Policy Act (1970), Endangered Species Act (1973), and Safe Drinking Water Act (1974). According to Stakhiv et al. (2016), these acts fostered and secured a bottom-up enabling institutional framework that focused on regulating, monitoring, and enforcing a suite of water quality and environmental laws passed in the 1970s.

5. *Links to human security: an area for future research*

According to a recent US National Intelligence Report, climate change and its resulting effects are likely to pose wide-ranging national security challenges for the United States and other countries over the next 20 years (NIC 2016). The pathways to insecurity outlined in this report include several drought-sensitive issues, such as:

- Adverse effects on food prices and availability
- Increased risks to human health
- Negative impacts on investments and economic competitiveness

While water scarcity and food insecurity have been shown to play roles in dislocation and unstable conditions, little is known about the relationships of these links to conflict (Erian et al. 2010). Hydroclimatic variability poses an important threat to human security through impacts on economies and livelihoods, independent of the conflict pathway (Kallis and Zografos 2014 ; Serageldin 2009). How drought and climate change may play into future fragility will be an area of increasing research and security interest.

Godschalck et al. (2009) noted that valuable lessons for mitigation planners and policymakers have emerged over the years, including the need to consider a portfolio of losses, integrating both qualitative and quantitative analyses, assessing benefits and value over a large number of projects, the need to explicitly acknowledge differing social values, and strengthening institutional mechanisms for collaboration, including impacts data collection in order to reduce vulnerability and enhance resilience. As Vayda and Walters (1999) caution, researchers should focus on human responses to environmental events without presupposing the impact of political processes on environmental events. Linking a deeper problem-oriented contextual analysis of an event (or events) within a broader political ecology of conditioning factors driving vulnerability is needed. As has been noted, critique by itself is not engagement (Walker 2007).

25.4 Final Thoughts

Drought results in widespread and complex impacts on society. Numerous factors influence drought vulnerability. As our population increases and becomes more urbanized, there are growing pressures on water and natural resource managers and policymakers to minimize these impacts. This also places considerable pressure on the science community to provide better tools, and credible and timely information to assist decision makers. The adaptive capacity of a community (defined here in the broadest terms) means little if available tools, data, and knowledge are not used effectively. Drought certainly exacerbates all of these problems and has significant cumulative impacts across all of these areas beyond the period of its climatological occurrence. Improving drought preparedness and management and its link to water management is one of the key challenges for the future.

The motivation for this book was to provide insights into these important issues and problems and, it is hoped, point toward some real and potential solutions, based on empirical evidence and experience. The contributors to this volume have addressed a wide range of topics that focus on integrating science, management, and policy issues in theory and practice. Building awareness of the importance of improved drought management today and investing in preparedness planning, mitigation, improved monitoring, and early warning systems and better forecasts will pay enormous dividends now and in the future.

Finding the financial resources to adopt risk-based drought preparedness plans and policies is always indicated as an impediment to changing the paradigm by policy and other decision makers. However, the solution seems clear: divert resources from reactive response programs that do little, if anything, to reduce vulnerability to drought (and, as has been demonstrated, may increase vulnerability) to a more proactive, risk-based management approach. For example, in the United States, billions of dollars have been provided as relief to the victims of drought in recent decades. The same is true in many drought-prone nations, both developed and developing, throughout the world. One can only imagine the advances that could have been made in predrought mitigation and preparedness strategies had a portion of those funds been invested in better monitoring networks, early warning and information delivery systems, decision support tools to improve decision-making, improved climate forecasts, drought planning and impact assessment methodologies. We are not here including those who are truly victims—that is, those whose lack of access and capacity to use these advances have left them at the mercy of those who exercise knowledge, decisions, and power (see Nakashima et al. 2012; Verver 2011). The key to invoking a new paradigm for drought management is educating the public—not only the recipients of drought assistance that have become accustomed to government interventions in times of crisis but also the rest of the public whose taxes are being used to compensate for losses. There will always be a role for emergency response, whether for drought or some other natural hazard, but it needs to be used sparingly and only when it does not conflict with preestablished drought policies that reflect sustainable resource management practices.

On the occasion of World Water Day in March 2004, the Secretary-General of the United Nations, Kofi Annan, stated:

> Water-related disasters, including floods, droughts, hurricanes, typhoons, and tropical cyclones, inflict a terrible toll on human life and property, affecting millions of people and provoking crippling economic losses. … However much we would wish to think of these as strictly natural disasters, human activities play a significant role in increasing risk and vulnerability. … Modern society has distinct advantages over those civilizations of the past that suffered or even collapsed for reasons linked to water. We have great knowledge, and the capacity to disperse that knowledge to the remotest places on earth. We are also the beneficiaries of scientific leaps that have improved weather forecasting, agricultural practices, natural resources management, and disaster prevention, preparedness, and management. New technologies will continue to provide the backbone of our efforts. But only a rational and informed political, social, and cultural response—and public participation in all stages of the disaster management cycle—can reduce disaster vulnerability, and ensure that hazards do not turn into unmanageable disasters.

In 2013, the Secretary-General of the United Nations, Ban Ki-moon, stated:

> Over the past quarter-century, the world has become more drought-prone, and droughts are projected to become more widespread, intense, and frequent as a result of climate change. The long-term impacts of prolonged drought on ecosystems are profound, accelerating land degradation and desertification. The consequences include impoverishment and the risk of local conflict over water resources and productive land. Droughts are hard to avert, but their effects can be mitigated. Because they rarely observe national borders they demand a collective response. The price of preparedness is minimal compared to the cost of disaster relief. Let us therefore shift from managing crises to preparing for

droughts and building resilience by fully implementing the outcomes of the High-level Meeting on National Drought Policy held in Geneva last March. (The complete statement from Ban Ki-moon is available at: http://www.un.org/sg/statements/?nid=6911)

We would argue that the complexities of drought and its differences from other natural hazards make drought more difficult to deal with than any other natural hazard, especially if the goal is to mitigate impacts. Special efforts must be made to address these differences as part of drought preparedness planning, or the differences will result in a failure of the mitigation and planning process. It is imperative that future drought management efforts consider the unique nature of drought, its natural and social dimensions, and the difficulties of developing effective early warning systems, reliable seasonal forecasts, accurate and timely impact assessment tools, comprehensive drought preparedness plans, effective mitigation and response actions, and drought policies that reinforce sustainable resource management objectives.

References

Erian, W., B. Katlan, and O. Babah. 2010. *Drought Vulnerability in the Arab Region. Special Case Study: Syria*. Technical submission to the UNISDR Global Assessment Report on Disaster Risk Reduction, GAR , U.N. Office of Disaster Risk Reduction, Geneva, Switzerland.

EU (European Union). 2000. *EU Water Framework Directive*. http://ec.europa.eu/environment/water/water-framework/index_en.html (accessed March, 2017).

FAO. 2014. *The Water-Energy-Food Nexus: A New Approach in Support of Food Security and Sustainable Agriculture*. FAO, Rome.

FAO. 2017. *The Future of Food and Agriculture: Trends and Challenges*. FAO, Rome.

Godschalck, D. R., A. Rose, E. Mittler, K. Porter, and C.T. West. 2009. Estimating the value of foresight: Aggregate analysis of natural hazard mitigation, benefits and costs. *Journal of Environmental Planning and Management* 52:739–756.

Kallis, G., and C. Zografos. 2014. Hydro-climatic change, conflict and security. *Climate Change* 123:69–82.

Kenney, D., A. Ray, B. Harding, R. Pulwarty, and B. Udall. 2010. Rethinking vulnerability on the Colorado River. *Journal of Contemporary Water Research & Education* 144:5–10.

Maskrey, A. 1997. *Report on National and Local Capabilities for Early Warning*. International Decade for Natural Disaster Reduction Secretariat, Geneva, pp. 33.

Merriam-Webster, 1977. *New Collegiate Dictionary*. G & C Merriam Company. Springfield, MA. p. 270.

Mittler, E., L. B. Bourque, M. M. Wood, and C. Taylor. 2009. How communities implement successful mitigation programs: Insights from the Multihazard Mitigation Council (MMC) Community Study. In *Multihazard Issues in the Central United States: Understanding the Hazards and Reducing the Losses*, J. E. Beavers, ed., pp. 51–73. American Society of Civil Engineers, Reston, VA.

Nakashima, D. J., K. Galloway McLean, H. D. Thulstrup, A. Ramos Castillo, and J. T. Rubis. 2012. *Weathering Uncertainty: Traditional Knowledge for Climate Change Assessment and adaptation*. UNESCO, Paris.

NDPC. 2000. *National Drought Policy Commission. National Drought Policy Act 1988*. https://govinfo.library.unt.edu/drought/finalreport/fullreport/pdf/reportfull.pdf (accessed June, 2017).

NIC. 2016. *Implications for US National Security of Anticipated Climate Change*. National Intelligence Council, Washington, DC.

NIDIS. 2007. *The National Integrated Drought Information System Implementation Plan*. NOAA, Boulder, CO. www.drought.gov (accessed June, 2017).

Pulwarty, R., and J. Verdin. 2013. Crafting early warning systems. In *Measuring Vulnerability to Natural Hazards: Towards Disaster Resilient Societies*, 2nd edn., J. Birkmann, ed., United Nations University Press, Tokyo, pp. 124-147.

Rogers, P. 1993. *America's Water: Federal Roles and Responsibilities*. MIT Press, Cambridge, MA.

Serageldin, I. 2009. Water wars? A talk with Ismail Serageldin. *World Policy Journal* 26(4):25–31.

Stakhiv E., W. Werick, and P. Brumbaugh. 2016. Evolution of drought management policies and practices in the United States. *Water Policy* 1–31. https://doi.org/10.2166/wp.2016.017.

Tanner, T. M., S. Surminski, E. Wilkinson, R. Reid, J. E. Rentschler, and S. Rajput. 2015. *The Triple Dividend of Resilience: Realizing Development Goals Through the Multiple Benefits of Disaster Risk Management*. Global Facility for Disaster Reduction and Recovery (GFDRR) at the World Bank and Overseas Development Institute (ODI), London. www.odi.org/tripledividend (accessed June, 2017).

UNDP Bureau of Crisis Prevention and Recovery. 2004. *Reducing Disaster Risk: A Challenge for Development (A Global Report)*. UNDP, New York.

Vayda, A. P., and B. B. Walters. 1999. Against political ecology. *Human Ecology* 21(1):167–179.

Verver, M.-T. 2012. The East African food crisis: Did regional early warning systems function? *Journal of Nutrition* 142(1):131–133. doi:10.3945/jn.111.150342.

Walker, P. 2007. Political ecology: Where is the politics? *Progress in Human Geography* 31:363–369.

Wilhite, D. A. (ed.) 2000. Drought: A Global Assessment, Volumes 1–2, In *Hazards and Disasters: A Series of Definitive Major Works*, A. Z. Keller, ed. Routledge, London, pp. 3–18.

Wilhite, D. A., and W. E. Easterling (eds.). 1987. *Planning for Drought: Toward a Reduction of Societal Vulnerability*. Westview Press, Boulder, CO.

Wilhite, D. A., M. V. K. Sivakumar, and R. Pulwarty. 2014. Managing drought risk in a changing climate: The role of national drought policy. *Weather and Climate Extremes* 3:4–13. http://dx.doi.org/10.1016/j.wace.2014.01.002.

Index

Printed and bound by CPI Group (UK) Ltd, Croydon, CR0 4YY

24/10/2024

01778306-0011